The Chemistry of Organophosphorus Compounds

The Chemistry of Organophosphorus Compounds

Edited by Deacon Palmer

STATES
ACADEMIC PRESS
www.statesacademicpress.com

States Academic Press,
109 South 5th Street,
Brooklyn, NY 11249, USA

Visit us on the World Wide Web at:
www.statesacademicpress.com

ISBN: 978-1-63989-737-7

Cataloging-in-Publication Data

The chemistry of organophosphorus compounds / edited by Deacon Palmer.
p. cm.
Includes bibliographical references and index.
ISBN 978-1-63989-737-7
1. Organophosphorus compounds. 2. Chemistry, Organic. I. Palmer, Deacon.
QD305.P46 C44 2023
547.07--dc23

Table of Contents

Preface

Organophosphorus compounds are derivatives of phosphoric, phosphinic or phosphonic acid in which the oxygen atoms directly bind with the phosphorus atom and can be replaced by the atoms of nitrogen or sulfur. They are primarily used as pesticides and insecticides, and serve as a substitute to chlorinated hydrocarbons persisting in the environment. Despite their extensive uses, certain organophosphorus compounds are toxic for human beings. These compounds are abundant in nature and due to their inherent chemical properties, play a vital role in various sectors. Asymmetry, multivalency, variable oxidation states and metal-binding properties are a few of the notable characteristics that enhance their identification as a distinct and adaptable family of chemicals. There has been rapid progress in organophosphorus chemistry and its applications are finding their way across multiple industries. This book aims to shed light on some of the unexplored aspects of organophosphorus chemistry and the recent researches in this field. It also aims to equip students and experts with the advanced topics and upcoming concepts in this area of chemistry.

This book has been the outcome of endless efforts put in by authors and researchers on various issues and topics within the field. The book is a comprehensive collection of significant researches that are addressed in a variety of chapters. It will surely enhance the knowledge of the field among readers across the globe.

It gives us an immense pleasure to thank our researchers and authors for their efforts to submit their piece of writing before the deadlines. Finally in the end, I would like to thank my family and colleagues who have been a great source of inspiration and support.

Editor

1

Recent Developments in Organophosphorus Flame Retardants Containing P-C Bond and their Applications

Sophie Wendels †, Thiebault Chavez †, Martin Bonnet, Khalifah A. Salmeia * and Sabyasachi Gaan *

Additives and Chemistry Group, Advanced Fibers, Empa, Swiss Federal Laboratories for Materials Science and Technology, Lerchenfeldstrasse 5, 9014 St. Gallen, Switzerland; sophie.wendels@empa.ch (S.W.); thiebault.chavez@empa.ch (T.C.); martin.bonnet@empa.ch (M.B.)

* Correspondence: khalifah.salmeia@empa.ch (K.A.S.); sabyasachi.gaan@empa.ch (S.G.)

† These two authors contributed equally to this work.

Abstract: Organophosphorus compounds containing P-C bonds are increasingly developed as flame retardant additives due to their excellent thermal and hydrolytic stability and ease of synthesis. The latest development (since 2010) in organophosphorus flame retardants containing P-C bonds summarized in this review. In this review, we have broadly classified such phosphorus compounds based on the carbon unit linked to the phosphorus atom i.e., could be a part of either an aliphatic or an aromatic unit. We have only considered those published literature where a P-C bond was created as a part of synthetic strategy to make either an intermediate or a final organophosphorus compound with an aim to use it as a flame retardant. General synthetic strategies to create P-C bonds are briefly discussed. Most popular synthetic strategies used for developing P-C containing phosphorus based flame retardants include Michael addition, Michaelis–Arbuzov, Friedels–Crafts and Grignard reactions. In general, most flame retardant derivatives discussed in this review have been prepared via a one- to two-step synthetic strategy with relatively high yields greater than 80%. Specific examples of P-C containing flame retardants synthesized via suitable synthetic strategy and their applications on various polymer systems are described in detail. Aliphatic phosphorus compounds being liquids or low melting solids are generally applied in polymers via coatings (cellulose) or are incorporated in the bulk of the polymers (epoxy, polyurethanes) during their polymerization as reactive or non-reactive additives. Substituents on the P atoms and the chemistry of the polymer matrix greatly influence the flame retardant behavior of these compounds (condensed phase vs. the gas phase). Recently, aromatic DOPO based phosphinate flame retardants have been developed with relatively higher thermal stabilities (>250 °C). Such compounds have potential as flame retardants for high temperature processable polymers such as polyesters and polyamides. A vast variety of P-C bond containing efficient flame retardants are being developed; however, further work in terms of their economical synthetic methods, detailed impact on mechanical properties and processability, long term durability and their toxicity and environmental impact is much needed for their potential commercial exploitations.

Keywords: P-C bond; substitution; addition; Michaelis-Arbuzov reaction; Michael addition; flame retardant; organophosphorus compounds; transformation; TGA; Cone calorimetry; UL 94

1. Introduction

P-C containing organophosphorus compounds find application as metal complex catalysts [1], reagents in organic chemistry [2,3], pesticides, insecticides, herbicides, medicine and biological systems [4], additives for polymers [5], surfactants, lubricants [6], surface treatments on metals [7], etc. In the past few decades, many review articles and book chapters regarding the creation of P-C bond and their potential applications have been published [6,8–13]. Organophosphorus compounds with P-C bonds are increasingly investigated as flame retardants due their high thermal and hydrolytic stability (P-C bond), ease of synthesis and suitability of processing at high temperature. One should pay attention to other properties like miscibility in the polymer matrix, reactivity with polymer or monomer, moisture absorption, tendency for blooming, toxicological properties, etc., which are very important and need to be considered before any commercial exploitation of these compounds can be realized. Unfortunately most published literature are academic and limited to aspects like synthesis, processability, flame retardancy and thermal properties of these compounds, thus it becomes very difficult to judge if any of these compounds discussed in this review can have any real commercial impact. The intention of this review is to summarize the work done in development flame retardants containing P-C bond without being critical of the shortcomings of the published literature.

P-C bonds are thermally quite stable and exhibit typical bond energies of ~272 kJ/mol [14]. Though the bond energies are lower than P-O (360 kJ/mol) and P-N (290 kJ/mol), the carbon substituents are not good leaving groups and thus resistant to possible nucleophilic attacks, which becomes relevant in high temperature polymer processing [15]. However, hydrolysis of P-C bonds can help in their biodegradation and reduce their possible harmful impact to the environment. Recent studies have provided insights into possible enzymatic hydrolysis of suitable P-C bonds [16–18]. Organophosphorus-based flame retardants are quite versatile in their flame retardant action and often exhibit both condensed and gas phase activity Depending on the substrate and phosphorus chemistry, there could be chemical interactions in the condensed phase at elevated temperatures, which lead to changes in the decomposition pathway of the polymer and possible formation of carbonaceous char residues on the surface of decomposing polymer, hence preventing its further oxidation. The char thus formed acts as a protective thermal barrier. In other instances, the phosphorus compounds and some of their decomposed products preferably volatilize from the polymer substrate when heated. These phosphorus species further decompose to release reactive phosphorus species, which then interact with the combustion intermediates in the gas phase as flame inhibitors [19,20]. In most cases, such interactions lead to recombination of the H and OH radicals and prevent their oxidation P-C bond containing phosphorus compounds like any organophosphorus flame retardants can exhibit either condensed phase or gas phase flame inhibition and is very specific to a polymer/flame retardant combination. For polymers (cellulose, wool), with relatively high concentration of hydroxyl or amino groups the phosphorus compounds primarily work in the condensed phase. In case of synthetic polymers containing oxygen and nitrogen atoms, catalytic hydrolysis of the ester or amide groups by phosphorus acids promotes an enhanced melt dripping and fast shrinkage from the flame. As far as olefin-based polymers are considered, the phosphorus compounds mainly act in the gas phase by recombining the key fuel species such as H and OH radicals and preventing their oxidation. Some minor physical effects due to volatilization of phosphorus compounds and dilution of the fuel can also occur [21].

In the following section, several analytical methods and techniques used by researchers to evaluate the thermal and fire properties of the new materials developed are mentioned. The choice of specific analytical technique depends mostly on the intended end use of the materials. TGA is generally employed to determine the thermal stability of new flame retardants and materials made from them. The degradation temperature of the flame retardant gives possible indication of its processing temperature. Higher char residue formed for the flame retardant polymer/formulation compared to the basis polymer indicates its possible condensed phase action. Similarly, higher char formation observed in other techniques such as PCFC and Cone calorimeter also indicate possible condensed based flame retardant action. Cone calorimeter and PCFC instrument is generally used to measure heat release of materials. Flame retardant action in condensed and gas phase leads in normally reduction in heat release rates. In cone calorimeter, one can also measure smoke and its composition. Flame retardants working in condensed phase usually leads to reduction in smoke toxicity. The ratio of CO/CO_2 ratio as measured in the cone calorimeter indicates the activity of flame retardant in gas phase. Higher ratio for flame retardant materials indicates possible gas phase flame inhibition [22]. LOI is a common technique, which was used for evaluating the flame retardancy of textile materials and films. Materials exhibiting LOI values of 25% and greater are considered to self-extinguishing. Usually, inherently flame retardant materials exhibit LOI values greater than 30%.

Several small case fire tests like UL 94 and vertical burning tests are also commonly used to evaluate the flame retardancy of materials. In these tests, the reaction of materials to small fire either in vertical and horizontal orientation measured.

2. General Principles of P-C Bond Formation

The P-C bond formation has been reviewed in the past by various researchers [6,8–10,12,13,23–30]. Conventional synthetic strategies like Michael addition, Arbuzov rearrangement reactions, Friedel-Crafts reactions, radical addition reactions, Pudovik reactions are the most common strategy for P-C bond creation. In the last decade, there has been an increased focus on development of new synthetic methodologies for P-C bond formation. As a result, new novel and green methodologies [24,28] in the organic synthesis gave impetus to replace toxic and harmful phosphorus starting materials with more benign alternatives [31].

Due to the electronic nature of the phosphorus atom, the organophosphorus compounds can be classified into different chemical classes based on the coordination number and the oxidation state of the phosphorus atom [23,32]. Accordingly, the chemical reactivity of the organophosphorus compounds can be influenced by their structural geometry and valency of the phosphorus atom. For example, *H*-phosphonates with P(O)H bonds are commonly used as starting materials for P-C bond formation and thus their chemical reactions have been investigated in detail [6,9,12,24,25,29–31,33,34]. It is reported that presence of the P=O bond is the driving force for their reactions [23]. In general, the phosphorus center of the *H*-phosphonates is electrophilic and can be modified to be nucleophilic [35]. Moreover, the P-H bond of *H*-Phosphonates is considered acidic and can go through acid base reactions [36]. On the other hand, the *H*-phosphonates are quinquevalent, tetra-coordinated phosphorus compounds, which exist in equilibrium in solution with their corresponding tervalent, tri-coordinated species (phosphite ester). The phosphite esters have a lone electron pair on the phosphorus center and thus can react with various electrophiles [23]. When these phosphite esters are protonated or bear electron withdrawing groups, they react with nucleophiles. Based on the chemical reactivity of the organophosphorus compounds, diverse synthesis approaches of P-C bond formation have been explored and can be categorized as illustrated in Tables 1–3.

Table 1. P-H transformation into P-C bond.

Reaction Type	General Examples	Selected Ref.
Pudovik reaction	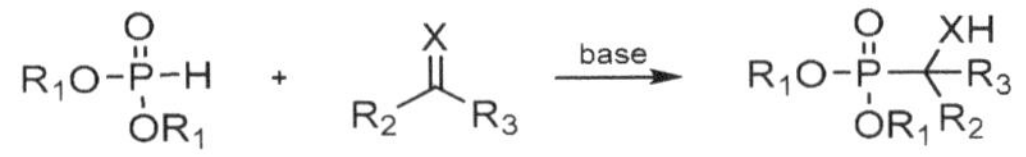 , X = NR_4, O	[6,30,37–39]
Kabachnik–Fields reaction	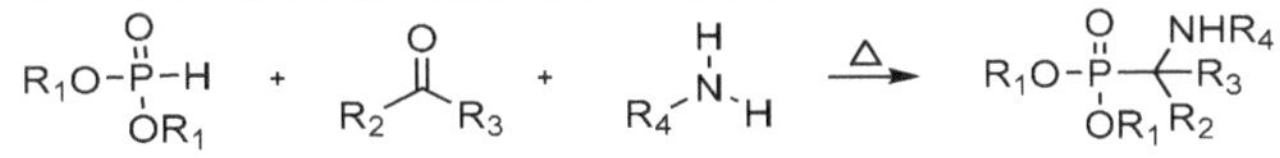	[6,30]
Michaelis–Becker reaction	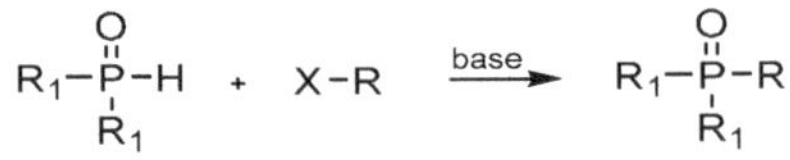 X = Halogen	[26]

Table 1. *Cont.*

Reaction Type	General Examples	Selected Ref.
Nucleophilic ring opening of epoxides	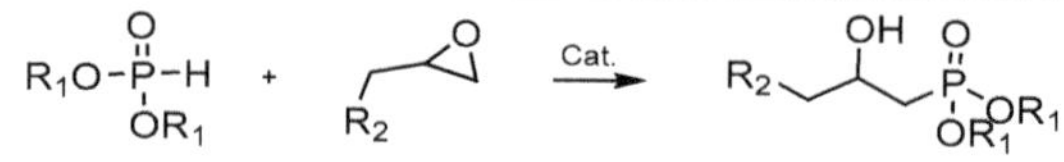	[6]
C-O/P-H cross coupling reaction	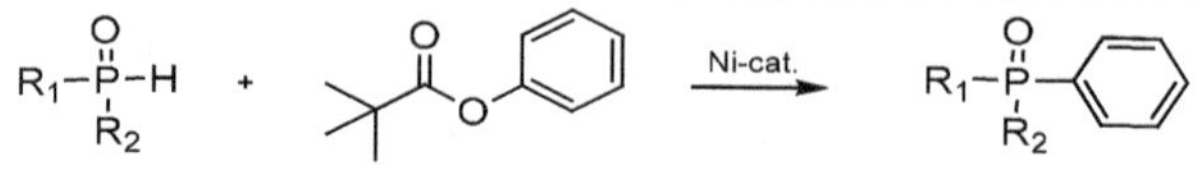	[40]
Dehydrogenative coupling with terminal alkynes	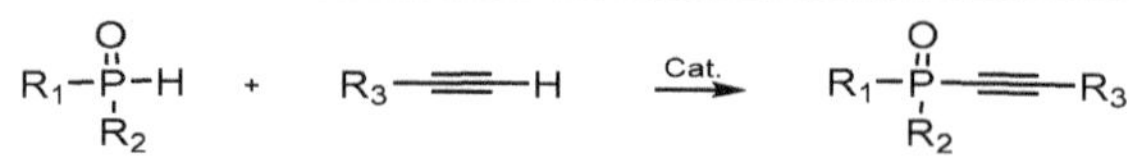	[6,12,41]
P-H insertion of α-Imino-carbenes	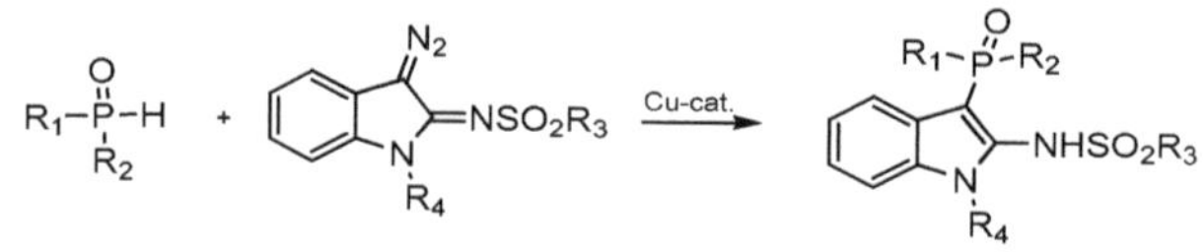	[42]
O_2-induced radical oxyphosphorylation reaction	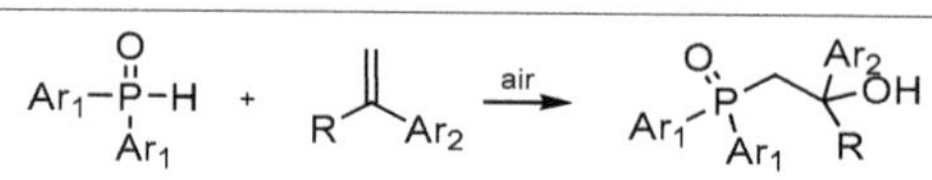	[43]
	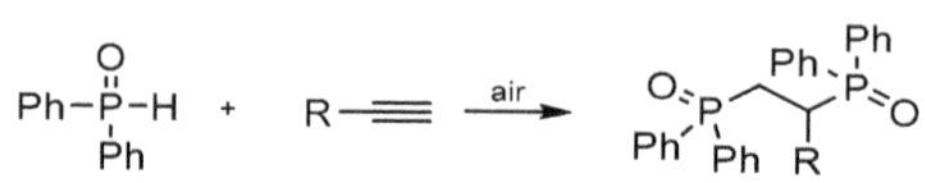	[44]
Phospha–Michael addition reaction	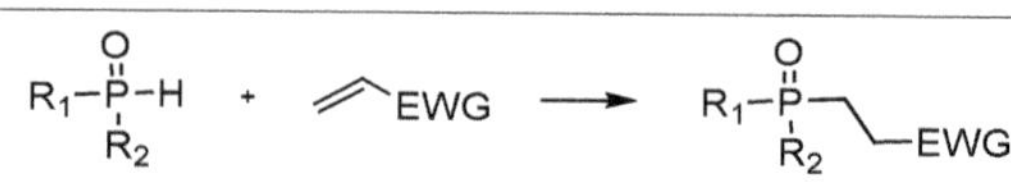	[5,15,29]

Table 2. Rearrangement/transformation Reactions.

Reaction Type	General Examples	Selected Ref.
Michaelis–Arbuzov reaction	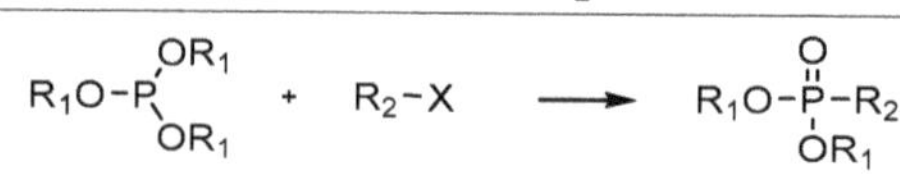	[5,27,28,45,46]
Sandmeyer-type reaction	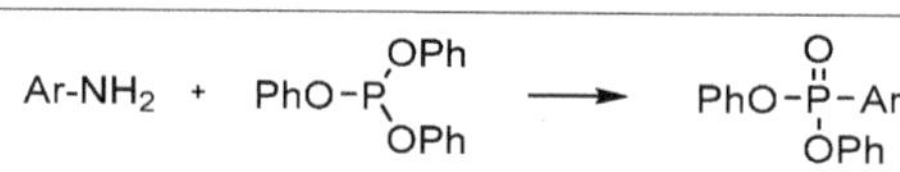	[47,48]
Decarbonylation of Acylphosphines	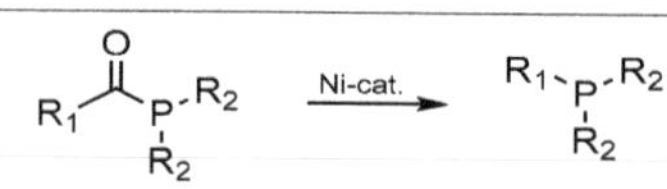	[49]
Abramov reaction	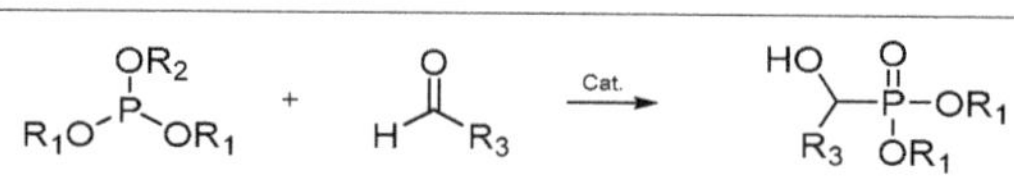 R_2: alkyl, trimethyl silyl	[30,35,50]

Table 3. C-H to C-P transformation.

Reaction Type	General Examples	Selected Ref.
Friedel–Crafts reaction	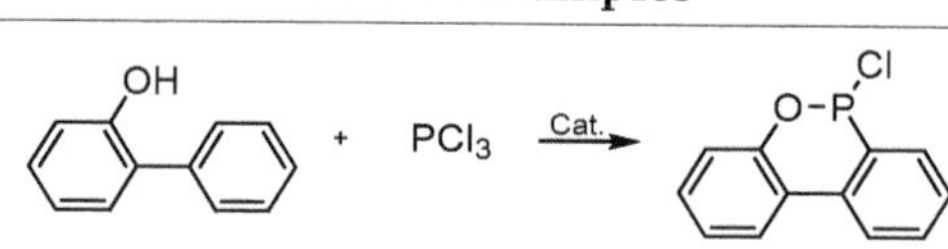	[5,8]
Via Aryne formation	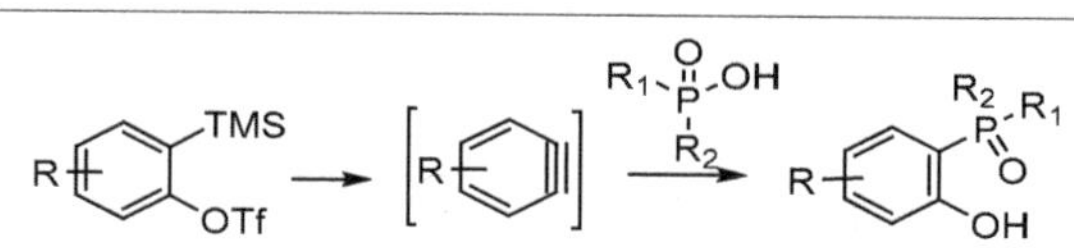	[51]
Nucleophilic substitution	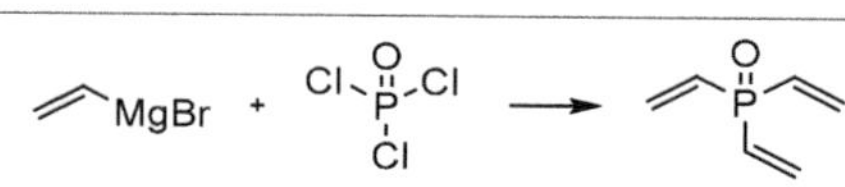	[8,52]

As seen in the table, there are numerous ways P-C bonds can be created and its potential academic research or industrial exploitations needs careful consideration. Flame retardant development is an applied research with a great potential for commercial exploitation and thus attention must be paid to the synthetic strategies for upscaling. In most cases, commercial availability of the starting chemicals and their cost determine their potential commercial exploitation. However, many other likely challenges can be encountered in industry, which hinders their industrial upscaling and utilities such as:

(1) Expensive synthesis approaches,
(2) Toxicity of the starting materials,
(3) Non-desirable wastes or by-product,
(4) Multiple purification steps,
(5) Difficulty handling of the chemical materials.

Currently only a few chemical reactions listed in Tables 1–3 are industrially exploited to manufacture flame retardants.

3. Flame Retardants with P-$C_{aliphatic}$ Bonds

In the following sections, we have summarized examples of P-$C_{aliphatic}$ bond containing organophosphorus compounds, which are classified based on the synthetic strategy for creation of P-C bond. For clarity, the specific P-C bonds that were created in the structure have been marked in bold.

3.1. Substitution Reactions

Substitution reactions are one of the most commonly used reactions used in organic chemistry and can be suitably exploited to create new bonds and eventually final products. Such reactions are commonly exploited commercially [53] and in research to create P-C bond containing useful flame retardants. In this section, we summarize development of organophosphorus compounds obtained by substitution reactions other than Grignard type reaction. The Grignard type reaction is described in detail later in the section.

The reaction of but-3-1-yl 4-methylbenzenesulfonate and butyl 4-methylbenzenesulfonate with triphenylphosphine afforded the phosphonium salts **1a** and **1b** respectively (Figure 1) [54]. The phosphonium salts **1a** and **1b** were obtained by precipitation in petroleum ether in 73% and 86% yield, respectively. Their thermal properties were studied using thermogravimetric analysis (TGA) under N_2 atmosphere and the temperatures for 5% weight loss ($T_{d5\%}$) of 335.8 °C and 208.9 °C were observed for compounds **1a** and **1b**, respectively. They were then incorporated in polycarbonate (PC) composites at a concentration of 5% together with an anti-dripping agent. TGA was used to study the thermal stability of these composites. Cone calorimeter, UL94 and limiting oxygen index (LOI) tests were used to evaluate the flame retardant performance of the composites. It was found that the composite containing flame retardant with the alkene group (**1a**) exhibited a significant increase of LOI value (32.8%) compared to composite with compound **1b** (30.8%) and a lower peak heat release rate (PHRR). The LOI value of the neat PC is 25%. The char yield at 700 °C was found to be higher for both compounds (17 wt % and 15.6 wt % for compounds **1a** and **1b**, respectively) than for the virgin PC (13.9 wt %), which means that the two compounds improved the flame retardancy of the PC and act in the condensed phase. It was also noted for both compounds that 5% in the composite was enough to obtain UL 94 V-0 rating.

The conjugated dienes **2** and **3** (Figure 2) were synthesized by reaction of mesityl oxide with phosphorus acid and hypophosphorus acid, respectively [55,56]. It was described in the patent that these compounds can be used as flame retardants, potentially in the plastic industry or in textile coatings. The same researchers have also studied the synthesis of acrylamide derivative (Compound **4**,

Figure 2) from phosphorus acid, mesityl oxide, acetic acid, acetic anhydride and phenothiazine and no additional fire performance of these compounds were reported [57].

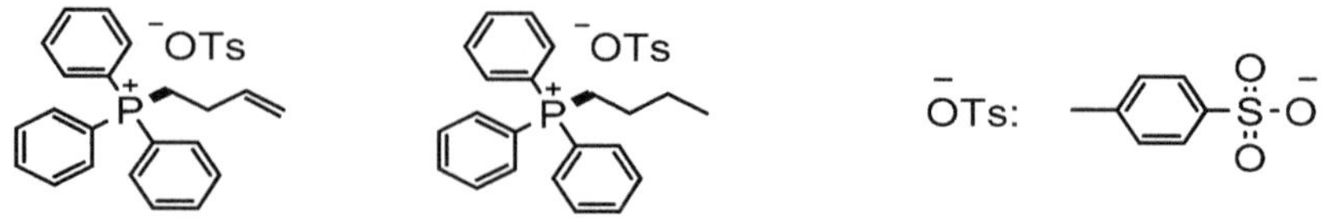

1a 1b

Figure 1. Chemical structures of **1a** and **1b** and their counter ion $^{-}$**OTs**.

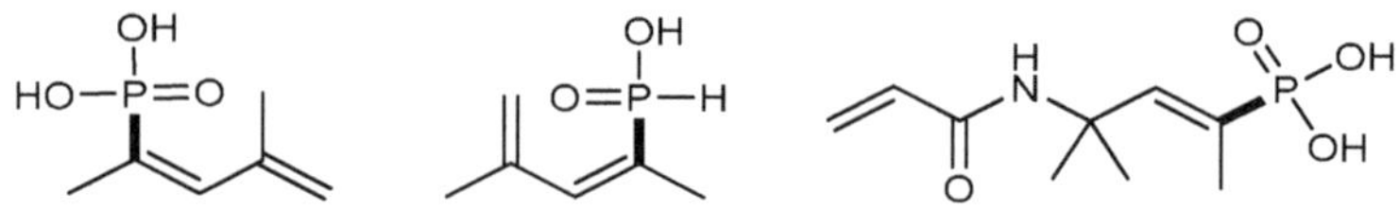

2 3 4

Figure 2. Chemical structures of **2**, **3** and **4**.

Dimethylpropyl phosphonate (**5a**, Figure 3) and dimethylallyl phosphonate (**5b**, Figure 3) were synthesized by the reaction of dimethyl phosphite in presence of inorganic base with propylbromide and allylbromide, respectively. They were obtained as liquids in 92% (Compound **5a**) and 93% (Compound **5b**) yield. Both compounds were characterized using their boiling point: 105 °C at 23 mbar for **5a** and 104 °C at 24 mbar for **5b** [14]. They were incorporated in flexible polyurethane (PU) foams at concentrations of 1, 2, 5 and 10 wt % based on polyol content and their flame retardant performance were evaluated. The thermal stability and flame retardancy were investigated by TGA and PCFC, which showed that the flame retardant additives act predominantly in the gas phase. The flammability of PU foams were further investigated by LOI, UL 94 HB and BKZ-VB Swiss standard fire test. It was found at a 10 wt % (based on weight of polyol) flame retardant loading, all flame retarded PU passed the BKZ-VB compared to the virgin PU foam. Moreover, the flame retarded PU with compound **5a** showed a HF2 rating in the UL 94 HB test while the treated PU with compound **5b** showed HF1 rating. LOI values of the PU foams increased with increasing the flame retardant concentrations. The LOI values of treated PU foams with the flame retardant additives **5a** and **5b** were 22.8% and 23.9%, respectively. PCFC measurements of the flame retarded PU foams did not show a char residue which confirms their gas phase mode of action.

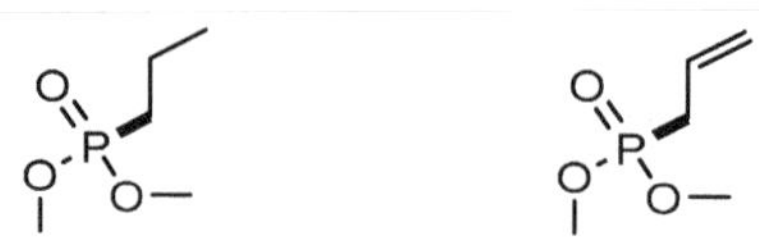

5a 5b

Figure 3. Chemical structures aliphatic phosphonates **5a** and **5b**.

Compounds **5a** and **5b** (Figure 3) were also compared to a commonly used flame retardant tris(1-chloro-2-propyl) phosphate (TCPP) [58]. It was concluded from both studies that compounds with an allyl group are more efficient flame retardants for PU foams than compounds with alkyl groups.

Allyltriphenylphosphonium bromide (Compound **6**, Figure 4) was synthesized from the reaction of triphenylphosphine and allyl bromide in dry toluene [59]. Subsequently, bentonite clay was modified with this phosphonium salt [60]. The surface modified bentonite clay was then used as a flame retardant additive (1, 2 and 3 wt %) for possible improvement in flame resistance and mechanical properties of polyester acrylate base UV-curable coatings. Tensile properties of films were evaluated from the stress–strain curve. The Young's modulus, tensile strength and elongation at break of the films were determined. It was found that the addition of the modified clay increased the Young's modulus but decreased the tensile strength. The elongation at break at first decreased but then increased with the increasing clay content. The coating became more brittle with addition of modified clay. The thermal properties and flammability properties were then investigated using TGA and LOI. The $T_{d10\%}$ of the neat coating was 373 °C while addition of clay first increased this value (the formulation with 1% modified clay showed a $T_{d10\%}$ of 387 °C), but then decreased to 361 °C for the formulation with modified clay content of 3 wt %. The char yield increased significantly from no char for the neat coating to 12% for the 3 wt % modified clay/coatings. The same trend was observed in the LOI measurements 17% and 22% for unmodified and 3 wt % clay/coatings. It was concluded from these results that the addition of flame retardant additive **6** to the clay had a significant impact on the mechanical as well as the flame retardancy of the polyester acrylate base UV-curable coatings acting dominantly in the condensed phase. All results were attributed to the crosslinking density of the allyl modified clay containing UV-cured coating.

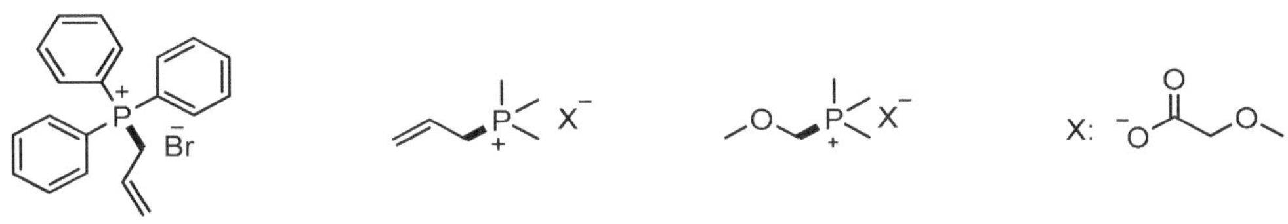

6 6a 6b

Figure 4. Chemical structures of phosphonium derivatives: **6**, **6a** and **6b**.

Ionic liquids are known to be nonflammable and have recently generated a lot of attention as environment friendly reagents. Compounds **6a** and **6b** (Figure 4) are ionic liquids and were synthesized from trimethylphosphine, methoxyacetic acid and allyl chloride (**6a**) or chloromethyl methyl ether (**6b**) [61]. Associated with methoxyacetate as counter anion, these compounds are described as the optimal ionic liquids for the dissolution of cellulose; however, no real application data was presented in the paper.

Recently, synthesis and application of different phosphinic acid derivatives has been reported [7]. Dibenzylphosphinic acid (**7a**, Figure 5) was synthesized from reaction of $H_2PO_2NH_4$ with benzyl bromide; diallylphosphinic acid (**7b**, Figure 5) was synthesized from reaction of $H_2PO_2NH_4$ with allyl bromide. These compounds were developed for application in epoxy resins for flame retardant and mechanical properties improvement. They were then treated with aluminium hydroxide to give aluminium–organophosphorus hybrids (AOPH) as follows: compound **7a** gave AOPH-NR and compound **7b** gave AOPH-C2. AOPH derivatives were then applied in epoxy resins to produce composites. LOI and TGA were performed for these composites and the analytical results showed that the AOPH-NR exhibits the best flame retardant properties: LOI for AOPH-NR was 28% and 23.6% for AOPH-C2. Moreover, the TGA analysis show that only AOPH-NR had an impact on the decomposition of the epoxy composites: the 5% and 50% decomposition temperatures decreased for AOPH-NR, whereas the other AOPH composites showed no real difference in thermal performance compared to the neat resin. TGA data indicated a slight increase in the char residue at 500 °C for both composites compared to the neat epoxy resin, showing a flame retardant mode of action in the condensed phase.

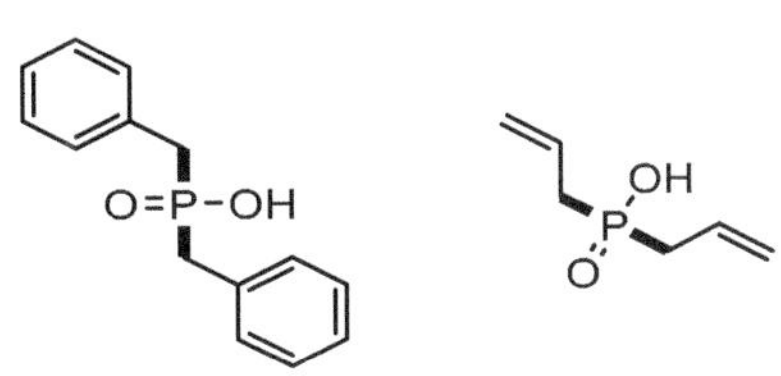

7a 7b

Figure 5. Chemical structure of **7a** and **7b**.

3.2. Addition Reactions

Addition reactions are also commonly used to create P-C bond containing flame retardants. The following section summarizes organophosphorus compounds obtained via addition reactions other than Michaelis–Arbuzov type reaction, which will be described in detail in the subsequent section.

Phosphinic acid derivatives bis(2-cyanoethyl)phosphinic acid (**7c**, Figure 6) was synthesized from reaction of $H_2PO_2NH_4$ with allyl cyanide; and bis(3-methoxy-3-oxopropyl)phosphinic acid (**7d**, Figure 6) was synthesized from reaction of $H_2PO_2NH_4$ with methyl acrylate [7]. These products were subsequently treated with aluminium hydroxide to give aluminium–organophosphorous hybrids (AOPH). The produced AOPH derivatives were also applied in epoxy resins.

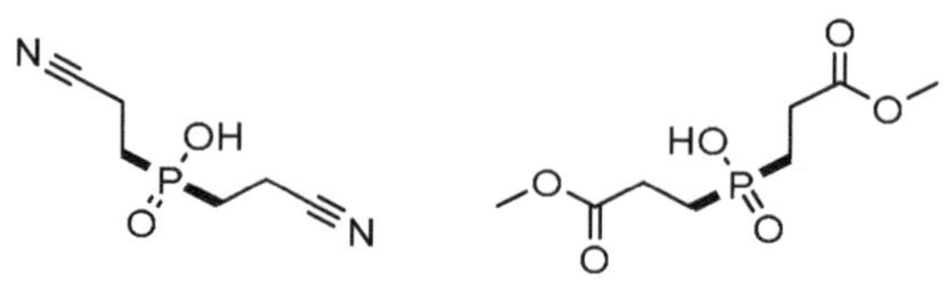

7c **7d**

Figure 6. Phosphinic acid derivatives **7c** and **7d**.

Diethylphosphinic acid (**8**, Figure 7) was obtained from a gas-liquid free-radical addition of sodium hypophosphite monohydrate and ethylene in acetic acid using benzoyl peroxide as initiator [62]. The thermal stability of compound **8** was investigated by TGA and the results showed two degradation peaks at 396 °C (small) and 424 °C (large), corresponding to the P-C bond breakage and combustion gas release, respectively. This compound was reported as a potential flame retardant; however, no further fire performance of this compound was reported.

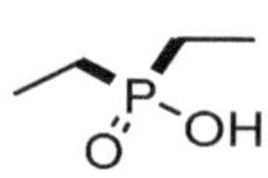

8

Figure 7. Chemical structure of **8**.

The synthesis of a cyclic *tert*-butyl phosphonate (Compound **9**, Figure 8) is patented. Compound **9** was obtained by reaction of *tert*-butyl phosphonic dichloride and neopenthyl glycol [63]. This compound was claimed as a flame retardant due to the presence of phosphorus atom and can be used as an additive for polymer resin like polyphenylene ether. The Izod impact strength as well as the melt viscosity index at 200 °C were measured for a formulation with a composition of 85 wt % aromatic vinyl resin, 15 wt % polyphenylene ether resin and 20 wt % compound **9** of the final resin. The results showed 8.8 kgf.cm/cm Izod impact strength and a melt viscosity index of 7.9. In addition, this formulation also exhibited UL 94 V-0 rating.

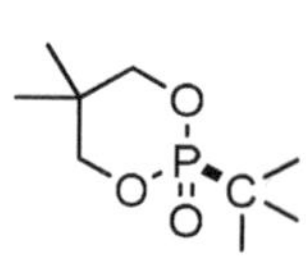

9

Figure 8. Cyclic *tert*-butyl phosphonate **9**.

The triazine-based DOPO compound (**10a**, Figure 9) was synthesized using Michael addition reaction of DOPO with 2-vinyl-4,6-diamino-1,3,5-triazine using DBU as a catalyst [15]. The product was collected as a white solid power in 93% yield with a m.p. of ~257 °C. It was dried overnight and then used as a PA6 additive in co-extrusion. Its impact on flammability, thermal, mechanical and rheological properties for PA6 were studied and compared to a commercially available flame retardant Exolit® OP 1230. PA6 formulations containing 17.1 wt % of compound **10a** and 16.8 wt % of Exolit® OP 1230 were prepared. Detailed thermal, mechanical and morphological characterizations using TGA, DSC, Scanning Electron Microscope (SEM), tensile tests, hardness and Charpy impact strength measurements were performed for these formulations. Flammability measurements and mechanistic studies were performed using PCFC, UL 94, vertical burning test and direct insertion probe mass spectroscopy (DIP-MS). It was found that in N_2 environment, neat PA6 showed a $T_{d5\%}$ of 388 °C and exhibited a one-step degradation process. In case of PA6 formulations, the $T_{d5\%}$ of the formulation decreased to 386 °C and 372 °C for Exolit® OP 1230 and compound **10a** respectively. In air, the decomposition took place in three steps for all formulations and the $T_{d5\%}$ followed the same trend as in N_2 environment. Thus, the addition of both flame retardants lowered the thermal stability of the PA6 polymer. Mechanical tests showed that the incorporation of compound **10a** in the PA6 formulation resulted in a more rigid and brittle material (higher Young's modulus but an impact strength of 4.53 kJ/m^2). Neat PA6 did not break in the Charpy test. PCFC measurements registered a PHRR of 738 W/g for PA6/**10a** formulation, followed by PA6 (590 W/g) and PA6/Exolit® OP 1230 formulation (538 W/g). This high PHRR for PA6/**10a** formulation may be attributed to a catalytic decomposition of PA6 formed from the decomposition of the DOPO-derivative at high temperature. However, based on the THR, compound **10a** showed a similar behavior compared to Exolit® OP 1230 (~27 kJ/g for both) and decreased the THR of the PA6 (28.5 kJ/g). The best UL 94 results were obtained for PA6 containing compound **10a** with a V-0 rating and afterflame times t_1/t_2 (time the flame stays after the first/the second flame) of 0 s/0 s respectively (2 s/3 s for Exolit® OP 1230 and 2 s/0 s for the neat PA6). The char yield under N_2 at 700 °C was found significantly higher for the PA6/**10a** formulation (2.4 wt %) compared to the PA6/Exolit® OP 1230 formulation (0.9 wt %) and the neat PA6 (0.28 wt %). Based on all analytical data, it was concluded that compound **10a** exhibit both gas and condensed phase active flame retardancy for PA6, and can be considered as a suitable alternative to halogen and halogen free flame retardants commercially available for PA6 application.

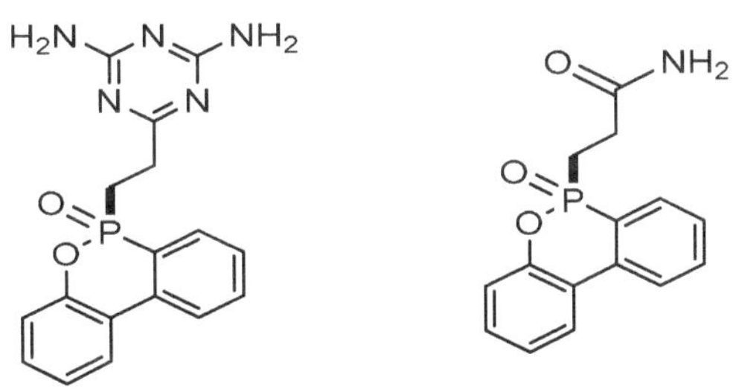

10a **10b**

Figure 9. DOPO derivatives of **10a** and **10b** structures.

The same group also synthesized DOPO-derivative **10b** (Figure 9) and compared its thermal and fire performances with compound **10a** [64]. Like in case of **10a**, **10b** was obtained using the synthetic procedure i.e., Michael addition of DOPO with acrylamide using 1,8-diazabicyclo(5.4.0)undec-7-ene (DBU) as a catalyst. It was obtained in a 75% yield and with m.p. ~170 °C. Both flame retardants were incorporated in rigid polyurethane foams (RPUF) with 1 wt % and 2 wt % phosphorus content and analyzed using UL 94 HB test, LOI, cone calorimetry, PCFC, TGA, DSC, SEM and Pyrolysis-Gas Chromatography Mass Spectroscopy (Py-GC-MS). Results showed that both flame retardants had HF-1 UL 94 HB rating while neat RPUF had a HBF rate. LOI values increased from 19.3% for the neat foam to 20.5% for PU foams containing compound **10b** and 20.9% for compound **10a**. Such increase in LOI are considered rather low compared to self-extinguishing systems which usually exhibit LOI > 25%. The cone calorimetry tests showed a significant decrease of the THR with the addition of flame retardant **10a** (23.1 MJ/m^2) and compound **10b** (19.3 MJ/m^2) compared to the neat PU foam

(24.9 MJ/m^2). TGA showed a slight decrease of the $T_{d10\%}$ for RPUF containing flame retardants **10a** and **10b** (300 °C for neat RPUF, 298 °C and 296 °C for RPUF with **10a** and **10b**, respectively). Evolved gas analysis, cone calorimetry and TGA experiments proved that DOPO derivatives-containing RPUF exhibit lower smoke and toxicant production and increased char residue, indicating the condensed phase mode of action.

3.3. Michaelis–Arbuzov Type Reactions and Rearrangement

Michaelis–Arbuzov reactions are one of the most common reactions used for the creation of phosphonate compounds. For example, Michaelis–Arbuzov reaction of 2-methoxy-5,5-dimethyl-1,3,2-dioxaphosphinane (**11**, Scheme 1) with allyl bromide produced 2-oxo-2-allyl-5,5-dimethyl-1,3,2-dioxaphosphinane (**11a**, Scheme 1). Epoxidation reaction of the later compound using *m*-CBPA, afforded the epoxy containing cyclic phosphonate **11b** [65]. Compound **11b** was reported as a white solid with 80% yield and a m.p. ~57–59 °C. It was then mixed with polyurethane (PU) at a 10 molar % in two different methods. The *Blend* method involved incorporation of the flame retardant as additive after the polymerization of the PU whereas the *Prep* method involved copolymerizing compound **11b** with PU. Pyrolysis combustion flow calorimetry (PCFC) test was used to evaluate the THR of the PU samples. It was found that the neat and the *Prep* PU had nearly the same THR (~22 kJ/g), which was lower than that for the *Blend* method (~25 kJ/g). The PCFC results confirmed that PU manufactured by both methods had no effect on THR of the PU. A possible explanation for this observation was attributed to the loss of compound **11b** during the final washing step in PU *Prep* method as well as to its weak incorporation in PU in the *Blend* method.

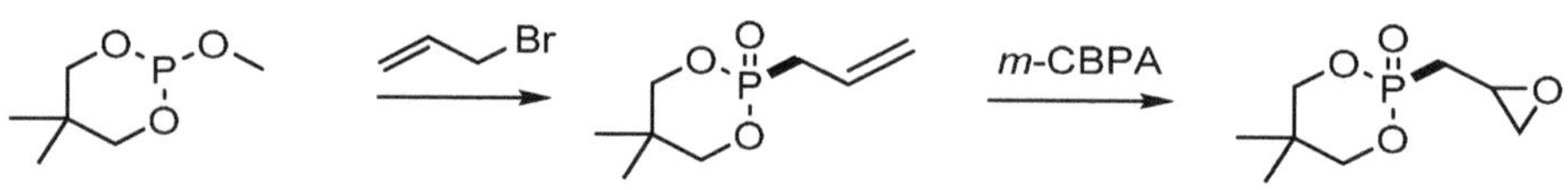

11 **11a** **11b**

Scheme 1. Chemical synthesis of compound **11b**.

Compound **14** (Scheme 2) was synthesized from reaction of compound **12** (Scheme 2) with DOPO derivative **13** (Scheme 2), via Michaelis–Arbuzov rearrangement [66]. The synthesized phosphorous compounds were then used as flame retardants additives in epoxy resins.

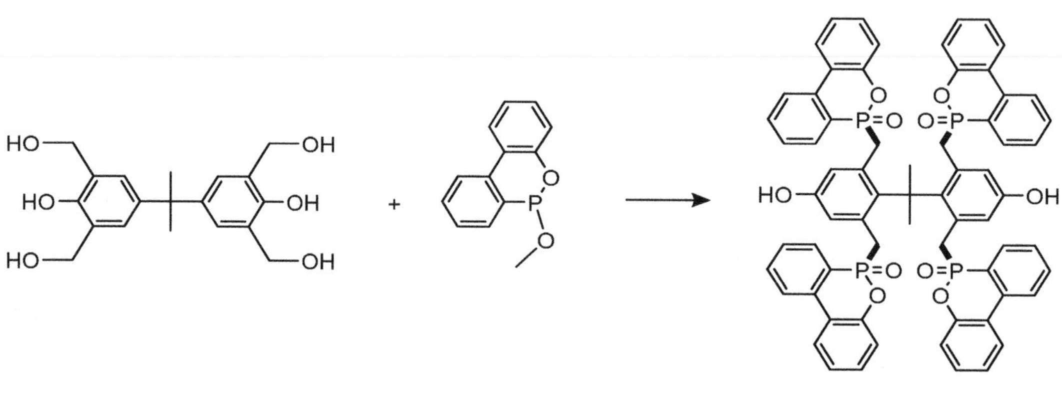

12 **13** **14**

Scheme 2. Synthetic route of compound **14**.

Diethyl allylphosphonate (**15**, Figure 10) has been synthesized from triethyl phosphite and allyl bromide and used in free radical precipitation polymerization in aqueous phase to produce poly(urethane-co-ester) (PMMD) copolymers [67].

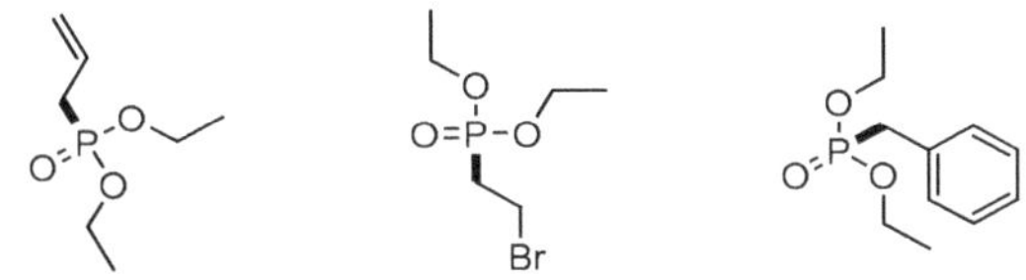

15 16 17

Figure 10. Flame retardant compounds **15**, **16** and **17**.

The PMMD copolymers were synthesized from poly(ethylene glycol), methacrylate, methyl methacrylate and monomer **15** in different molar ratios 1:6:2 (PMMD-162) and 1:6:3 (PMMD-163). BMMD copolymers were blended with ABS to improve its flame retardancy. Improvements in flame retardancy have been observed using LOI and UL 94 tests. LOI values were 20.8% for 20 wt % PMMD-162 and 20.3% for 20 wt % PMMD-163, while neat ABS showed an LOI value of 17.8%. UL 94 tests also showed better fire results with increased amount of FR.

Diethyl bromoethylphosphonate (**16**, Figure 10) has been synthesized from 1,2-dibromoethane and triethyl phosphite to improve flame resistance of several polymers [68]. Compound **16** was used to synthesize other monomers for ROMP homopolymerization. Dimethyl benzylphosphonate (**17**) was synthesized from trimethyl phosphite and benzyl bromide and was applied in PU foam to improve the flame retardancy of PU foams. The structural influence of the organophosphorous compounds on its flame retardancy was also studied [14]. It was reported that applying compound **17** in PU foams improves the flame retardancy of PU foams as confirmed by LOI, BKZ-VB and UL 94 HB tests. The concentrations of compound **17** used in these experiments were 1, 2, 5 and 10%. LOI values for the formulations ranged from 21.2% for the lowest to 22.5% for the highest concentrate respectively. Burning velocity decreased from 5.9 mm/s for the least concentration of additive to 4.1 mm/s for the highest concentration. UL 94 HB tests confirmed HBF rating for the PU foam with 1% to 5% concentration of the flame retardant additive **17** and it improved to an HF2 rating for 10% concentration.

Different DOPO-based derivatives (Figure 11) have been synthesized using the Michaelis–Arbuzov reaction with the aim of studying their flame retardant mechanism in PU foams [69]. For example, methyl-DOPO (**18**) was synthesized in 97% yield and incorporated in polyol to obtain the PU foam. Three different concentrations were tested: 5, 7.5 and 10 phpp (parts per hundred parts polyol). Burning test FMVSS 302 has been performed on the produced foams. With increase in the concentration of **18,** the total heat release (THR) of the foam as measured in cone calorimeter decreased, but the total smoke release (TSR) increased, indicating a strong gas flame inhibiting effect of compound **18**.

Methyl-DOPO (**3.18**) and compounds **19** to **28** (Figure 11) can also be synthesized using Michaelis–Arbuzov rearrangement [70]. For example, a new synthetic route involves using the transesterification of phosphonous acid diesters with the corresponding aliphatic alcohols and subsequent catalyzed Michaelis–Arbuzov rearrangement reaction, which leads to the formation of DOPO derivatives **19** to **28.** These structures were developed as halogen-free flame retardant alternatives for application in various polymers. The melting points were measured for compounds **18** (131 °C), **21** (106 °C), **23** (59–61 °C), **27** (108 °C) and **28** (105 °C). No further analytical studies were performed on these compounds. Such compounds can find potential applications in polymers with processing temperature below 200 °C. The synthesis of DOPO derivatives **18** and **28** were also investigated in other studies [71].

Compound **30** (Scheme 3) was synthesized via Michaelis–Arbuzov rearrangement reaction with a yield of 82% and applied as flame retardant additive in epoxy resin [72]. TGA and UL 94 test were performed on the resin samples. It was observed in TGA measurements that there was only a slight difference in $T_{d5\%}$ between the neat epoxy resin (320 °C) and resin with the compound **30** (340 °C), which indicates a possible gas phase action for this flame retardant. V-0 rating was obtained for epoxy formulation containing compound **30.** It was inferred in this work that the lower the oxidation state of the phosphorus compound, the better is its flame retardancy.

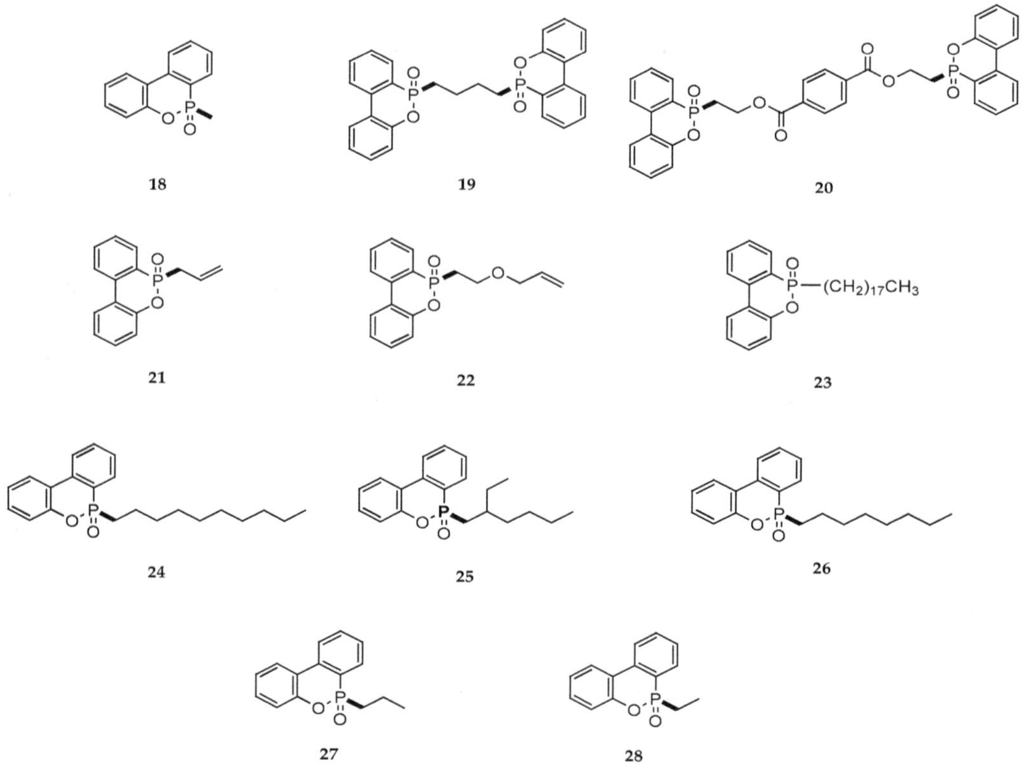

Figure 11. Various DOPO derivatives.

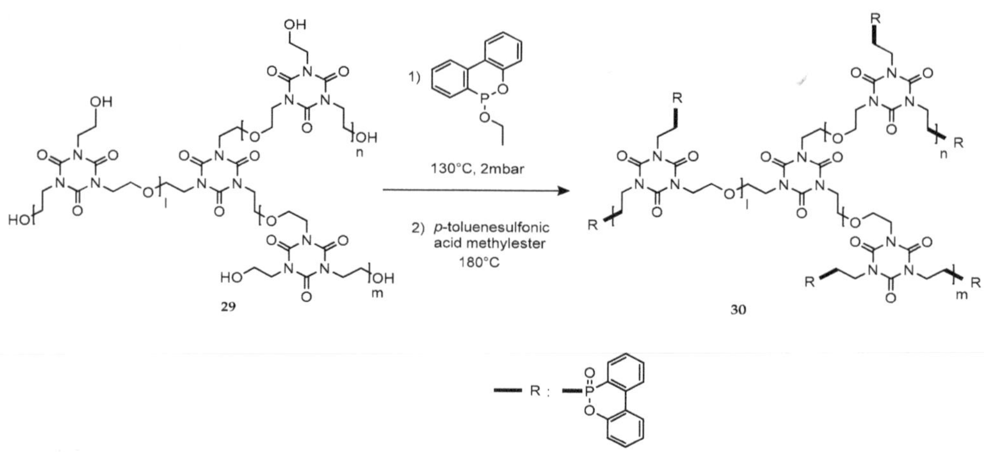

Scheme 3. Synthetic route for oligomer **30**.

The spiro-cyclic phosphorus-containing flame retardant **32b** was synthesized from reaction of 5-hydroxymethyl-5-methyl-2-phenyl-1,3,2-dioxaphosphorinane-2-oxide with compound **32a** (Scheme 4), in 70% yield [73]. P-C bond in compound **32a** was formed via Michaelis–Arbuzov reaction using the starting material **31** followed by chlorination reaction using thionyl chloride. The final product **32b** was then blended with ABS in 20 wt % and 30 wt % loading and the thermal decomposition and flammability of the formulations were investigated using TGA and UL 94 tests. A 20 wt % loading of **32b** in ABS didn't show any improvement in flame retardancy. However, for a higher concentration of **32b** (30 wt %), the temperature of 10% weight loss ($T_{d10\%}$) was 402 °C and 384 °C under N_2 and air, respectively. For the neat ABS, a $T_{d10\%}$ of 326 °C and 306 °C in N_2 and air were observed, respectively. For the 30 wt % **32b** formulation, a PHRR of 268.4 W/g and a THR of 26.9 kJ/g (the neat ABS having showed a PHRR of 505 W/g and a THR of 32.7 kJ/g) was observed.

It was suggested that this flame retardant acts in a condensed phase when incorporated in PC and in a gas phase when incorporated in ABS.

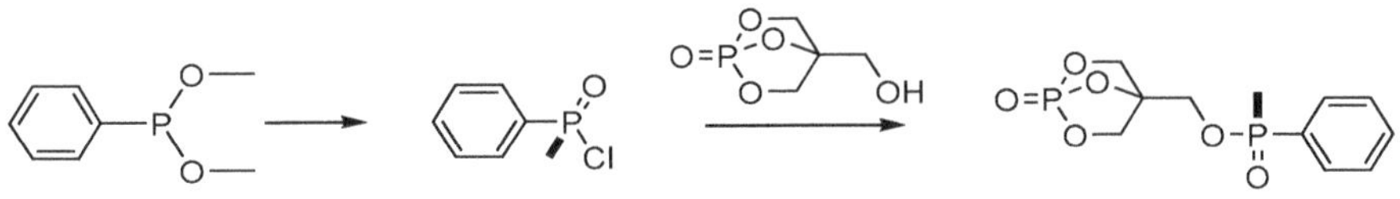

31 32a 32b

Scheme 4. Schematic route for compound **32b** synthesis.

For a formulation of 30 wt % of compound **32b** in ABS resin, one could achieve a V-0 rating. V-0 classification is the highest classification in terms of fire safety in UL 94 test. However, for the PC blend, a UL 94 V-0 rating was obtained with a much lower concentration (4 wt %) of compound **32b**.

The norbornene-based phosphorus containing compound **33a–33e** (Figure 12) were synthesized using ring opening metathesis (ROMP) and investigated as flame retardants [68]. However, the main block monomer **33** has been synthesized in multistep reactions including Michaelis–Arbuzov rearrangement in 68% yield. The polymers **33a–33e** were then applied by coating to folded printed paper and subsequent burning tests were performed. The TGA results of these polymers showed that most of them (**33a**, **33b**, **33d**, **33e**) work in condensed phase: the residual weights were 31 wt %, 34 wt %, 27 wt %, 38 wt %, respectively, whereas, for **33c**, a residual weight of 21 wt % was observed. These polymers were also subjected to photolithography using "thiol-ene click" chemistry, which allowed the phosphonated polymer to get a promising negative photoresist. TGA was performed on the polymer samples and homopolymer of **33a** exhibited a two-step thermal decomposition. A $T_{d10\%}$ of 331 °C was observed, which was 50 °C lower than for other polymers studied. However, the obtained coating from compound **33a** showed high flame resistance. This was attributed to higher residual weight (37 wt %) for the treated paper as observed in the TGA experiments. LOI value of the blank paper was 21% with a flame speed of 1 cm/s. With 40 wt % of **33a** homopolymer coating on paper, the LOI value was increased to 25% with a burning speed <0.1 cm/s. At a oxygen concentration 35%, the homopolymer **33a** coated paper showed a burning speed <0.5 cm/s, which was half of that obtained for the blank paper. The norbornene derived phosphonated polymer was tested for the first time as flame retardant and the study confirmed its self-extinguishing ability when applied on paper.

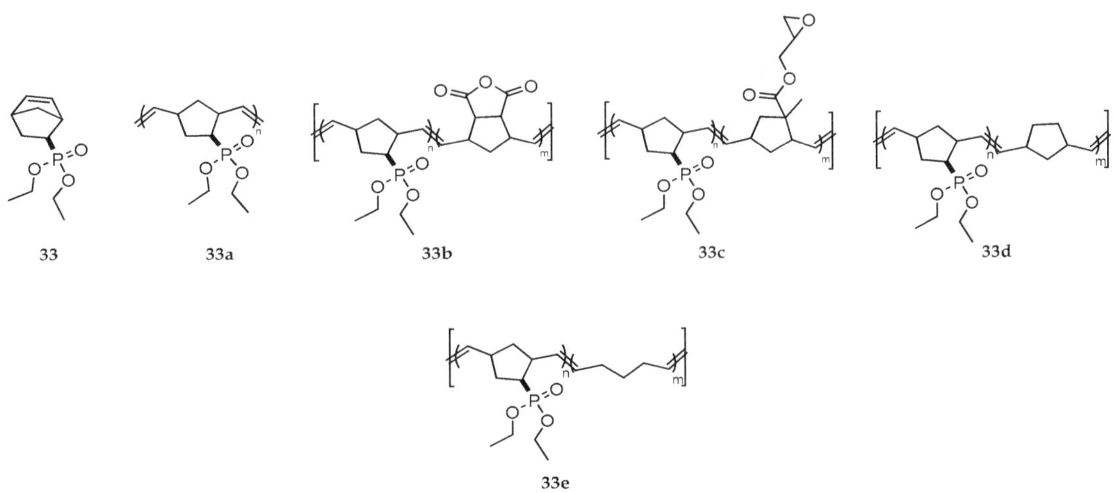

33 33a 33b 33c 33d

33e

Figure 12. Monomer **33** and its corresponding polymers.

Methylphenylphosphinic acid (**34a**, Figure 13) was obtained from hydrolysis of methyl methylphenylphosphinate (MMPP) in hydrochloric acid in 90% yield. MMPP was synthesized in a two-step synthetic strategy. It involved reactions of dichlorophenylphosphine with methanol to form dimethyl phenylphosphonite and subsequent rearrangement reaction with methyl iodide to yield MMPP [74], in 65% overall yield. It was observed that the thermal decomposition of compound **34a** occurred in one step compared to other structurally analogues flame retardants studied (**34b**, **34c** and **34d**, Figure 13). In addition, the compound **34a** showed a $T_{d10\%}$ < 300 °C in the TGA experiment. The flame retardant additives were also copolymerized with acrylonitrile-butadiene-styrene (ABS) and ethylene-vinyl acetate (EVA) separately with a 20 wt % loading (~4 wt % phosphorus). Based on the residue obtained in TGA, it was reported that all these flame retardants act in the gas phase in EVA formulation, compounds **34a** and **34b** act in gas phase in ABS formulations and compounds **34c** and **34d** act in both gas and condensed phases in ABS formulations. A UL 94 V-0 rating for formulation containing 20 wt % of compound **3.34a** in ABS was observed. Compounds **34a**, **34b**, **34c** and **34d** were compared for their flame retardant performance. It was observed that compound **34b** exhibited the best flame resistance; however, it was considered not suitable as real flame retardant because of its high solubility in water. Compound **34a** exhibited the second best result in the UL 94 test, was also the most volatile, and followed by **34c** and **34d**. A general conclusion on the influence of the structure on the flame retardancy was made i.e., flame retardant having a methyl group attached directly to the phosphorus atom showed greater flame retardancy compared to compounds having a phenyl, hydrogen or a hydroxyl group.

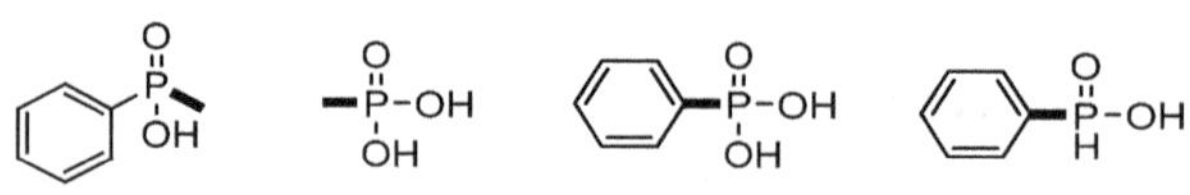

34a 34b 34c 34d

Figure 13. Flame retardant **34a–34d** structures.

3.4. Grignard Type Reaction

Grignard reactions can also be used for the creation of new P-C bonds [75]. Hexaallyltriphosphazene (HAPP) (**35**, Figure 14) was prepared from the reaction of allylmagnesium bromide with hexachlorocyclotriphosphazene (in 94% yield) for application in epoxy system to improve its flame retardancy. HAPP was further reacted in two steps to give hexa(3-triglycidyloxysilylpropyl) triphosphazene (**36**) in 92% yield, which was then used in epoxy formulations. LOI tests were performed for the epoxy system with and without flame retardant additive **36**. The results showed that compound **36** played a major role in improving flame retardancy. For example, the LOI value of the neat resin was 22.5%, whereas, with 15 wt % of **36**, the LOI value increased to 32.1%. Moreover, the TGA data revealed that the higher the concentration of compound **36** in the epoxy system, the higher the $T_{d5\%}$ and $T_{d50\%}$ temperatures. The TGA results also showed that, for the most concentrated resin (15 wt % of compound **36**), the char residue under was 10.2%. Therefore, it was reported that this flame retardant acts predominantly in the gas phase.

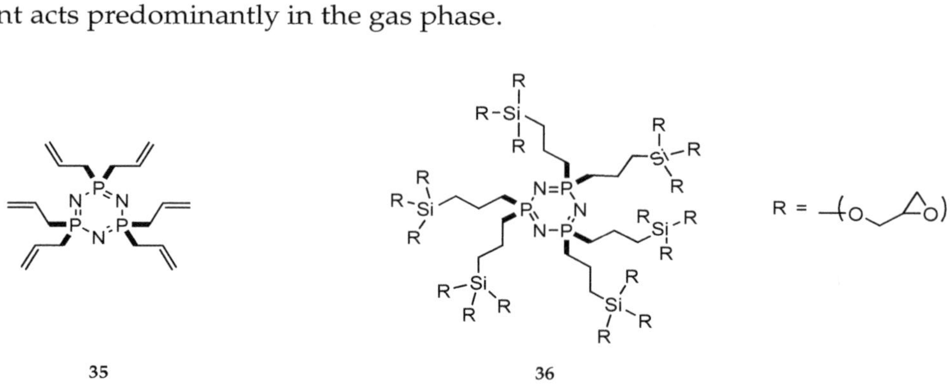

35 36

Figure 14. Chemical structures of **35** and **36**.

Diallylphenylphosphine oxide (**38**, Figure 15) was prepared from reaction of allylmagnesium bromide with phenylphosphonic dichloride (**37**, Figure 15), in about 55% yield [76]. The light yellow product obtained was applied in UV-cured organic–inorganic hybrid coatings on corona treated Plexiglass® panels. The formulations contained tetraethyl orthosilicate (TEOS), diallylphenylphosphine oxide (5 to 20 wt %), an aliphatic urethane diacrylate resin and a silane-coupling agent. Free films on Teflon™ wells were also prepared and cured with UV-light. Mechanical tests were performed on the cured samples with and without addition of compound **38** and evaluated for tensile properties like the Young's modulus and the elongation at break, the contact angle, the pendulum hardness and the cross-cut adhesion. It was found that the addition of diallylphenylphosphine oxide increased the Young's modulus but decreased both the tensile strength and the elongation at break compared to the formulation without the compound **38**. The contact angle also decreased with the addition of compound **38** to the formulation due to its high polarity. The pendulum hardness increased with addition of the flame retardant due to a higher cross-linking density and the cross-cut adhesion (which ranged from 0 for the best and 5 the poorest adhesion) showed a result of 0 for a 15 wt % and 20 wt % flame retardant content. The thermal stability of the samples was also investigated by TGA and it was found that addition of 20 wt % of compound **38** lowered the $T_{d5\%}$ by 40 °C compared to the sample without the flame retardant. The char content as measured in TGA for the most concentrated resin (20 wt % of **38**) is lesser for lower (10% of **38**) concentration formulation. Normally for a flame retardant working in the condensed phase, the char content in TGA analysis increases with increase in its concentration in the polymer. Thus, one can infer from the TGA data that flame retardant **38** work mostly in the gas phase. The photo-DSC test was also performed, and it was concluded that the enthalpy of polymerization increased with the increase in concentration of compound **38**.

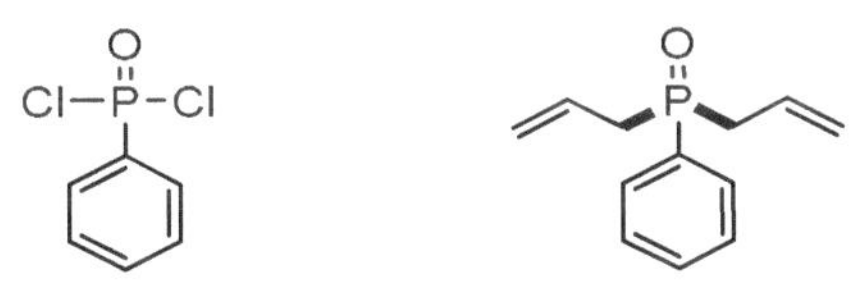

37 38

Figure 15. Phosphine oxide derivative **38** and its precursor **37**.

4. Flame Retardants with P-$C_{aromatic}$ Bonds

In the following sections, we have summarized specific examples of P-$C_{aromatic}$ bond containing organophosphorus compounds, which are classified based on the synthetic strategy for creation of P-C bond. For clarity, the specific P-C bonds that were created in the structure have been marked in bold.

4.1. Friedel–Crafts Type Reactions

One of the main routes to phosphorus-aromatic carbon bonds formation is a modified Friedel–Crafts reaction usually involving phosphorus trichloride and a Lewis acid. DOP-Cl (Compound **39**, Figure 16) has reportedly been synthesized starting from *o*-phenylphenol using phosphorus trichloride in presence of zinc chloride catalyst in 95% yield [66,77]. DOP-Cl is an important precursor for DOPO derivatives due to its high chemical reactivity towards nucleophiles [5]. As an example, it is shown in the literature that DOP-OMe reacts with hydroxymethyl substituted phenols via a Michaelis–Arbuzov type rearrangement to give compound **14** (Scheme 2) in 32% yield. Compound **14** has a glass transition temperature of 108.5 °C. In the published patent, it was claimed that this family of compounds can be used as flame retardant for a wide range of polymers such as PU, epoxy and other thermosetting resins or thermoplastics in general. It is further claimed that these compounds are particularly adapted to protective coating formulations, electrical laminates and molded thermoplastic products [77].

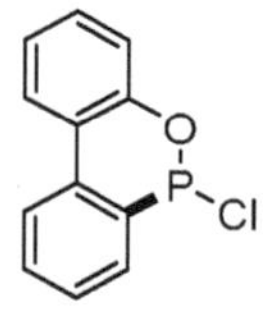

39

Figure 16. Structure of DOP-Cl **39**.

The same reagents as described earlier were also used in another patent to prepare compound **39** (Figure 16) and was subsequently oxidized [78]. Next, esterification with phenols or polyhydroxybenzenes was performed to develop several flame retardants and one such example is compound **40** (Figure 17) (obtained from bisphenol A). No further information on characterization, flame retardancy or applications were provided.

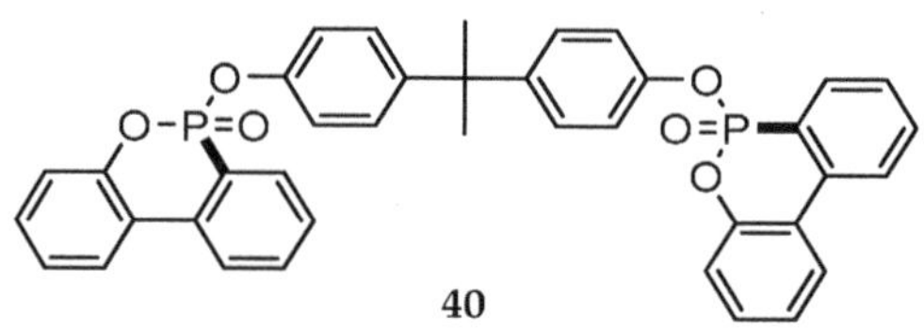

40

Figure 17. Chemical structure of bisphosphonate **40**.

Friedel–Crafts reactions can also be performed using aluminum chloride as Lewis acid. It was reported that dichlorophenylphosphine sulfide reacts with two equivalents of toluene in the presence of $AlCl_3$ catalyst to give **41a** (Scheme 5, 84% yield) [79]. The compound was further oxidized to give BCPPO (**41b**, Scheme 5). Several polycondensations of this flame-retardant monomer and terephthalic acid (TPA) with bisphenol A were performed using different TPA/**41b** ratios, yielding six different polyarylates. Films were prepared from these phosphorus-containing polyarylates for further characterization. Mechanical properties of the polymers were evaluated, and it was found that tensile strength and elongation at break decreased with introduction of **41b** in the polyarylate. Crystallinity was lost when **41b** was incorporated in the polymer. Glass transition temperature, $T_{d5\%}$ temperature and char yield followed the same positive trend when phosphorus content was increased, showing tangible evidence that **41b** introduction can improve the thermal stability of the resulting polymer. To confirm improved flame retardancy of the new materials, LOI value was measured and was shown to increase from 30.7% to 34.5% with increasing phosphorus content. Further study of the thermal decomposition behavior of the polymers concluded that increasing the proportion of compound **41b** in the polymer increased the char formation, indicating that compound **41b** acts in both the gas and condensed phase. Finally, an additional goal of this work was to study the solubility of the synthesized polyarylenes. Good solubility was achieved in a wide selection of organic solvents for formulations with 20% or more of compound **41b**. This was attributed to the increase of free volume induced by the introduction of bulky pendant phenyl units in the chain. It was thus concluded in this work that introduction of compound **41b** enhanced the thermal stability while improving the polyarylates' processability.

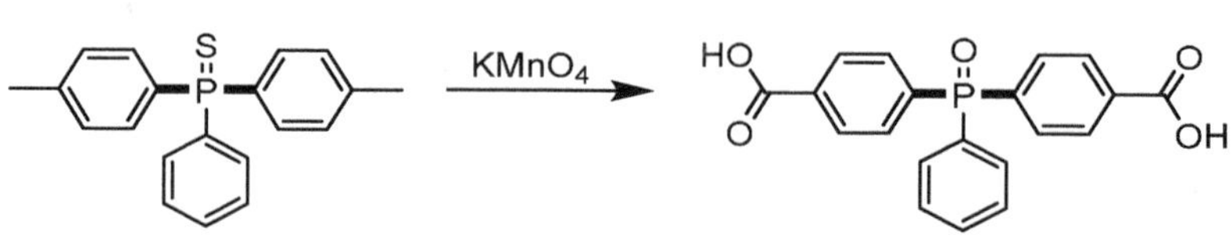

Scheme 5. Oxidation of **41a** to **41b**.

DAP-Cl (compound **42a**, Figure 18), a precursor to the nitrogen-containing DAPO (compound **42b**, Figure 18), was synthesized in 95% yield using 2-aminobiphenyl, PCl_3 and $AlCl_3$ as catalyst [80,81]. It was reported that DAPO exists in the solution in a tautomeric equilibrium with its corresponding phosphonamidic acid [82]. Another DOPO derivative, DOPS (compound **42c**, Figure 18), was synthesized by reacting DOP-Cl with hydrogen sulfide at room temperature (89% yield) [83]. Epoxy phenol novolac resins were functionalized with DAPO and DOPS, followed by a curing step. Phosphorus content in these formulations ranged from 0.6 to 1.6 wt %. DSC measurements showed that DAPO-modified resins had higher T_g compared to the reference DOPO-modified resins. This was attributed to the presence of an amine group that can also act as a cross-linking reagent, resulting in a more rigid network. On the contrary, DOPS-modification induced a sharp decrease of T_g, which can result in lower crosslinking density due to the bulky sulfur atom. Flame retardancy of these modified resins was investigated and compared to DOPO-modified resins. At 1.6 wt % P loading, DOPS- and DAPO-modified resins achieved UL 94 V0 rating. Char yield in TGA experiments reached 35% for DAPO and 32% for DOPS based resins. Nitrogen-phosphorus synergistic effect was considered to be the reason behind increased char yield, and was also presented as a possible explanation for their slower decomposition rate. Sulfur-phosphorus synergy hasn't yet been extensively studied so no conclusion could be drawn. Increased char yield indicate a high probability that DOPS and DAPO may work in condensed and gas phase, whereas DOPO acts mainly in gas phase.

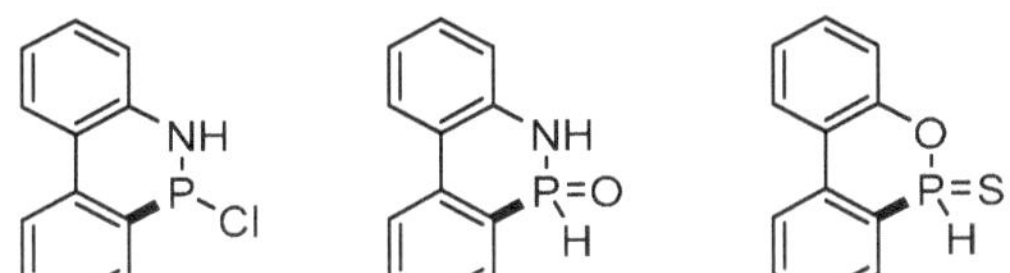

42a 42b 42c

Figure 18. DOPO analogues.

Two similar compounds were prepared using aluminum chloride and phosphorus trichloride reagents and di-*p*-tolylether [84] or diphenylamine [85], affording DPPO (compound **42d**, 91% yield) and DPPA (compound **42e**, 61% yield), respectively (Scheme 6). Compound **42d** was then oxidized with hydrogen peroxide or reacted with formaldehyde, affording DPPO-OH and DPPO-CH_2OH, respectively. These two intermediates along with unmodified **42d** were then used to functionalize epoxy novolac resin DEN 438 by their reaction with the epoxy groups, resulting in DPPO-OH-DEN, DPPO-CH_2OH-DEN and DPPO-DEN. The same procedure was applied to DOPO, affording DOPO-OH, DOPO-CH_2OH-DEN and DOPO-DEN. Each flame-retardant prepolymer was prepared with phosphorus content ranging between 0.5% and 2.0%. 4,4′-Diaminodiphenylmethane (DDM) was used as hardener during the curing step.

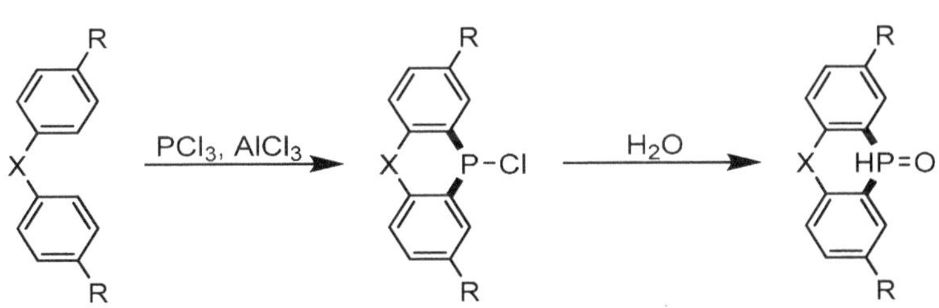

X = O ; R = CH_3 (42d)
X = N ; R = H (42e)

Scheme 6. Synthetic route for phosphine oxides DPPO (**42d**) and DPPA (**42e**).

Compound **42e** was directly reacted to epoxy groups of DGEBA resin (via the phosphorus and nitrogen atoms) to obtain DPPA-DEN prepolymers [85]. The phosphorus content of the formulations varied from 0 to 3.20 wt %. Curing was then performed using DDM or 3,5-diethyltoluene-2,6-diamine (DETDA) as hardeners. High level flame retardant properties were achieved for all the DOPO- and DPPO-based epoxy resins. At a 0.81 wt % phosphorus content, all of them acehieved UL 94 V-0 rating. For this same phosphorus content, DOPO-based resins exhibited LOI value between 31% and 31.5%, whereas, for DPPO-based resins, LOI values ranged between 27.4% and 28.3%, both considered really good values for flame retardants. In case of compound **42d**, PO radicals were shown to form in the gaseous phase, indicating the gas phase mode of action [84]. DPPA-DEN resins exhibited lower flame retardant performance, needing a phosphorus content of 3.20% to attain UL 94 V-0 rating as well achieving LOI values 28.5% (DDM-cured) and 29.3% (DETDA-cured). Thus, it was concluded that DPPO-, DPPA- and DOPO-functionalization proved to be beneficial in improving flame retardants in epoxy resins.

4-carboxyphenylphenylphosphinic acid (**43**, Figure 19) was prepared by addition of toluene to dichlorophenylphosphine and anhydrous aluminum chloride followed by hydrolysis and the oxidation reaction [86], achieving a yield of 33.1%. Its solubility in water was investigated from 26.70 °C to 94.52 °C and experimental data was correlated using a second-order polynomial with an absolute average deviation of 2.1%. These solubility measurements were considered of interest because of the important role of water in the industrial production and purification of compound **43**. Being a reactive phosphorus-containing monomer, compound **43** can be used as a co-monomer for manufacturing functional polyesters.

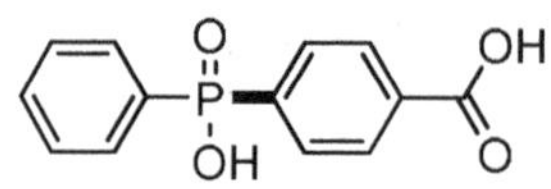

43

Figure 19. Chemical structure of compound **43**.

4.2. General Substitution Reactions

Compound **44** (Figure 20) was reported to be synthesized by adding chlorodiphenylphosphine to a mixture of *p*-bromostyrene and butyllithium, followed by an oxidation step, resulting in 65% yield [87]. Cationic copolymerization was investigated with soybean oil for different ratios of styrene to substance **44**, using divinylbenzene as crosslinker. Due to a problem of heterogeneity before and after the polymerization reaction, an alternative phosphorus-containing styrene substance was investigated. This alternative compound underwent cross-metathesis with a vegetable-oil derivative and was copolymerized, affording homogeneous thermoset polymers. Thermal-oxidative degradation was investigated by TGA, showing that 1 wt % phosphorus incorporation improved the thermal stability of the thermoset slightly. Condensed phase action was inferred from the increased char formation observed in the TGA. Phosphorus-free reference polymer achieved 19.2% LOI value, and phosphorus-containing polymer was reported to achieve a maximum LOI value of 24%.

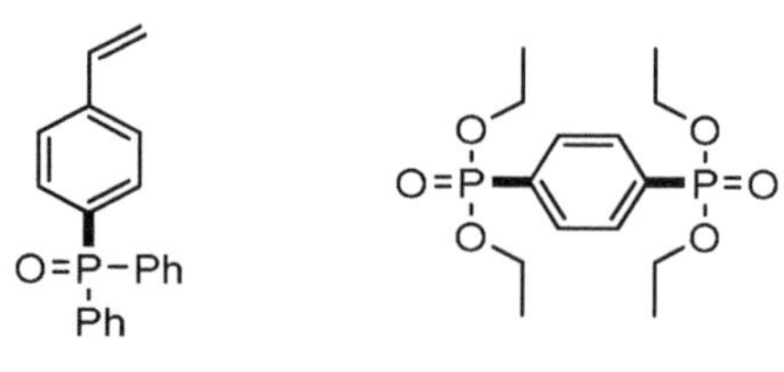

44 **45**

Figure 20. Phosphine oxide derivative **44** and phosphonate derivative **45**.

1,4-bis(diethylphosphoro)benzene (**45**, Figure 20) was synthesized from the reaction of 1,4-dibromobenzene with triethyl phosphite [88]. Compound **45** was used in battery electrolytes compositions to enhance flame retardancy properties and to reduce impact of possible thermal runaway.

Phosphorus-carbon bonds formation were reported with good yields for substitution reactions involving triethyl phosphite and cyanuric chloride. These compounds are considered very efficient as charring agents due the presence of nitrogen in the triazine ring and thus expected to exhibit P-N synergism.

In a recent work [89], diethyl phosphite-functionalized cyanuric chloride (86.1% yield) was dimerized using neopentyl glycol as a bridging agent, yielding compound **46** (Figure 21, 92.5% yield). The target flame retardant can bind covalently to cotton fabrics by reaction of the cellulose hydroxyl groups via a substitution reaction. The best curing method was established from several trials and was applied to treat cotton fabric samples with 10 to 30 wt % of the flame retardant. TGA analysis showed an improvement of thermal stability and increase of the char yield above 30% for all treated samples, indicating the condensed phase mode of action of the flame retardant. During vertical flammability testing, char length became shorter with increasing flame retardant content, and a B-1 rating was achieved for samples treated with more than 20 wt % flame retardant. Improved flame retardancy was confirmed from LOI tests, which showed a LOI value of 30.3% and 18% for 30 wt % of flame retardant treated fabric and untreated fabric, respectively. Washing with water of treated cotton showed a rapid decrease in LOI values. Initial LOI of 28.4% dropped to 22% after 10 times washing indicating not all flame retardants reacted with the cellulose. Such flame retardants primarily act in the condensed phase in cellulose based materials.

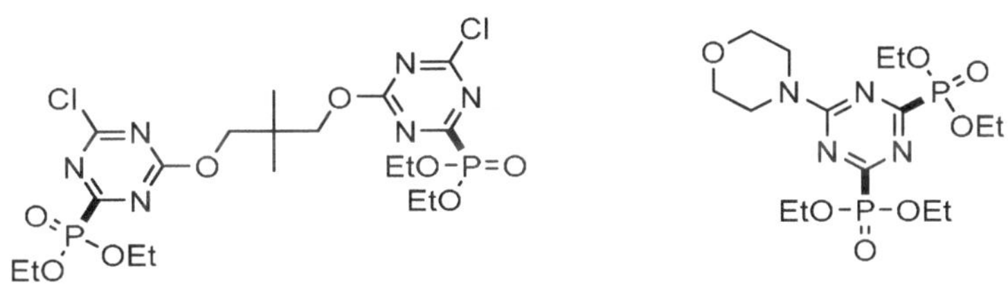

46 47

Figure 21. Triazine derivatives **46** and **47**.

A reaction of three equivalents of triethyl phosphite with cyanuric chloride resulted in 2,4,6-trisubstituted 1,3,5-triazine (86.1% yield). One of the diethyl phosphate groups on the triazine ring was then substituted by morpholine, producing another substituted triazine-based compound **47** (Figure 21, 92.5% yield) [90]. An oligomer was synthesized by transesterification reaction of compound **47** with pentaerythritol in 93.6% yield. It was then incorporated in polypropylene at 10 and 30 wt %, which respectively achieved UL 94 V-2 and V-0 ratings, while LOI values of 23.4% and 28.9% were observed for the same formulations, respectively. A polymer formulation was further prepared with 30 wt % ammonium polyphosphate/pentaerythritol/**47** (ratio 3:1:2), which achieved a higher LOI of 29.4% and UL 94 V-0 rating. This improvement in flame retardancy was attributed to the possible synergistic effect between the three additives.

4.3. *Addition Reactions*

Recently in the literature, a large amount of research work can be found on addition reactions of DOPO derivatives on α,β-unsaturated ketones. This is mainly due to the versatility and high yields allowed by this type of reaction. DOPO addition on 1,4-benzoquinone is one of the most common uses of this addition reaction. DOPO-BQ (compound **48a**, Scheme 7) was synthesized following a previously reported method in 93% yield [91] as an intermediate leading to six flame retardant additives for lithium ion batteries. They were then characterized for their thermal properties using TGA and DSC [92]. From these six compounds, the compound **48b** (Scheme 7) that achieved a $T_{d5\%}$ of 210 °C was selected for further flame retardant evaluation. It was mixed with conventional Li-ion electrolytes, with weight fractions ranging from 2.5 to 25 wt %. Self-extinguishing time (SET) measurements were performed, showing a sharp decrease in SET, whereas ionic conductivity underwent a 50% decrease when increasing the weight fraction of the additive from 2.5 wt % to 25 wt %. The flame retardant efficiency of the electrolyte formulations was confirmed, but lower battery performances are to be expected due to lower ionic conductivity results. Char formation indicated a condensed phase action. Studies on actual battery performances along with electrochemical properties are underway.

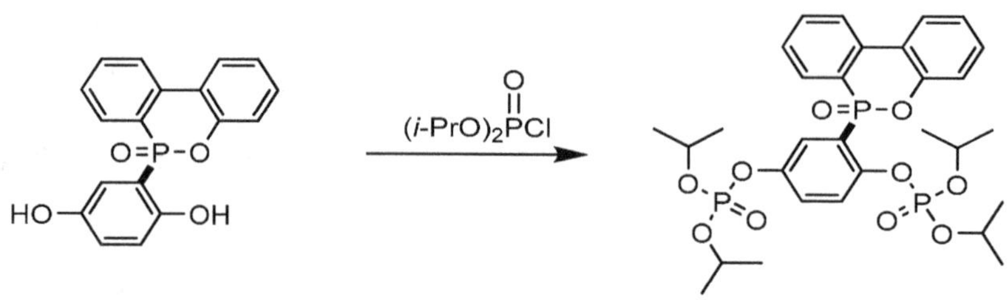

48a **48b**

Scheme 7. Synthesis of DOPO based phosphinate derivative **48b** from **48a**.

Compound **48a** was reacted with acryloyl chloride to obtain a phosphorus-containing acrylate monomer (compound **49**, Figure 22) [93]. Various concentrations (0 to 20 wt %) of this compound were introduced into the formulation of unsaturated polyester resins synthesized by radical copolymerization. A high LOI value of 24% was achieved for formulations containing ≥5 wt % phosphorus-containing monomer. $T_{d5\%}$ peaked at 260 °C in air, at 300 °C in nitrogen, for the highest flame retardant-containing formulation, and its peak heat release rate (PHRR) was reduced to 288.7 W/g. Its flame retardant properties were confirmed both in condensed and gas phase, due to charring and flame inhibition phenomena, respectively.

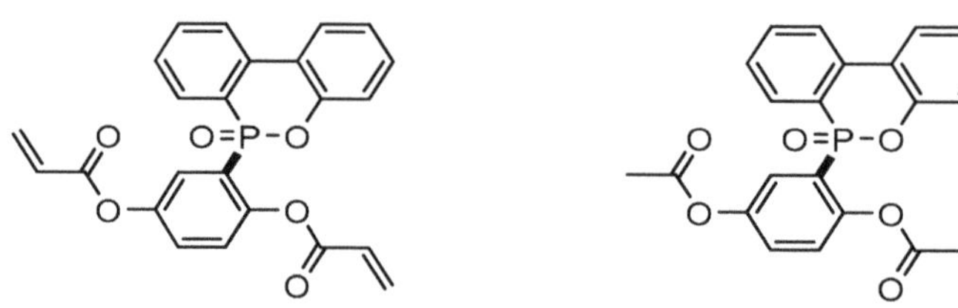

49 **50**

Figure 22. DOPO based phosphinate derivatives **49** and **50**.

Substance **48a** was also reacted with acetic anhydride, affording **50** (Figure 22, 75% yield) [94]. Substance **50** was then copolymerized with *p*-acetoxybenzoic acid, terephthalic acid and the condensation product of terephthalic acid with 1,8-octanediamine, producing a new flame retarded thermotropic liquid crystal polymer. DSC experiment showed lower T_g and T_m for it compared to usual liquid crystal polymers, namely, 127 °C and 147 °C. It adopted nematic liquid crystalline behavior in the temperature range 180–330 °C. Thermal stability of the polymers was investigated by TGA and $T_{d5\%}$ was estimated at 385 °C, showing satisfying thermal stability and char yield, allowing for very good flame resistance.

Diglycidyl ether of bisphenol-A epoxy resin (DGEBA) was modified by compound **48a**, affording the phosphorus-containing epoxy resin **51** (Figure 23) [95]. Four cured epoxy resin/polyhedral oligomeric silsesquioxane (POSS) hybrids were synthesized, and all samples contained 2 wt % phosphorus but with varying silicon content (from 0 to 4 wt %). The thermal-oxidative stability of the polymers was evaluated by TGA. With increasing the silicon content, TGA analysis showed a decrease of $T_{d5\%}$ and an increase of $T_{d50\%}$ and char yield, which was attributed to the phosphorus-silicon synergism. Further flame retardant properties were estimated by PCFC, giving evidence that PHRR decreased and occurred at higher temperatures with increasing silicon content. T_g of the polymer was shown to increase with increase in phosphorus and silicon content. SEM and XPS were used to analyze char residues and results obtained reinforce hypotheses on phosphorus-silicon synergistic impact on epoxy resins flame retardancy.

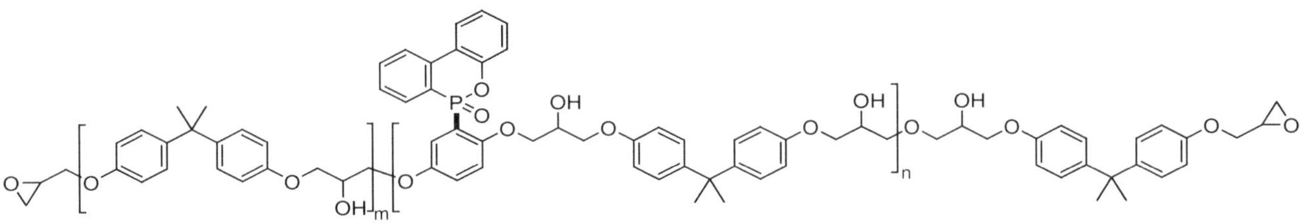

51

Figure 23. DOPO-BQ (**48a**) functionalized DGEBA resin **51**.

Similar to the previous study, different epoxy resin **51** were synthesized from reaction of DGEBA and compound **48a** in different molar ratios [96] and several dicyanate esters were used as curing agents. TGA data under nitrogen showed evidence that degradation temperature and char yield for the resin increased with the increasing phosphorus content, showing a condensed phase mode of action. UL 94 V-0 rating was attained for formulations containing around 2 wt % phosphorus, when UL 94 V-1 rating was achieved for the resins containing around 1 wt % phosphorus, the curing dicyanate ester having no major influence. Potential applications as flame retardant laminates or protective coatings were suggested for these resins.

Previously synthesized compound **48a** [91] and DPPO-HQ, (compound **52**, Figure 24) (obtained by reacting compound **42d** (Scheme 6) with *p*-benzoquinone) [84,97] were used as diols to form aromatic polysulfones (PSU) by nucleophilic aromatic substitution polycondensation with 4,4′-difluorodiphenylsulfone (DFDPS) [98]. Two polysulfones were thus obtained, having around 5 wt % phosphorus content. Both exhibited significantly higher T_g than phosphorus-free, bisphenol A-based PSU (reference), and a two-step decomposition was observed instead of one-step for the reference polymer in TGA. TGA and Py/GC-MS data were used to formulate a hypothesis concerning the thermal degradation process of these formulations. It was also concluded that DOPO-based PSU combined gaseous and solid-state flame-retarding mechanisms, whereas DPPO-based PSU favored solid-state action. Thus, DOPO-based PSU was considered as a slightly better in terms of flame retardancy.

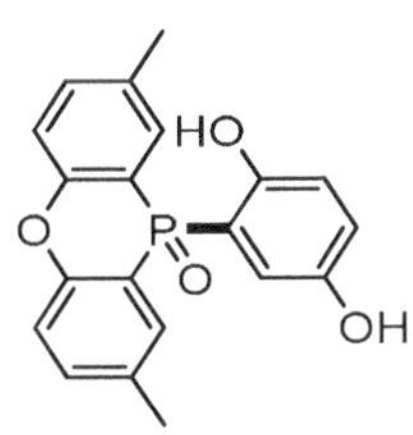

52

Figure 24. Phosphine oxide derivative **52**.

Following the same reaction principle as described earlier, DOPO can also react with 1,4-naphthoquinone to give DOPO-NQ (compound **53a**, Scheme 8, 70% yield) [99], which was then functionalized by trimellitic anhydride chloride and 4-aminophenol to give a bisphenol derivative (compound **53b**, Scheme 8, 71.8% yield) [100]. Several liquid crystalline co-poly(ester imide)s were prepared by polycondensation of an aromatic diacid chloride with various ratios of the compound **53b** and aliphatic diols. Thermal analysis of the polymers was studied by TGA, and it was observed that all polymers followed a two-step decomposition process. The least stable polymer exhibited a $T_{d5\%}$ of 340 °C, whereas the most stable showed a $T_{d5\%}$ of 395 °C. The thermal degradation behavior of the polymers was investigated by FTIR-TGA and SEM-TGA analysis. Phosphorus-nitrogen synergy is expected to contribute to flame retardancy, suggesting gas and condensed phase activity for the phosphorus compounds. Solubility in various solvents was investigated. Liquid crystalline properties were confirmed by XRD and DSC analysis.

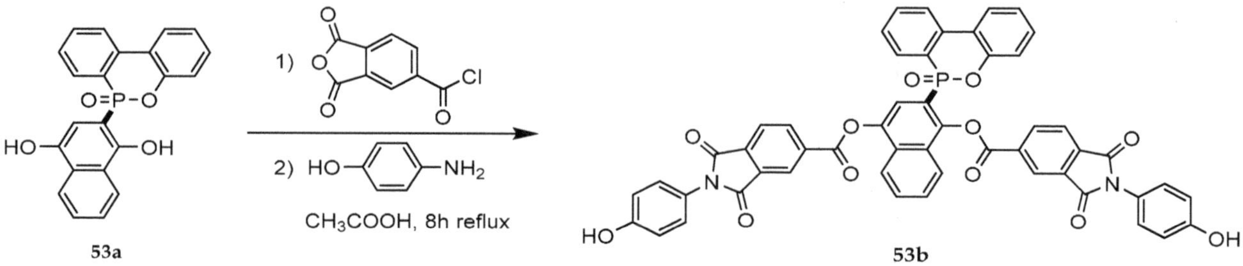

Scheme 8. Synthetic route for DOPO based phosphinate **53b**.

In a second paper [101], the same group synthesized three different poly(ester imide)s by solution polycondensation of equimolar compound **53b** (Scheme 8) and aromatic diacids (either terephthalic acid, isophthalic acid or 4,4′-dicarboxydiphenylether) yielding PEI-1, PEI-2 and PEI-3, respectively. TGA experiments were run and concluded that these polymers exhibited 55% char and were stable up to 375 °C. Their T_g were respectively measured at 226 °C, 200 °C and 190 °C, making them good candidates for thermoforming. Energy-dispersive X-ray spectroscopy (EDX) confirmed the presence of phosphorus in the char, proving condensed phase action. Liquid crystalline properties were confirmed by WAXD, optical properties and DSC studies.

Diphenylphosphine oxide was reported to react with benzoquinone to give compound **54** (DPO-BQ, Figure 25, 78% yield) [102]. Its thermal stability was shown to be slightly better than the reference compound **48a**, having a $T_{d5\%}$ of 315 °C and a slower decomposition in air. The compound **54** was then reacted to epoxy resin, followed by curing with 4,4′-diaminodiphenylsulfone to achieve thermoset resins having 1 and 2 wt % phosphorus. TGA investigations showed few differences between the two formulations containing various concentrations of compound **54**, $T_{d5\%}$, being around 391 °C in nitrogen and 384 °C in air. Polymer formulation with 2 wt % phosphorus achieved a UL 94 V-0 rating and LOI of 32.4%, whereas the one containing only 1 wt % phosphorus reached a V-1 rating and a LOI value of 29.2%. Good flame retardance was achieved at low phosphorus loadings, although TGA results showed little improvements compared to phosphorus-free epoxy resins, except for higher char yields, indicating a condensed phase mode of action.

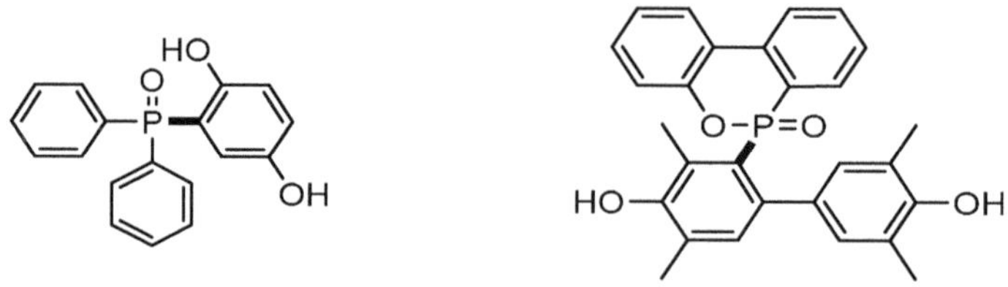

Figure 25. Chemical structures of compounds **54** and **55**.

A recent patent reported that DOPO addition on 3,3′,5,5′-tetramethyldiphenoquinone gave a yellow compound **55** (Figure 25) in 73% yield [103]. The aim of this work was to synthesize phosphorus-containing compounds with low steric hindrance for application in flame retardant epoxy resins.

According to previous studies, a disubstituted phosphorus compound **56a** and a monosubstituted phosphorus compound **56b** (Figure 26) were obtained as a mixture using modified separation and purification steps. Melting points for these compounds were 258–260 °C and 184–187 °C, respectively [104]. Further characterizations were performed and were compared with other phosphorus compounds described later in the text.

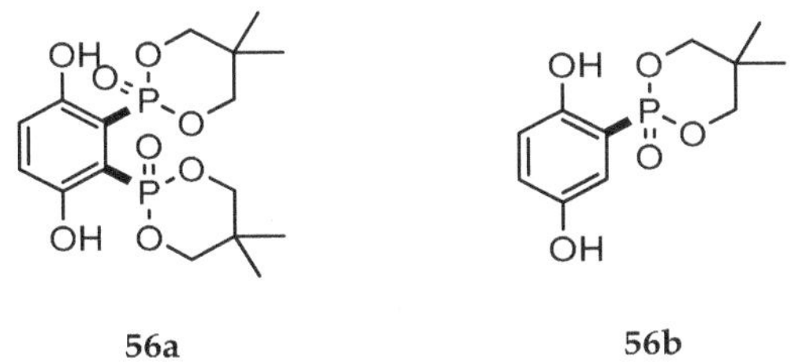

Figure 26. Phosphonate derivatives **56a** and **56b**.

4.4. Michaelis–Arbuzov Rearrangement

In the same work [104], phosphonoterephthalic acid derivative **56c** (Figure 27) was synthesized using a Michaelis–Arbuzov rearrangement from a previously published synthesis protocol [105]. Compounds **56d** and **56e** (Figure 27) syntheses were also based on a previous study from the same group (exhibiting m.p. of 296–298 °C and 310 °C, respectively). The syntheses involved iodoester reagents with palladium catalyst in a yield of around 80%. PU foams were prepared with compounds **56a** to **56e** according to two different processing methods: *Preps* (reactive flame retardant copolymerized during PU foam formation) and *Blends* (10 wt % of flame retardant additive used with already prepared PU). These foams were analyzed for thermal and flame retardant behavior using TGA and PCFC, respectively. It was found that compound **56e** did decrease the heat release and inhibited melting and flowing of the PU in the *Blend* form. In the *Prep* form, the addition of this compound did not affect the heat release. The resulting PU formulation also had a two-step thermal degradation and increased char yield at 800 °C (neat PU around 9% and PU with 10 wt % flame retardant around 29%). For compound **56d**, the *Prep* form didn't affect either the heat release, but, in the *Blend* form, it induced a significant decrease of the THR. One possible explanation of this phenomenon was the incomplete incorporation of the flame retardant in the PU foam. Compound **56c** showed the same flame retardant behavior as **56e** for PU prepared by *Blend* as well as in *Prep* forms. The previously mentioned phosphonoterephthalic acid derivatives **56a** and **56b** have shown similar efficiency in decreasing the total heat release in both processing methods. It was also found that they didn't inhibit melting and flowing of the PU. In conclusion, compounds **56c** and **56e** showed the best flame retardancy properties of all the investigated structures.

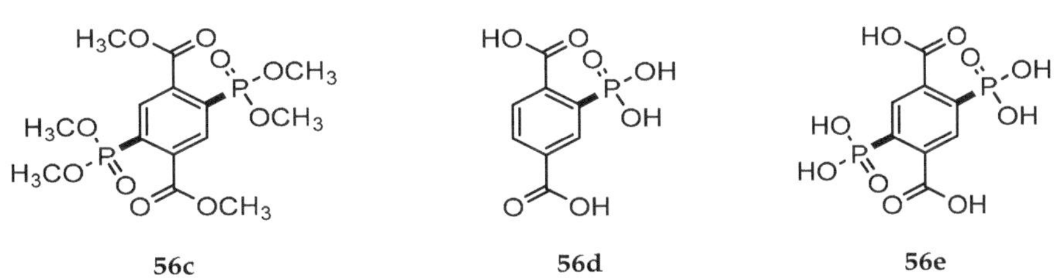

Figure 27. Chemical structures of compounds **56c** to **56e**.

4.5. [1,3]-Sigmatropic Rearrangement

Two polyphenols were transformed into corresponding phosphates using diethyl phosphite following Atherton–Todd reaction [106]. Further treatment with lithium diisopropylamine allowed breakage of a P-O bond and migration of the phosphorus to create a new P-C bond. This process known as [1,3]-sigmatropic rearrangement yielded two different compounds, **57a** (62% yield, m.p. 169–171 °C) and **57b** (65% yield, m.p. 216–217 °C), and both were obtained as white crystals. Their structures are shown in Figure 28. Although no flame retardant tests were performed, it was concluded that these phosphorus-containing polyphenols could be introduced as flame retardant monomers in high-performance polymers including PEEK, PEES, PEK, PES or PC.

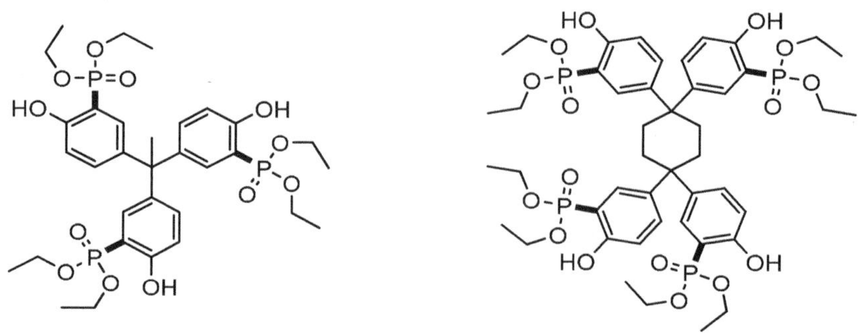

57a **57b**

Figure 28. Phosphonate derivatives of polyphenols **57a** and **57b**.

According to the same mechanism, the phosphonate product (diethyl-[2-hydroxy-5-vinylphenyl) phosphonate (compound **58a**, Scheme 9) was synthesized starting from diethyl-(4-vinylphenyl) phosphate and obtained in a yield of 82% [107]. Diethyl phosphite was then used to react with the hydroxyl group using Atherton–Todd reaction, yielding compound **58b** (Scheme 9, 85.7% yield). Compound **58b** was copolymerized with styrene using three different compositions (5, 10 and 20 wt % of monomer **58b**) yielding copolymers that T_g decreases from 101 to 95 °C when flame-retardant monomer mass fraction is increased in the polymer formulation. TGA showed a two-step degradation behavior explained by the presence of diethyl phosphate and diethyl phosphonate, allowing the main degradation step to happen at a temperature at least 60 °C higher than the neat polystyrene. Copolymer incorporation lowered the temperature of the first decomposition step but increased the temperature of the second degradation stage. The char yield increased from 1.1% to 7.3% with increasing the phosphorus-containing monomer, showing evidence of a condensed phase mechanism. PCFC measurements followed a similar trend, the total heat release decreasing from 40.1 to 38.2 kJ/g with increasing phosphorus content. As a conclusion, efficient flame retardant vinyl polymers were obtained by copolymerizing phosphorus-functionalized styrene with styrene units.

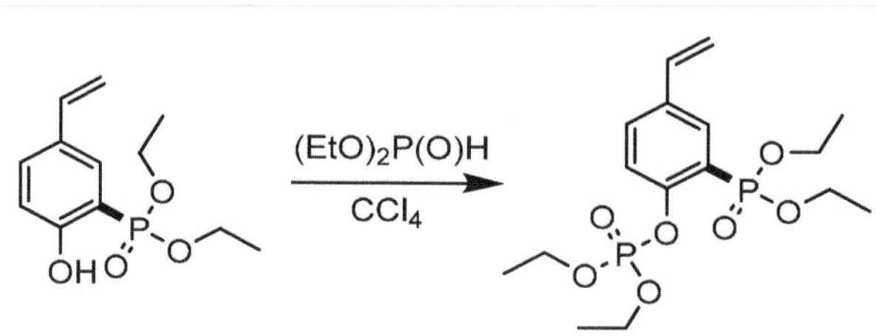

58a **58b**

Scheme 9. Synthesis of the phosphorus-containing vinyl monomer **58b** from **58a**.

4.6. Phospho-Fries Rearrangement

A phosphorus version of the Fries rearrangement was reported by treating tetraethyl-1,2-phenylene bis(phosphate) with a mixture of butyllithium and lithium diisopropylamide to produce diphosphonate compounds **59** (Figure 29, 67% yield) [108]. This procedure was a modified version of a synthesis reported in a previous publication [109] explaining that phosphate esters in presence of an organolithium compound form anions that undergo phosphorus migration from oxygen to carbon. Compound **59** was an intermediate to four lithium salts: three boron-centered anions and one phosphorus-centered. Pyrolysis combustion flow calorimetry (PCFC) results showed evidence that lithium boron phosphorus complexes reached high char yield (up to 49.60%), which lowers the peak heat release rate temperature below 312 °C. These high char yield values indicate condensed phase activity, and could even indicate phosphorus-boron synergism. TGA analysis completed this study by proving these new lithium salts were stable up to 200 °C. It was concluded these new compounds should be effective flame retardants for lithium ion batteries thanks to their reduced flammability, either as electrolyte additives or as possible alternative to existing lithium salts.

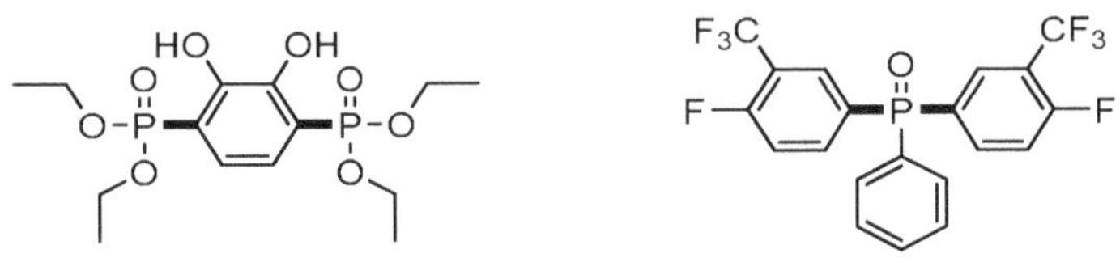

59 **60**

Figure 29. Chemical structure of compounds **59** and **60**.

4.7. Grignard Type Reactions

Grignard reagents represent an additional method of P-C bonds synthesis in which the nucleophilic carbon can react as a base. 5-bromo-2-fluorobenzotrifluoride was added to magnesium turnings, forming the associated organomagnesium compound. Phenylphosphonic dichloride was then added to yield compound **60** (Figure 29) as a pale yellow solid (80% yield) [110]. Polymerization reactions were carried out using an equimolar ratio of compound **60** and different diphenol derivatives. Six novel poly(arylene ether phosphine oxide)s were obtained, with T_g in the range of 173–252 °C, and $T_{d5\%}$ values being between 457 °C and 490 °C. PCFC investigation showed great influence of the diphenol structure, and the peak heat release rate varied from 184 to 67 W/g. Films were prepared, resulting in transparent, brittle or flexible films depending on the diphenol monomer used. The mechanical properties of the flexible films were evaluated, tensile stress reaching up to 54 MPa, Young's modulus up to 0.97 GPa and elongation at break up to 47%. This study brought to light compounds with interesting thermal stabilities, and choosing the right diphenol monomer can allow to finely tune the poly(arylene ether phosphine oxide) properties.

In a similar approach, diphenylphosphinic chloride was reacted with 4-(trifluoromethyl) phenylmagnesium bromide to yield compound **61a** (Scheme 10) [111] having a melting point of around 90.5–91.2 °C. This product underwent nitration followed by hydrogenation to give the amino derivative **61b** (Scheme 10). This phosphorus-containing diamino compound can be used to prepare polyimides, polyamides that exhibit very good fire retardancy, thermal stability, mechanical properties, adhesion as well as low dielectric constant. Possible applications would be adhesives for metals, semiconductor packaging or optical materials.

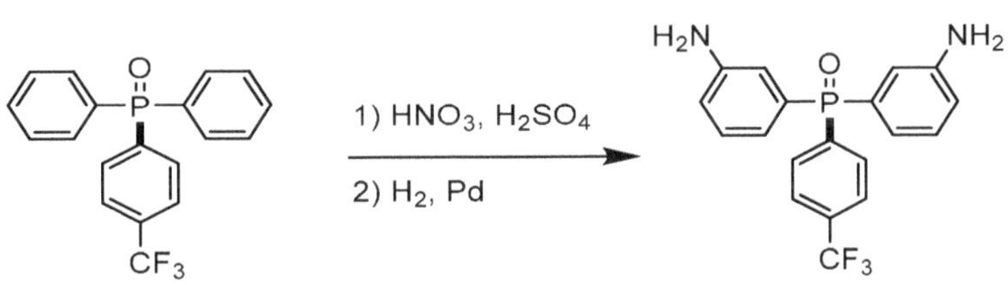

61a **61b**

Scheme 10. Two-step amination of **61a** to yield **61b**.

5. Conclusions

P-C bond creation and its use in synthetic chemistry offer a wide range of possibility of design and development of exciting organophosphorus flame retardants. Conventional synthetic methodologies

Michael addition, Michaelis–Arbuzov, Friedels–Crafts and Grignard reactions are the most commonly used in academics for development of P-C bonded flame retardants. Use of green reagents, catalytic reactions, microwave technology, photo initiated reactions, etc. are increasingly used by researchers to synthesize P-C containing organophosphorus compounds; however, these techniques are seldom utilized in development of flame retardant compounds. Though economics for these new routes for synthesis is important for future industrial exploitation, concerns for the environment and new regulations will help their acceptance in industrial exploitation. In our opinion, some synthetic strategies like Michael addition, Friedel–Crafts reaction and Michaelis–Arbuzov reaction can be industrially exploited in industry, taking in their advantages like higher yields, selectivity and simplicity of reaction. However, other reactions discussed in this review are also commercially feasible if the properties of the flame retardant significantly outweigh its cost of production.

One can find a vast amount of literature in development of P-C containing flame retardants with detailed investigations of their flame retardant properties and potential applications in various polymers. In most cases, such flame retardants exhibit satisfactory to excellent fire performance in the polymers, and questions regarding their impact on their processability, mechanical performance, and long-term serviceability of the final products need more attention. As organophosphorus compounds are also toxic (used as pesticides, herbicides, insecticides, nerve gasses), one has to pay attention while developing new structures. Toxicity screening of the new compounds should be carried out early in their development to prevent accidents and potential issues relate to toxicity. Research on ways to hydrolyze P-C bonds will help reduce their potential environmental impact and biocompatibility.

Author Contributions: Sophie Wendels, Thiebault Chavez, Martin Bonnet, Khalifah A. Salmeia and Sabyasachi Gaan contributed to the writing of the review manuscript. Khalifah A. Salmeia and Sabyasachi Gaan led and structured the manuscript.

Abbreviations

ABS	Acrylonitril butadiene styrene
AOPH	Aluminium–organophosphorus hybrids
BKZ-VB	Swiss standart fire test (vertical specimen)
CPPPA	4-Carboxyphenylphenylphosphinic acid
DBU	1,8-Diazabicyclo(5.4.0)undec-7-ene
DDM	Diaminodiphenylmethane
DETDA	Diethyldiamino toluene
DFDPS	4,4-Difluorodiphenylsulfone
DGEBA	Diglycidyl ether of bisphenol-A epoxy resin
DIP-MS	Direct insertion probe mass spectroscopy
DOPO	10-Dihydro-9-oxa-10-phosphaphenanthrene 10-oxide
DPPO	2,8-Dimethyl-phenoxaphosphine-10-oxide
DSC	Differential scanning calorimetry
EDX	Energy-dispersive X-ray spectroscopy
EVA	Ethylene-vinyl acetate
FPUF	Flexible polyurethane foams
FR	Flame retardant
FTIR	Fourier Transform Infrared Spectroscopy
HAPP	Hexaallyltriphosphazene
KMnO4	Potassium permanganate
LOI	Limiting oxygen index
MCC	Microscale Combustion Calorimetry
m-CPBA	meta-Chloroperoxybenzoic acid
MMPP	Methyl methylphenylphosphinate
m.p.	Melting point
N_2	Nitrogen
PA6	Polyamide 6

PC	Polycarbonate
PCFC	Pyrolysis-combustion flow calorimetry
PEEK	Polyether ether ketone
PEK	Polyether ketones
PES	Poly(ethersulfone)
PHRR	Lower peak heat release rate
PMMD	Poly(ethylene glycol) methacrylate/methyl methacrylate/diethyl allylphosphonate
PN-TLCP	Thermotropic liquid crystal polymer
POSS	Polyhedral oligomeric silsesquioxane
PSU	Polysulfones
PU	Polyurethane
Py-GC-MS	Pyrolysis-gas chromatography mass spectroscopy
ROMP	Ring opening metathesis polymerisation
RPUF	Rigid polyurethane foams
SEM	Scanning electron microscope
SET	Self-extinguishing time
TCPP	Tris(1-chloro-2-propyl) phosphate
Tdx%	Temperature for x% weight loss
TEOS	Tetraethyl orthosilicate
TGA	Thermogravimetric analysis
THR	Total heat release
TPA	Terephthalic acid
TSR	Total smoke release
UL94 HB	Standard for Tests for Flammability of Plastic Materials for Parts in Devices and Appliances (horizontal specimen)
UL94 VA or UL94 VB	Standard for Tests for Flammability of Plastic Materials for Parts in Devices and Appliances (vertical specimen)
UV	Ultraviolet
WAXD	Wide-angle X-ray scattering
XPS	X-ray photoelectron spectroscopy

References

1. Demchuk, O.M.; Jasinski, R. Organophosphorus ligands: Recent developments in design, synthesis, and application in environmentally benign catalysis. *Phosphorus Sulfur Silicon Relat. Elem.* **2016**, *191*, 245–253. [CrossRef]
2. Majoral, J.-P. *Topics in Current Chemistry*; Springer: Berlin, Germany, 2002; p. 244.
3. Majoral, J.-P. *Topics in Current Chemistry*; Springer: Berlin, Germany, 2003; p. 263.
4. Demkowicz, S.; Rachon, J.; Dasko, M.; Kozak, W. Selected organophosphorus compounds with biological activity. Applications in medicine. *RSC Adv.* **2016**, *6*, 7101–7112. [CrossRef]
5. Salmeia, K.A.; Gaan, S. An overview of some recent advances in dopo-derivatives: Chemistry and flame retardant applications. *Polym. Degrad. Stab.* **2015**, *113*, 119–134. [CrossRef]
6. Michal Sobkowski, A.K.; Stawinski, J. Recent advances in h-phosphonate chemistry. Part 2. Synthesis of c-phosphonate derivatives. In *Topics in Current Chemistry*; Montchamp, J.-L., Ed.; Springer: Berlin, Germany, 2015; Volume 361, pp. 179–216.
7. Ma, J.; Yang, J.; Huang, Y.; Ke, C. Aluminum-organophosphorus hybrid nanorods for simultaneously enhancing the flame retardancy and mechanical properties of epoxy resin. *J. Mater. Chem.* **2012**, *22*, 2007–2017.
8. Montchamp, J.-L. Carbon-hydrogen to carbon-phosphorus transformations. In *Topics in Current Chemistry*; Montchamp, J.-L., Ed.; Springer: Berlin, Germany, 2015; Volume 361, pp. 217–252.
9. Menigaux, D.; Belmont, P.; Brachet, E. Light on unsaturated hydrocarbons—"Gotta heterofunctionalize them all". *Eur. J. Org. Chem.* **2017**, *2017*, 2008–2055. [CrossRef]
10. Pan, X.-Q.; Zou, J.-P.; Yi, W.-B.; Zhang, W. Recent advances in sulfur- and phosphorous-centered radical reactions for the formation of s–c and p–c bonds. *Tetrahedron* **2015**, *71*, 7481–7529. [CrossRef]

11. Wauters, I.; Debrouwer, W.; Stevens, C.V. Preparation of phosphines through c–p bond formation. *Beilstein J. Org. Chem.* **2014**, *10*, 1064–1096. [CrossRef] [PubMed]
12. Chen, T.; Zhang, J.-S.; Han, L.-B. Dehydrogenative coupling involving p(o)-h bonds: A powerful way for the preparation of phosphoryl compounds. *Dalton Trans.* **2016**, *45*, 1843–1849. [CrossRef] [PubMed]
13. Pullarkat, S.A. Recent progress in palladium-catalyzed asymmetric hydrophosphination. *Synthesis* **2016**, *48*, 493–503. [CrossRef]
14. Liang, S.; Neisius, M.; Mispreuve, H.; Naescher, R.; Gaan, S. Flame retardancy and thermal decomposition of flexible polyurethane foams: Structural influence of organophosphorus compounds. *Polym. Degrad. Stab.* **2012**, *97*, 2428–2440. [CrossRef]
15. Butnaru, I.; Fernandez-Ronco, M.P.; Czech-Polak, J.; Heneczkowski, M.; Bruma, M.; Gaan, S. Effect of meltable triazine-dopo additive on rheological, mechanical, and flammability properties of pa6. *Polymer* **2015**, *7*, 1541–1563. [CrossRef]
16. Agarwal, V.; Borisova, S.A.; Metcalf, W.W.; van der Donk, W.A.; Nair, S.K. Structural and mechanistic insights into c-p bond hydrolysis by phosphonoacetate hydrolase. *Chem. Biol.* **2011**, *18*, 1230–1240. [CrossRef] [PubMed]
17. Kamat, S.S.; Williams, H.J.; Raushel, F.M. Intermediates in the transformation of phosphonates to phosphate by bacteria. *Nature* **2011**, *480*, 570–573. [CrossRef] [PubMed]
18. Ternan, N.G.; Hamilton, J.T.; Quinn, J.P. Initial in vitro characterisation of phosphonopyruvate hydrolase, a novel phosphate starvation-independent, carbon-phosphorus bond cleavage enzyme in burkholderia cepacia pal6. *Arch. Microbiol.* **2000**, *173*, 35–41. [CrossRef] [PubMed]
19. Liang, S.; Hemberger, P.; Levalois-Grützmacher, J.; Grützmacher, H.; Gaan, S. Probing phosphorus nitride (p≡n) and other elusive species formed upon pyrolysis of dimethyl phosphoramidate. *Chemistry* **2017**, *23*, 5595–5601. [CrossRef] [PubMed]
20. Liang, S.; Hemberger, P.; Neisius, N.M.; Bodi, A.; Grützmacher, H.; Levalois-Grützmacher, J.; Gaan, S. Elucidating the thermal decomposition of dimethyl methylphosphonate by vacuum ultraviolet (vuv) photoionization: Pathways to the po radical, a key species in flame-retardant mechanisms. *Chemistry* **2015**, *21*, 1073–1080. [CrossRef] [PubMed]
21. Salmeia, K.; Gaan, S.; Malucelli, G. Recent advances for flame retardancy of textiles based on phosphorus chemistry. *Polymers* **2016**, *8*, 319. [CrossRef]
22. Salmeia, K.; Fage, J.; Liang, S.; Gaan, S. An overview of mode of action and analytical methods for evaluation of gas phase activities of flame retardants. *Polymers* **2015**, *7*, 504–526. [CrossRef]
23. Stawinski, J.; Kraszewski, A. How to get the most out of two phosphorus chemistries. Studies on h-phosphonates. *Acc. Chem. Res.* **2002**, *35*, 952–960. [CrossRef] [PubMed]
24. Keglevich, G.; Grun, A.; Balint, E.; Kiss, N.Z.; Bagi, P.; Toke, L. Green chemical syntheses and applications within organophosphorus chemistry. *Struct. Chem.* **2017**, *28*, 431–443. [CrossRef]
25. Luo, K.; Yang, W.-C.; Wu, L. Photoredox catalysis in organophosphorus chemistry. *Asian J. Org. Chem.* **2017**, *6*, 350–367. [CrossRef]
26. Sharova, E.V.; Artyushin, O.I.; Odinets, I.L. Synthetic routes to carbamoylmethylphosphoryl compounds—Extractants for the processing of spent nuclear fuels. *Russ. Chem. Rev.* **2014**, *83*, 95–119. [CrossRef]
27. Haegele, G. Michaelis-arbuzov reactions of perhalogenated cyclobutenes with trialkyl phosphites. *Phosphorus Sulfur Silicon Relat. Elem.* **2015**, *190*, 2094–2109. [CrossRef]
28. Keglevich, G.; Kiss, N.Z.; Balint, E.; Bagi, P.; Grun, A.; Kovacs, T.; Henyecz, R.; Abranyi-Balogh, P. Milestones in microwave-assisted organophosphorus chemistry. *Phosphorus Sulfur Silicon Relat. Elem.* **2016**, *191*, 1416–1420. [CrossRef]
29. Enders, D.; Saint-Dizier, A.; Lannou, M.-I.; Lenzen, A. The phospha-michael addition in organic synthesis. *Eur. J. Org. Chem.* **2005**, *2006*, 29–49. [CrossRef]
30. Herrera, R.P. Organocatalytic hydrophosphonylation reaction of carbonyl groups. *Chem. Rec.* **2017**. [CrossRef] [PubMed]
31. Montchamp, J.-L. Phosphinate chemistry in the 21st century: A viable alternative to the use of phosphorus trichloride in organophosphorus synthesis. *Acc. Chem. Res.* **2014**, *47*, 77–87. [CrossRef] [PubMed]
32. Troev, K.D. Structure and spectral characteristics of h-phosphonates. In *Chemistry and Application of H-Phosphonates*; Elsevier Science Ltd.: Amsterdam, The Netherlands, 2006; pp. 11–22.

33. Yu, X.; Ge, B.; Zhang, Y.; Wang, X.; Wang, D. Silver-mediated phosphonylation of c(sp^2)−h bonds with p−h bonds: Direct c−h functionalization of ferrocenyl anilides and dialkyl phosphites under palladium- and copper-free conditions. *Asian J. Org. Chem.* **2016**, *5*, 1253–1259. [CrossRef]
34. Renaud, J.-L.; Gaillard, S. Iron-catalyzed carbon-nitrogen, carbon-phosphorus, and carbon-sulfur bond formation and cyclization reactions. In *Iron Catalysis ii*; Bauer, E., Ed.; Springer: Cham, Switzerland, 2015; pp. 83–144.
35. Guin, J.; Wang, Q.; van Gemmeren, M.; List, B. The catalytic asymmetric abramov reaction. *Angew. Chem. Int. Ed.* **2015**, *54*, 355–358. [CrossRef] [PubMed]
36. Troev, K.D. Methods for preparation and physical properties of h-phosphonates. In *Chemistry and Application of H-Phosphonates*; Elsevier Science Ltd.: Amsterdam, The Netherlands, 2006; pp. 1–9.
37. Balint, E.; Tajti, A.; Adam, A.; Csontos, I.; Czugler, M.; Keglevich, G.; Karaghiosoff, K.; Abranyi-Balogh, P. The synthesis of α-aryl-α-aminophosphonates and α-aryl-α-aminophosphine oxides by the microwave-assisted pudovik reaction. *Beilstein J. Org. Chem.* **2017**, *13*, 76–86. [CrossRef] [PubMed]
38. Subramanyam, C.; Thaslim Basha, S.; Madhava, G.; Nayab Rasool, S.; Adam, S.; Durga Srinivasa Murthy, S.; Naga Raju, C. Synthesis, spectral characterization and bioactivity evaluation of novel α-aminophosphonates. *Phosphorus Sulfur Silicon Relat. Elem.* **2017**, *192*, 267–270. [CrossRef]
39. Kamath, P.; Rajan, R.; Deshpande, S.H.; Montgomery, M.; Lal, M. An efficient synthesis of 3-phosphorylated benzoxaboroles via the pudovik reaction. *Synthesis* **2017**. [CrossRef]
40. Yang, J.; Xiao, J.; Chen, T.; Han, L.-B. Nickel-catalyzed phosphorylation of phenol derivatives via c–o/p–h cross-coupling. *J. Org. Chem.* **2016**, *81*, 3911–3916. [CrossRef] [PubMed]
41. Yang, J.; Chen, T.; Zhou, Y.; Yin, S.-F.; Han, L.-B. Mechanistic studies on the palladium-catalyzed cross dehydrogenative coupling of p(o)–h compounds with terminal alkynes: Stereochemistry and reactive intermediates. *Organometallics* **2015**, *34*, 5095–5098. [CrossRef]
42. Wang, L.; Wu, Y.; Liu, Y.; Yang, H.; Liu, X.; Wang, J.; Li, X.; Jiang, J. Copper-catalyzed p–h insertions of α-imino carbenes for the preparation of 3-phosphinoylindoles. *Org. Lett.* **2017**, *19*, 782–785. [CrossRef] [PubMed]
43. Peng, P.; Lu, Q.; Peng, L.; Liu, C.; Wang, G.; Lei, A. Dioxygen-induced oxidative activation of a p-h bond: Radical oxyphosphorylation of alkenes and alkynes toward β-oxy phosphonates. *Chem. Commun.* **2016**, *52*, 12338–12341. [CrossRef] [PubMed]
44. Guo, H.; Yoshimura, A.; Chen, T.; Saga, Y.; Han, L.-B. Air-induced double addition of p(o)-h bonds to alkynes: A clean and practical method for the preparation of 1,2-bisphosphorylethanes. *Green Chem.* **2017**, *19*, 1502–1506. [CrossRef]
45. Rafikov, S.R.; Chelnokova, G.N.; Dzhilkibaeva, G.M. Michaelis-arbuzov reaction kinetics and the use of this reaction for polymer synthesis. *Tr. Inst. Khim. Nauk Akad. Nauk Kaz. SSR* **1969**, *23*, 80–97.
46. Shaikh, R.S.; Düsel, S.J.S.; König, B. Visible-light photo-arbuzov reaction of aryl bromides and trialkyl phosphites yielding aryl phosphonates. *ACS Catal.* **2016**, *6*, 8410–8414. [CrossRef]
47. Wang, S.; Qiu, D.; Mo, F.; Zhang, Y.; Wang, J. Metal-free aromatic carbon–phosphorus bond formation via a sandmeyer-type reaction. *J. Org. Chem.* **2016**, *81*, 11603–11611. [CrossRef] [PubMed]
48. Peng, H.; Cai, R.; Xu, C.; Chen, H.; Shi, X. Nucleophile promoted gold redox catalysis with diazonium salts: C-br, c-s and c-p bond formation through catalytic sandmeyer coupling. *Chem. Sci.* **2016**, *7*, 6190–6196. [CrossRef]
49. Yu, R.; Chen, X.; Martin, S.F.; Wang, Z. Differentially substituted phosphines via decarbonylation of acylphosphines. *Org. Lett.* **2017**, *19*, 1808–1811. [CrossRef] [PubMed]
50. Nakanishi, K.; Kotani, S.; Sugiura, M.; Nakajima, M. First asymmetric abramov-type phosphonylation of aldehydes with trialkyl phosphites catalyzed by chiral lewis bases. *Tetrahedron* **2008**, *64*, 6415–6419. [CrossRef]
51. Zhang, P.; Fan, H.; Tian, S.; Chen, Y.; Yan, J. Synergistic effect of phosphorus-nitrogen and silicon-containing chain extenders on the mechanical properties, flame retardancy and thermal degradation behavior of waterborne polyurethane. *RSC Adv.* **2016**, *6*, 72409–72422. [CrossRef]
52. Bisaro, F.; Gouverneur, V. Desymmetrization by direct cross-metathesis producing hitherto unreachable p-stereogenic phosphine oxides. *Tetrahedron* **2005**, *61*, 2395–2400. [CrossRef]

53. Horrocks, A.R.; Zhang, S. Enhancing polymer flame retardancy by reaction with phosphorylated polyols. Part 2. Cellulose treated with a phosphonium salt urea condensate (proban cc) flame retardant. *Fire Mater.* **2002**, *26*, 173–182. [CrossRef]
54. Hou, S.; Zhang, Y.J.; Jiang, P. Phosphonium sulfonates as flame retardants for polycarbonate. *Polym. Degrad. Stab.* **2016**, *130*, 165–172. [CrossRef]
55. Panmand, D.S.; Tiwari, A.D.; Panda, S.S.; Monbaliu, J.-C.M.; Beagle, L.K.; Asiri, A.M.; Stevens, C.V.; Steel, P.J.; Hall, C.D.; Katritzky, A.R. New benzotriazole-based reagents for the phosphonylation of various n-, o-, and s-nucleophiles. *Tetrahedron Lett.* **2014**, *55*, 5898–5901. [CrossRef]
56. Liu, Z.; De Campo, F. Conjugated Diene Phosphinate Compounds, Preparation Method and Use Thereof. U.S. Patent 20120208975 A1, 2014.
57. Liu, Z.; De Campo, F. N-Substituted Acrylamides, Preparation Method and Use Thereof. Patent WO2011050534A1, 2011.
58. Neisius, N.M.; Liang, S.; Gaan, S.; Mispreuve, H.; Nascher, R.; Rentsch, D. Flame retardants for flexible polyurethane foams: Structure-property-relationship studies. In Proceedings of the 243rd ACS National Meeting & Exposition, San Diego, CA, USA, 2012.
59. Çakmakçı, E.; Mülazim, Y.; Kahraman, M.V.; Apohan, N.K. Flame retardant thiol-ene photocured coatings. *React. Funct. Polym.* **2011**, *71*, 36–41. [CrossRef]
60. Zeytuncu, B.; Çakmakçı, E.; Kahraman, M.V. Allyl phosphonium salt-modified clay for photocured coatings: Influence on the properties of polyester acrylate-based coatings. *Polym. Compos.* **2015**, *36*, 946–954. [CrossRef]
61. Mai, N.L.; Koo, Y.-M. Computer-aided design of ionic liquids for high cellulose dissolution. *ACS Sustain. Chem. Eng.* **2016**, *4*, 541–547. [CrossRef]
62. Yang, L.; Han, X.Y.; Li, L.L.; Bi, C.L.; Tang, X.J.; Zhang, B.G. Synthesis of aluminium diethylphosphinate by gas-liquid free radical addition reaction under atmospheric pressure. *Adv. Mater. Res.* **2011**, *194–196*, 2237–2240. [CrossRef]
63. Lee, M.S.; Ju, B.J.; Lee, B.G.; Hong, S.H. Non-Halogen Flameproof Resin Composition. U.S. Patent 20100152345, 2008.
64. Liu, X.; Salmeia, K.A.; Rentsch, D.; Hao, J.; Gaan, S. Thermal decomposition and flammability of rigid pu foams containing some dopo derivatives and other phosphorus compounds. *J. Anal. Appl. Pyrolysis* **2017**, *124*, 219–229. [CrossRef]
65. Benin, V.; Cui, X.; Morgan, A.B.; Seiwert, K. Synthesis and flammability testing of epoxy functionalized phosphorous-based flame retardants. *J. Appl. Polym. Sci.* **2015**, *132*. [CrossRef]
66. Worku, A.Z.; Mullins, M.J. Phosphorus-Containing Compounds Useful for Making Halogen-Free, Ignition-Resistant Polymers. Patent WO2012161926, 2012.
67. Li, S.; Wang, J.; Li, Y.; Wu, G.; Wang, Y.; Wang, W.; Guo, J. Preparation and applications of the tertiary copolymer poly(ethylene glycol) methacrylate/methyl methacrylate/diethyl allylphosphonate. *J. Appl. Polym. Sci.* **2016**, *133*. [CrossRef]
68. Zhang, X.; Zhang, Q.; Xie, C.; Gao, A.; Chang, Z.; Kwon Oh, J.; Yang, P.; Li, P. Phosphonated homopolymers and copolymers via ring opening metathesis polymerization: T_g tuning, flame resistance, and photolithography. *J. Polym. Sci. Part A* **2016**, *54*, 1396–1408. [CrossRef]
69. König, A.; Kroke, E. Flame retardancy working mechanism of methyl-dopo and mppp in flexible polyurethane foam. *Fire Mater.* **2012**, *36*, 1–15. [CrossRef]
70. Artner, J.; Ciesielski, M.; Ahlmann, M.; Walter, O.; Döring, M.; Perez, R.M.; Altstädt, V.; Sandler, J.K.W.; Schartel, B. A novel and effective synthetic approach to 9,10-dihydro-9-oxa-10-phosphaphenanthrene-10-oxide (dopo) derivatives. *Phosphorus Sulfur Silicon Relat. Elem.* **2007**, *182*, 2131–2148. [CrossRef]
71. Dittrich, U.; Just, B.; Doring, M.; Ciesielski, M. Process for the Preparation of 9, 10-Dihydro-9-oxa-10-organylphosphaphenanthrene-10-oxide and Derivatives of the Same Substituted on the Phenyl Groups. U.S. Patent 20050038279, 2005.
72. Zang, L.; Wagner, S.; Ciesielski, M.; Müller, P.; Döring, M. Novel star-shaped and hyperbranched phosphorus-containing flame retardants in epoxy resins. *Polym. Adv. Technol.* **2011**, *22*, 1182–1191. [CrossRef]
73. Hoang, D.; Kim, J. Flame-retarding behaviors of novel spirocyclic organo-phosphorus compounds based on pentaerythritol. *Macromol. Res.* **2015**, *23*, 579–591. [CrossRef]

74. Nguyen, C.; Lee, M.; Kim, J. Relationship between structures of phosphorus compounds and flame retardancies of the mixtures with acrylonitrile-butadiene-styrene and ethylene-vinyl acetate copolymer. *Polym. Adv. Technol.* **2011**, *22*, 512–519. [CrossRef]
75. Kang, N.; Du, Z.; Li, H.; Zhang, C. Synthesis and characterization of p/si flame retardant and its application in epoxy systems. *Polym. Adv. Technol.* **2012**, *23*, 1329–1334. [CrossRef]
76. Karataş, S.; Hoşgör, Z.; Kayaman-Apohan, N.; Güngör, A. Preparation and characterization of phosphine oxide containing organosilica hybrid coatings by photopolymerization and sol–gel process. *Prog. Org. Coat.* **2009**, *65*, 49–55. [CrossRef]
77. Worku, A.Z.; Mullins, M.J. Phosphorus-Containing Compounds Useful for Making Halogen-Free, Ignition-Resistant Polymers. U.S. Patent 20160319107 A1, 2016.
78. Su, W.C.; Sheng, C.S. Method for Preparing a Biphenylphosphonate Compound. U.S. Patent 20050101793, 2005.
79. Tian, C.; Liu, P.; Ye, G.; Xu, J. Soluble polyarylates containing phosphorus in main chains with thermal stability. *J. Appl. Polym. Sci.* **2013**, *130*, 3521–3529. [CrossRef]
80. Campbell, I.G.M.; Way, J.K. Synthesis and stereochemistry of heterocyclic phosphorus compounds. Part I. Preparation of (+)- and (-)-10-*p*-dimethylaminophenyl-9,10-dihydro-9-aza-10-phosphaphenanthrene. *J. Chem. Soc.* **1960**, 5034–5041. [CrossRef]
81. Dewar, M.J.S.; Kubba, V.P. New heteroaromatic compounds. Part vi. Novel heterocyclic compounds of phosphorus. *J. Am. Chem. Soc.* **1960**, *82*, 5685–5688. [CrossRef]
82. Rakotomalala, M.; Ciesielski, M.; Zevaco, T.; Doering, M. New phosphacyclic molecules and their application as flame retardants for epoxy resins. *Phosphorus Sulfur Silicon Relat. Elem.* **2011**, *186*, 989–996. [CrossRef]
83. Rakotomalala, M.; Wagner, S.; Zevaco, T.; Ciesielski, M.; Walter, O.; Doring, M. Synthesis and reactivity of 6h-dibenzo[c,e][1,2]oxaphosphinine 6-sulfide, a novel thiophosphacyclic molecule. *Heterocycles* **2011**, *83*, 743–753.
84. Schäfer, A.; Seibold, S.; Lohstroh, W.; Walter, O.; Döring, M. Synthesis and properties of flame-retardant epoxy resins based on dopo and one of its analog dppo. *J. Appl. Polym. Sci.* **2007**, *105*, 685–696. [CrossRef]
85. Schäfer, A.; Seibold, S.; Walter, O.; Döring, M. Novel high t_g flame retardancy approach for epoxy resins. *Polym. Degrad. Stab.* **2008**, *93*, 557–560. [CrossRef]
86. Wang, Z.-W.; Sun, Q.-X.; Wu, J.-S.; Wang, L.-S. Solubilities of 2-carboxyethylphenylphosphinic acid and 4-carboxyphenylphenylphosphinic acid in water. *J. Chem. Eng. Data* **2003**, *48*, 1073–1075. [CrossRef]
87. Sacristán, M.; Ronda, J.C.; Galià, M.; Cádiz, V. Phosphorus-containing soybean-oil copolymers: Cross-metathesis of fatty acid derivatives as an alternative to phosphorus-containing reactive flame retardants. *J. Appl. Polym. Sci.* **2011**, *122*, 1649–1658. [CrossRef]
88. Mandal, B.; Filler, R. Thermal Runaway Inhibitors. U.S. Patent 20030003358 A1, 2003.
89. Li, X.; Chen, H.; Wang, W.; Liu, Y.; Zhao, P. Synthesis of a formaldehyde-free phosphorus-nitrogen flame retardant with multiple reactive groups and its application in cotton fabrics. *Polym. Degrad. Stab.* **2015**, *120*, 193–202. [CrossRef]
90. Zuo, J.-D.; Liu, S.-M.; Sheng, Q. Synthesis and application in polypropylene of a novel of phosphorus-containing intumescent flame retardant. *Molecules* **2010**, *15*, 7593–7602. [CrossRef] [PubMed]
91. Chun Shan, W.; Ching Hsuan, L. Synthesis and properties of phosphorus containing polyarylates derived from 2-(6-oxido-6h-dibenzoxaphosphorin-6-yl)-1,4-dihydroxy phenylene. *Polymer* **1999**, *40*, 4387–4398. [CrossRef]
92. Zheng, J.; Yu, Y.; Zhang, L.; Zhen, X.; Zhao, Y. Synthesis of flame-retardant phosphaphenanthrene derivatives with high phosphorus contents. *Aust. J. Chem.* **2014**, *67*, 1688–1692. [CrossRef]
93. Bai, Z.; Song, L.; Hu, Y.; Yuen, R.K.K. Preparation, flame retardancy, and thermal degradation of unsaturated polyester resin modified with a novel phosphorus containing acrylate. *Ind. Eng. Chem. Res.* **2013**, *52*, 12855–12864. [CrossRef]
94. Xia, Y.; Ge, X.; Hou, C.; Wang, C.; Ma, W. Synthesis and characterization of a new kind of flame-retardant thermotropic liquid crystal containing 1,8-octanediamine softsegment. *Polym.-Plast. Technol. Eng.* **2013**, *52*, 1–6. [CrossRef]
95. Wang, X.; Hu, Y.; Song, L.; Xing, W.; Lu, H. Thermal degradation behaviors of epoxy resin/poss hybrids and phosphorus-silicon synergism of flame retardancy. *J. Polym. Sci. Part B* **2010**, *48*, 693–705. [CrossRef]

96. Ho, T.-H.; Hwang, H.-J.; Shieh, J.-Y.; Chung, M.-C. Thermal, physical and flame-retardant properties of phosphorus-containing epoxy cured with cyanate ester. *React. Funct. Polym.* **2009**, *69*, 176–182. [CrossRef]
97. Seibold, S.; Schäfer, A.; Lohstroh, W.; Walter, O.; Döring, M. Phosphorus-containing terephthaldialdehyde adducts—Structure determination and their application as flame retardants in epoxy resins. *J. Appl. Polym. Sci.* **2008**, *108*, 264–271. [CrossRef]
98. Hoffmann, T.; Pospiech, D.; Häußler, L.; Sahre, K.; Komber, H.; Harnisch, C.; Auf Der Landwehr, M.; Schäfer, A.; Döring, M. Phosphorus-containing polysulfones—A comparative study. *High Perform. Polym.* **2010**, *22*, 715–741. [CrossRef]
99. Sun, Y.-M.; Wang, C.-S. Synthesis and luminescent characteristics of novel phosphorus containing light-emitting polymers. *Polymer* **2001**, *42*, 1035–1045. [CrossRef]
100. Serbezeanu, D.; Vlad-Bubulac, T.; Hamciuc, C.; Aflori, M. Synthesis and properties of novel phosphorus-containing thermotropic liquid crystalline copoly(ester imide)s. *J. Polym. Sci. Part A* **2010**, *48*, 5391–5403. [CrossRef]
101. Serbezeanu, D.; Vlad-Bubulac, T.; Hamciuc, C.; Doroftei, F. Synthesis and characterization of new phosphorus-containing poly(ester imide)s. *High Perform. Polym.* **2010**, *22*, 916–929. [CrossRef]
102. Tian, X.; Wang, Z.; Yu, Q.; Wu, Q.; Gao, J. Synthesis and property of flame retardant epoxy resins modified with 2-(diphenylphosphinyl)-1,4-benzenediol. *Chem. Res. Chin. Univ.* **2014**, *30*, 868–873. [CrossRef]
103. Chen, G.-C.H.; Huang, C.J.; Tu, A.P.; Hwang, K.Y. Phosphorus-Containing Compound and Method for Preparing the Same. US20100105939, 2009.
104. Benin, V.; Gardelle, B.; Morgan, A.B. Heat release of polyurethanes containing potential flame retardants based on boron and phosphorus chemistries. *Polym. Degrad. Stab.* **2014**, *106*, 108–121. [CrossRef]
105. Ivan, N.; Benin, V.; Morgan, A.B. Preparation of phosphonoterephthalic acids via palladium-catalyzed coupling of aromatic iodoesters. *Synth. Commun.* **2013**, *43*, 1831–1836. [CrossRef]
106. Finocchiaro, P.; Consiglio, G.A.; Imbrogiano, A.; Failla, S. Synthesis and characterization of new organic phosphonates monomers as flame retardant additives for polymers. *Phosphorus Sulfur Silicon Relat. Elem.* **2007**, *182*, 1689–1701. [CrossRef]
107. Dumitrascu, A.; Howell, B.A. Flame-retarding vinyl polymers using phosphorus-functionalized styrene monomers. *Polym. Degrad. Stab.* **2011**, *96*, 342–349. [CrossRef]
108. Rectenwald, M.F.; Gaffen, J.R.; Rheingold, A.L.; Morgan, A.B.; Protasiewicz, J.D. Phosphoryl-rich flame-retardant ions (frions): Towards safer lithium-ion batteries. *Angew. Chem. Int. Ed.* **2014**, *53*, 4173–4176. [CrossRef] [PubMed]
109. Dhawan, B.; Redmore, D. *O*-hydroxyaryl diphosphonic acids. *J. Org. Chem.* **1984**, *49*, 4018–4021. [CrossRef]
110. Ghosh, A.; Bera, D.; Wang, D.-Y.; Komber, H.; Mohanty, A.K.; Banerjee, S.; Voit, B. Synthesis, characterization and properties of new semifluorinated poly(arylene ether phosphine oxide)s. *Macromol. Mater. Eng.* **2012**, *297*, 145–154. [CrossRef]
111. Yoon, T.H.; Jeong, K.; Jo, Y.J. Triarylphosphine Oxide Derivatives Containing Fluorine Substituent and Preparing Method Thereof. U.S. Patent 20020095057 A1, 2002.

2

Insight into the Synthesis and Characterization of Organophosphorus-Based Bridged Triazine Compounds

Khalifah A. Salmeia [1,*], Antonia Neels [2,*], Dambarudhar Parida [1], Sandro Lehner [1], Daniel Rentsch [3] and Sabyasachi Gaan [1]

1 Additives and Chemistry, Advanced Fibers, Empa Swiss Federal Laboratories for Materials Science and Technology, Lerchenfeldstrasse 5, 9014 St. Gallen, Switzerland

2 Center for X-ray Analytics, Swiss Federal Laboratories for Materials Science and Technology, Empa, Überlandstrasse 129, 8600 Dübendorf, Switzerland

3 Laboratory for Functional Polymers, Empa Swiss Federal Laboratories for Materials Science and Technology, Überlandstrasse 129, 8600 Dübendorf, Switzerland

* Correspondence: khalifah.salmeia@empa.ch (K.A.S.); antonia.neels@empa.ch (A.N.)

Abstract: In this article, we report the synthesis of 2,4,6-substituted *s*-triazine-based organophosphorus compounds via a two-step process, which enables their production in high yields, and with a high purity as solids. In the first step, a Michaelis–Arbuzov rearrangement of cyanuric chloride with triethyl phosphite afforded 2,4,6-trisdiethoxyphosphinyl-1,3,5-triazine (HEPT). Subsequently, the nucleophilic substitution reaction on the triazine carbon was achieved, owing to the electron-withdrawing ability of the phosphonate groups. This characteristic of HEPT facilitated its derivatization with bi-functional amines, producing novel P–C containing bridged triazine organophosphorus compounds. The molecular structures of all of the compounds were confirmed by NMR spectroscopy, CHN elemental analysis, and single crystal X-ray analysis. In the thermogravimetric analysis in an N_2 environment, >33% char formation was observed for the bridged compounds. The chemical composition analysis of the char obtained under the oxidative thermal decomposition of the bridged compounds confirmed the presence of phosphorus- and nitrogen-enriched species, which indicate their function in the condensed phase. Comparatively, the detection of HPO and H–C≡P in the gas phase during the pyrolysis of the bridged compounds can act as a source for $PO^{\bullet}$, which is known for its gas phase flame inhibition reactions. The synergy of significant char formation and the generation of intermediates leading to $PO^{\bullet}$ during pyrolysis makes these molecules promising flame-retardant additives.

Keywords: Michaelis–Arbuzov rearrangement; cyanuric chloride; triazine derivatives; organophosphorus compounds

---

1. Introduction

Nitrogen-containing heterocycles (azaheterocycles) are common organic compounds that play a crucial role in different applications. Although the symmetric 1,3,5-triazine, which is referred to as s-triazine, is considered monotonous [1], new synthetic approaches were investigated to produce new substituted 1,3,5-triazine derivatives [2]. The most common chemical reaction hereof is the substitution of the chlorine atom of the commercially available 2,4,6-trichloro-1,3,5-triazine (cyanuric chloride) with nucleophiles [3]. Cyanuric chloride has been known since the early 1800, and has a similar reaction profile to acyl chlorides, rather than alkyl chlorides; since then, it has received great attention as a chemical reagent for different chemical reactions, as shown in Scheme 1 [4,5].

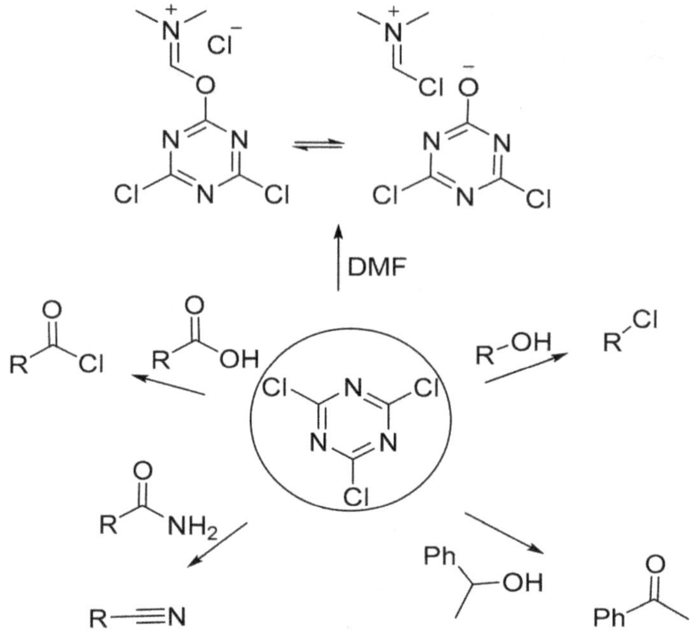

Scheme 1. General schematic example for using cyanuric chloride as a chemical reagent.

Moreover, the ability to replace the chlorine atom (Cl) in the presence of an acid scavenger was found to produce mono-, di-, and tri-substituted s-triazines [3]. The first displacement of a Cl atom happens between 0 °C to ambient temperature, and the subsequent replacement of Cl occurs at higher reaction temperatures [2,6]. The phosphorus-bearing azaheterocyclic compounds have received great attention because of their different applications in medicinal chemistry and agrochemicals [7]. Industrially, more than 70% of cyanuric chloride production is used for the production of pesticides and herbicides [8]. Substituted 1,3,5-triazine derivatives could find an application as flame retardant additives [9,10], and their flame retardant performance was investigated extensively on cotton [11–15], epoxy resins [16–18], and polypropylene [19–22].

The nitrogen-containing flame retardants work either in gas phase or in condensed phase. These flame-retardant derivatives, which release a large amount of non-flammable gases upon decomposition, act predominantly in the gas phase. On the other hand, the additives that lead to the formation of acidic derivatives, act in the condensed phase [23]. It is known that organophosphorus flame-retardant additives act predominantly in the gas phase, producing $PO^{\bullet}$ species during thermal decomposition, which quench the radical species ($H^{\bullet}$ and $^{\bullet}OH$), thereby inhibiting the cycle of fire [24,25]. Phosphorus-containing additives liberate phosphoric acid during their decomposition, and act synergistically together with nitrogen-containing additives in improving the char formation [26]. The synergism of P,N-containing flame retardants was extensively investigated, for enhancing both the gas- and condensed-phase mode of action [17,27]. Triazine-based flame-retardant additives can act in both the gas and condensed phases simultaneously [6,9]. Notwithstanding, the triazine compound itself was found to be a good char forming agent [12,17,19,21,26,28,29]. Despite the increase in the number of synthetic methodologies, popular methods for phosphonylation of cyanuric chloride include the Michaelis–Arbuzov rearrangement and Michaelis–Becker reaction. The Michaelis–Arbuzov rearrangement of cyanuric chloride with trialkyl phosphite was investigated to produce 2,4,6-trisdialkoxyphosphinyl-1,3,5-triazine [30]. This type of reaction can occur vigorously with neat trialkyl phosphite at ambient conditions, and can take place in boiling solvents. The partial substitution of the chlorine atom of cyanuric chloride via the Michaelis–Arbuzov rearrangement has also been investigated [31]. However, the ability of the nucleophilic substitution of the 2-chloro-atom of 4,6-substituted 1,3,5-triazine decreases with the lessening electron-withdrawing effect of its substituents. The electron-withdrawing substituents reduce the electron density on the carbon atoms of the triazine ring, making it more susceptible to a nucleophilic reaction. A new class of phosphino-derivatives of s-triazine was also synthesized using three different synthetic methodologies [32].

Herein, we report a two-step synthetic of P-containing bridged triazine compounds using cyanuric chloride as a starting material. This synthetic process enabled the preparation of bridged triazine compounds in high yields and with a high purity as solids. The synthesis of the ethylene diamine-based bridged triazine compound is already reported in the literature, however, its detailed chemical and thermal characterizations are missing [33]. Bridged compounds, being larger structures, are expected to be thermally more stable. Additionally, the bridged characteristic of the triazine structures could decrease their chemical reactivity, which is an important requirement for the high-temperature processing of polymers. Moreover, the addition of a nitrogen bridge in the structure alters the P:N ratio, which has a strong influence on their flame retardant behavior [34]. The chemical structures of these compounds were characterized by NMR spectroscopy, elemental analysis, and single crystal X-ray analysis. Furthermore, we have performed their detailed thermal analysis using thermogravimetric analysis (TGA) and direct insertion probe mass spectrometry (DIP-MS).

2. Results and Discussion

2.1. Synthesis of Chemicals

The synthetic methodology to obtain the bridged triazine derivatives is shown in Scheme 2. The Michaelis–Arbuzov rearrangement of cyanuric chloride with triethyl phosphite afforded 2,4,6-trisdiethoxyphosphinyl-1,3,5-triazine (HEPT; Scheme 2, step 1). It was then used as an intermediate to synthesize *N,N′*-Bis[4,6-bis(diethylphosphono)-1,3,5-Triazin-yl]-1,2-diaminoethane (EDA-bis-TEPT) and *N,N′*-Bis[4,6-bis(diethylphosphono)-1,3,5-Triazin-yl]-piperazine (Pip-bis-TEPT) (Scheme 2, step 2).

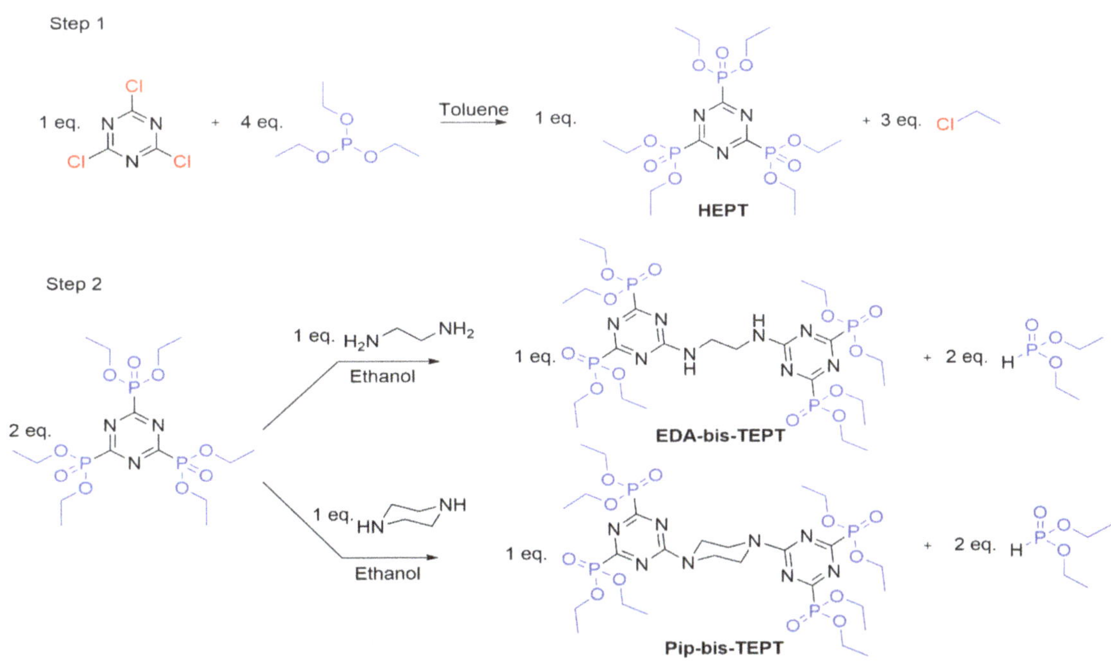

Scheme 2. The synthesis scheme of triazine bridged derivatives of 2,4,6-trisdiethoxyphosphinyl-1,3,5-triazine (HEPT) and the final products *N,N′*-Bis[4,6-bis(diethylphosphono)-1,3,5-Triazin-yl]-1,2-diaminoethane (EDA-bis-TEPT) and *N,N′*-Bis[4,6-bis(diethylphosphono)-1,3,5-Triazin-yl]-piperazine (Pip-bis-TEPT).

It is noteworthy to mention that only three equivalents of triethyl phosphite, with respect to cyanuric chloride, are enough to produce the precursor HEPT. However, in order to obtain a high product yield, four equivalents of HEPT were used. Attempts to partially replace the three Cl-atoms via the Michaelis–Arbuzov rearrangement to prepare di-substituted triazine derivatives in a high yield and with a high purity were unsuccessful. This may be attributed to the electron-withdrawing effect of the phosphonate group, which reduces the electron density of the triazine ring and enhances the activity of the Cl-atom for further substitution reactions, such as preferring the tri-substituted product [31].

The substitution of one phosphonate group of HEPT with the amine group, to produce EDA-bis-TEPT and Pip-bis-TEPT, was performed following a previously reported procedure [10,30,35]. It is reported that the substitution of the phosphonate group is affected by the reaction solvent, using ammonia as a nucleophile [35]. By using ethanol as a reaction solvent, two phosphonate groups were replaced by the amine group, whilst one group was replaced when the reaction was performed in benzene. In the case of this work, the substitution reaction was carried out in ethanol at ambient conditions, to which only one phosphonate group of HEPT was selectively replaced with an amine group in a high yield, and no by-products were detected.

It can be assumed that by replacing one phosphonate group with an electron-donating atom, such as nitrogen, the electron density of the triazine ring increases, thus decreasing the reactivity of the ring for further substitution reactions at ambient conditions [31]. On the contrary to the previous report [33], this methodology led to producing pure products that were isolated as solids. Hence, the correlation between the NMR spectroscopy and single crystal X-ray analysis provided an insight into the chemical structure of these products.

2.2. Characterization of the Molecular Structure of Chemicals

All of the synthesized chemicals were analyzed by 1D and 2D ^{1}H, ^{13}C, and ^{31}P-NMR spectroscopy. The ^{31}P-NMR data of the target materials showed high-frequency shifts of the ^{31}P-NMR resonances as soon as one of the phosphonate groups of the precursor HEPT was replaced by an amine group, producing EDA-bis-TEPT and Pip-bis-TEPT (Figures S1, S2A, and S3A). From the chemical structures, all of the phosphorus atoms were expected to have the same chemical environment, and the presence of one signal for each compound in their corresponding ^{13}P-NMR spectra was expected. However, the ^{31}P-NMR spectrum of EDA-bis-TEPT showed two resonances assigned to an AB spin system, with a ^{31}P–^{31}P coupling constant of about 8.8 Hz. Owing to the presence of the electron-donating NH groups next to the triazine ring, the two phosphorous atoms became inequivalent, and consequently, two different signals were observable at ambient temperatures [33].

The electron-donating ability of the NH group could influence the electron densities around the triazine carbons of EDA-bis-TEPT that bind directly to the phosphorus atoms, as confirmed by X-ray analysis (Table 1 and Figure 1). Hence, the electron densities around these carbon atoms are inequivalent, and one carbon atom has more electron density than the other carbon atom. Accordingly, the ^{13}C-NMR spectrum of EDA-bis-TEPT showed two doublet of doublets for the carbon atoms that directly bind to the phosphorus atoms in the same ring (Figure S2A). In contrast, each phosphonate group on one ring was found to be identical to the other group on the other triazine ring. The ^{13}C-NMR analysis showed a low-frequency chemical shift of the triazine carbon when a phosphonate group was replaced by a nitrogen atom, from 171.2 ppm observed for HEPT, to 163.2 ppm and 161.8 ppm for EDA-bis-TEPT and Pip-bis-TEPT, respectively (Figures S1, S2A, and S3A). This can be attributed to the increase of electron density on the triazine carbon atom that binds to the amine group, because of the electron-donating ability of the nitrogen atom.

Table 1. Selected bond distances (Å) for the synthesized compounds 2,4,6-trisdiethoxyphosphinyl-1,3,5-triazine (HEPT); *N,N′*-Bis[4,6-bis(diethylphosphono)-1,3,5-Triazin-yl]-1,2-diaminoethane (EDA-bis-TEPT); *N,N′*-Bis[4,6-bis(diethylphosphono)-1,3,5-Triazin-yl]-piperazine (Pip-bis-TEPT).

HEPT		EDA-bis-TEPT		Pip-bis-TEPT	
Atoms	**Distances (Å)**	**Atoms**	**Distances (Å)**	**Atoms**	**Distances (Å)**
P1–C1	1.846 (4)	P1–C3	1.827 (5)	P1–C1	1.826 (3)
N1–C1	1.337 (6)	P2–C4	1.834 (6)	P2–C2	1.830 (3)
		N3–C3	1.355 (8)	N1–C1	1.344 (4)
		N3–C4	1.354 (7)	N1–C2	1.347 (4)
		N4–C4	1.323 (8)	N2–C2	1.323 (4)
		N4–C2	1.361 (8)	N2–C3	1.353 (4)
		N2–C3	1.309 (8)	N3–C1	1.326 (4)
		N2–C2	1.366 (7)	N3–C3	1.354 (4)
		N1–C2	1.335 (8)	N4–C3	1.343 (4)

The solid-state chemical structures of all of the compounds were confirmed by the single crystal X-ray analysis (Figure 1). HEPT crystallizes in a centrosymmetric hexagonal cell, in space group $P6_3/m$. The molecular formula corresponding to the crystallographic asymmetric unit is ($C_{15}H_{30}N_3O_9P_3$). The molecule is on a mirror plane, resulting in a positional disorder with 50% atomic position occupation of all P and O atoms; the C1 and N1 atoms are precisely positioned on the mirror plane, and the remaining C atoms are also 50% occupied, related to the positional disorder. EDA-bis-TEPT crystallizes in a centrosymmetric triclinic cell, in space group P-1. The molecular formula corresponds to ($C_{24}H_{46}N_8O_{12}P_4$), and all of the molecules lie on symmetry centers. The crystallographic asymmetric unit is composed out of three half molecules. Ten out of twelve ethyl groups have been disordered. The positional disorder has been refined, with a 50% atomic position occupation of all of the concerned C and H atoms, and the thermal vibration has been constrained to be equal. The positional disorder encountered for this structure results in a weak diffraction power of the crystal. However, the crystal structure has been refined, with reasonable standard uncertainties on atomic positions, bond lengths, and angles. Pip-bis-TEPT crystallizes in a centrosymmetric triclinic cell, space group P-1. The molecular formula is ($C_{26}H_{48}N_8O_{12}P_4$); the asymmetric unit is one half of the molecule, as the molecule is positioned on an inversion center. All of the structures were solved by direct methods using the program SHELXS [36], and refined by full-matrix least-squares on F^2 with SHELXL [37]. The hydrogen atoms were included in the calculated positions, and were treated as riding atoms using SHELXL-97 default parameters. All of the non-hydrogen atoms were refined anisotropically, and no absorption correction was applied.

It was found that the P–$C_{triazine}$ bond distance of HEPT (1.846 (4) Å) became shorter when one phosphonate group was replaced with the corresponding amine functional group, affording either EDA-bis-TEPT or Pip-bis-TEPT (Table 1).

It is noteworthy to mention that the P–$C_{triazine}$ bond distance within either EDA-bis-TEPT or Pip-bis-TEPT are not relatively equal. For example, the bond distances of P2–C4 and P1–C3 in EDA-bis-TEPT (Figure 1) are 1.834 (6) and 1.827 (5) Å, respectively. The same was observed in Pip-bis-TEPT, as the bond distance of P1–C1 is 1.826 (3) Å, and for P2–C2 is 1.830 (3) Å. The decrease of the P–$C_{triazine}$ bond distance in EDA-bis-TEPT and Pip-bis-TEPT in comparison to HEPT indicates that the bond of the phosphonate group to the triazine ring became stronger by replacing one phosphonate group with the amine group.

The correlation of the NMR spectroscopy and single-crystal X-ray analysis shows that the electron density around the phosphonate groups decreased, which can be attributed to the electron-donating nature of nitrogen, which enhances the electron density of the triazine ring. Therefore, the electron-withdrawing ability of the phosphonate groups decreased, which may, in turn, reduce the eligibility of the triazine ring for further substitution reactions at ambient conditions [31]. This can be observed in the NMR spectra, because of the difference in the chemical shifts of the phosphorus and carbons atoms in EDA-bis-TEPT and Pip-bis-TEPT, in comparison to the precursor HEPT. The X-ray data show that the lengths of the C–N bonds in the triazine ring in compounds EDA-bis-TEPT and Pip-bis-TEPT were found to be non-equal with respect to the precursor, HEPT. For example, the bond distances of C–N of the triazine ring, which bind to the bridged amine groups, are relatively longer than those that are close to the phosphonate groups. The nature of the amine group as an electron-donor by resonance leads to the donation of its free-electron to the triazine ring, which increases the electron density of the triazine ring, but deteriorates the distribution of the electron density, as shown in Scheme 3.

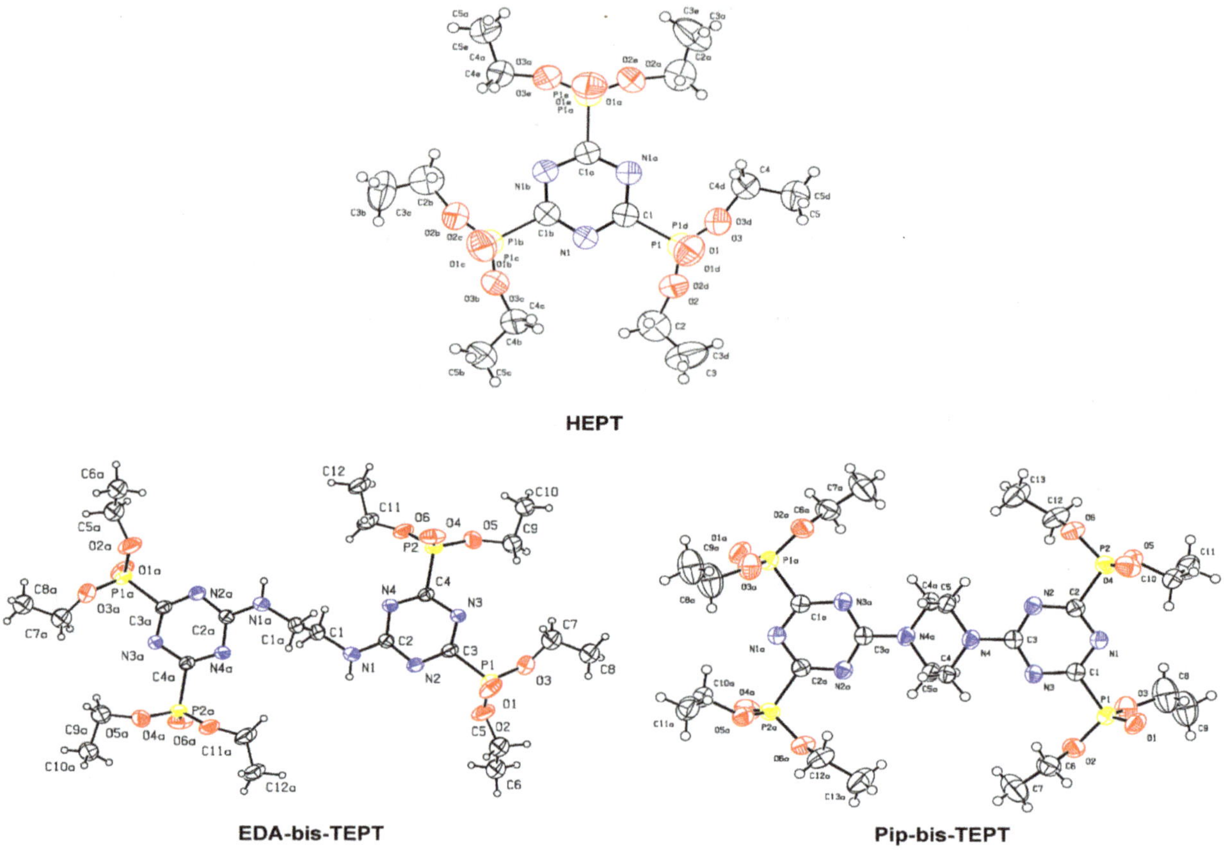

Figure 1. Molecular structures of HEPT, EDA-bis-TEPT, and Pip-bis-TEPT. Thermal ellipsoids are drawn with a 50% probability level [37]. Phosphorus (yellow), oxygen (red), and nitrogen (blue).

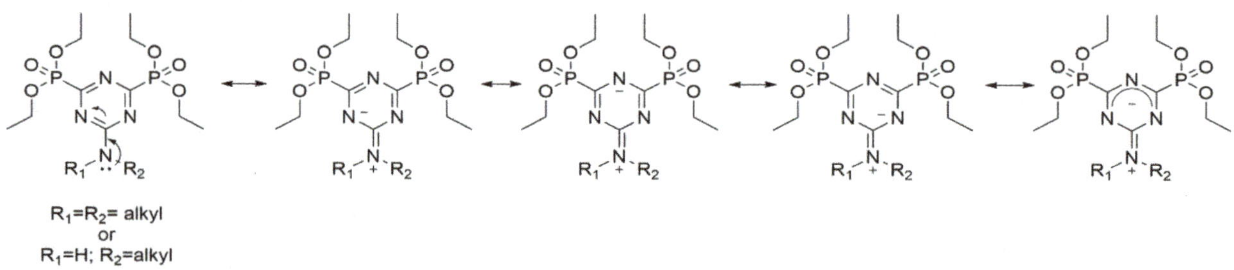

Scheme 3. Representative scheme showing the electronic resonance in a triazine ring because of the electron-donating ability of the amine group.

2.3. *Thermal Characteristics and Chemical Composition of the Resulting Chars*

The thermal gravimetric analysis (TGA) is an established technique that provides a quantitative physical and chemical phenomena, including the thermal decomposition and char formation in milligram sample quantities [24]. The thermal stability of HEPT, EDA-bis-TEPT, and Pip-bis-TEPT was measured in TGA (Figure 2), and the data are summarized in Table 2.

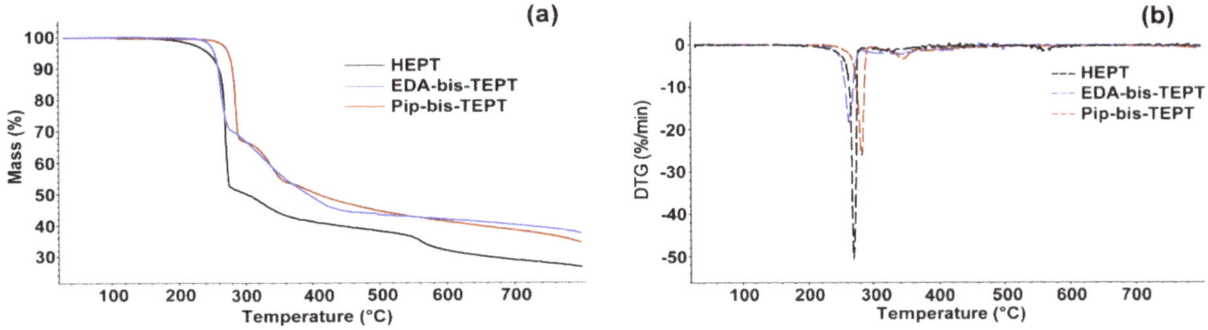

Figure 2. Thermal gravimetric analysis (TGA) (**a**), and derivative thermogravimetry (DTG) (**b**) plots of the organophosphorus-based bridged triazine derivatives under an N_2 atmosphere.

Table 2. Thermal gravimetric analysis (TGA) data of the synthesized compounds.

Sample	T_{onset} (°C)		T_{max} [a] (°C)		Char Residue (wt%) at 800 °C		N:P (wt%)
	N_2	O_2	N_2	O_2	N_2	O_2	
HEPT	265 ± 1	265 ± 0.5	270	271	27 ± 0.5	20 ± 0.5	0.45
EDA-bis-TEPT	253 ± 0.5	252 ± 0.5	263	263	36 ± 1	25 ± 0.5	0.90
Pip-bis-TEPT	277 ± 3	261 ± 1	283	272	33 ± 1	25 ± 1	0.90

[a] T_{max}: the maximum-rate decomposition temperature.

As shown in Figure 2 and Table 2, all three of the compounds exhibited a significant residual char upon their pyrolysis under the N_2 atmosphere. It is noticed that, by increasing the percentage ratio of nitrogen to phosphorus atoms in the corresponding molecules, the char residue increases. The elemental composition of residue after the oxidative decomposition of the samples, imitating the oxidative decomposition that can happen upon combustion, was investigated using energy-dispersive X-ray analysis (EDX; Figure 3a–c). All of the data are summarized in Table 3.

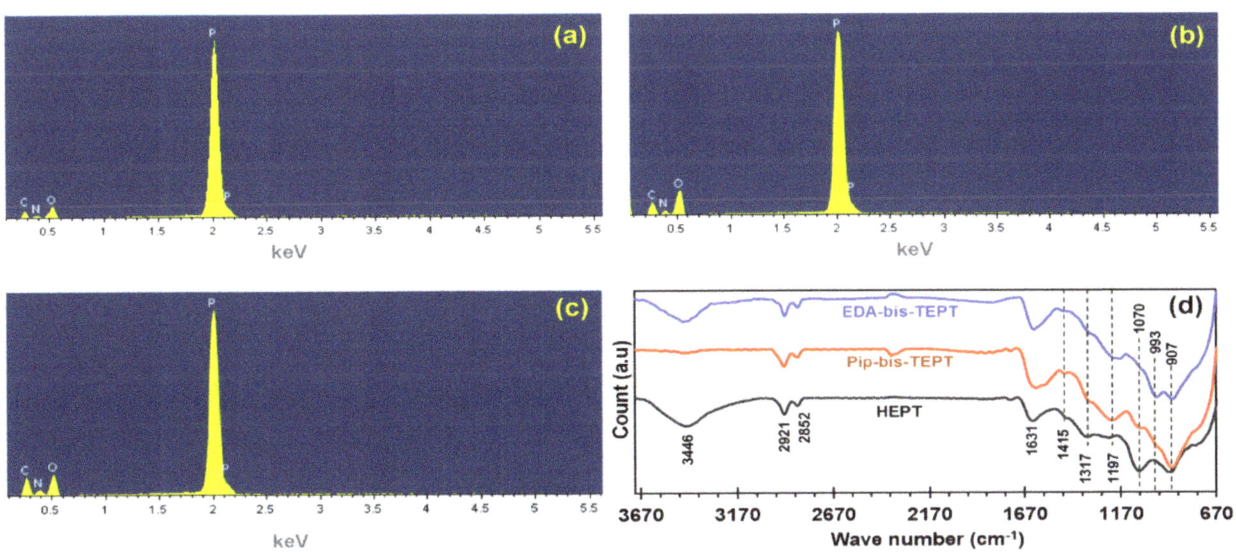

Figure 3. Energy-dispersive X-ray analysis (EDX) elemental composition of the char residue of HEPT (**a**), EDA-bis-TEPT (**b**), Pip-bis-TEPT (**c**), and the Fourier-Transform Infrared Spectroscopy (FTIR) spectra for their corresponding char residues (**d**). All of the char residues were obtained after the oxidative pyrolysis of the corresponding compounds.

Table 3. Elemental analysis of the char residue after TGA under air. EDX—energy-dispersive X-ray analysis.

Sample	EDX (wt%)			Elemental Loss (wt%) [a]		
	P	C	N	P	C	N
HEPT	37 ± 2	24 ± 1	15 ± 1	61	87	65
EDA-bis-EPT	28 ± 2	26 ± 1	15 ± 0.5	57	83	74
Pip-bis-TEPT	27 ± 1	29 ± 2	18 ± 3	57	82	68

[a] The elemental loss (wt%) was calculated based on the EDX analysis, from the following equation: (1 − ((char percentage x element percentage in char)/(100 × element percentage of the compound))) × 100.

As shown in Table 3, the char residues of each compound retained about 40% of the original phosphorus content, with respect to their corresponding compounds. The retained phosphorus could play a crucial role in the condensed phase, and could contribute towards the char formation. The loss in phosphorus content can be attributed to the formation of $PO^{\bullet}$ species in the gas phase. Identical results were also observed from the percentage mass loss of nitrogen content and the retained amount in the residual char. On the other hand, the percentage mass loss of carbon content was more than 80%. This can be attributed to the accelerated formation of CO_2 and H_2O, upon the oxidative pyrolysis of the hydrocarbon contents of the molecules. The retained amount of carbon was enough for the formation of voluminous char.

The chemical composition of the residual chars obtained by the oxidative pyrolysis of all of the compounds were qualitatively studied by Fourier-Transform Infrared Spectroscopy (FTIR) analysis (Figure 3d). The bands at 907 cm^{-1} and 993 cm^{-1} were attributed to the existence of P–O–P and P–O–C bonds, respectively [19]. A high intensity of 907 cm^{-1} and 993 cm^{-1} suggests that the main composition of char is pyrophosphate and polyphosphates, which is consistent with the EDX analysis. The band around 1317 cm^{-1} can be attributed to the P=O stretch [38], and at 1415 cm^{-1} it is because of the presence of P–N structures in the char [39]. Evidence of a carbonization reaction (C=C formation), which promotes char formation, can be seen as the presence of a strong band at 1631 cm^{-1} [39]. The bands at 2852 cm^{-1} and 2921 cm^{-1} were corresponding to the stretching vibration of the C–H bond, and the broad band at 3446 cm^{-1} could be for the stretching vibration of the N–H bond [40]. Surprisingly, the IR spectrum of the Pip-bis-TEPT char showed a relatively low intensity for the N–H bond in comparison to other char residues.

2.4. Direct Insertion Probe Mass Spectrometry (DIP-MS) Measurements

The above-mentioned SEM-EDX and FTIR study on residual char shed light on the possible mode of action of HEPT, EDA-bis-TEPT, and Pip-bis-TEPT in the condensed phase. However, about 75% to 80% of the products were lost in the gas phase. To identify the volatile products that evolved during their pyrolysis, DIP-MS measurements were performed, and the total ion chromatogram for the decomposition was investigated as a function of the temperature. The total ion chromatogram of each product revealed the release of diethyl phosphite (m/z 138) for all of the compounds, in addition to the volatilization of the main compounds, namely HEPT (m/z 489), EDA-bis-TEPT (m/z 763), and Pip-bis-TEPT (m/z 789), as illustrated in Figure 4a–c. The change of intensities between each product and its corresponding diethyl phosphite illustrate its gas phase mode of action. For example, comparing the intensities of all of the products with their corresponding release of diethyl phosphite (Figure 4a–c) can clarify their eligibility to act in the gas phase as flame retardant additives. A comparison of HEPT (Figure 4a) and EDA-bis-TEPT (Figure 4b) with their corresponding diethyl phosphite shows a relatively high intensity of the latter compound. That is to say, both compounds can have efficacy in the gas phase. The low intensity of EDA-bis-TEPT regarding diethyl phosphite (Figure 4b) indicates that its pyrolysis can happen faster than its evaporation. On the contrary to HEPT and EDA-bis-TEPT, Pip-bis-TEPT showed more thermal stability, as its evaporation is in concurrent with its pyrolysis.

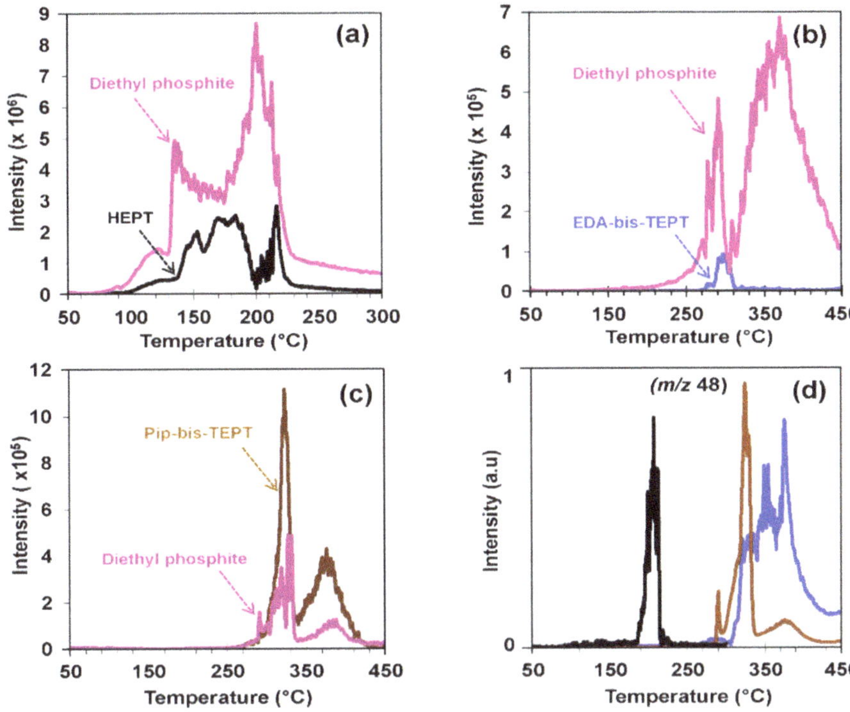

Figure 4. Extracted ion chromatogram of the synthesized compounds in comparison to their release of diethyl phosphite (**a**–**c**) and production of HPO (*m/z* 48) (**d**).

Diethyl phosphite is considered an efficient flame retardant that acts in the gas phase, with the potential to produce $PO^{\bullet}$. Despite the difficulty of tracking this species using DIP-MS, a minor product was detected for all of the compounds at *m/z* 48, which can be attributed to HPO (Figure 4d), which upon its pyrolysis, may produce $PO^{\bullet}$. Moreover, another minor peak was detected for all of the compounds at *m/z* 44, which can be attributed to the production of CO_2. However, as all of the measurements in DIP-MS were performed in an inert atmosphere, which may exclude the production of CO_2, the production of H–C≡P instead of CO_2 can be expected [41].

3. Materials and Methods

3.1. Materials

Cyanuric chloride (99% purity), triethyl phosphite (98% purity), ethylenediamine (*ReagentPlus*®, ≥99% purity), and Piperazine (*ReagentPlus*®, ≥99% purity) were purchased from Sigma-Aldrich (Buchs, Switzerland). Toluene (HPLC grade) was purchased from Biosolve BV (Valkenswaard, Netherlands), and absolute ethanol was purchased from Honeywell Fluka (Reinach, Switzerland). All of the chemicals and solvents were used as received, without any further purification.

3.2. Thermal Characterization, DIP-MS, SEM-EDX, Elemental Analysis, and NMR Spectrometer

The thermogravimetric analysis (TGA) was performed on a NETZSCH TG 209 F1 instrument (NETZSCH-Gerätebau GmbH, Selb, Germany). Tests were conducted on 2–5 mg samples at a heating rate of 10 °C/min, from 25 °C to 800 °C. The TGA curves represent the average of two measurements. Differential scanning calorimetry (DSC) analyses were performed on the DSC 214 Polyma instrument (NETZSCH-Gerätebau GmbH, Selb, Germany). All of the experiments were conducted at a controlled heating rate of 10 °C/min, under a nitrogen flow of 50 mL/min. A direct insertion probe mass spectrometry (DIP-MS) analysis was conducted for a 1–2 µg sample, using ThermoQuest FINNIGAN apparatus (Austin, TX, USA). The probe was heated from 30 °C to 300 °C for HEPT, and from 30 °C to 450 °C for EDA-bis-TEPT and Pip-bis-TEPT, at a rate of 10 K/min and

10^{-6} mbar pressure. Energy-dispersive X-ray spectroscopy (EDX) was applied in combination with a scanning electron microscope (SEM), using an Inca X-sight device from Oxford Instruments (EDX) and a S-4800 microscope from Hitachi (SEM) (Tokyo, Japan). For each sample, the EDX analysis was performed at two different areas on completely charred samples obtained from TGA, and an average value was obtained from the resulting elemental compositions. An estimated area of 100 μm × 100 μm was scanned for the EDX measurements using an acceleration voltage of the electron beam of 20 kV, an emission current of 15μA, and a working distance of 15mm. Under these experimental conditions, the information depth is in the order of a couple of microns. The micro-elemental analysis (CHN) was performed by the molecular and biomolecular analysis service (MoBiAS) of the Laboratory of Organic Chemistry at Swiss Federal Institute of Technology in Zurich (ETHZ). The ^{1}H, ^{13}C, and ^{31}P-NMR data were recorded on a Bruker Avance III 400 NMR spectrometer (Bruker Biospin AG, Fällanden, Switzerland) at 400.2, 100.6, and 162.0 MHz, respectively. The 1D and the 2D ^{1}H–^{13}C HSQC and ^{1}H–^{13}C HMBC correlation NMR experiments used for the full resonance assignments were performed on a 5 mm CryoProbe™ Prodigy probe equipped with a z-gradient at 298 K, using the Bruker standard pulse programs and parameter sets, applying 90° pulse lengths of 11.4 μs (^{1}H), 10.0 μs (^{13}C), and 12.0 μs (^{31}P). The ^{1}H and ^{13}C chemical shifts (δ) in ppm are calibrated to residual solvent peaks (DMSO-d_6: δ = 2.49 and 39.5 ppm), the ^{31}P chemical shifts were referenced to an external sample with neat H_3PO_4 at 0.0 ppm. Wherever possible, the ^{1}H–^{31}P and ^{31}P–^{13}C coupling constants are reported in Hz. For ^{13}C-NMR data multiplicities, *s* = quaternary carbon, *d* = CH, *t* = CH_2, and *q* = CH_3 are shown.

3.3. Single Crystal X-Ray Structure Determination

The crystals of HEPT, EDA-bis-TEPT, and Pip-bis-TEPT were grown by slow evaporation of their corresponding ethanol solution. Suitable crystals were selected and mounted on a Stoe Mark II-Imaging Plate Diffractometer System (Stoe and Cie, 2015), equipped with a graphite-monochromator (Darmstadt, Germany). Data collection was performed at −100 °C using Mo-Kα radiation (λ = 0.71073 Å). There were 180 exposures (5 min per exposure) that were obtained, at an image plate distance of 135 mm, $\varphi = 0°$ and $0 < \omega < 180°$, with the crystal oscillating through 1° in ω. The resolution was D_{min}–D_{max} 24.00–0.82 Å. The structures were solved by direct methods using the program SHELXS (Sheldrik, Göttingen, Germany) [36], and refined by full-matrix least-squares on F^2 with SHELXL (Sheldrik, Göttingen, Germany) [37]. The hydrogen atoms were included in the calculated positions, and treated as riding atoms using SHELXL-97 default parameters. All of the non-hydrogen atoms were refined anisotropically. No absorption correction was applied. More refinement details for HEPT, EDA-bis-TEPT, and Pip-bis-TEPT are given in the supporting information. CCDC 1893600-1893602 contain the supplementary crystallographic data for compounds HEPT, EDA-bis-TEPT and Pip-bis-TEPT, respectively. These data can be obtained free of charge from The Cambridge Crystallographic Data Centre, via www.ccdc.cam.ac.uk/data_request/cif. The crystal data and refinements are summarized in Table 4.

Table 4. Crystal data and structure refinement of the synthesized substances.

Identification Code	HEPT	EDA-bis-TEPT	Pip-bis-TEPT
Empirical formula	$C_{15}H_{30}N_3O_9P_3$	$C_{24}H_{46}N_8O_{12}P_4$	$C_{26}H_{48}N8O_{12}P_4$
Formula weight	489.33	762.57	788.60
Temp/K	293	173	293
Crystal system	Hexagonal	Triclinic	Triclinic
Space group	P63/m	P-1	P-1
a (Å)	13.5150 (15)	7.463 (5)	7.3620 (9)
b (Å)	13.5150 (15)	21.171 (13)	11.8117 (16)
c (Å)	7.6602 (10)	21.200 (12)	12.4732 (16)
α (°)	90	118.49 (4)	106.602 (10)
β (°)	90	95.68 (5)	103.435 (10)
γ (°)	120	97.27 (5)	104.434 (10)
V (Å^3)	1211.7 (4)	2858 (3)	951.2 (2)
Z	2	3	1
D_{calc} (g.cm^{-3})	1.341	1.329	1.377
Absorption coefficient (mm^{-1})	0.293	0.262	0.264
F(000)	516	1206	416
Crystal size /mm^3	0.4 × 0.45 × 0.5	0.5 × 0.45 × 0.4	0.5 × 0.45 × 0.4
Radiation (Å)	MoKα (λ = 71073)	MoKα (λ = 0.71073)	MoKα (λ = 0.71073)
2θ range for data collection (°)	1.7 to 25.7	1.931 to 26.273	3.606 to 51.326
Index ranges	$-16 \le h \le 14$, $-16 \le k \le 16$, and $-9 \le l \le 9$	$-8 \le h \le 9$, $-25 \le k \le 25$, and $-25 \le l \le 25$	$-8 \le h \le 8$, $-14 \le k \le 12$, and $-15 \le l \le 15$
Reflections collected	5013	19228	8103
Independent reflections	5013 (R_{int} = 0.071)	10234 (R_{int} = 0.0906)	3528 (R_{int} = 0.0964)
Data/restraints/parameters	820/0/85	10,234/18/582	3528/0/225
Goodness of fit on F^2	1.145	1.024	1.093
Final R indexes (I ≥ 2σ (I))	R_1 = 0.0672, wR_2 = 0.2142	R_1 = 0.0947, wR_2 = 0.2612	R_1 = 0.0573, wR_2 = 0.1588
Final R indexes (all data)	$R_1{}^a$ = 0.0853, $wR_2{}^b$ = 0.2379	$R_1{}^a$ = 0.1662, $wR_2{}^b$ = 0.3192	$R_1{}^a$ = 0.0700, $wR_2{}^b$ = 0.1804
Largest diff. peak/hole (e Å^{-3})	0.64/−0.27	0.632/−0.793	0.97/−0.88

[a] $R_1 = \sum(||F_o| - |Fc||)/\sum|F_o|$. [b] $wR_2 = \{[\sum w(F_o{}^2 - F_c{}^2)^2]/[\sum w(F_{o2})^2]\}^{1/2}$, $w = 1/[\sigma^2(F_o{}^2) + (aP)^2 + bP$, where $P = [\max(F_o{}^2),0) + 2(F_c{}^2)]/3$.

3.4. Synthesis

3.4.1. Synthesis of 2,4,6-Trisdiethoxyphosphinyl-1,3,5-Triazine (HEPT)

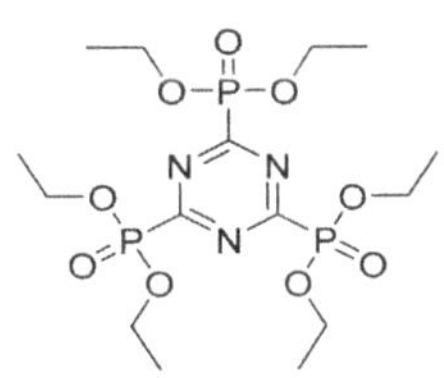

A five-neck double-jacket reactor (5 L) was equipped with a heating system, mechanical stirrer, reflux condenser, dropping funnel, and connected to a cooling trap. The reactor was purged with N_2, then charged with cyanuric chloride (268.9 g; 1.46 mol) and anhydrous toluene (2.5 L). The solution was then heated to 60 °C, and triethyl phosphite (969.0 g; 5.83 mol) was added dropwise in a rate that the reaction temperature did not exceed 80 °C. After the complete addition, the reaction solution was stirred at 70 °C for 4 h, and then allowed to cool down to ambient temperature. The volatiles were then completely removed in a vacuum, affording a pale-yellow solid that was washed thoroughly with heptane. The solid was then collected by filtration, and dried overnight in a vacuum oven at 60 °C, affording white crystals. The yield was as follows: 691.4 g (97%); m.p. 96 °C. ^{1}H-NMR (400.2 MHz, DMSO-d_6) δ (ppm) was as follows: 4.28 (qt, $J_{H,H}$ = 7.1 Hz, $J_{H,P}$ = 7.3 Hz, 12H, O–$\mathbf{CH_2}$–CH_3); 1.32 (t, $J_{H,H}$ = 7.1 Hz, 18H, O-CH_2-$\mathbf{CH_3}$). ^{13}C-NMR (100.6 MHz, DMSO-d_6) δ (ppm) was as follows: 171.2 (sdt, $J_{C,P}$ = 257, 12.5 Hz, Cq); 64.4 (t [a]), O-$\mathbf{CH_2}$–CH_3); 16.14 (q [a]), O-CH_2–$\mathbf{CH_3}$). ^{31}P-NMR (162.0 MHz, DMSO-d_6) δ (ppm) was as follows: 1.4.

[a)] Signals exhibit higher-order multiplet splitting because of J(P,C).

3.4.2. Synthesis of *N,N'*-Bis[4,6-Bis(diethylphosphono)-1,3,5-Triazin-yl]-1,2-Diaminoethane (EDA-bis-TEPT)

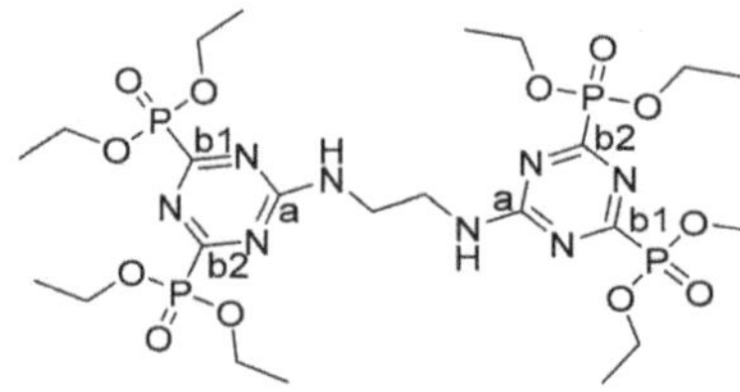

A two-neck round-bottomed flask (500 mL) was charged with HEPT (30.0 g; 61.3 mmol) in absolute ethanol (100 mL). Ethylenediamine (1.82 g; 30.4 mmol) was then dropwise added at ambient temperature. After the complete addition, the yellow solution was allowed to stir at ambient temperature. After 4 h, the volatiles were completely removed in a vacuum, affording a viscous yellow oil. Toluene (200 mL) was then added with vigorous stirring, affording a white precipitate. The solid was then collected by filtration, and dried overnight in a vacuum oven at 80 °C. The yield was as follows: 19.0 g (81%); m.p. 150 °C. ^{1}H-NMR (400.2 MHz, DMSO-d_6) δ (ppm) was as follows: 8.99 (m, 2H, **NH**); 4.21 (m, 16H, O–$\mathbf{CH_2}$–CH_3); 3.56 (m, 4H, N–$\mathbf{CH_2}$–$\mathbf{CH_2}$–N); 1.28 (m, 24H, O–CH_2–$\mathbf{CH_3}$). ^{13}C-NMR (100.6 MHz, DMSO-d_6) δ (ppm) was as follows: 170.9 (sdd [a], $J_{C,P}$ = 260, 14.6 Hz, 2C, C-b1); 170.2 (sdd [a], $J_{C,P}$ = 261, 15.2 Hz, 2C, C-b2); 163.2 (st, $J_{C,P}$ = 18.1 Hz, 2C, C-a); 63.7 (tdd, $J_{C,P}$ = 22.6, 6.1 Hz, 8C, O–CH_2–CH_3); 39.1 (t, 2C, N–CH_2–CH_2–N); 16.1(tdd, $J_{C,P}$ = 5.8, 2.0 Hz, 8C, O–CH_2–CH_3). ^{31}P-NMR (162.0 MHz, DMSO-d_6) δ (ppm): 3.1 (d [a], $J_{P,P}$ = 8.8, P-b1); 2.9 (d [a], $J_{P,P}$ = 8.8, P-b2)).^{1}H–^{13}C HMBC: **NH** → N–CH_2–CH_2–N, **C**-a; O–$\mathbf{CH_2}$$CH_3$ → O–CH_2–$\mathbf{CH_3}$; N–$\mathbf{CH_2}$–$\mathbf{CH_2}$–N → N–CH_2–CH_2–N, **C**-a; O–$CH_2$$\mathbf{CH_3}$ → O–CH_2–CH_3. The following analysis was calculated for $C_{24}H_{46}N_8O_{12}P_4$: C, 37.80; H, 6.08; N, 14.69. The following was found: C, 37.60; H, 5.99; N, 14.57.

[a] Because the NH group positions "b1" and "b2" are not chemically equivalent, resonance assignments to b1 or b2 arbitrarily.

3.4.3. Synthesis of *N,N'*-Bis[4,6-Bis(diethylphosphono)-1,3,5-Triazin-yl]-Piperazine (Pip-bis-TEPT)

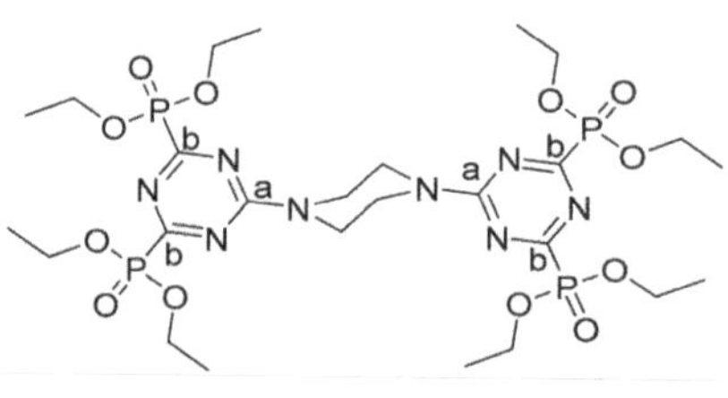

A two-neck round-bottomed flask (500 mL) was charged with HEPT (30.0 g; 61.3 mmol) in absolute ethanol (100 mL). Piperazine (2.61 g; 30.4 mmol) was dissolved in absolute ethanol (50 mL), and added dropwise at ambient temperature. After the complete addition, the yellow solution was allowed to stir at ambient temperature. After 4 h, the volatiles were completely removed in a vacuum, affording a viscous yellow oil. After standing at ambient conditions, a white crystalline product formed that was collected by filtration and washed with heptane. The solid was then collected and dried overnight in a vacuum oven at 70 °C. The yield was as follows: 18.4 g (76%); m.p. 125 °C. ^{1}H-NMR (400.2 MHz, DMSO-d_6) δ (ppm) was as follows: 4.24 (m, 16H, O–$\mathbf{CH_2}$–CH_3); 3.99 (s, 8H, N–($\mathbf{CH_2}$–$\mathbf{CH_2}$)$_2$–N); 1.30 (t, $J_{H,H}$ = 7.1 Hz, 24H, O–CH_2–$\mathbf{CH_3}$). ^{13}C-NMR (100.6 MHz, DMSO-d_6) δ (ppm) was as follows: 170.5 (sdd, $J_{C,P}$ = 260, 14.9 Hz, 4C, C-b); 161.8 (st, $J_{C,P}$ = 18.3 Hz, 2C, C-a); 63.9 (tm [a], 8C, O–CH_2–CH_3); 42.4 (t, 4C, N–(CH_2–CH_2)$_2$–N); 16.2(tm [a], 8C, O–CH_2–CH_3). ^{31}P-NMR (162.0 MHz, DMSO-d_6) δ (ppm) was as follows: 2.9. ^{1}H-^{13}C HMBC: O–$\mathbf{CH_2}$–CH_3 → O–CH_2–CH_3; N–($\mathbf{CH_2}$–$\mathbf{CH_2}$)$_2$–N → C-a, N–(CH_2–CH_2)$_2$–N; O–$CH_2$$\mathbf{CH_3}$ → O–CH_2–CH_3. The following analysis was calculated for $C_{26}H_{48}N_8O_{12}P_4$: C, 39.60; H, 6.14; N, 14.21. The following was found: C, 39.58; H, 6.04; N, 14.11.

[a] Because of steric reasons, a rather complicated signal pattern was observed for the $P(=O)(OEt)_2$ carbons.

4. Conclusions

Cyanuric chloride was used as a starting material to synthesize bridged triazine-based organophosphorus compounds. The Michaelis–Arbuzov rearrangement of triethyl phosphite with cyanuric chloride afforded intermediate HEPT, which was used for the further substitution reactions. Because of the electron-withdrawing effect of the phosphonate group, the HEPT group showed a relative positive charge in the triazine ring, making the ring eligible for nucleophilic substitution reactions. This nature of HEPT was utilized to achieve a substitution reaction with ethylenediamine and piperazine at ambient temperature, affording EDA-bis-TEPT and Pip-bis-TEPT, respectively. Diethyl phosphite is the only by-product that was detected, and could be isolated by precipitating EDA-bis-TEPT and Pip-bis-TEPT in toluene. Single crystals for X-ray analysis provided compelling evidence regarding the chemical structures of the target compounds and the electronic distribution in a triazine ring. The correlation between the NMR spectroscopy and single crystal X-ray analysis could confirm the increase in the electron density of the triazine ring once one phosphonate group of HEPT was replaced with the amine group, resulting in either EDA-bis-TEPT or Pip-bis-TEPT. This change led to a decrease in the reactivity of the latter compounds to any further substitution reaction at ambient conditions, and prevented the formation of a mixture of substituted product or polymeric materials. However, more investigations are in progress in order to investigate the possibility of further substitution reactions, either at high temperatures or in different chemical solvents. The pyrolysis of the synthesized products showed the formation of the residual char, which is considered as a barrier to protect the sub-layers from heat and fire. An analysis of the volatiles that evolved upon pyrolysis showed the presence of diethyl phosphite, HPO, and H–C≡P, which are considered a suitable source for $PO^{\bullet}$. Such radical scavenging of the $O^{\bullet}$ and $^{\bullet}OH$ species does result in an additional protection of the matrix from fire. Therefore, these compounds have promising flame retardant characteristics [34]. Accordingly, the efficacy of these compounds as flame retardant additives will be further investigated in our future work.

Supplementary Materials: The following additional information is available online: Figure S1: ^{1}H, ^{13}C, and ^{31}P-NMR spectra of HEPT (DMSO-d_6); Figure S2A: ^{1}H, ^{13}C, and ^{31}P-NMR spectra of EDA-bis-TEPT (DMSO-d_6); Figure S2B: Regions of interest of ^{1}H-^{13}C HSQC (A) and ^{1}H-^{13}C HMBC (B) NMR spectra of EDA-bis-TEPT (DMSO-d_6); Figure S3A: ^{1}H, ^{13}C, and ^{31}P-NMR spectra of Pip-bis-TEPT (DMSO-d_6); Figure S3B: Regions of interest of ^{1}H-^{13}C HSQC (A) and ^{1}H-^{13}C HMBC (B) NMR spectra of Pip-bis-TEPT (DMSO-d_6); Figure S4: DSC curve of HEPT; Figure S5: DSC curve of EDA-bis-TEPT; Figure S6: DSC curve of Pip-bis-TEPT; Table S1: Final coordinates and equivalent isotropic displacement parameters of the non-hydrogen atoms for HEPT; Table S2: Hydrogen atom positions and isotropic displacement parameters for HEPT; Table S3: Anisotropic displacement parameters for HEPT; Table S4: Bond angles for HEPT; Table S5: Atomic coordinates and equivalent isotropic displacement parameters of the non-hydrogen atoms for EDA-bis-TEPT; Table S6: Hydrogen coordinates and isotropic displacement parameters for EDA-bis-TEPT; Table S7: Anisotropic displacement parameters for EDA-bis-TEPT; Table S8: Bond angles for EDA-bis-TEPT; Table S9: Fractional atomic coordinates and equivalent isotropic displacement parameters of the non-hydrogen atoms for Pip-bis-TEPT; Table S10: Hydrogen atom coordinates and isotropic displacement parameters for Pip-bis-TEPT; Table S11: Anisotropic displacement parameters for Pip-bis-TEPT; Table S12: Bond angles for Pip-bis-TEPT.

Author Contributions: Conceptualization, K.A.S. and S.G.; methodology, K.A.S.; software, K.A.S., S.L., D.R., D.P. and A.N.; validation, K.A.S. and S.G.; formal analysis, K.A.S., A.N., D.R. and D.P.; investigation, K.A.S and S.G.; resources, S.G.; data curation, K.A.S.; writing—original draft preparation, K.A.S. and D.P.; writing—review and editing, K.A.S., D.P., A.N., D.R. and S.G.; visualization, K.A.S. and S.G.; supervision, K.A.S.; project administration,

Acknowledgments: The authors wish to thank Wheeldon Dean Edward for his assistance in proof reading the manuscript.

References

1. Prokhorov, A.M.; Prokhorova, P.E. Triazines and tetrazines. *Prog. Heterocycl. Chem.* **2015**, *27*, 451–464. [CrossRef]
2. Banerjee, R.; Brown, D.R.; Weerapana, E. Recent developments in the synthesis of bioactive 2,4,6-trisubstituted 1,3,5-triazines. *Synlett* **2013**, *24*, 1599–1605. [CrossRef]
3. Blotny, G. Recent applications of 2,4,6-trichloro-1,3,5-triazine and its derivatives in organic synthesis. *Tetrahedron* **2006**, *62*, 9507–9522. [CrossRef]
4. Probst, D.A.; Hanson, P.R.; Barda, D.A. Cyanuric chloride. In *e-EROS Encyclopedia of Reagents for Organic Synthesis*; John Wiley & Sons, Ltd.: Hoboken, NJ, USA, 2004; pp. 2874–2878.
5. Senier, A. Contributions to the history of cyanuric chloride and cyanuric acid. *J. Chem. Soc. Trans.* **1886**, *49*, 311–313. [CrossRef]
6. Bretterbauer, K.; Schwarzinger, C. Melamine derivatives. A review on synthesis and application. *Curr. Org. Synth.* **2012**, *9*, 342–356. [CrossRef]
7. Van der Jeught, S.; Stevens, C.V. Direct Phosphonylation of Aromatic Azaheterocycles. *Chem. Rev.* **2009**, *109*, 2672–2702. [CrossRef] [PubMed]
8. Huthmacher, K.; Most, D. Cyanuric acid and cyanuric chloride. In *Ullmann's Encyclopedia of Industrial Chemistry*; Wiley-VCH Verlag GmbH & Co. KGaA: Weinheim, Germany, 2012; Volume 11.
9. Liu, X.; Hao, J.; Gaan, S. Recent studies on the decomposition and strategies of smoke and toxicity suppression for polyurethane based materials. *RSC Adv.* **2016**, *6*, 74742–74756. [CrossRef]
10. Cipolli, R.; Masarati, E.; Rossi, C.; Oriani, R.; Nucida, G. Self-extinguishing polymeric compositions. EP551154A1, 14 July 1993.
11. Li, X.; Chen, H.; Wang, W.; Liu, Y.; Zhao, P. Synthesis of a formaldehyde-free phosphorus-nitrogen flame retardant with multiple reactive groups and its application in cotton fabrics. *Polym. Degrad. Stab.* **2015**, *120*, 193–202. [CrossRef]
12. Chang, S.C.; Condon, B.; Nguyen, T.-M.; Graves, E.; Smith, J. Antiflammable properties of capable phosphorus-nitrogen-containing triazine derivatives on cotton. *ACS Symp. Ser.* **2012**, *1118*, 123–137. [CrossRef]
13. Chang, S.C.; Condon, B.; Graves, E.; Uchimiya, M.; Fortier, C.; Easson, M.; Wakelyn, P. Flame retardant properties of triazine phosphonates derivative with cotton fabric. *Fibers Polym.* **2011**, *12*, 334–339. [CrossRef]
14. Nguyen, T.-M.D.; Chang, S.; Condon, B.; Slopek, R. Synthesis of a novel flame retardant containing phosphorus-nitrogen and its comparison for cotton fabric. *Fibers Polym.* **2012**, *13*, 963–970. [CrossRef]
15. Easson, M.; Condon, B.; Yoshioka-Tarver, M.; Childress, S.; Slopek, R.; Bland, J.; Nguyen, T.-M.; Chang, S.C.; Graves, E. Cyanuric chloride derivatives for cotton textile treatment-synthesis, analysis, and flammability testing. *AATCC Rev.* **2011**, *11*, 60–66.
16. Schmidt, C.; Ciesielski, M.; Greiner, L.; Doering, M. Novel organophosphorus flame retardants and their synergistic application in novolac epoxy resin. *Polym. Degrad. Stab.* **2018**, *158*, 190–201. [CrossRef]
17. Yang, S.; Wang, J.; Huo, S.; Wang, M.; Wang, J. Preparation and flame retardancy of a compounded epoxy resin system composed of phosphorus/nitrogen-containing active compounds. *Polym. Degrad. Stab.* **2015**, *121*, 398–406. [CrossRef]
18. You, G.; Cheng, Z.; Tang, Y.; He, H. Functional Group Effect on Char Formation, Flame Retardancy and Mechanical Properties of Phosphonate–Triazine-based Compound as Flame Retardant in Epoxy Resin. *Ind. Eng. Chem. Res.* **2015**, *54*, 7309–7319. [CrossRef]
19. Chen, H.; Wang, J.; Ni, A.; Ding, A.; Han, X.; Sun, Z. The effects of a macromolecular charring agent with gas phase and condense phase synergistic flame retardant capability on the properties of PP/IFR composites. *Materials* **2018**, *11*, 111. [CrossRef] [PubMed]
20. Ma, D.; Li, J. Synthesis of a bio-based triazine derivative and its effects on flame retardancy of polypropylene composites. *J. Appl. Polym. Sci.* **2019**. [CrossRef]
21. Ye, X.; Wang, Y.; Zhao, Z.; Yan, H. A novel hyperbranched poly(phosphorodiamidate) with high expansion degree and carbonization efficiency used for improving flame retardancy of APP/PP composites. *Polym. Degrad. Stab.* **2017**, *142*, 29–41. [CrossRef]
22. Zuo, J.-D.; Liu, S.-M.; Sheng, Q. Synthesis and application in polypropylene of a novel of phosphorus-containing intumescent flame retardant. *Molecules* **2010**, *15*, 7593–7602. [CrossRef]

23. Klatt, M. Nitrogen-based flame retardants. In *Non-Halogenated Flame Retardant Handbook*; Morgan, A.B., Wilkie, C.A., Eds.; Scrivener Publishing LLC: Beverly, MA, USA, 2014; pp. 143–168.
24. Salmeia, K.A.; Fage, J.; Liang, S.; Gaan, S. An Overview of Mode of Action and Analytical Methods for Evaluation of Gas Phase Activities of Flame Retardants. *Polymers* **2015**, *7*, 504–526. [CrossRef]
25. Salmeia, K.A.; Gaan, S. An overview of some recent advances in DOPO-derivatives: Chemistry and flame retardant applications. *Polym. Degrad. Stability* **2015**, *113*, 119–134. [CrossRef]
26. Hu, X.-P.; Li, W.-Y.; Wang, Y.-Z. Synthesis and characterization of a novel nitrogen-containing flame retardant. *J. Appl. Polym. Sci.* **2004**, *94*, 1556–1561. [CrossRef]
27. Liu, Y.; Ren, Y.; Dong, E. Development of environmental friendly flame retardants for wood plastic composites (WPC). *Adv. Mater. Res.* **2011**, *332–334*, 1880–1883. [CrossRef]
28. Mikroyannidis, J.A. Synthesis, physical, and thermal properties of linear poly(dialkoxyphosphinyl-s-triazines). *J. Polym. Sci. Part A Polym. Chem.* **1988**, *26*, 583–593. [CrossRef]
29. Nguyen, M.M.; Al-Abdul-Wahid, M.S.; Fontenot, K.R.; Graves, E.E.; Chang, S.C.; Condon, B.D.; Grimm, C.C.; Lorigan, G.A. Understanding the mechanism of action of triazine-phosphonate derivatives as flame retardants for cotton fabric. *Molecules* **2015**, *20*, 11236–11256. [CrossRef] [PubMed]
30. Morrison, D.C. sym-Triazinetriphosphonic acid esters. *J. Org. Chem.* **1957**, *22*, 444. [CrossRef]
31. Hewertson, W.; Shaw, R.A.; Smith, B.C. 313. 1,3,5-Triazines. Part III. Arbuzov reactions of chlorotriazines. *J. Chem. Soc.* **1963**, 1670–1675. [CrossRef]
32. Hewertson, W.; Shaw, R.A.; Smith, B.C. 197. 1,3,5-Triazines. Part IV. Phosphino-1,3,5-triazines. *J. Chem. Soc.* **1964**, 1020–1026. [CrossRef]
33. Kreher, T.; Costisella, B.; Kirschke, K.; Bartoszek, M.; Quaiser, S.; Fischer, M. Dynamic NMR studies of phosphorylated diamine-coupled bis-1,3,5-triazines. *Phosphorus Sulfur Silicon Relat. Elem.* **1998**, *141*, 135–146. [CrossRef]
34. Salmeia, K.; Gaan, S.; Bruehwiler, M. Flame retardants. WO2018069249A1, 19 April 2018.
35. Moreau, J.P.; Chance, L.H. New method for preparing alkyl (amino-s-triazinyl) phosphonates. *J. Chem. Eng. Data* **1970**, *15*, 581–583. [CrossRef]
36. Sheldrick, G.M. A short history of SHELX. *Acta. Crystallogr. A* **2008**, *64*, 112–122. [CrossRef] [PubMed]
37. Spek, A.L. Single-crystal structure validation with the program PLATON. *J. Appl. Crystallogr.* **2003**, *36*, 7–13. [CrossRef]
38. Zhou, S.; Song, L.; Wang, Z.; Hu, Y.; Xing, W. Flame retardation and char formation mechanism of intumescent flame retarded polypropylene composites containing melamine phosphate and pentaerythritol phosphate. *Polym. Degrad. Stability* **2008**, *93*, 1799–1806. [CrossRef]
39. Su, X.; Yi, Y.; Tao, J.; Qi, H.; Li, D. Synergistic effect between a novel triazine charring agent and ammonium polyphosphate on flame retardancy and thermal behavior of polypropylene. *Polym. Degrad. Stability* **2014**, *105*, 12–20. [CrossRef]
40. Mahapatra, S.S.; Karak, N. s-Triazine containing flame retardant hyperbranched polyamines: Synthesis, characterization and properties evaluation. *Polym. Degrad. Stability* **2007**, *92*, 947–955. [CrossRef]
41. Liang, S.; Hemberger, P.; Neisius, N.M.; Bodi, A.; Gruetzmacher, H.; Levalois-Gruetzmacher, J.; Gaan, S. Elucidating the Thermal Decomposition of Dimethyl Methylphosphonate by Vacuum Ultraviolet (VUV) Photoionization: Pathways to the PO Radical, a Key Species in Flame-Retardant Mechanisms. *Chem. Eur. J.* **2015**, *21*, 1073–1080. [CrossRef] [PubMed]

3

Synthesis, Spectral Characterization of Several Novel Pyrene-Derived Aminophosphonates and their Ecotoxicological Evaluation using *Heterocypris incongruens* and *Vibrio fisheri* Tests

Jarosław Lewkowski [1,*], Maria Rodriguez Moya [1], Marta Chmielak [1,†], Diana Rogacz [2], Kamila Lewicka [2] and Piotr Rychter [2,*]

[1] Department of Organic Chemistry, Faculty of Chemistry, University of Łódź, Tamka 12, 91-403 Łódź, Poland; MRM_chem@wp.pl (M.R.M.); martachmielak1@gmail.com (M.C.)

[2] Institute of Chemistry, Environmental Protection and Biotechnology, Jan Długosz University in Częstochowa, 13/15 Armii Krajowej Av., 42-200 Częstochowa, Poland; diana.rogacz@gmail.com (D.R.); lewickakamilla@gmail.com (K.L.)

* Correspondence: jlewkow@uni.lodz.pl (J.L.); p.rychter@ajd.czest.pl (P.R.)

† M.Sc. student at the Faculty of Chemistry, University of Łódź (Academic Year 2015–2016).

Academic Editor: György Keglevich

Abstract: Four diphenyl pyrene-derived aminophosphonates were synthesized. Attempts were made to synthesize diphenyl *N*-(*R*)-α-methylbenzylamino(pyren-1-yl)methylphosphonate (**3e**) in order to obtain the chiral aminophosphonate bearing a pyrene moiety. Because these attempts failed, dimethyl and dibenzyl *N*-(*R*)-α-methylbenzyl substituted aminophosphonates **4** and **5** were synthesized and the predominant diastereoisomer of dimethyl aminophosphonate **4** was isolated. The resolution of the diastereomeric mixture of **5** failed. Aminophosphonates **3a–d** and the predominant diastereoisomer of **4** were investigated in terms of their ecotoxicity using tests performed on the ostracode *Heterocypris incongruens* and the fluorescent bacterium *Vibrio fisheri*. The tests confirmed the moderate-to-high ecotoxicity of aminophosphonates **3a–d** and **4**, but no evident correlation between the structure and toxicity has been found.

Keywords: pyrene-1-carboxaldehyde; aminophosphonates; ecotoxicity; ostracode test; *Vibrio fisheri* test

1. Introduction

Aminophosphonic derivatives are well known for their biological activity. They are described as effective inhibitors of various enzymes and, therefore they find applications as crop protective [1], antibacterial [2] or anticancer [3] agents. Analogues of *C*-phenylphosphonoglycine, have been demonstrated to show activity in various fields as herbicides and plant growth regulators [4], agrochemical fungicides [5,6] and glutamate receptor agonists and antagonists [7].

In a sort of logical consequence, a lot of compounds from this group have been synthesized, including those bearing benzene and its polycyclic aromatic analogues [8], five- and six-membered heteroaromatic rings [9,10], and even the ferrocene [11] moiety. Their syntheses are generally based on the Kabachnik-Fields and the aza-Pudovik reactions [12–14].

In contrast, aminophosphonic derivatives of 1-pyrene, i.e., amino(1-pyrene)methylphosphonic compounds have been described only twice in the chemical literature. Hudson et al. [15] reported the synthesis of diethyl *N*-benzhydrylamino(1-pyrene)methyl phosphonate, and Jayaprakash and co-workers [16] described the synthesis of a series of *N*-aryl substituted, diethyl amino(1-pyrene)methyl phosphonates and results of preliminary studies on their fluorescence properties.

We have tried to fill this void and very recently have published results of our studies on the synthesis, photophysical properties and cytotoxic action of a series of α-aminophosphonates bearing pyren-1-yl moieties [17]. They demonstrated moderate-to-satisfactory cytotoxicities against two colon cancer cell lines. Such interesting biological properties may result in potential future applications and their biological hazards should be evaluated using typical biotoxicological tests.

Herein, we wish to report the synthesis of a new series of *N*-substituted-*C*-(pyren-1-yl)-phosphonoglycine derivatives and their ecotoxicological properties evaluated using ostracode (*Heterocypris incongruens*) and fluorescence bacteria *Vibrio fisheri* tests.

Nowadays, ecotoxicology faces the challenge of evaluating and predicting the effects of a variety of chemicals on aquatic species and ecosystems. Since a chemical pollutant may follow various routes from a source to the target, there is an urgent need to follow their potential ecotoxicological impact using different biotests [18–21].

Toxkit microbiotests are presently used in many countries worldwide for biomonitoring of contaminated waters and soils. These small-scale bioassays are becoming more and more popular in laboratories all over the world because of their various useful advantages, such as high degree of standardization and precision, simple testing procedures which are very clearly documented down to the last detail, operation with small sample volumes, the repeatability as well as giving comparable results to equivalent ISO and OECD full-scale tests. However, the major advantage of Toxkit microbiotests in comparison to conventional bioassays is that the tested microorganisms are incorporated in the kits in a dormant or immobilized form, from which they can be hatched or activated "on demand" prior to carrying out the toxicity tests [22].

The Microtox® system test with the luminescent bacterium *Vibrio fischeri* and the crustacean *Heterocypris incongruens*, when used as bioindicator allows one to monitor potential toxicity of soil or water contaminants in a rapid, simple and inexpensive way. This bioanalytical technique allows one to overcome some disadvantages of the traditional physicochemical methods such as concentration of toxic substances below analytical limits or potential chemical interactions leading to additive, synergistic or antagonistic effect which cannot be identified using the abovementioned methods. The *Vibrio fischeri* bioassay can be applied in all types of water, including surface and ground water, wastewater, drilling sump fluids and many other aqueous solutions [23]. Both *Vibrio fischeri* and *Heterocypris incongruens* bioassay tests have been already successfully used for ecotoxicity determination of various pharmaceuticals, pesticides, ionic liquids, detergents, heavy metals, road dust, eluates and organic extracts of wastes, etc. [24–30].

As it was mentioned above, the aim of the study was the preliminary evaluation of the potential toxicity of newly synthesized aminophosphonates **3a–d** and **4**, if introduced to the environment. In order to perform it, the Microtox® test using the bioluminescent bacterium *Vibrio fischeri* was employed to examine the acute toxicity of sediment eluates loaded with tested compounds. Due to the cost effective, simple operation and time-savings, the biotest based on the bioluminescence inhibition of *Vibrio fischeri* is probably the most widely applied bacterial test used for environmental pollution monitoring.

However, due to fact that the obtained substances are almost insoluble in water, the Microtox® solid phase test (MSPT), as a one of the most popular tests for sediment toxicity assessments due to its simplicity, reproducibility, ecological relevance and sensitivity, has been used [26,31–33].

Additionally, the standardized crustacean biotest with *Heterocypris incongruens* was selected as a representative benthic invertebrate species for screening the toxicity of sediments containing the tested substances. Crustaceans are promising representative species of the benthic environment as they spend their lives swimming in water and in direct contact with the sediment. Therefore, the Ostracodtoxkit® test has been used for toxicity evaluation. The toxicity test using mentioned freshwater ostracodes is a sub-chronic static test that exposes individuals to whole sediments over a period of 6 days, the endpoints being mortality and growth changes [34].

2. Results and Discussion

2.1. *Synthesis of Aminophosphonic Derivatives* **3a–e**, **4** *and* **5** *Bearing 1-Pyrene Moieties*

Schiff bases **1a–e** were synthesized following the published and commonly known procedure by simple mixing pyrene-1-carboxaldehyde with an appropriate amine in hexane (in a case of *p*-anisidine, *p*-toluidine and aniline) or methanol (for benzylamine and (*R*)-α-methylbenzylamine) and stirring them at room temperature for 24 h; the reaction with *p*-anisidine required 48 h [14]. This procedure gave imines **1a–e**, which were isolated and used for further conversions without any purification. ^{1}H-NMR spectra were recorded only to verify the identity, based on the diagnostic ^{1}H-NMR singlet corresponding to the proton of the azomethine group (–CH=N–) above 8 ppm [17].

Aminophosphonates **3a–d**, **4** and **5** were synthesized basing on the aza-Pudovik reaction, i.e., the addition of diphenyl phosphite to imines **1a–d**. However, important modifications were necessary to be introduced, whereby reactions were carried out in a small amount of dichloromethane and diphenyl phosphite was used in three-fold excess. The reaction mixtures were heated for 5 min then stirred for 24 h at room temperature. After workup as described in the Experimental Section, the resulting aminophosphonates **3a–d** were obtained in 60%–70% yields (Scheme 1). They were purified by column chromatography on neutral aluminum oxide, because diphenyl amino(1-pyrene)methylphosphonates **3a–d** decomposed on silica gel.

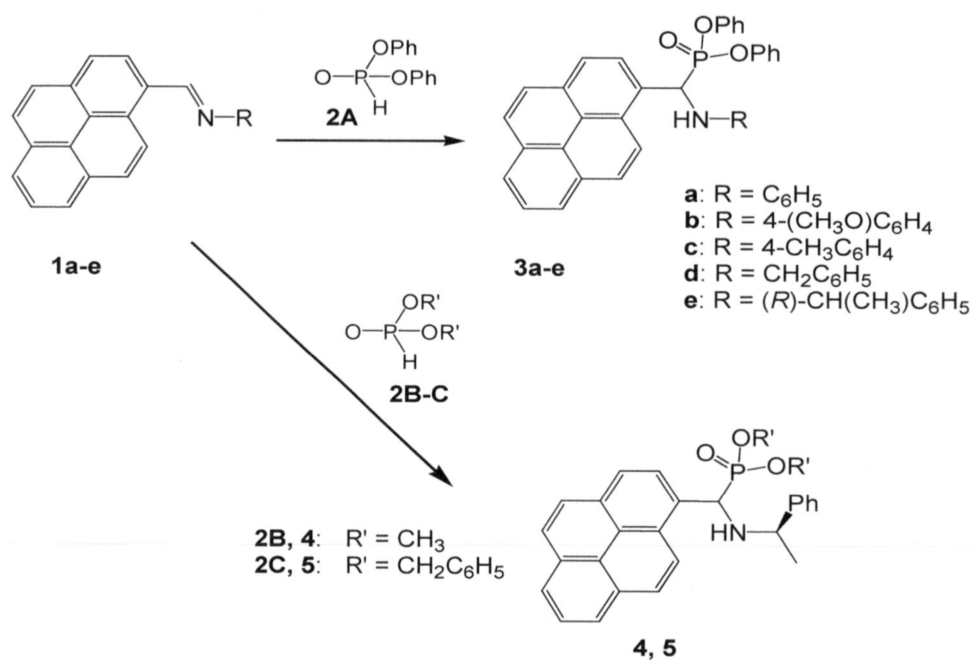

Scheme 1. Synthesis of *N*-substituted amino(1-pyrene)methylphosphonates **3a–e**, **4** and **5**.

The identity of synthesized compounds was verified by means of ^{1}H-, ^{31}P- and ^{13}C-NMR as well as by FT-IR spectroscopy. The proton NMR spectra of diphenyl pyrene-derived aminophosphonates **3a–d** showed the characteristic signals of all compounds, i.e., the doublet corresponding to the proton from the CH-P bond and the system of signals attributable to the pyrene ring protons. The ^{13}C-NMR spectra revealed the presence of two doublets at around 150 ppm with a coupling constant close to 10 Hz, which are attributed to the phenyl ring carbons from the P-O-C groups. Doublets in the 50–55 ppm range with coupling constants between 150 and 160 Hz are easily recognizable as C-P bond signals. IR bands at around 1250 cm^{-1} correspond to the bending vibrations of P=O bonds, while the presence of aromatic carbon-carbon bonds was confirmed by characteristic bands at around 1600, 1570, 1500 and 1450 cm^{-1}. Scans of NMR spectra are collected in the Supplementary Materials as Figures S1–S7.

In order to obtain the chiral, non-racemic diphenyl ester of a phosphonoglycine derivative bearing a pyren-1-yl moiety, the synthesis of diphenyl *N*-(*R*)-α-methylbenzylamino(pyren-1-yl)-methylphosphonate (**3e**) was performed using the method described above. The resulting mixture contained two diastereomers in a 100:15 ratio, but all attempts to purify the product by column chromatography on neutral aluminum oxide caused extensive decomposition of the product. From among available adsorbents, silica gel was totally inconvenient (*vide supra*), and chromatography on microcrystalline cellulose with chloroform allowed us to get rid of impurities only roughly, i.e., crude diphenyl *N*-(*R*)-α-methylbenzylamino(pyren-1-yl)methylphosphonate (**3e**) was obtained.

Thus, in order to investigate the stereochemistry of phosphite addition to the azomethine bond of pyrene-1-carboxaldimine **1e**, addition of dimethyl and dibenzylphosphites has been performed. Dimethyl *N*-(*R*)-α-methylbenzylamino(pyren-1-yl)methylphosphonate (**4**) was obtained as a mixture of two diastereoisomers in a 100:13 ratio. We succeeded in separation of the predominant diastereoisomer using column chromatography on silica gel. Its absolute configuration however could not be determined, because the predominant isomer of aminophosphonate **4** is a dense oily liquid, thus an X-ray study could not be performed. Isolation of the minor diastereoisomer of **4** failed, it was always obtained in a mixture with the major one.

Dibenzyl *N*-(*R*)-α-methylbenzylamino(pyren-1-yl)methylphosphonate (**5**) was obtained as a mixture of two diastereoisomers in a 100:26 ratio. Unfortunately, the obtained mixture of diastereoisomers was not separable by column chromatography, as all applied solvent systems led to mixtures with different ratios of isomers. However, the crude diastereomeric mixture of **5** was purified by column chromatography on silica gel to give satisfactory elemental analysis results.

Thus the obtained diphenyl aminophosphonates **3a–d** and the isolated predominant diastereoisomer of **4** were tested in terms of their ecotoxicity using two popular ecotoxicological tests, the ostracode *Heterocypris incongruens* and fluorescence bacteria *Vibrio fisheri*.

2.2. Ecotoxicological Properties of Aminophosphonates **3a–d** *and* **4**

2.2.1. Microtox Toxicity Assay

Bioassays are the complementary tool for characterizing the biological effects and hazards of contaminated sediments or soils. During the solid phase Microtox® assay the bioluminescent bacteria are exposed to a sediment suspension and their response reflects totally the action of toxic agents along with synergists and antagonists present in a given sample. The effects on light emission are evaluated in the liquid phase that remains after removal of the sediments by filtration. Effective concentration (EC_{50}) values (Table 1) were calculated from the concentration–response curves (Figure 1).

Obtained EC_{50} values ranged between 92 and 533 mg/kg of dry weight of soil, depending on the tested compound. Among the compounds **3a–c** with aromatic substituents on the amine groups, diphenyl *N*-(4-methoxyphenyl)amino(pyren-1-yl)methylphosphonate (**3b**) was found to be the most toxic, with the lowest EC_{50} value (92.63 mg/kg). Much less harmful action against *Vibrio fischeri* was found for the *N*-(4-methylphenyl) derivative **3c** (EC_{50} = 214.5 mg/kg), while diphenyl *N*-phenylamino(pyren-1-yl)methylphosphonate (**3a**) demonstrated nearly a lack of toxicity. Thus, the presence of a substituent at the phenyl group linked to nitrogen seems to be crucial in terms of toxicological properties of the studied compounds **3a–c**. The results also showed that a methoxy substituent increases the toxicity of the compound **3b** in comparison with the methyl substituted aminophosphonate **3c**.

Among the aminophosphonates **3d** and **4** having the amino groups substituted with non-aromatic groups, results were indeed surprising. Diphenyl *N*-benzylamino(pyren-1-yl)methylphosphonate (**3d**) with a *N*-benzyl substituent demonstrated five times higher toxicity (EC_{50} = 102.7 mg/kg) as compared to the *N*-(*R*)-α-methylbenzylamino derivative **4** (EC_{50} = 533.1 mg/kg). Therefore we can conclude that the presence of a methyl substituent in the benzyl moiety dramatically decreases the toxicity of compound **4**.

Table 1. Microtox® EC_{50} values (mg/kg of soil dry weight) of exposure of the luminescent marine bacteria *Vibrio fischeri* to tested aminophosphonates **3a–d** and **4**, with respective 95% confidence limits (in brackets) obtained in the fit of the data.

Compounds	EC_{50} (Lower Limit; Upper Limit)	Coefficient of Determination (R^2)
3a	495.9 (203.4; 1209)	0.7632
3b	92.63 (65.27; 131.4)	0.8912
3c	214.5 (165.0; 279.0)	0.9452
3d	102.7 (81.83; 128.9)	0.9571
4	533.1 (439.4; 646.8)	0.9516

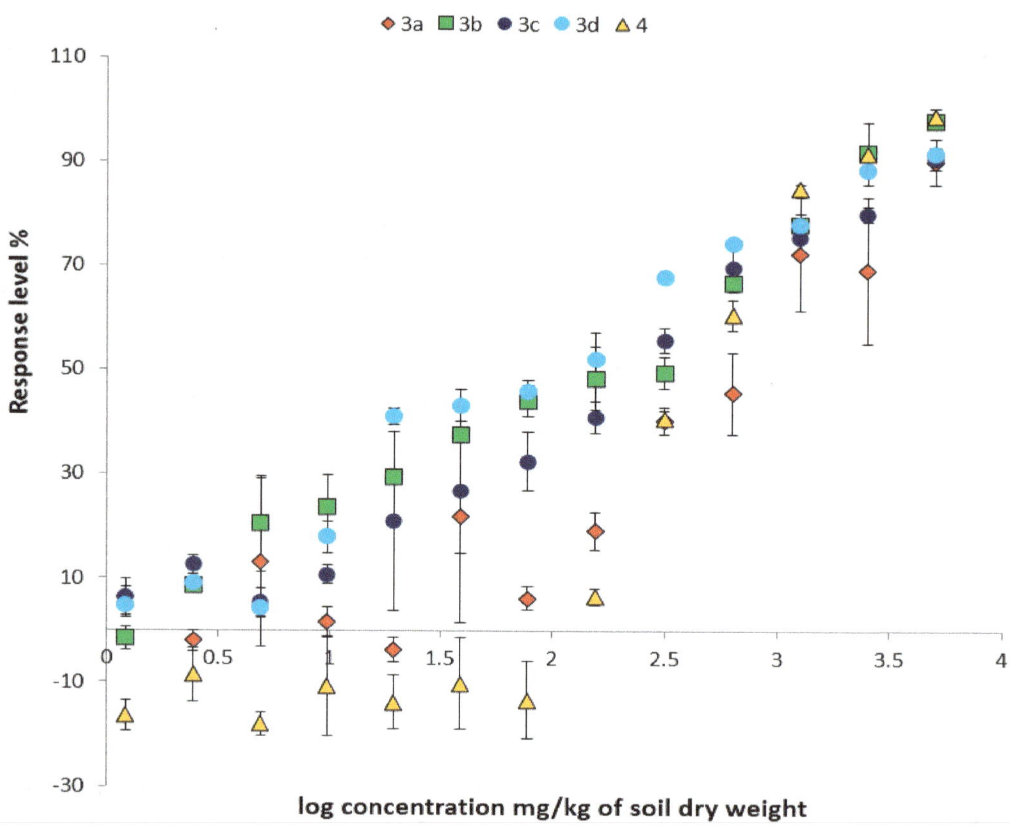

Figure 1. The EC_{50} graded dose-response curves for the five tested compounds **3a–d** and **4**. Overlapped curves are plotted by the Microtox® Analyzer 500 software. Error bars represent standard deviation error (SD, n = 3 determinations).

2.2.2. Ostracod Test Kit

Ostracodtoxkit F™ is the very first sediment-contact microbiotest with a crustacean test species for the assessment of the total toxicity of sediments, hence covering the toxic hazards of both dissolved and not dissolved pollutants. Considering the very low solubility of synthesized aminophosphonates **3a–d** and **4**, it seemed justified to evaluate reliably their potential toxicity.

The first criterion of the effect on tested crustaceans is the percent of mortality. Toxicity was expressed in % of mortality, i.e., the lethal effect of the sample on the number of ostracods compared with the control plate. As it was expected, the mortality of ostracodes increased with growing concentration of tested substance in soil (Figure 2). The concentration 500 mg/kg of soil was found to be lethal for all samples. However, among all tested compounds, *N*-benzylamino (pyren-1-yl)methylphosphonate (**3d**) demonstrated the highest toxicity at every concentration when compared to other samples (with the sole exception of aminophosphonate **4** at a concentration of

100 mg/kg of soil). Even at a concentration of 10 mg/kg of dry weight of soil, compound **3d** caused 10% mortality of crustaceans, while other samples at this concentration did not show any mortal effect.

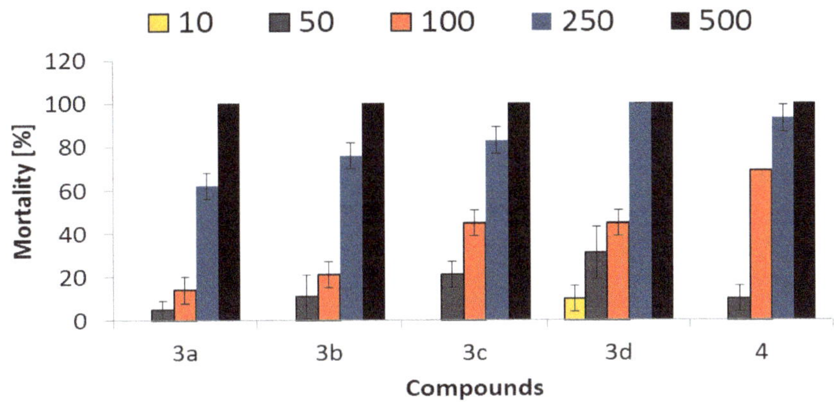

Figure 2. The mortality (%) of *Heterocypris incongruens* living in soil contaminated with pyrene-derived aminophosphonates **3a–d** and **4** at concentrations: 10, 50, 100 and 500 mg/kg of soil dry weight. Error bars represent standard deviation error (SD, $n = 6$ determinations).

Growth inhibition is the second criterion of the toxic effect indicated by the Ostracodtoxkit F™ microbiotest. This criterion allows one to evaluate the sub-lethal toxicity of sediments. The growth inhibition is determined by comparing the size of surviving ostracods living in the test sediment with the size of ostracods living in the reference sediment at the end of the test. Determination of the sub-lethal impact of sediment toxicants is justified only for sediments which do not cause a high ostracod mortality. It is advised that the growth inhibition should only be determined for sediments which mortality was found to be less than 30%. Therefore, the growth inhibition was chosen as a criterion of sub-lethal effects to determine toxicity not inducing substantial mortality in the test organisms. Consequently, measurements of lengths were only performed when the mortality was less than 30%. In this respect, the growth inhibition rates have not been measured for aminophosphonates **3c**, **3d** and **4** at concentrations of 100, 250 and 500 mg per kg of soil dry weight. Crustaceans were more resistant against compounds **3a** and **3b** and their growth inhibition reached ca. 20% for a concentration of 100 mg/kg of soil, nevertheless it was not measured at 250 and 500 mg per kg of soil dry weight. Values of growth inhibition for compounds **3c**, **3d** and **4** at 50 mg/kg of soil concentration varied between 16% and 23% (Figure 3).

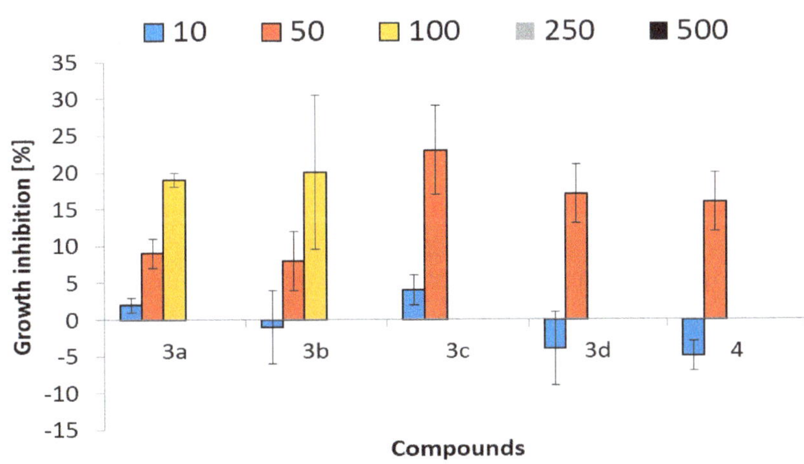

Figure 3. Growth inhibition (%) of *Heterocypris incongruens* living in soil contaminated with pyrene-derived aminophosphonates **3a–d** and **4** at concentrations of 10, 50, 100 and 500 mg/kg of soil dry weight. Error bars represent standard deviation error (SD, $n = 6$ determinations).

Comparison of results of luminescence inhibition for *Vibrio fischeri* tests with growth inhibition of crustaceans for all concentration of compounds did not show any reasonable relationship. Only diphenyl *N*-benzylamino(pyren-1-yl)methylphosphonate (**3d**), which was strongly toxic for *Vibrio fischeri* (EC_{50} ~103 mg/kg of soil) also revealed the highest mortality of ostracods even at a concentration of 10 mg/kg of soil.

However, diphenyl *N*-(4-methoxyphenyl)amino(pyren-1-yl) methylphosphonate (**3b**) that induced a very strong luminescence inhibition in *V. fischeri* was only moderately toxic for crustaceans considering both the mortality and the growth inhibition index. On the other hand, diphenyl *N*-phenylamino(pyren-1-yl)methylphosphonate (**3a**), being nearly non-toxic for *V. fischeri* (EC_{50} close to 500 mg/kg of soil) is also very weakly toxic for ostracods (with respect to both criteria, the mortality and the growth inhibition).

Thus, based on the obtained results, *Vibrio fischeri* bacteria were very sensitive to aminophosphonates **3b** and **3d**, since relatively low effective concentration (EC_{50}) values were determined, while for ostracods, the highest negative impact considering both mortality and inhibition growth have been found for diphenyl *N*-benzylamino(pyren-1-yl)methylphosphonate (**3d**). Therefore, being the most toxic against both tested organisms, the aminophosphonate **3d** should be considered as a potentially hazardous compound for environment, which may be toxic against many other living species. Its high toxicity may be related with its *N*-benzyl group, which is more flexible than other substituents and may bind more easily to enzymes or receptors of the tested organisms. Nevertheless, this is only a very preliminary hypothesis, which requires further verification.

The usefulness of bioassays for the assessment of areas affected by phosphate industry wastes has already been reported and discussed [35,36]. Authors performing risk assessments in such contaminated areas have proven that the development of sensitive and observable organisms as bio-indicators for monitoring and management of the environmental pollution is a very useful tool to complement chemical analyses. Biological analysis allows one to reflect the harmful effects of contaminants on organisms and provides data to reveal the mechanisms of the toxic effects including formation, development and removal of contaminants.

Since the soil is generally the reservoir for agrochemical residues, the evaluation of any xenobiotics including pesticides using a simple battery of assay tests is strongly needed to control level and fate of their residues. In a successful study focused on the ecotoxicological evaluation of select pesticides (chlorpyrifos, glyphosate, vinclozolin, endosulfan) present in agricultural soils, Antunes et al have reported the use of standard aquatic bioassays for testing of soil eluates [30].

Our obtained effective concentration values are very close to the inhibition concentration (IC_{50}) values for herbicides examined by Joly et al. [28]. Toxicity values of the pesticides *S*-metolachlor, nicosulfuron and benoxacor were 178.4, 167.8 and 93.3 mg/L, respectively. In this respect, the toxicity of the first two herbicides are close to that of the tested aminophosphonate **3c**, while the last IC_{50} value matches the values of compounds **3b** and **3d**. Nevertheless, the action of aminophosphonates **3a–d** and **4** against *V. fischeri* demonstrated much lower toxicity when compared to the herbicides tested by the mentioned authors.

Ostracodtoxit tests using *Heterocypris incongruens* are also a very useful tool for environmental toxicity determination. Results related to mortality and growth inhibition of ostracods correlated with various types of polycyclic aromatic hydrocarbons (PAHs) have been reported by Cvancarova et al. [37]. They found that *H. incongruens* was strongly sensitive to bioavailable PAHs, while luminescence inhibition of *V. fischeri* varied widely between ca. 15%–90% depending on the type of soil examined.

Oleszczuk and Hollert [38], in their studies devoted to determination of the influence of different soils on sewage sludge toxicity, have found that the mortality values of *Heterocypris incongruens* ranged from 0.26% to 11.5% depending on the sludge tested [39]. Growth inhibition values ranged from 10.7% to 36.2% depending on the type of sludge. Płaza et al. [39] applied Microtox®, Ostracodtoxkit F™, to assay bioremediation processes in soils heavily contaminated with petroleum. The test species demonstrated varying sensitivity to soils and the effects on test organisms exposed to tested soils correlated with the soil contaminants' concentrations [39].

Results of toxicity against *H. incongruens* obtained for the studied aminophosphonates **3a–d** and **4** were at a similar level. The mortality varied from ca. 5% until 70% for the concentration range 10–100 mg per kg of soil. At higher concentrations, the tested compounds were practically lethal for crustaceans. The growth inhibition varied from 16% up to 23% for a concentration of 50 mg per kg of soil. All the data presented above lead us to stress the necessity of handling pyrene-derived aminophosphonic derivatives with special care, as they may be potentially hazardous for the environment.

3. Materials and Methods

3.1. Chemistry

All solvents (POCh, Gliwice, Poland) were applied routinely and dried prior to use. Commercial reagents (Aldrich, Poznań, Poland) were generally used as received. NMR spectra were recorded on an Avance III 600 MHz apparatus (Bruker, Billerica, MA, USA) operating at 600 MHz (^{1}H-NMR), 150 MHz (^{13}C-NMR) and 243 MHz (^{31}P-NMR). IR spectra were recorded on a Nexus FT-IR spectrometer (Thermo Nicolet, Waltham, MA, USA). Rotations were measured using a 241 MC polarimeter (Perkin-Elmer, Waltham, MA, USA). Melting points were measured using a MelTemp II apparatus (Bibby Scientific Limited, Staffordshire, UK), in a capillary and were not corrected. Elemental analyses were performed in the Laboratory of Microanalysis, the Center of Molecular and Macromolecular Studies PAS in Łódź and the Laboratory of Molecular Spectroscopy at the Faculty of Chemistry, University of Łódź.

3.1.1. General Procedure for Synthesis of Phosphonates **3a–d**, **4** and **5**

Equimolar quantities of 1-pyrenecarboxaldehyde and amine (2 mmol each) were dissolved in hexane (**1a–c**) or methanol (**1d–e**) (20 mL) and stirred at reflux for 24 h (**1b–c**, **1d–e**) or 48 h (**1a**). The reaction was monitored by ^{1}H-NMR and after the completion, the solvent was removed under reduced pressure, and obtained products were used directly for further conversion. Imine (1 mmol) and the appropriate phosphite **2A–C** (3 mmol or 5–6 mmol in a case of dimethyl phosphite due to the low solubility of the imine in it) were placed in a 25 mL round-bottom flask, which was heated with stirring on a water bath for 5 min until all of the imine was dissolved in the phosphite. In the case of derivatives **3a–d**, dichloromethane (2 mL) was added in order to improve the solubility. Then the mixture was stirred at room temperature for 24 h. The crude aminophosphonates were purified by column chromatography on neutral aluminum oxide using chloroform as eluent.

Diphenyl N-phenylamino(pyren-1-yl)methylphosphonate (**3a**). Y = 0.41 g (76%), yellow solid, m.p. 184–185 °C. ^{1}H-NMR ($CDCl_3$): δ 8.38 (d, $^3J^{HH}$ = 9.3 Hz, H_{pyr}, 1H); 8.28 (dd, $^3J^{HH}$ = 8.0 and $^4J_{HH}$ = 2.6 Hz, H_{pyr}, 1H); 8.20 (t, H_{pyr}, $^3J^{HH}$ = 8.0 Hz, 2H); 8.12 (d, $^3J^{HH}$ = 8.0 Hz, H_{pyr}, 1H); 8.10 (d, $^3J^{HH}$ = 9.2 Hz, H_{pyr}, 1H); 8.07–8.06 (m, H_{pyr}, 1H); 8.04–8.01 (m, H_{pyr}, 2H); 7.30–7.29 (m, H_{Ph}, 5H); 6.94 (t, $^3J^{HH}$ = 7.4 Hz, H_{Ph}, 1H); 6.84 (dd, $^3J^{HH(1)}$ = $^3J^{HH(2)}$ = 7.8 Hz, H_{Ph}, 2H); 6.79 (app. d, $^3J^{HH}$ = 8.5 Hz, *p*-C_6H_4, 2H); 6.67 (d, $^3J^{HH}$ = 7.4 Hz, H_{Ph}, 2H); 6.46 (app. d, $^3J^{HH}$ = 8.5 Hz, *p*-C_6H_4, 2H); 5.92 (d, $^2J_{PH}$ = 24.1 Hz, CHP, 1H); 5.10–5.08 (m, $POCH_2$, 2H); 4.65 and 4.22 (Part of AMX system, $^2J_{HH}$ = 11.6 Hz and $^3J_{PH}$ = 7.8 and 8.5 Hz, $POCH_2$, 2H). ^{13}C-NMR ($CDCl_3$): δ 150.7 (d, $^2J_{CP}$ = 9.9 Hz, C_{POC}); 150.3 (d, $^2J_{CP}$ = 9.5 Hz, C_{POC}); 146.1 (d, $^3J_{CP}$ = 15.27 Hz, C_{Ph}); 131.6 (C_{pyr}); 131.5 (d, $^4J_{CP}$ = 3.1 Hz, C_{pyr}); 130.8 (C_{pyr}); 130.2 (C_{pyr}) 130.0 (C_{Ph}); 129.6 (C_{pyr}); 129.5 (d, $^4J_{CP}$ = 2.0 Hz, C_{Ph}); 128.7 (C_{pyr}); 128.6 (d, $^4J_{CP}$ = 2.1 Hz, C_{pyr}); 127.9 (C_{pyr}); 127.7 (C_{pyr}); 127.4 (C_{pyr}); 127.0 (C_{pyr}); 126.8 (C_{pyr}); 126.3 (C_{pyr}); 125.8 (C_{Ph}); 125.7 (C_{Ph}); 125.6 (d, $^4J_{CP}$ = 3.4 Hz, C_{pyr}); 125.5 (C_{pyr}); 125.2 (C_{pyr}); 125.0 (C_{pyr}); 122.1 (C_{pyr}); 120.9 (d, $^3J_{CP}$ = 4.0 Hz, C_{Ph}); 119.9 (d, $^3J_{CP}$ = 4.4 Hz, C_{Ph}); 119.1 (C_{Ph}); 114.2 (C_{Ph}); 52.2 (d, $^1J_{CP}$ = 154.6 Hz, PC). ^{31}P-NMR ($CDCl_3$): δ 16.16. IR (KBr): 3313 (νCH); 3037; 2923 (CH_3); 1600, 1581; 1524, 1483 (C=C); 1249 (P=O); 1214; 1185; 1062; 1021 (P-O); 941; 853;

824; 622.Elemental analysis: Calcd. for $C_{35}H_{26}NO_3P$: C, 77.91; H, 4.86; N, 2.60. Found: C, 77.76, H, 4.86, N, 2.79.

Diphenyl N-(4-methoxyphenyl)amino(pyren-1-yl)methylphosphonate (**3b**). Y = 0.3 g (52%), yellow solid, m.p. 161 °C. ^{1}H-NMR ($CDCl_3$): δ 8.49 (d, $^3J^{HH}$ = 9.2 Hz, H_{pyr}, 1H); 8.34 (dd, $^3J^{HH}$ = 7.7 and $^4J_{HH}$ = 1.9 Hz, H_{pyr}, 1H); 8.24-8.21 (m, H_{pyr}, 3H); 8.14 (d, H_{pyr}, $^3J^{HH}$ = 8.1 Hz, 1H); 8.08 (d, $^3J^{HH}$ = 8.9 Hz, H_{pyr}, 1H); 8.04 (d, $^3J^{HH}$ = 7.7 Hz, H_{pyr}, 1H); 8.02–8.01 (m, H_{pyr}, 2H); 7.31–7.28 (m, H_{Ph}, 2H); 7.19–7.16 (m, H_{Ph}, 3H); 6.95–6.92 (m, H_{Ph}, 2H); 6.89–6.87 (m, H_{Ph}, 1H); 6.62–6.58 (m, H_{Ph}, *p*-C_6H_4, 6H); 6.23 (d, $^2J_{PH}$ = 24.4 Hz, CHP, 1H); 3.61 (s, CH_3, 3H). ^{13}C-NMR ($CDCl_3$): δ 153.2 (d, $^2J_{CP}$ = 5.4 Hz, C_{para}); 150.7 (d, $^2J_{CP}$ = 9.9 Hz, C_{POC}); 150.3 (d, $^2J_{CP}$ = 9.7 Hz, C_{POC}); 131.5 (C_{pyr}); 131.2 (d, $^4J_{CP}$ = 3.2 Hz, C_{pyr}); 130.8 (C_{pyr}); 130.0 (C_{Ph}); 129.7 (C_{pyr}); 129.6 (d, $^3J_{CP}$ = 7.2 Hz, C_{pyr}); 129.5 (C_{Ph}); 128.8 (C_{pyr}); 128.6 (C_{pyr}); 127.9 (C_{pyr}); 127.7 (C_{pyr}); 126.3 (C_{Ph}); 125.8 (C_{Ph}); 125.6 (C_{pyr}); 125.6 (C_{pyr}); 125.5 (C_{pyr}); 125.2 (d, $^4J_{CP}$ = 1.8 Hz, C_{pyr}); 125.0 (C_{pyr}); 124.9 (C_{pyr}); 122.2 (C_{pyr}); 121.0 (d, $^3J_{CP}$ = 4.2 Hz, C_{Ph}); 120.0 (d, $^4J_{CP}$ = 4.4 Hz, C_{Ph}); 115.7 (d, $^4J_{CP}$ = 9.7 Hz, C_{Ph}); 115.1 (C_{para}); 55.7 (C_{Ar}-OC); 53.0 (d, $^1J_{CP}$ = 154.3 Hz, PC). ^{31}P-NMR ($CDCl_3$): δ 16.32. IR (KBr): 3383 (νNH); 3294 (νCH); 3037; 2958 (CH_3); 1591, 1511, 1486; 1451 (C=C); 1245 (P=O); 1211; 1068; 1024 (P-O); 941; 834; 815; 618. Elemental analysis: Calcd. for $C_{36}H_{28}NO_4P$: C, 75.91; H, 4.95; N, 2.46. Found: C, 75.76, H, 5.06, N, 2.54.

Diphenyl N-(4-methylphenyl)amino(pyren-1-yl)methylphosphonate (**3c**). Y = 0.44 g (79%), yellow solid, m.p. 181–182 °C. ^{1}H-NMR ($CDCl_3$): δ 8.51 (d, $^3J^{HH}$ = 9.3 Hz, H_{pyr}, 1H); 8.34 (dd, $^3J^{HH}$ = 8.2 and $^4J_{HH}$ = 2.5 Hz, H_{pyr}, 1H); 8.25-8.21 (m, H_{pyr}, 3H); 8.13 (d, H_{pyr}, $^3J^{HH}$ = 8.2 Hz, 1H); 8.08 (d, $^3J^{HH}$ = 8.9 Hz, H_{pyr}, 1H); 8.04 (t, $^3J^{HH}$ = 7.6 Hz, H_{pyr}, 1H); 8.01 (d, $^3J^{HH}$ = 8.8 Hz, H_{pyr}, 2H); 7.31–7.29 (m, H_{Ph}, 2H); 7.20–7.15 (m, H_{Ph}, 3H); 7.05–7.02 (m, H_{Ph}, 2H); 6.94–6.91 (m, H_{Ph}, 2H); 6.88-6.86 (m, H_{Ph}, 1H); 6.68-6.64 (m, H_{Ph}, 3H); 6.58–6.57 (m, H_{Ph}, 2H); 6.31 (d, $^2J_{PH}$ = 24.6 Hz, CHP, 1H). ^{13}C-NMR ($CDCl_3$): δ 150.7 (d, $^2J_{CP}$ = 9.8 Hz, C_{POC}); 150.3 (d, $^2J_{CP}$ = 9.7 Hz, C_{POC}); 143.7 (d, $^2J_{CP}$ = 15.9 Hz, C_{para}); 131.6 (C_{pyr}); 131.2 (d, $^4J_{CP}$ = 3.2 Hz, C_{pyr}); 130.8 (C_{pyr}); 130.0 (C_{Ph}); 129.8 (C_{pyr}); 129.6 (d, $^3J_{CP}$ = 7.3 Hz, C_{pyr}); 129.5 (C_{para}); 128.8 (d, $^4J_{CP}$ = 2.1 Hz, C_{pyr}); 128.6 (C_{pyr}); 128.3 (C_{pyr}); 127.8 (C_{para}); 127.7 (d, $^4J_{CP}$ = 1.6 Hz, C_{pyr}); 126.3 (C_{pyr}); 125.8 (C_{pyr}); 125.6 (C_{Ph}); 125.5 (d, $^4J_{CP}$ = 2.3 Hz, C_{pyr}); 125.5 (C_{Ph}); 125.2 (d, $^4J_{CP}$ = 1.9 Hz, C_{pyr}); 125.0 (C_{pyr}); 124.9 (C_{pyr}); 122.2 (C_{pyr}); 121.0 (d, $^3J_{CP}$ = 4.0 Hz, C_{Ph}); 120.0 (d, $^3J_{CP}$ = 4.4 Hz, C_{Ph}); 114.3 (C_{para}); 52.5 (d, $^1J_{CP}$ = 155.1 Hz, PC); 20.5 (C_{Ar}-OC). ^{31}P-NMR ($CDCl_3$): δ 15.23. IR (KBr): 3303 (νCH); 2923 (CH_3); 1591, 1489, 1458 (C=C); 1268 (P=O); 1024 (P-O); 926, 840; 770; 688. Elemental analysis: Calcd. for $C_{36}H_{28}NO_3P$: C, 78.11; H, 5.10; N, 2.53. Found: C, 77.86; H, 5.31; N, 2.47.

Diphenyl N-benzylamino(pyren-1-yl)methylphosphonate (**3d**). Y = 0.43 g (77%), yellow solid, m.p. 118–119 °C. ^{1}H-NMR ($CDCl_3$): δ 8.51–8.49 (m, H_{pyr}, 1H); 8.26 (d, $^3J^{HH}$ = 8.0 Hz, H_{pyr}, 1H); 8.24–8.21 (m, H_{pyr}, 3H); 8.13–8.09 (m, H_{pyr}, 3H); 8.04 (t, $^3J^{HH}$ = 7.6 Hz, H_{pyr}, 1H); 7.29–7.25 (m, H_{Ph}, 5H); 7.20–7.19 (m, H_{Ph}, 2H); 7.15–7.13 (m, H_{Ph}, 3H); 7.02–7.00 (m, H_{Ph}, 2H); 6.93 (app. t, $^3J^{HH}$ = 7.4 Hz, H_{Ph}, 1H); 6.73 (app. d, $^3J^{HH}$ = 8.2 Hz, H_{Ph}, 2H); 5.58 (d, $^2J_{PH}$ = 20.8 Hz, CHP, 1H); 3.93 (d, $^3J^{HH}$ = 13.3 Hz, NCH_2, 1H); 3.66 (d, $^3J^{HH}$ = 13.3 Hz, NCH_2, 1H). ^{13}C-NMR ($CDCl_3$): δ 151.0 (d, $^2J_{CP}$ = 9.9 Hz, C_{POC}); 150.5 (d, $^2J_{CP}$ = 9.7 Hz, C_{POC}); 139.1 (C_{Ph}); 131.6 (C_{pyr}); 131.4 (d, $^4J_{CP}$ = 3.0 Hz, C_{pyr}); 130.8 (C_{pyr}); 130.4 (C_{pyr}); 130.3 (C_{pyr}) 130.0 (C_{pyr}); 129.9 (C_{pyr}); 129.8 (C_{Ph}); 129.5 (C_{Ph}); 128.8 (C_{Ph}); 128.6 (C_{Ph}); 128.2 (C_{pyr}); 128.0 (C_{pyr}); 127.7 (d, $^4J_{CP}$ = 1.5 Hz, C_{pyr}); 127.5 (C_{pyr}); 126.3 (C_{pyr}); 125.7 (C_{Ph}); 125.5 (d, $^4J_{CP}$ = 2.8 Hz, C_{pyr}); 125.4 (C_{Ph}); 125.2 (C_{pyr}); 125.0 (C_{pyr}); 122.7 (C_{pyr}); 120.8 (d, $^3J_{CP}$ = 4.3 Hz, C_{Ph}); 120.3 (d, $^3J_{CP}$ = 4.3 Hz, C_{Ph}); 51.6 (d, $^3J_{CP}$ = 17.8 Hz, PCNC). ^{31}P-NMR ($CDCl_3$): δ 16.22. IR (KBr): 3386 (νNH); 3041, 2917 (CH_3); 1616, 1587, 1518, 1451 (C=C); 1245 (P=O); 1066; 1024 (P-O); 941, 843; 777; 618. Elemental analysis: Calcd. for $C_{36}H_{28}NO_3P$: C, 78.11; H, 5.10; N, 2.53. Found: C, 77.81; H, 5.24; N, 2.40.

3.1.2. *Dimethyl N-(R)-α-Methylbenzylamino(pyren-1-yl)methylphosphonate* (**4**)

The reaction was carried out as it is described above, but the post-reaction mixture was dissolved in dichloromethane and washed three times with a saturated aqueous sodium bicarbonate solution. The organic layer was dried, the solvent was removed in vacuo and the residual oil was purified by column chromatography on silica gel with chloroform as eluent. The procedure gave 440 mg of **4** as a dense yellow oil. *Isolated predominant diastereomer*: Y = 0.4 g (92%, after column chromatography), dense yellow oil. $[\alpha]_D^{20} = -48.3°$ ($c = 0.1$, $CHCl_3$). ^{1}H-NMR ($CDCl_3$): δ 8.29–8.27 (m, H_{pyr}, 1H); 8.22 (d, $^3J^{HH} = 8.0$ Hz, H_{pyr}, 1H); 8.21–8.18 (m, H_{pyr}, 3H); 8.10–8.06 (m, H_{pyr}, 3H); 8.02 (t, $^3J^{HH} = 7.6$ Hz, H_{pyr}, 1H); 7.25–7.23 (m, PhH, 2H); 7.22–7.20 (m, PhH, 2H); 7.19–7,16 (m, PhH, 1H); 5.33 (d, $^2J_{PH} = 21.1$ Hz, CHP, 1H); 3.91 (q, $^3J^{HH} = 6.5$ Hz, *CH*Ph, 1H); 3.86 (d, $^3J_{PH} = 10.5$ Hz, $POCH_3$, 3H); 3.26 (d, $^3J_{PH} = 10.5$ Hz, $POCH_3$, 3H); 1.39 (d, $^3J^{HH} = 6.5$ Hz, $CHCH_3$, 3H). ^{13}C-NMR ($CDCl_3$): δ 145.2 (C_{Ph}); 131.6 (C_{pyr}); 131.1 (d, $^4J_{CP} = 2.3$ Hz, C_{pyr}); 130.8 (C_{pyr}); 130.3 (C_{pyr}); 129.4 (d, $^3J_{CP} = 7.7$ Hz, C_{pyr}); 128.6 (C_{Ph}); 128.2 (C_{pyr}); 127.8 (C_{pyr}); 127.7 (C_{pyr}); 127.3 (C_{Ph}); 127.0 (C_{Ph}); 126.2 (C_{pyr}); 126.1 (C_{pyr}); 125.6 (C_{pyr}); 125.4 (d, $^4J_{CP} = 2.8$ Hz, C_{pyr}); 125.3 (C_{pyr}); 125.1 (C_{pyr}); 125.0 (C_{pyr}); 55.8 (d, $^3J_{CP} = 11.9$ Hz, PCNC); 54.0 (d, $^2J_{CP} = 7.4$ Hz, POC); 53.6 (d, $^2J_{CP} = 7.2$ Hz, POC); 29.9 (C_{aliph}); 27.8 (C_{aliph}). ^{31}P-NMR ($CDCl_3$): δ 26.70. IR (KBr): 3440 (νNH); 3309 (νCH); 2949 (CH_2); 1600, 1584, 1511, 1454 (C=C); 1242 (P=O); 1055; 1027 (P-O); 846; 815; 761; 609. Elemental analysis: Calcd. for $C_{27}H_{26}NO_3P$: C, 73.12; H, 5.91; N, 3.16. Found: C, 73.15, H, 5.72, N, 2.96. *Mixture of diastereomers—the minor diastereomer is denoted with an asterisk.* ^{1}H-NMR ($CDCl_3$): δ 8.29–8.28 (m, H_{pyr}, 1H); 8.23–8.22 (m, H_{pyr}, 1H); 8.21–8.17 (m, H_{pyr}, 3H); 8.17–8.16* (m, H_{pyr}, 1H); 8.10–8.06 (m, H_{pyr}, 3H); 8.03–8.00 (m, H_{pyr}, 1H); 7.95–7.94* (m, H_{pyr}, 1H); 7.30–7.26* (m, PhH); 7.24–7.22 (m, PhH, 2H); 7.22–7.20 (m, PhH, 2H); 7.19–7.16 (m, PhH, 1H); 7.12–7.10 (m, PhH); 5.33 (d, $^2J_{PH} = 21.0$ Hz, CHP, 1H); 4.98* (d, $^2J_{PH} = 21.7$ Hz, CHP, 1H); 3.91 (q, $^3J^{HH} = 6.5$ Hz, *CH*Ph, 1H); 3.86 (d, $^3J_{PH} = 10.6$ Hz, $POCH_3$, 3H); 3.83* (d, $^3J_{PH} = 10.5$ Hz, $POCH_3$, 3H); 3.50* (q, $^3J^{HH} = 6.6$Hz, *CH*Ph, 1H); 3.26 (d, $^3J_{PH} = 10.4$ Hz, $POCH_3$, 3H); 3.23* (d, $^3J_{PH} = 10.4$ Hz, $POCH_3$, 3H); 1.39 (d, $^3J^{HH} = 6.5$ Hz, $CHCH_3$, 3H); 1.34* (d, $^3J^{HH} = 6.6$ Hz, $CHCH_3$, 3H). ^{31}P-NMR ($CDCl_3$): δ 26.70 and 26.18* (10:4). Elemental analysis: Calctd for $C_{27}H_{26}NO_3P$: C, 73.12; H, 5.91; N, 3.16. Found: C, 73.17, H, 5.85, N, 3.10.

3.1.3. *Dibenzyl N-(R)-α-Methylbenzylamino(pyren-1-yl)methylphosphonate* (5)

The procedure was identical as in a case of aminophosphonate **4**. *Mixture of diastereoisomers—the predominant diastereoisomer is denoted with an asterisk* (^{13}C-NMR data are quoted with two decimal digits due to the large number of signals situated very close one to another). Y = 0.12 g, (20%, after column chromatography), dense yellow oil (mixture). ^{1}H-NMR ($CDCl_3$): δ 8.21–8.20 (m, H_{pyr}, 2H); 8.16* (d, $^3J^{HH} = 7.5$ Hz, H_{pyr}, 1H); 8.15–8.12 (m, H_{pyr}); 8.10–8.06 (m, H_{pyr}, 4H); 8.03–8.01 (m, H_{pyr}, 1H); 7.98* (d, $^3J^{HH} = 9.3$ Hz, H_{pyr}, 1H); 7.38–7.36 (m, PhH, 2H); 7.34–7.32 (m, PhH, 3H); 7.22–7.17 (m, PhH, 5H); 6.97–6.94* (m, PhH, 2H); 6.92–6.90 (m, PhH); 6.89–6.86* (m, PhH, 2H); 6.83–6.81 (m, PhH); 6.76–6.74 (m, PhH, 2H); 6.68–6.66 (m, PhH); 5.38* (d, $^2J_{PH} = 21.0$ Hz, CHP, 1H); 5.26* and 5.10* (Part of ABX system, $^2J_{HH} = 11.8$ and $^3J_{PH} = 8.6$ and 7.0 Hz, OCH_2Ph, 2H); 5.24–5.22 (m, CHP); 5.13–5.03 (m, OCH_2Ph); 4.72* and 4.39* (part of AMX system, $^2J_{HH} = 11.8$ and $^3J_{PH} = 8.4$ and 7.3 Hz, OCH_2Ph, 2H); 4.67-4.64 (m, OCH_2Ph); 4.40–4.35 (m, OCH_2Ph); 3.93* (q, $^3J^{HH} = 6.4$ Hz, *CH*Ph, 1H); 1.33* (d, $^3J^{HH} = 6.4$ Hz, $CHCH_3$, 3H); 1.30–1.29 (m, $CHCH_3$). ^{13}C-NMR ($CDCl_3$): δ 145.21 (C_{Ar}); 144.28 (C_{Ar}); 136.83 (C_{Ar}); 136.79 (C_{Ar}); 136.01 and 135.98 (C_{Ar}); 131.58 (C_{Ar}); 131.06 and 131.05 (C_{Ar}); 130.87 (C_{Ar}); 130.41 (C_{Ar}); 129.57 and 129.52 (C_{Ar}); 128.67 and 128.66 (C_{Ar}); 128.63 (C_{Ar}); 128.45 (C_{Ar}); 128.41 (C_{Ar}); 128.12 (C_{Ar}); 128.03 (C_{Ar}); 127.98 (C_{Ar}); 127.85 (C_{Ar}); 127.75, 127.74, 127.71 and 127.70 (C_{Ar}); 127.49 and 127.48 (C_{Ar}); 127.43 (C_{Ar}); 127.29 (C_{Ar}); 126.98 (C_{Ar}); 126.13 (C_{Ar}); 125.49 (C_{Ar}); 125.40 (C_{Ar}); 125.32 and 125.30 (C_{Ar}); 125.19 (C_{Ar}); 125.16 (C_{Ar}); 125.09 (C_{Ar}); 125.00 (C_{Ar}); 124.97 (C_{Ar}); 68.79 and 68.75 (POC); 68.64 and 68.60 (POC); 68.34 and 68.30 (POC); 68.10 and 68.06 (POC); 55.94 (C_{aliph}); 55.86 (C_{aliph}); 55.58 (C_{aliph}); 55.46 (C_{aliph}); 24.92 (C_{aliph}); 22.85 (C_{aliph}). ^{31}P-NMR ($CDCl_3$): δ 25.31*; 24.64 (4:1).Elemental analysis: Calcd. for $C_{39}H_{34}NO_3P$: C, 78.64; H, 5.75; N, 2.35. Found: C, 78.48, H, 5.96, N, 2.16.

3.1.4. *Diphenyl N-(R)-α-Methylbenzylamino(pyren-1-yl)methylphosphonate* (3e)

To a solution of the imine **1e** (1mmol) and diphenyl phosphite (1 mmol) in dichloromethane (ca. 2 mL) was slowly added $BF_3 \cdot OEt_2$ (24 mg, 0.67 mmol) via syringe. The mixture was stirred at room temperature for 24 h while monitoring the course of the reaction using ^{31}P-NMR spectroscopy or TLC. The solvent was removed in vacuo and the residual oil was roughly purified by column chromatography on microcrystalline cellulose with chloroform as eluent to give **3e** as a brown oil in 0.9 g (72%) yield of a crude product. Further attempts to purify and to analyze the product resulted in its decomposition.

3.2. Toxicity Tests

3.2.1. Microtox® Toxicity Assay

The method is based on the analysis of light emission reduction of luminescent bacteria (*Vibrio fischeri*) under toxic stress. The tests were carried out in a Microtox® M500 analyzer according to the 1992 Microtox® Manual. The Microtox® Solid-Phase Test (MSPT) was adopted to report of Doe et al. [31]. The MSPT procedure allows the test organisms to come direct contact with the solid sample in an aqueous suspension of the test sample. Thus it is possible to detect toxicity which is due to the insoluble solids that are not in the solution.

All materials and reagents were purchased from MODERNWATER (New Castle, DE, USA). Toxicity was determined by using the marine luminescent bacterium, *Vibrio fischeri*, naturally adapted to a saline environment. Briefly, bacteria were regenerated with 1 mL of Reconstitution Solution (0.01%) and placed in the reagent well of the Microtox®. A suspension of 7 g of the sediment was prepared in 35 mL of a Solid Phase Diluent (3.5% NaCl) and was magnetically stirred for 10 min. Then a series of dilutions were made and bacteria (approx. 1×10^6 cell$\cdot$ mL^{-1} per assay) were exposed to these dilutions and to a blank (3.5% NaCl solution) for 20 min. Next, after filtration, the light output of supernatants containing exposed bacteria was measured after 5 min with a Microtox® Analyzer 500. Inhibition was calculated as the concentration of compound loaded to sediment (mg/L) that caused a 50% reduction in the light emitted by the bacteria, and EC_{50} along with 95% confidence limit determined by the software provided with the Analyzer.

3.2.2. Ostracod Test Kit

Ecotoxicity evaluation of synthesized compounds was performed in a short term contact test using Ostracodtoxkit F™ provided by MicroBiotests Inc. (Gent, Belgium). This direct sediment contact bioassay was performed in multiwell test plates using neonates of the benthic ostracod crustacean *Heterocypris incongruens* hatched from cysts [35]. After 6 days of contact with the sediment (or soil) the percentage mortality and the growth of the crustaceans were determined and compared to the results obtained in a non-treated reference sediment (soil).

Briefly, according to manual of Ostracodtoxkit test, the cysts (*Heterocypris incongruens*) were transferred into a Petri dish filled with 10 mL standard fresh water (reconstituted water) and were incubated at 25 °C for 52 h under continuous illumination (approx. 3000–4000 lux).

After 48 h of cysts incubation, pre-feeding of the freshly hatched ostracods was performed with algae (spirulina-powder) provided in the test kit. Next, after hatching, before feeding with algal food suspension, the length measurements of ostracod neonates has been done. Algae (*Selenastrum capricornutum*) used as feed in the test plate were reconstituted according to the manufacturer's procedure. Each well of a test plate was filled in the following order: 2 mL standard freshwater, 2×500 µL of sediment (soil) treated and non-treated for comparison (blank), 2 mL already prepared algal suspension, 10 ostracods. The test plates were covered with Parafilm® and closed by a lid. Then multiwall plates were incubated at 25 °C in darkness for 6 days. After 6 days of exposure, the ostracods have been recovered from the multiwells to determine the percentage mortality. To calculate the growth inhibition of survived organisms, their length measurements have been also done. Mortality of test organisms was determined in six replicates. The measurement of length was carried out by means of a micrometric strip placed on the bottom of a glass microscope plate. Growth inhibition (GI) of *H. incongruens* in the test sediment was calculated as follows:

$$\% \text{growth inhibition} = 100 - [(\text{growth in test sed.}/\text{growth in ref. sed.}) \times 100] \quad (1)$$

Statistical differences between variables were analyzed with ANOVA.

4. Conclusions

To conclude, a series of diphenyl pyrene-derived aminophosphonates **3a–d**, **4** and **5** has been prepared. Our previous study [17] revealed that pyrene-derived aminophosphonates have interesting cytotoxic properties against colon cancer cell lines, therefore, the ecotoxicity of the newly synthesized compounds has been evaluated using the crustacean *Heterocypris incongruens* and bacterium *Vibrio fischeri* biotests. The evaluation demonstrated clearly that all compounds **3a–d** and **4** are toxic in a moderate to high degree and that diphenyl *N*-benzylamino(pyren-1-yl) methylphosphonate (**3d**) shows the highest degree of ecotoxicity. It has been also revealed that a weak correlation between the toxicity against *H. incongruens* and against *V. fischeri* has been found and the structure-toxicity correlation is rather problematic.

Although no evident correlations between the structure of the tested compounds and their toxicity have been observed, the combination of the different bioassays—the Microtox® and Ostracodtoxkit F™ tests—has been proved to be a very useful tool for determining and comparing the potential toxicities of examined substances **3a–d** and **4** as well as for the assessment of their toxicity interactions.

Acknowledgments: We thank the Polish National Centre of Science (NCN) for funding the work in the framework of the grant no. 2014/13/B/NZ9/02418. Within the framework of this grant, funds for covering the costs to publish in open access were scheduled.

Author Contributions: Jarosław Lewkowski and Piotr Rychter conceived and designed the experiments; Maria Rodriguez Moya and Marta Chmielak performed the chemical experiments; Diana Rogacz and Kamila Lewicka performed the ecotoxicological experiments; Jarosław Lewkowski and Piotr Rychter analyzed the data (Jarosław Lewkowski chemistry, Piotr Rychter. ecotoxicity); Jarosław Lewkowski and Piotr Rychter wrote the paper.

References

1. Kafarski, P.; Lejczak, B.; Tyka, R.; Koba, L.; Pliszczak, E.; Wieczorek, P. Herbicidal activity of phosphonic, phosphinic, and phosphonous acid analogues of phenylglycine and phenylalanine. *J. Plant Growth Regul.* **1995**, *14*, 199–203. [CrossRef]
2. Kafarski, P.; Lejczak, B. Biological Activity of Aminophosphonic Acids. *Phosphorus Sulfur Silicon Relat. Elem.* **1991**, *63*, 193–215. [CrossRef]
3. Kafarski, P.; Lejczak, B. Aminophosphonic Acids of Potential Medical Importance. *Curr. Med. Chem. Anti-Cancer Agents* **2001**, *1*, 301–312. [CrossRef] [PubMed]
4. Hudson, H.R. Aminophosphonic and Aminophosphinic Acids and their Derivatives as Agrochemicals. In *Aminophosphonic* and Aminophosphinic Acids: Chemistry and Biological Activity; Kukhar, V.P., Hudson, H.R., Eds.; Wiley and Sons: Chichester, UK, 2000.
5. Maier, L.; Diel, P.J. Organic Phosphorus Compounds 94. Preparation, Physical and Biological Properties of Aminoarylmethylphosphonic- and -Phosphonous Acids. *Phosphorus Sulfur Silicon Relat. Elem.* **1991**, *57*, 57–64. [CrossRef]
6. Hudson, H.R. Phosphorus-Containing Fungicides: A Review of Current Research and Prospects. *Phosphorus Sulfur Silicon Relat. Elem.* **1999**, *144–146*, 441–444. [CrossRef]
7. Jane, D.E. Neuroactive aminophosphonic and aminophosphinic acid derivatives. In *Aminophosphonic and Aminophosphinic Acids: Chemistry and Biological Activity*; Kukhar, V.P., Hudson, H.R., Eds.; Wiley and Sons: Chichester, UK, 2000.
8. Mohd-Pahmi, S.H.; Hussein, W.M.; Schenk, G.; McGeary, R.P. Synthesis, modelling and kinetic assays of potent inhibitors of purple acid phosphatase. *Bioorg. Med. Chem. Lett.* **2011**, *21*, 3092–3094. [CrossRef] [PubMed]

9. Klimczak, A.A.; Kuropatwa, A.; Lewkowski, J.; Szemraj, J. Synthesis of new *N*-arylamino(2- furyl) methylphosphonic acid diesters, and in vitro evaluation of their cytotoxicity against esophageal cancer cells. *Med. Chem. Res.* **2013**, *22*, 852–860. [CrossRef]
10. Boduszek, B. The Acidic Cleavage of Pyridylmethyl(amino)phosphonates. Formation of the Corresponding Amines. *Tetrahedron* **1996**, *52*, 12483–12494. [CrossRef]
11. Lewkowski, J.; Rzeźniczak, M.; Skowroński, R.; Zakrzewski, J. The first synthesis of ferrocenyl aminophosphonic esters. *J. Organomet. Chem.* **2001**, *631*, 105–109. [CrossRef]
12. Keglevich, G.; Bálint, E. The Kabachnik–Fields Reaction: Mechanism and Synthetic Use. *Molecules* **2012**, *17*, 12821–12835. [CrossRef] [PubMed]
13. Ali, T.E.; Abdel-Kariem, S.M. Methods for the synthesis of α-heterocyclic/heteroaryl-α-aminophosphonic acids and their esters. *ARKIVOC* **2015**, *6*, 246–287. [CrossRef]
14. Kafarski, P.; Górny vel Górniak, M.; Andrasiak, I. Kabachnik-Fields Reaction under Green Conditions—A Critical Overview. *Curr. Green Chem.* **2015**, *2*, 218–222. [CrossRef]
15. Hudson, H.R.; Lee, R.J.; Matthews, R.W. 1-Amino-1-aryl- and 1-Amino-1-heteroaryl-methanephosphonicacids and Their *N*-Benzhydryl–Protected Diethyl Esters: Preparation and Characterization. *Phosphorus Sulfur Silicon Relat. Elem.* **2004**, *179*, 1691–1709. [CrossRef]
16. Jayaprakash, S.H.; Uma Maheswara Rao, K.; Satheesh Krishna, B.; Siva Prasad, S.; Syama Sundar, C.; Suresh Reddy, C. PAA-SIO2 Catalyzed Synthesis, Uv Absorption, and Fluorescence Emission Studies of Diethyl (aryl/hetero aryl amino)(pyren-1-yl)-methylphosphonates. *Phosphorus Sulfur Silicon Relat. Elem.* **2015**, *190*, 449–460. [CrossRef]
17. Lewkowski, J.; Rodriguez Moya, M.; Wrona-Piotrowicz, A.; Zakrzewski, J.; Kontek, R.; Gajek, G. Synthesis, fluorescent properties and the promising cytotoxicity of pyrene-derived aminophosphonates. *Beilstein J. Org. Chem.* **2016**, *12*, 1229–1235. [CrossRef]
18. Ma, X.Y.; Wang, X.C.; Ngo, H.H.; Guo, W.; Wu, M.N.; Wang, N. Bioassay based luminescent bacteria: Interferences, improvements, and applications. *Sci. Total. Environ.* **2014**, *468–469*, 1–11. [CrossRef] [PubMed]
19. Connon, R.E.; Geist, J.; Werner, I. Effect-Based Tools for Monitoring and Predicting the Ecotoxicological Effects of Chemicals in the Aquatic Environment. *Sensors* **2012**, *12*, 12741–12771. [CrossRef] [PubMed]
20. Lewkowski, J.; Malinowski, Z.; Matusiak, A.; Morawska, M.; Rogacz, D.; Rychter, P. The Effect of New Thiophene-Derived Aminophosphonic Derivatives on Growth of Terrestrial Plants: A Seedling Emergence and Growth Test. *Molecules* **2016**, *21*, 694. [CrossRef] [PubMed]
21. Matusiak, A.; Lewkowski, J.; Rychter, P.; Biczak, R. Phytotoxicity of New Furan-derived Aminophosphonic Acids, N^{-}Aryl Furaldimines and 5^{-}Nitrofuraldimine. *J. Agric. Food Chem.* **2013**, *61*, 7673–7678. [CrossRef] [PubMed]
22. MicroBioTests. Toxkit Advantages/Assets. Available online: http://www.microbiotests.be/information/toxkit-advantagesassets/ (accessed on 19 July 2016).
23. Baran, A.; Tarnawski, M. Phytotoxkit/Phytotestkit and Microtoxs as tools for toxicity assessment of sediments. *Ecotox. Environ. Safe.* **2013**, *98*, 19–27. [CrossRef] [PubMed]
24. Weltens, R.; Deprez, K.; Michiels, L. Validation of Microtox as a first screening tool for waste classification. *Waste Manage.* **2014**, *34*, 2427–2433. [CrossRef] [PubMed]
25. Yu, X.; Zuo, J.; Tang, X.; Li, R.; Li, Z.; Zhang, F. Toxicity evaluation of pharmaceutical wastewaters using the alga *Scenedesmus obliquus* and the bacterium *Vibrio fischeri*. *J. Hazard. Mat.* **2014**, *266*, 68–74. [CrossRef] [PubMed]
26. Rosado, D.; Usero, J.; Morillo, J. Assessment of heavy metals bioavailability and toxicity toward *Vibrio fischeri* in sediment of the Huelva estuary. *Chemosphere* **2016**, *153*, 10–17. [CrossRef] [PubMed]
27. Montalban, M.G.; Hidalgo, J.M.; Collado-Gonzalez, M.; Banos, F.G.D.; Víllora, G. Assessing chemical toxicity of ionic liquids on *Vibrio fischeri*: Correlation with structure and composition. *Chemosphere* **2016**, *155*, 405–414. [CrossRef] [PubMed]
28. Joly, P.; Bonnemoy, F.; Charvy, J.-C.; Bohatier, J.; Mallet, C. Toxicity assessment of the maize herbicides S-metolachlor, benoxacor, mesotrione and nicosulfuron, and their corresponding commercial formulations, alone and in mixtures, using the Microtox test. *Chemosphere* **2013**, *93*, 2444–2450. [CrossRef] [PubMed]
29. Pedrazzani, R.; Ceretti, E.; Zerbini, I.; Casale, R.; Gozio, E.; Bertanza, G.; Gelatti, U.; Donato, F.; Feretti, D. Biodegradability, toxicity and mutagenicity of detergents: Integrated experimental evaluations. *Ecotoxicol. Environ. Saf.* **2012**, *84*, 274–281. [CrossRef] [PubMed]

30. Antunes, S.C.; Pereira, J.L.; Cachada, A.; Duarte, A.C.; Goncalves, P.; Sousa, J.P.; Pereira, R. Structural effects of the bioavailable fraction of pesticides in soil: Suitability of elutriate testing. *J. Hazard. Mater.* **2010**, *184*, 215–225. [CrossRef] [PubMed]
31. Burga Perez, K.F.; Charlatchka, R.; Sahli, L.; Ferard, J.-F. New methodological improvements in the Microtox® solid phase assay. *Chemosphere* **2012**, *86*, 105–110. [CrossRef] [PubMed]
32. Stronkhorst, J.; Ciarelli, S.; Schipper, C.A.; Postma, J.F.; Dubbeldam, M.; Vangheluwe, M.; Brils, J.M.; Hooftman, R. Inter-laboratory comparison of five marine bioassays for evaluating the toxicity of dredged material. *Aquat. Ecosyst. Health Manag.* **2004**, *7*, 147–159. [CrossRef]
33. Doe, K.; Scroggins, R.; Mcleay, D.; Wohlgeschaffen, G. Solid-phase test for sediment toxicity using the luminescent bacterium *Vibrio fischeri*. In *Small-Scale Freshwater Toxicity Investigations*; Blaise, C., Férard, J.F., Eds.; Springer: Dordrecht, Holland, 2005; Volume 1, pp. 107–136.
34. ISO. *Water Quality—Determination of Fresh Water Sediment Toxicity to Heterocyprisincongruens (Crustacea, Ostracoda)*, 1st ed.; ISO 14371:2012; The International Organization for Standardization: Geneva, Switzerland, 2012.
35. Martínez-Sánchez, M.J.; Pérez-Sirvent, C.; García-Lorenzo, M.L.; Martínez-López, S.; Bech, J.; García-Tenorio, R.; Bolívar, J.P. Use of bioassays for the assessment of areas affected by phosphate industry wastes. *J. Geochem. Explor.* **2014**, *147*, 130–138. [CrossRef]
36. Gouider, M.; Feki, M.; Sayadi, S. Bioassay and use in irrigation of untreated and treated wastewaters from phosphate fertilizer industry. *Ecotoxicol. Environ. Saf.* **2010**, *73*, 932–938. [CrossRef] [PubMed]
37. Cvancarova, M.; Kresinova, Z.; Cajthaml, T. Influence of the bioaccessible fraction of polycyclic aromatic hydrocarbons on the ecotoxicity of historically contaminated soils. *J. Hazard. Mater.* **2013**, *254–255*, 116–124. [CrossRef] [PubMed]
38. Oleszczuk, P.; Hollert, H. Comparison of sewage sludge toxicity to plants and invertebrates in three different soils. *Chemosphere* **2011**, *83*, 502–509. [CrossRef] [PubMed]
39. Płaza, G.; Nałęcz-Jawecki, G.; Ulfig, K.; Brigmon, R.L. The application of bioassays as indicators of petroleum-contaminated soil remediation. *Chemosphere* **2005**, *59*, 289–296. [CrossRef] [PubMed]

4

Synthesis, Properties and Stereochemistry of 2-Halo-1,2λ⁵-oxaphosphetanes

Anastasy O. Kolodiazhna and Oleg I. Kolodiazhnyi *

Institute of Bioorganic Chemistry and Petrochemistry, National Academy of Sciences of Ukraine, 1, Murmanska Street, Kiev 02094, Ukraine; anastasik7@rambler.ru
* Correspondence: olegkol321@gmail.com

Academic Editors: György Keglevich and Derek J. McPhee

Abstract: Results of research into four-membered 2-halo-1,2λ^5-oxaphosphetane phosphorus(V)-heterocycles are presented. The preparation of 2-halo-1,2λ^5-oxaphosphetanes by reaction of *P*-haloylides with carbonyl compounds is described. The mechanism of asynchronous [2+2]-cycloaddition of ylides to aldehydes was proposed on the base of low-temperature NMR investigations. 2-Halo-1,2λ^5-oxaphosphetanes were isolated as individual compounds and their structures were confirmed by ^{1}H-, ^{13}C-, ^{19}F- and ^{31}P-NMR spectra. These compounds are convenient reagents for preparing of various organic and organophosphorus compounds hardly available by other methods. Chemical and physical properties of the 2-halo-1,2λ^5-oxaphosphetanes are reviewed. The 2-chloro-1,2λ^5-oxaphosphetanes, rearrange with formation of 2-chloroalkyl-phosphonates or convert into *trans*-phosphorylated alkenes depending on the substituents at the α-carbon atom. Prospective synthetic applications of 2-halo-1,2λ^5-oxaphosphetanes are analyzed. The 2-halo-1,2λ^5-oxaphosphetanes may be easily converted to various alkenylphosphonates: allyl- or vinylphosphonates, phosphorus ketenes, thioketenes, ketenimines.

Keywords: 2-halo-1,2λ^5-oxaphosphetanes; allylphosphonates; vinylphosphonates; phosphorus ketenes

1. Introduction

One of the most interesting and intriguing classes of organophosphorus compounds are the 1,2-oxaphosphetanes—four-membered heterocycles containing pentacoordinated phosphorus [1–7]. Since 1,2-oxaphosphetanes are well-known intermediates in the Wittig reaction, a number of efforts have been made for their structural characterization both in solution and the solid state [8–13].

In 1967, Birum and Matthews had already reported the structural characterization (NMR and X-ray study) of the first isolated 1,2-oxaphosphetane **1**. Compound **1** was prepared in 76% yield by allowing hexaphenylcarbodiphosphorane to react with hexafluoroacetone in dry diglyme [14] (Scheme 1).

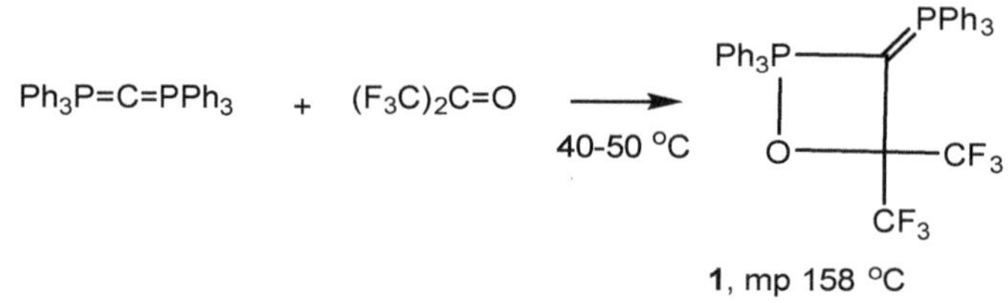

Scheme 1. First representative of stable 1,2-oxaphosphetanes isolated by Birum and Matthews.

Vedejs [3,4] succeeded in detecting of 1,2-oxaphosphetanes **2–4** by low temperature NMR spectroscopy during typical Wittig reactions and observed that these intermediates readily decompose upon warming to room temperature into alkenes and phosphine oxides (Scheme 2).

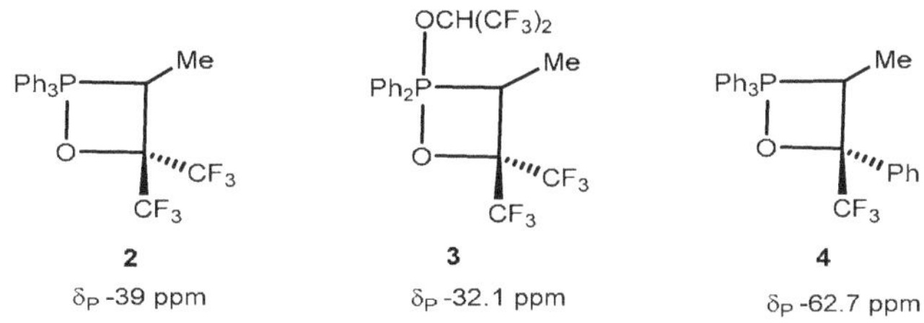

Scheme 2. 1,2-oxaphosphetanes **2–4** registered by low temperature NMR spectroscopy.

Schmutzler and co-workers reported several stabilized bis(trifluoromethylated) oxaphosphetanes **5**, **6** (Scheme 3) [15] which were characterized by NMR, MS spectra and X-ray analysis. At room temperature Berry-pseudorotation was fast on the NMR time scale, impeding one from distinguishing apical and equatorial P-CF_3 groups. Decreasing the temperature to -60 °C in toluene-d_8 allowed resolving the signals for all CF_3 groups of molecule **7** [16].

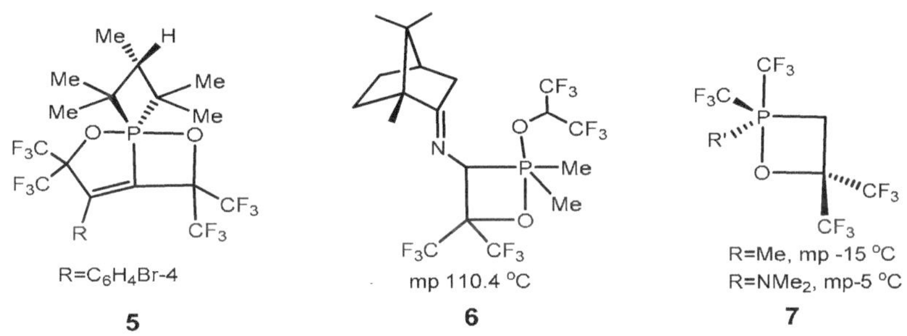

Scheme 3. Stabilized bis(trifluoromethylated) oxaphosphetanes **5**,**6**.

Kojima [17] reported the interesting anti-apicophilic spirophosphorane **8** bearing an oxaphosphetane ring. The structure of compounds **8** was confirmed by X-ray diffraction. Crystallization from hexane gave the pure anti-apicophilic derivative. Stereomutation of compound **8** was observed in the presence of acids and slowed down when DBU was present, suggesting that the isomerization into **9** is rather the result of a P–O bond breaking-recombination process. Evidently, this conversion represents an example of a thermodynamically stable oxaphosphetanes, in which pseudorotation is faster than alkene formation (Scheme 4).

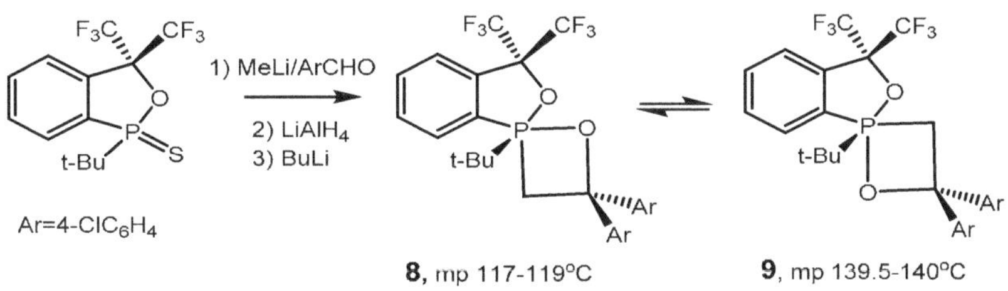

Scheme 4. Stereomutation of anti-apicophilic oxaphosphetane **8**.

Most of the previously reported stable oxaphosphetane structures contain fluorine-bearing or bicyclic phosphole-type ligands either at the phosphorus position or at the 4 position in the oxaphosphetane ring **10**, **11** [15–20]. Recently, Streubel and coworkers have prepared the first 1,2-oxaphosphetane complexes **12** formally similar to traditional oxaphosphetanes, using low-temperature ring expansion of epoxides with a Li/Cl phosphinidenoid complex [18] (Scheme 5).

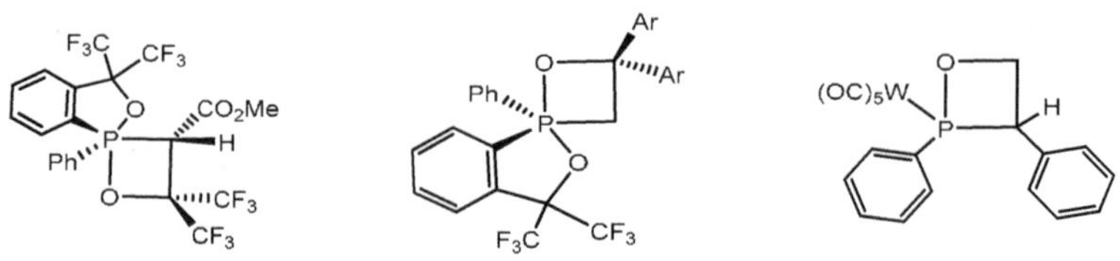

10 **11 [1a]** mp 179 °C **12**

Scheme 5. Oxaphosphetanes stabilized by CF_3 groups **10**,**11** and 1,2-oxaphosphetane complex **12**.

Gilheany studied oxaphosphetane intermediates in the Wittig reaction by variable-temperature NMR spectroscopy [9–11]. Compound **13** was obtained by low-temperature acid quenching of the Wittig reaction of ylide with benzaldehyde, a suitable representative aromatic aldehyde (see Scheme 6). The major diastereomer was the *syn*-**13** on the basis that the unquenched Wittig reaction gives the (Z)-alkene as the major product. In this manner, the *syn/anti* ratio of **13** was 89:11.

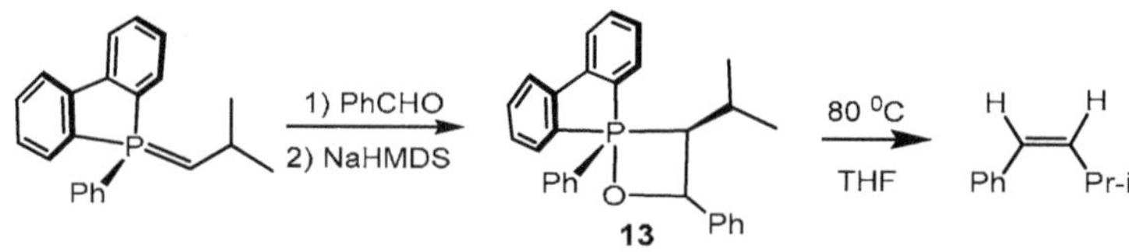

Scheme 6. Stereospecific decomposition of oxaphosphetane **13**.

Keglevich reported detection of enantiomers of *P*-stereogenic pentacoordinated phosphorus compounds [20]. Detailed ^{31}P-NMR investigations of oxaphosphetes in optically active solvents have clearly shown that the most electronegative substituents (e.g., oxygen) prefers the apical position in a trigonal bipyramidal structure and that the pentacoordinated phosphorus atom is in a dynamic condition due to pseudorotation. Berger and coworkers found that 2-furyl groups on the phosphorus atom increase the thermal stabilities of oxaphosphetanes and succeeded in isolation and determination of the X-ray structure of tris(2-furyl) substituted oxaphosphetane, the stability of which is attributed to the electron-withdrawing properties of the 2-furyl group [21].

Among the stable oxaphosphetanes [1–22], 2-halo-1,2λ^5-oxaphosphetanes **14**, which possess relatively high stability and diverse reactivity, attract particular interest [23–38] (Scheme 7). These oxaphosphetanes containing fluorine, chlorine or bromine atoms bonded to phosphorus are an interesting class of pentacoordinated phosphorus heterocycles possessing peculiar properties. The chemical properties of 2-halo-1,2λ^5-oxaphosphetanes, first of all of *P*-chloro- and *P*-fluoroylides, due to the presence of a labile halogen atom on phosphorus, are very specific and differ from the properties of triphenylphosphonium ylides. For example, reactions and conversions of 2-halo-1,2λ^5-oxaphosphetanes proceed with preservation of the P-C bond and leads to the formation of different organophosphorus compounds. In addition, this type of compounds exhibit also physico-chemical properties uncharacteristic for traditional *P,P,P*-triorganosubstituted oxaphosphetanes. For the first time the 2-halo-1,2λ^5-oxaphosphetanes were prepared in our laboratory almost twenty years ago and up to today we and some other authors are still studying their chemistry. In this article, we summarize the synthesis and properties of this type of organophosphorus compounds. The chemistry of 2-halo-1,2λ^5-oxaphosphetanes was not previously analyzed, generalized or reviewed.

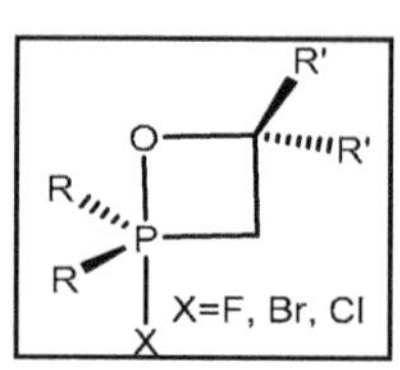

14

Scheme 7. Examples of stable oxaphosphetanes.

2. Results and Discussion

2.1. General Description

Since pentacoordinated phosphoranes formally have 10 electrons in the valence shell, they display a specific bonding model. Therefore pentacoordinated phosphoranes take a trigonal bipyramidal structure and there are two ligating sites, apical and equatorial sites. The apical bond consists of three-center four-electron bond using the *p* orbital of the central phosphorus atom, while the equatorial bond is a typical *s* bond using sp^2 hybrid orbital of the phosphorus atom. This three-center four-electron bond forms three molecular orbitals.

It is generally known that pentacoordinated phosphoranes rapidly undergo intramolecular positional isomerization without bond cleavage. A very rapid non-dissociative intramolecular site exchange is usually explained by the Berry pseudorotation mechanism [19,22,23].

The 2-halo-1,2λ^5-oxaphosphetanes (halogen = chlorine, bromine, fluorine) are the most stable representatives of this type of compounds. They can be purified by distillation under vacuum and stored in a refrigerator. At the same time they possess interesting chemical properties and participate in various chemical transformations [24–28].

The stability of 2-halo-oxaphosphetanes changes in the same sequence of substituents R^4 at the endocyclic carbon atoms at position 4. The most stable are compounds containing strong electron-accepting groups at C(4), drawing off electron density from the oxygen atom as a result of which the three-centre apical bond O—P—Hal is strengthened. Chloro-oxaphosphetanes, having less electron-accepting substituents at C(4), are dissociated to a large extent and correspondingly are converted into vinylphosphine oxides at room temperature.

2.2. Synthesis of 2-Halo-1,2λ^5-oxaphosphetane

Available methods for the synthesis of 2-halo-1,2λ^5-oxaphosphetanes can be used for investigation of the reaction mechanism of phosphorus ylides with carbonyl compounds as well as for preparing stable oxaphosphetanes that can be used as reactants for organic synthesis. The 2-halo-1,2λ^5-oxaphosphetanes were prepared by reaction of *P*-fluoro-, chloro- or bromoylides with carbonyl compounds. *P*-Chloro- and *P*-bromoylides react with active ketones, containing a trifluoromethyl group, with the formation of stable [2+2]-cycloaddition products, 2-chloro- or 2-bromo-1,2λ^5-oxaphosphetenes **14** were isolated in yields close to quantitative as crystalline substances or as liquids distillable in vacuum (Scheme 8, Table 1) [29–42].

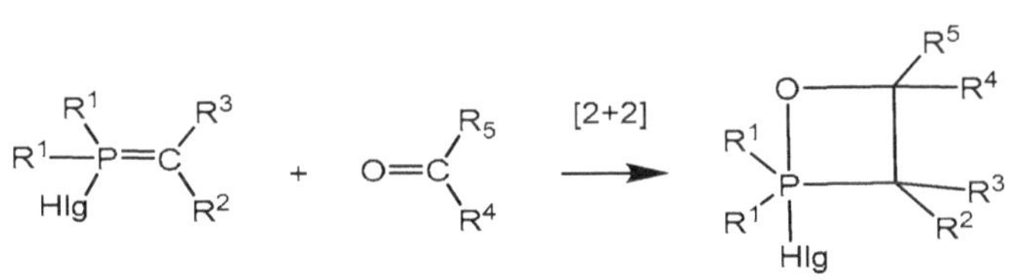

14

Scheme 8. Synthesis of 2-halo-1,2λ^5-oxaphosphetanes. Hlg = F, Cl, Br; R^1 = Alk, Ph; CR^2R^3 = CH_2, CHAlk, $CAlk_2$, CHPh, $CHSiMe_3$, CCl_2, CBr_2; CR^4R^5 = C=O, CNPh, CHAlk, $CAlk_2$, CHPh, CPh_2, $CH_2CH = CH_2$.

The addition of *P*-halogen-ylides to ketones proceeded stereoselectively and led predominantly to the formation of one of the possible 2-halo-1,2λ^5-oxaphosphetane diastereomers. 2-Halo-1,2λ^5-oxaphosphetanes dissociate at the *P*-halogen bond in solution with the formation of cyclic phosphonium salts, as a result of which an equilibrium is established between the forms with five- and four-coordinate phosphorus atoms. Dissociation of 2-halo-1,2λ^5-oxaphosphetanes is enhanced by reducing the electron-accepting properties of substituents and also by increasing priority solvent. The 2-halo-1,2λ^5-oxaphosphetanes containing electron-accepting groups at C(4) are distinctly stabler than oxaphosphetanes with alkyl groups in this position [25–27].

The reaction of *P*-bromomethylides with fluorinated acetophenone afforded in high yield oxaphosphetanes **15**, which were isolated as crystalline compounds (Scheme 9 and Table 1, entries 10–13). The compounds **15** exist in solution as cyclic phosphonium salt.

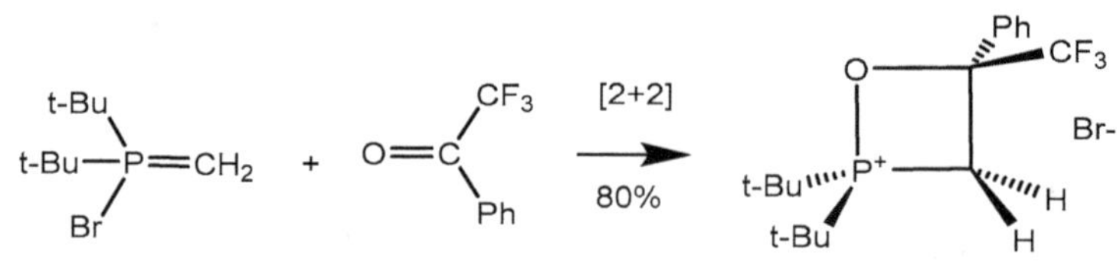

15

δ_P +46 ppm
mp 132 °C (hexane)

Scheme 9. Ionized form of 2-bromo-1,2λ[5]-oxaphosphetanes **15**.

At the same time the 2-chloro-1,2λ^5-oxaphosphetanes **16** exist as mixture of P(IV) and P(V)-forms. These compounds can be distilled under vacuum and dissolved in non-polar solvents (benzene) (Scheme 10 and Table 1, entries 1–9). Reaction of *P*-Fluoroylids with aldehydes and ketones proceeds in ether or pentane at −40–−20 °C and leads to the formation of stable 2-fluoro-1,2λ^5-oxaphosphetanes **17** (Scheme 11 and Table 2).

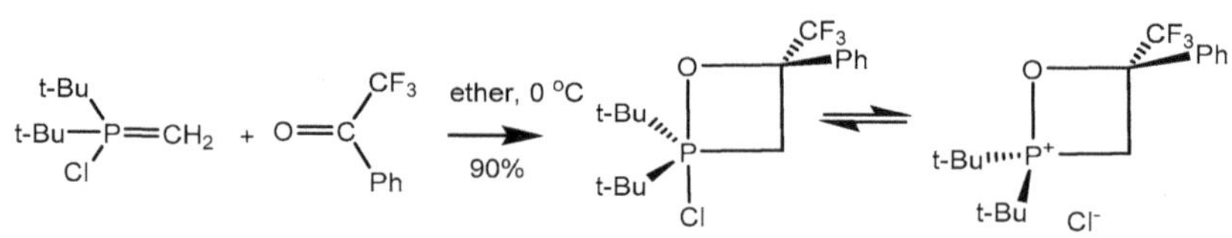

16

mp 82-84 °C, δ_P 9.94 ppm

Scheme 10. P(IV) and P(V)-forms of 2-chloro-1,2λ^5-oxaphosphetane **16**.

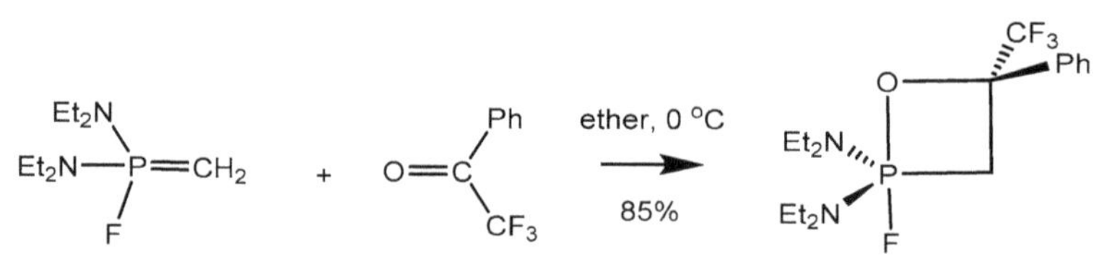

17

bp 110 °C (0.06 mmHg), mp 20 °C,
δ_P -38.96 ppm

Scheme 11. 2-Fluoro-1,2λ^5-oxaphosphetane **17**.

The compounds **17** are liquids distilling in vacuo, the structure of which was proved by means of mass and NMR spectra. The ^{31}P-NMR spectra of 2-fluoro-oxaphosphetanes **17** present doublets with 800 Hz $^1J_{PF}$ constants in the high magnetic field of a NMR spectrum at −37–−8 ppm. This corresponds to a pentacoordinate state of compounds **17** [24,26–31]. Tetracoordinated forms of 2-fluoro-oxaphosphetanes **17** were not registered by ^{19}F- and ^{31}P-NMR spectroscopy (Scheme 11).

C-Silyl-*P*-chloroylides **18** react with carbonyl compounds to afford 2-chloro-1,2λ^5-oxaphosphetanes **19**. The oxaphosphetanes **19** bearing an electronegative CF_3 substituent at C-4 are relatively stable and can be isolated and analyzed by NMR (Scheme 12). The NMR spectra of these compounds reveal signals at 0.2 ppm, singlet (Me_3Si), at 2 ppm, doublet, $^3J_{PH}$ 18.0 Hz (C^3H), and at 4.5 ppm (C^4H). ^{13}C-NMR signals of C-3 and C-4 carbons were found at 30 and 90 ppm, correspondingly. The ^{31}P-NMR signals of 2-chloro-1,2λ^5-oxaphosphetanes **19** at δ_P +48 ppm (R^1 = *i*-PrO) and at δ_P = 60 ppm (R^1 = Et_2N) correspond to tetracoordinate phosphorus included in four-membered phosphetane cycle (Table 1) [25,32].

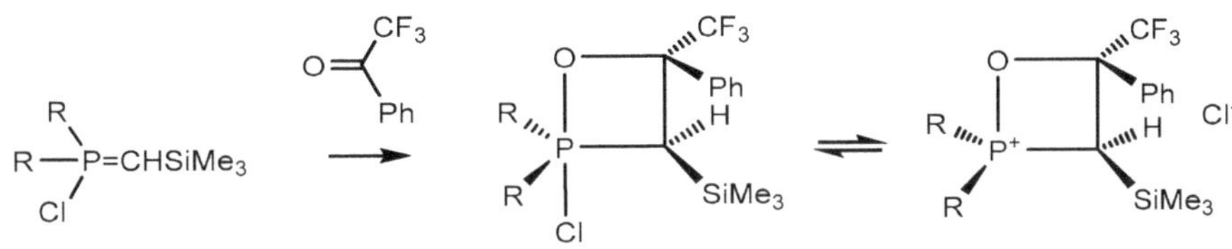

18 19

Scheme 12. Silylated 2-halogene-1,2λ^5-oxaphosphetanes **19**.

Table 1. 2-Chloro- and 2-bromo-1,2λ^5-oxaphosphetanes.

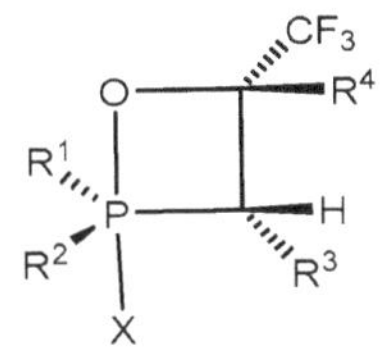

Entry	R^1	R^2	R^3	R^4	X	δ_P	δ_F	References
1	*t*-Bu	*t*-Bu	H	C_6H_4F-4	Cl	+ 9.0	+7.2 c (CF_3); −32.9 (C_6H_4F)	[36,42]
2	*i*-Pr	*i*-Pr	Ph	Ph	Cl	+1.23; +0.53	+1.08; +1.64	[42]
3	*t*-Bu	*t*-Bu	H	Ph	Cl	+ 9.94	+1.0	[42]
4	*t*-Bu	*t*-Bu	H	An-4	Cl	+13.6	+7.35 c	[36,42]
5	*t*-Bu	*t*-Bu	H	CF_3	Cl	+0.25	6.0	[36,42]
6	*t*-Bu	*t*-Bu	Me	Ph	Cl	+13.8; +11.65	+6.1	[42]
7	*t*-Bu	Et_2N	H	Ph	Cl	+5.01; +9.13	+6.06; +6.09	[42]
8	*t*-Bu	*t*-Bu	Me	CF_3	Cl	+2.53	+5.38;	[42]
9	*t*-Bu	Et_2N	H	CF_3	Cl	−2.6	+2.65 q; +2.94 q, $^1J_{FF}$ 8	[42]
10	*t*-Bu	*t*-Bu	H	CF_3	Br	+24.4	+6.25	[42]
11	*t*-Bu	*t*-Bu	H	Ph	Br	+46	+7.35	[42]
12	*t*-Bu	*t*-Bu	H	C_6H_4F-4	Br	+46	+5.6 (CF_3); −34.2 (C_6H_4F)	[42]
13	*t*-Bu	*t*-Bu	H	An-4	Br	+60	+6.1	[42]
14	*t*-Bu	*t*-Bu	H	Ph	OMe	−14.6	+1.15	[42]
15	*t*-Bu	*t*-Bu	H	Ph	OPh	−12.8	+2.0	[42]

Sotiropulos and Bertrand [33] reported the addition of phosphacumulene ylides **20** bearing the diazo group to isocyanates, leading to the formation of products **22**. It was proposed that initial nucleophilic attack of the ylide carbon atom on the carbonyl carbon gives a oxaphosphetane **21**, which depending on the relationship of the oxygen atom (or NR group) to nitrogen or phosphorus rearranges into products of 1,4- or 1,5-cyclisation **22** (Scheme 13). Benzaldehyde gives the 2-chlorooxaphosphetane **23** with the ylide **20** which readily eliminates hydrogen chloride and a nitrogen molecule being converted into an acetylene phosphonate **24** (Scheme 14).

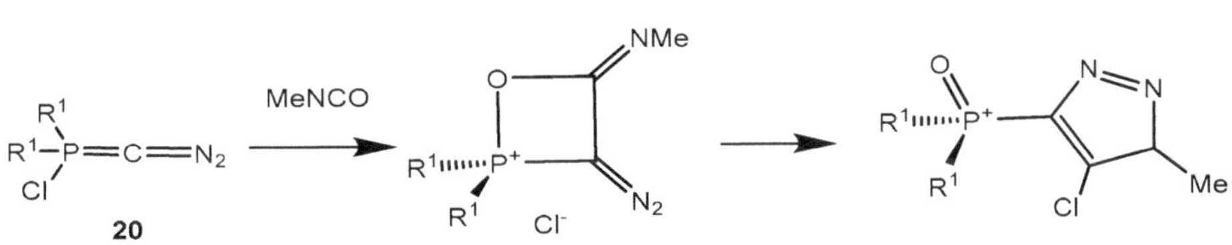

20 21 22

R=i-Pr$_2$N

Scheme 13. Adduct of phosphacumulene-ylide **20** with isocyanate.

A number of 2-fluoro-1,2λ^5-oxaphosphetanes **25** were prepared by reaction of *P*-fluoroylides with aldehydes and ketones (Scheme 15, Table 2). The 2-fluoro-1,2λ^5-oxaphosphetanes **25** are stable compounds, which can be isolated and purified by distillation under vacuum or by crystallization from non-polar solvents. These compounds are much distinguished from unstable adducts of carbonyl compounds with triphenylphosphonium ylides. Cycloadducts of *P*-fluoroylides with carbonyl compounds, 2-fluoro-1,2λ[5]-oxaphosphetanes, are also much more stable than 2-chloro- or 2-bromo-1,2λ[5]-oxaphosphetanes [25–31].

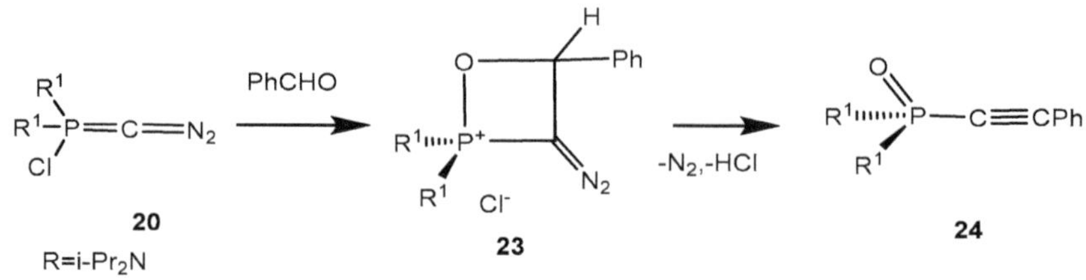

Scheme 14. Reaction of phosphacumulene-ylide **20** with benzaldehyde.

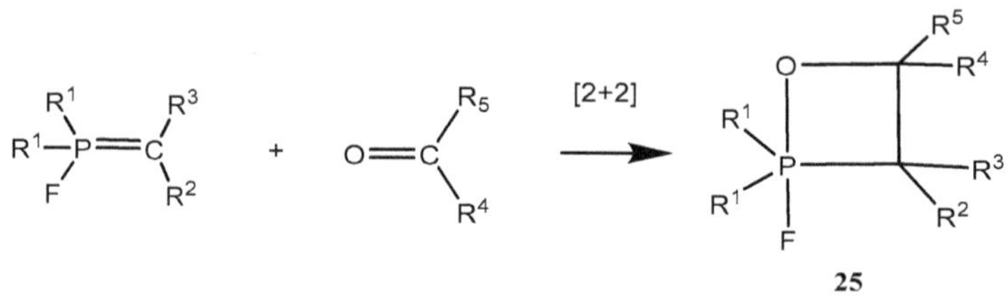

Scheme 15. Synthesis of 2-fluoro-1,2λ[5]-oxaphosphetanes **25**. R^1 = Alk, Ph; CR^2R^3 = CH_2, CHAlk, $CAlk_2$, CHPh, $CHSiMe_3$, CCl_2, CBr_2; CR^4R^5 = C=O, CNPh, CHAlk, $CAlk_2$, CHPh, CPh_2, $CH_2CH = CH_2$.

The stability of 2-fluoro-oxaphosphetanes is explained by the high electronegativity of the fluorine atom, compared to the electronegativities of chlorine and bromine. The P-F bond in 2-fluoro-1,2λ[5]-oxaphosphetanes is very strong, and, therefore, these compounds do not dissociate with formation of cyclic phosphonium salts, what is observed, for example, with 2-chloro-1,2λ[5]-oxaphosphetanes. Various stable 2-fluoro-1,2λ[5]-oxaphosphetanes were synthesized, isolated as pure specimens, and characterized (Table 2). Typical representatives of such compounds **26–29** are shown in Schemes 16–18 [26,29].

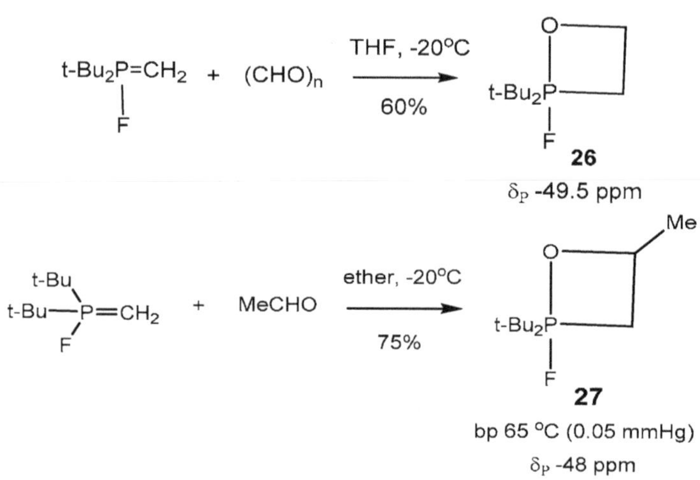

Scheme 16. The simplest representatives of 2-Fluoro-1,2λ[5]-oxaphosphetanes **26,27**.

The ^{31}P-NMR spectra of compounds **25** show a doublet at −6 to −45 ppm with 760–850 Hz PF coupling constants appropriate for axial fluorine atoms. The ^{31}P-NMR spectra of 2-fluoro-1,2λ[5]-oxaphosphetanes exhibit a doublet in the range −6 to −45 ppm, belonging to the five-coordinate phosphorus with corresponding coupling constants on axial fluorine atoms (760–850 Hz). The ^{19}F-NMR spectra of 2-fluoro-1,2λ[5]-oxaphosphetanes exhibit doublets with the same PF coupling constants.

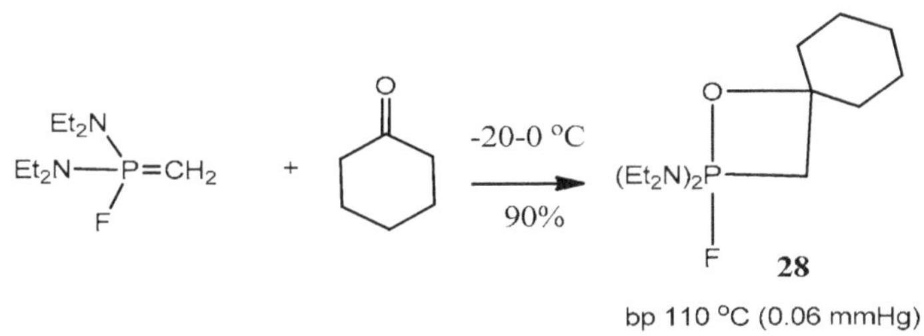

Scheme 17. 2-Fluoro-1,2λ^5-oxaphosphetane **28**.

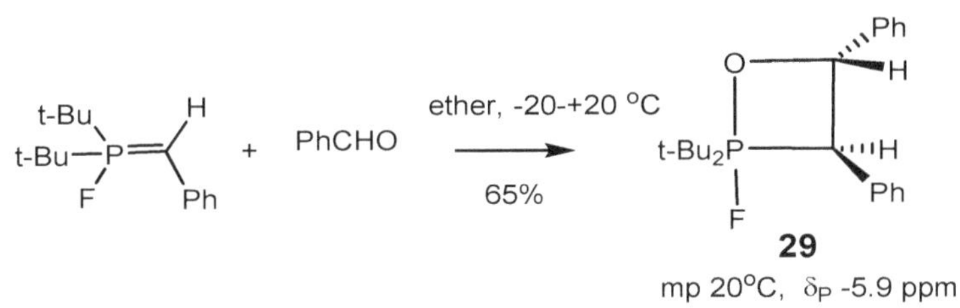

Scheme 18. 2-Fluoro-1,2λ^5-oxaphosphetane **29**.

The *P,P*-difluoroylides **30** react with aldehydes and active ketones to afford the stable oxaphosphetanes **31** bearing two fluorine atoms at the phosphorus (Scheme 19) [34–36]. The compounds **31** (R=CF_3) were distilled under reduced pressure without decomposition and were characterized by means of NMR spectroscopy.

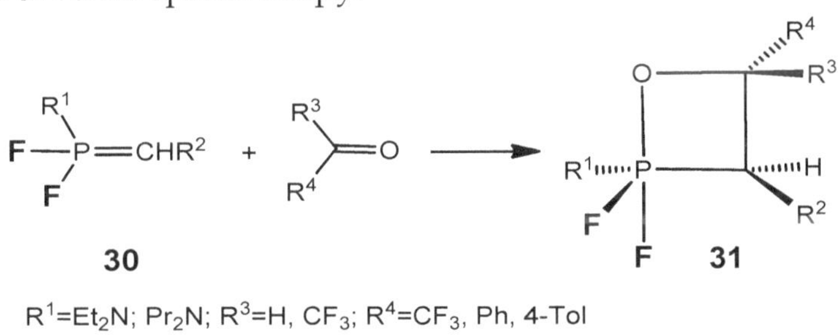

Scheme 19. 2,2-Difluoro-1,2λ^5-oxaphosphetanes **31**.

The ^{19}F-NMR spectra contain two double doublets at −47 and −65 ppm with expected coupling constants for the axial and equatorial fluorine atoms: $^1J_{PFa}$ = 915 Hz, $^1J_{PFe}$ = 1025 Hz, and $^2J_{FaFe}$ = 62 Hz. Apparently [2+2]-cycloaddition of the C=O group to the ylide **15** proceeds with high stereoselectivity, because the ^{19}F- and ^{31}P-NMR spectra show the signals belonging to the single diastereomer of the compounds **31**. The signals of the second diastereomer, the existence of which one can suppose as a consequence of the presence of two asymmetric endocyclic C-3 and C-4 carbon atoms are absent. The ^{13}C-NMR spectra reveal the presence of the signals at 62.5 ppm ($^1J_{CP}$ = 150 Hz, $^2J_{CF}$ = 53 Hz) and 75.6 ppm due to the C-3 and C-4 carbon atoms in complete accordance with the assigned structure of oxaphosphetane (Figures 1 and 2).

P-Fluoroylides add easily isocyanates, carbon dioxide and carbon disulfide with formation of oxaphosphetanes **32–35**. For example, the reaction of *P*-fluoroylides with phenyl isocyanate resulted in the formation of 2-fluorooxaphosphetane 34 which was stable during several hours at ambient temperature. Compound **34** was purified by crystallization in pentane and isolated as colorless crystalline matter. In IR spectra of compounds was found the strong band at 1730 cm^{-1} belonging to the C=O bond in a four-membered cycle. The ^{31}P-NMR spectra disclosed a doublet with constant $^1J_{PF}$ = 780 Hz and a doublet with the same constant in ^{19}F-NMR spectra. The oxaphosphetanes **32–35** convert at room temperature slowly and at heating quickly into phosphorylated heterocumulenes (ketenes, thioketenes, ketenimines, see Scheme 20) [29,37–39].

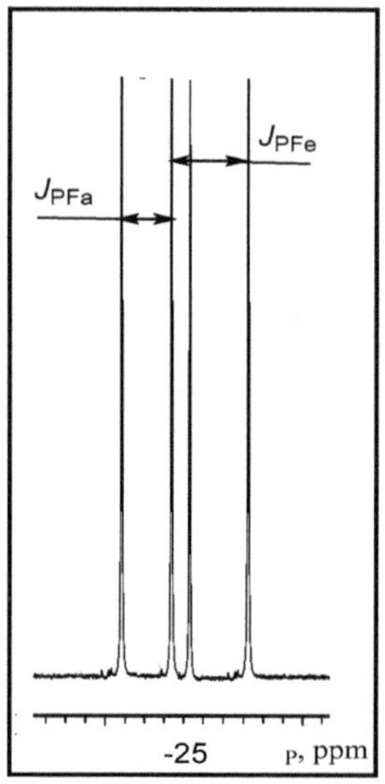

Figure 1. ^{31}P-NMR spectrum of 2-(diethylamino)-2,2-difluoro-4-tolyl-3-propyl-4-(trifluorophenyl-1,2λ^5-oxaphosphetane **31** (See Table 2, entry 15).

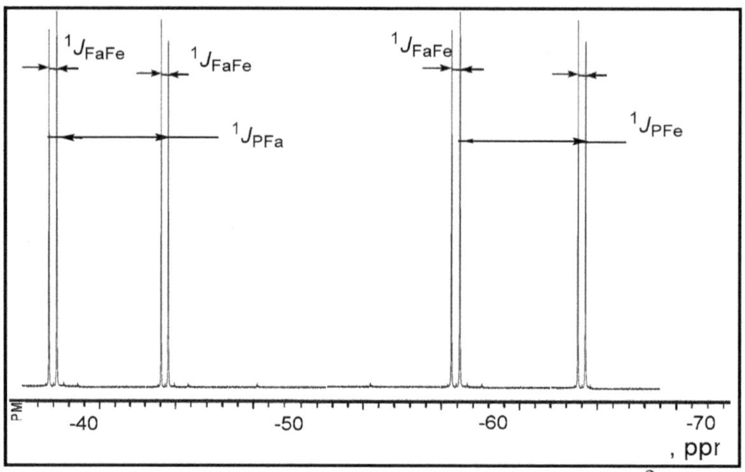

Figure 2. ^{19}F-NMR spectra of 2-(diethylamino)-2,2-difluoro-4-tolyl-3-propyl-4-(trifluorophenyl-1,2λ^5-oxaphosphetane **31** (see Table 2, entry 15).

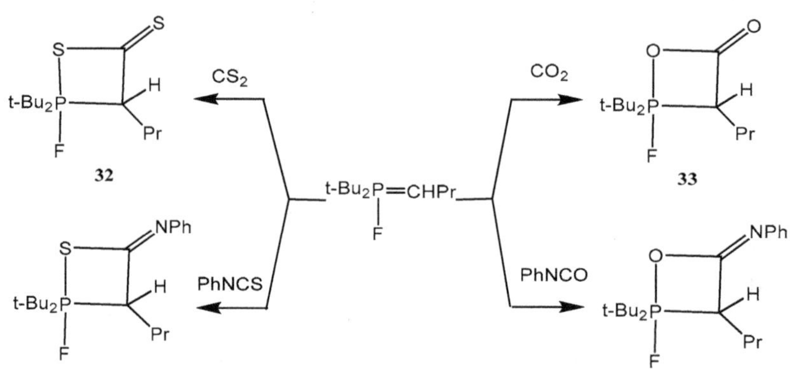

Scheme 20. Addition of heterocumulenes (X=C=Y) to *P*-fluoroylides.

Table 2. 2-Fluoro-1,2λ^5-oxaphosphetanes [26,32,37].

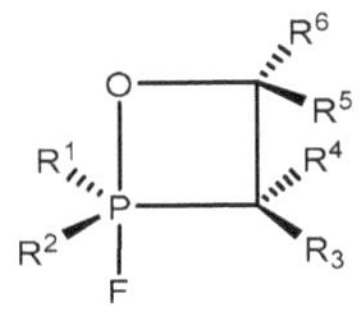

Entry	R^1	R^2	R^3	R^4	R^5	R^6	Yield, %	b.p., °C (P, mmHg)	δ_P, ppm.	δ_F, ppm.	$^1J_{PF}$, Hz
1	Et_2N	Et_2N	H	H	H	Bu	85	90 (0.06)	−42.15	31.30	767
2	Et_2N	Et_2N	H	H	H	C_6H_{13}	85	100 (0.06)	−42.25	31.12	766
3	Et_2N	Et_2N	H	H	Me	Me	70	75 (0.02)	−47.53	31.04	765
4	Et_2N	Et_2N	H	H	Me	Et	85	85 (0.06)	−49.20	31.04	765
5	Et_2N	Et_2N	H	H	Me	Ph	99	[a]	−44.00	32.00	766
6	Et_2N	Et_2N	H	H	$(CH_2)_5$		90	110 (0.06)	−44.60	34.90	766
7	Et_2N	Et_2N	H	Me	$(CH_2)_5$		99	[a]	−44.00	21.00	853
8	Et_2N	Et_2N	H	H	Ph	Ph	99	[a]	−41.60	34.90	766
9	Et_2N	Et_2N	H	H	H	Ph	99	[a]	−42.13	31.18	768
10	Et_2N	Et_2N	Cl	Cl	H	Pr	90	[a]	−47.00	5.40	853
11	Et_2N	Et_2N	Br	Br	H	Pr		[a]	−56.03	5.80	842
12	Et_2N	Et_2N	H	Me	CF_3	Ph	80	120 (0.06)	−36.39; −38.27 (7:1) [b]	−0.4	795; 795
13	Ph	Ph	Ph	H	CH_3	H	95	[a]	−43.50	47.8	670
14	Et_2N	F	H	Pr	CF_3	Ph	85	105 (0.06)	−26.7 (99% dr)	−46.9, −65 (CF_3)	915; 1025
15	Et_2N	F	H	Pr	CF_3	4-Tol	80	108 (0.06)	−26.5 (99% dr)	−46.6, −65 (CF_3)	915; 1025
16	Et_2N	Et_2N	H	Ph	CF_3	Ph	85	<20 [c]	−38.96	27	792.6
17	*t*-Bu	*t*-Bu	H	Pr	CF_3	Ph	70	125 (0.06)	−7.73, −8.79 (9:1) [b]	−1.03	768; 770
18	*t*-Bu	*t*-Bu	H	Pr	H	Ph	75	120 (0.08)	−10.1; −9.62 (15:1) [b]	9.9	762; 762
19	*t*-Bu	*t*-Bu	H	H	H	Ph	70	<20 [c]	−5.9	13	750
20	Et_2N	Et_2N	H	*i*-Pr	H	Ph	70	110 (0.06)	−33.7; −33.54 (15:1) [b]	7.2	827; 827
21	*t*-Bu	*t*-Bu	H	Pr	C=NPh		50	84–86 [c]	−18.85	5.55	820
22	*t*-Bu	*t*-Bu	H	Pr	C=O			[a]	20.01	−26.23	785
23	Ph	Ph	Ph	$SiMe_3$	Ph	H	85	[a]	−43.5	28.76	670
24	Et_2N	Et_2N	H	$SiMe_3$	Ph	H	90	[a]	−39.18; −38.83 (3:2) [b]	−40.19; −8.84	670
25	*t*-Bu	*t*-Bu	H	$SiMe_3$	Ph	H	90	[a]	−11.0	-	768
26	Et_2N	Et_2N	H	$SiMe_3$	Pr	H	90	[a]	−39.27; −42.11 (4:1) [b]	−43.71; −42.61	768
27	Et_2N	Et_2N	H	$SiMe_3$	Bu	H	95	[a]	−39.22; −42.32 (4:1) [b]	−43.71; −42.61	768
28	Et_2N	Et_2N	H	$SiMe_3$	C_6H_{13}	H	90	[a]	−39.20; −42.10 (3:1) [b]	−43.71; −42.61	763
29	Et_2N		H	$SiMe_3$	C_8H_{17}	H	90	[a]	−39.88; 42.20 (4:1) [b]	−42.85; −42.07	763
30	Et-N		H	$SiMe_3$	CF_3	Ph	95	[a]	−38.60; −37.45 (3:1) [b]	−41.11 −49.01	777

[a] Oil; [b] Diastereomers; [c] melting point (m.p.).

2.3. *Properties of Oxaphosphetanes*

The cycloaddition of *P*-halogenylides to aldehydes and ketones proceeds stereoselectively and leads predominantly to the formation of one of the possible diastereomers of 2-halo-1,2λ^5-oxaphosphetanes (Tables 1 and 2). *P*-Haloylides (chloro- and bromo) react stereoselectively with aldehydes to give predominantly single diastereoisomers of 2-halo-oxaphosphetanes, bearing asymmetric endocyclic C-3 and C-4 carbon atoms (dr ~7:1–15:1). It was established by NMR that the ratio of diastereoisomers of 2-halooxaphosphetanes containing asymmetric atoms at C(3) and C(4) was within the limits ~99:1–90:10. The ratio of diastereomers after completion of the reaction was approximately 6:4–9:1. However on heating, as a result of permutational changes in the molecule, the ratio of diastereomers grew in favour of the thermodynamically more stable diastereomer. 2-Fluoro- and 2-chloroxaphosphetanes containing chiral phosphorus or endocyclic carbon atoms exist as mixtures of diastereomers whose ratio depends on the nature of the starting reagents (Scheme 21). The diastereomeric purity of the compounds, assessed by NMR spectroscopy, was 80%–96%. The reaction of α,α,α-trifluoroacetophenone with *P,P*-difluoroylides proceeded with very high stereo-selectivity to furnish only one diastereomer of compound. At the same time 2-halo-1,2λ^5-oxa-phosphetanes bearing asymmetric phosphorus atom are formed with low stereoselectivity (dr~3:1).

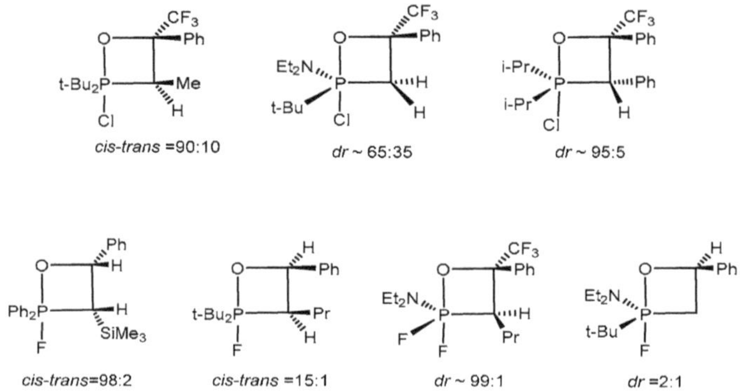

Scheme 21. Diastereomeric composition of 2-chloro- and 2-fluoro-1,2λ^5-oxaphosphetanes.

The reaction of *P*-halogenylides with aldehydes leads to the formation of an *erythro*-oxaphosphetanes with the oxygen in the apical position, because of the tendency of electron-acceptor atoms to occupy this position [40]. The preferred orientation of the transition state leading to the *erythro*-oxaphosphetane is stereoselective approach of the ylide nucleophilic center to the carbonyl group at an angle of 107°, with the double bonds of both reagents arranged in one plane. This was associated with the Burgi-Dunitz trajectory concept (Scheme 22) [41]. The first step of reaction stereoselectively leads to the formation of the betaine **B**. The conversion of reagents into oxaphosphetane requires minimum energy when it proceeds with the non-synchronous formation of bonds between the carbonyl and ylidic carbon atoms. On the second step the betaine converts into oxaphosphetane **C** with low stereoselectivity, because of free rotation around of the P-C bond [28,42]. This mechanism of asynchronous addition of *P*-haloylides to carbonyl compounds was confirmed by low temperature NMR investigations. The reaction of di-*tert*-butylchlorphosphonium methylide with $Ph(CF_3)C{=}O$ was studied by ^{31}P-NMR at low temperature (−70–0 °C) in diethyl ether solution. At −70 °C was found only a decreasing signal of *P*-chloroylide (+114 ppm) and a growing signal of 2-chloro-1,2λ^5-oxaphosphetane (+9.9 ppm). The signal which could be referred to betaine was not found, probably, because of the high speed of betaine cyclisation into 2-chloroxaphosphetane. However the reaction of *tert*-butyl(diethylamino)chlorophosphonium methylide with benzaldehyde began at −70 °C and led to the formation of *erythro*-betaine which was registered by increasing signal δ_P +106 ppm, located in more weaker field than the signal of initial *P*-chloroylide (+102 ppm) (Figure 3). The betaine formation was proceeded stereoselectively as in the ^{31}P-(^{1}H-)NMR spectrum only one diastereomer signal was registered. The rise of temperature to −50 °C led to the formation of two signals of *threo* and *erythro*-oxaphosphetane diastereomers δ_P + 92.6 and +84.6 ppm in the ratio of 2:1. Evidently, the formation of diastereomers proceeded via $S_N2@P$ substitution with inversion of configuration at the phosphorus atom [43]. The formation of isomers indicates on the nucleophilic substitution at asymmetric tetracoordinate phosphorus atom, proceeding with partial inversion of configuration. At −50–−40 °C the oxaphosphetane was converted into 2-chloralkylphosphine oxide, which was also formed as two *threo-* and *erythro*-diastereomers in the ratio of 2:1 (δ_P +48.7 and +49.2 ppm). The conversion was completed at −20 °C. The *threo-* and *erythro* diastereomers were separated and isolated in pure state by chromatography and crystallization (m.p. 100 and 123 °C, see Table 3, entries 13, 14). Although at present many of the puzzling features about the reaction mechanism of ylides phosphorus with carbonyl compounds have been clarified, it seems there is no general mechanism, which could explain the progress of the reaction, transition states and stereochemistry. Nevertheless in case of *P*-chloroylide addition to aldehydes, on the basis of presented above stereochemical researches the asynchronous [2+2]-cycloaddition seems to be the most likely mechanism.

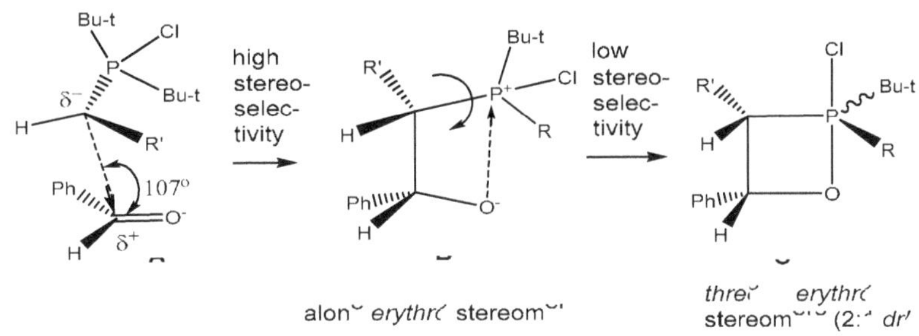

Scheme 22. Mechanism of asynchronous [2+2]-cycloaddition of *P*-chloroylide to benzaldehyde.

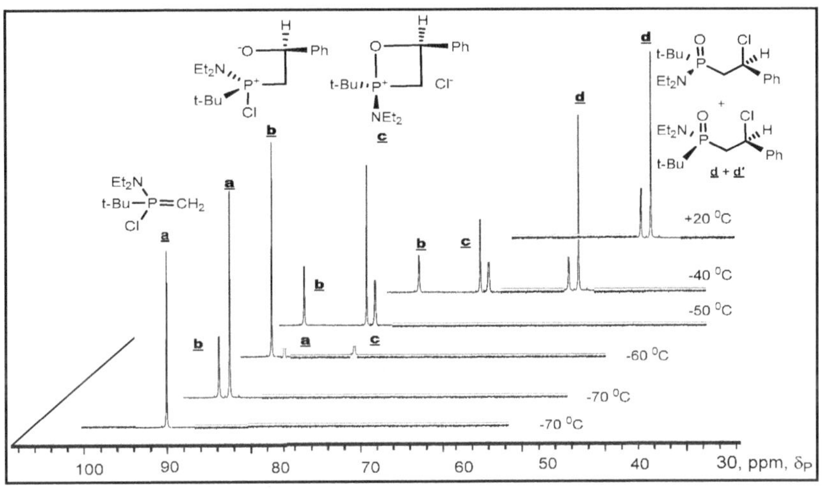

Figure 3. ^{31}P-(^{1}H-)NMR monitoring of the reaction of ylide **a** with PhCHO leading via the formation of betaine **b**, and oxaphosphetane **c** to the formation of β-chloroalkylphosphonates **d** and **d′**.

2-Halo-oxaphosphetanes exist in pentacoordinate form, but ionize under certain conditions with formation of tetracoordinate forms. In the mass spectra of 2-chlorooxaphosphetanes the peak of molecular ion was observed that indicated the covalent character of the P–Cl bond. All oxaphosphetanes show methyl carbons with $^{1}J_{PC}$ values characteristic of equatorial placement in the trigonal bipyramid. The resonances of the methylene carbons of the phosphetane ring system lay downfield of the methyl carbons and the large measured $^{1}J_{PC}$ values indicate equatorial placement. The four-membered ring is occupying an apical-equatorial plane where the O-P-C angle is 90°, and never a diequatorial plane or a diapical plane where the O-P-C angle would have to be 120 or 180°, respectively.

In the ^{1}H-NMR spectra of the 2-halooxaphosphetanes possessing asymmetric carbon atom C^4 the magnetic nonequivalence of protons CH^aH^b in the four-membered ring became apparent because the geminal spin-spin interaction arose between them. Each of the signals CH^a and CH^b was the double doublet with the coupling constant with phosphorus nucleus $^{2}J_{PH}$ = 20–22 Hz and a constant of geminal interaction $^{2}J_{HH}$ = 16–17 Hz. The chemical shifts ^{31}P of 2-halogenoxaphosphetanes depended on the polarity of the solvent. Thus, in nonpolar solvents (diethyl ether, pentane) the signals of δ_P are located in the strong field of the ^{31}P-NMR spectrum (from −3 to 10 ppm) that corresponds to the pentacoordinated state of the phosphorus atom. In polar solvents and especially in the presence of Lewis acids, like $AlCl_3$, the values of δ_P were shifted downfield. For example, for 2-chloro-oxaphosphetanes the chemical shift of phosphorus, δ_P was as follows (ppm): 9 (pentane), 22 (CH_2C1_2), 30 ($CHC1_3$), 35 (CH_3CN), 45 ($CHC1_3$ + $A1C1_3$ (traces)). With the increase in amount of aluminum chloride up to equimolar level the value of δ_P 100 ppm was registered for 2-chlorooxaphosphetanes that was in accordance with the values of chemical shifts for the known alkoxyphosphonium salts and apparently indicated complete ionization of chlorophosphorane with the formation of phosphonium salt. The dependence of the chemical shift value of phosphorus in 2-chlorooxaphosphetanes on the solvent polarity and on the presence of Lewis acid is in accordance with the published data showing that chlorophosphoranes can be ionized with the formation of phosphonium structures. Herewith in the ^{31}P-NMR spectrum the resultant signal is registered due to fast exchange in the phosphorus coordination **15A**$\rightleftharpoons$**15B**. The value of δ_P is shifted downfield proportionally to the increase in the phosphonium structure content, which in its turn depends on the solvent polarity. Substitution of electron withdrawing CF_3 group at C4 atom by hydrogen atom destabilizes the oxaphosphetane cycle and considerably reinforces the ionization of P-C1 bond. The stability of 2-halogenoxaphosphetanes decreases, and positive values δ_P and ionization of P-C1 bond increases in the sequence of substituents at C4 atom: CF_3 > C_6H_4F-4 >-C_6N_5 > C_6H_4OMe > H (Table 4) [25,30,31]. Chemical shifts of phosphorus in 2-bromo-1,2λ^5-oxaphosphetanes (δ_P = +20–+75 ppm) were in the weaker fields relatively to chemical shifts of chloro-oxaphosphetanes. Evidently high positive values δ_P of 2-bromoxaphosphetanes is undoubtedly explained by bigger, than in case of 2-chlorooxaphosphetanes, contribution of phosphonium forms **15B** in the equilibrium 15**A**$\rightleftharpoons$15**B**.

Table 3. 2-Chloroalkylphosphine oxides **35** (Scheme 23) [36,42].

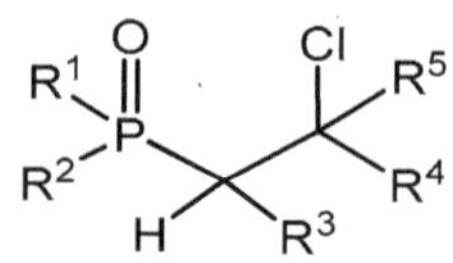

N	R^1	R^2	$R^{3\prime}$	R^4	R^5	Yield, %	m.p./b.p. °C	δ_P
1	*t*-Bu	*t*-Bu	H	H	CF_3	60	62 (pentane)	57.8
2	*t*-Bu	*t*-Bu	H	Me	CF_3	60	86 (hexane)	57.0
3	*t*-Bu	*t*-Bu	H	Pr	CF_3	60	88 (hexane)	57.0
4	*t*-Bu	*t*-Bu	H	Ph	CF_3	75	112.5 (benzene)	57.2
5	*t*-Bu	*t*-Bu	H	$C_6H_4NO_2$	CF_3	70	139 (hexane)	59
6	*t*-Bu	*t*-Bu	Me	$C_6H_4NO_2$	CF_3	70	131 (hexane)	60.1
7	*t*-Bu	*t*-Bu	Pr	$C_6H_4NO_2$	CF_3	60	143.5 (hexane)	60.0
8	Me_2N	Me_2N	H	Pr	CF_3	60	b.p. 95 (0.06 mmHg)	45.8
9	Et_2N	Et_2N	H	Ph	CF_3	40	123 (ether-pentane)	46.0
10	Et_2N	Et_2N	H	Ph	CF_3	20	160 (hexane)	46.3
11	*i*-Pr_2N	*i*-Pr_2N	Me	Ph	H	50	107–111	34.4
12	*t*-Bu	Ph	H	Ph	H	75	oil	57.9; 58.3 (10:1)
13	*t*-Bu	Et_2N	H	H	Ph	20	100 (hexane)	46.3
14	*t*-Bu	Et_2N	H	Ph	H	40	123 (ether-pentane)	46.0

Table 4. Effect of solvent on δ_P of 2-chloro and 2-bromo-1,2λ^5-oxaphosphetanes.

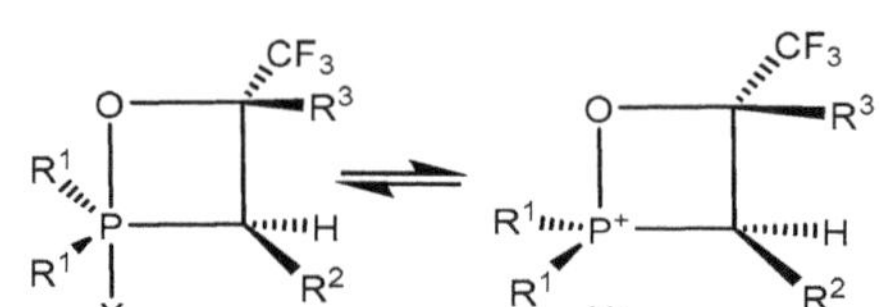

15A **15B**

R^3	CF_3	C_6H_4F	C_6H_5	C_6H_4OMe
δ_P ($CHCl_3$), Hlg = C1	+0.25	+9.0	+10.0	+13.6
δ_P (ether), Hlg = Br	+25.0	+60.0	+65.0	+75.0

2-Chloro-1,2λ^5-oxaphosphetanes enter into a number of interesting chemical conversions. Thus, 2-chloro-1,2λ^5-oxaphosphetanes as a result of [1,3]-migration of chlorine atom to carbon atom underwent 2-chlorooxaphosphetane-2-chlorooxyphosphine oxide rearrangement to convert into the 2-chlorooxyphosphine oxides **35**. Thermal stability of 2-chloro-1,2λ^5-oxaphosphetanes was decreased in case of oxaphosphetanes which not contain at C4 atom strong acceptor substituents, which rearranged into chloroalkylphosphonates at room temperature (Table 3). At heating 2-chloralkylphosphine oxides yield vinylphosphine oxides **38**. Hydrolysis of 2-chloro- and bromo-oxaphosphetanes led to the formation of 2-hydroxyalkylphosphine oxides **36** (Table 5). The chlorine atom of 2-chloro-1,2λ^5-oxaphosphetanes is easily substituted on methoxy- or phenoxy groups by reaction with methanol or phenol in the presence of triethylamine with formation of 2-alkoxyoxaphosphetanes **37**, which at heating were converted into alkenes (Scheme 23, Table 1, entries 14, 15) [27,42].

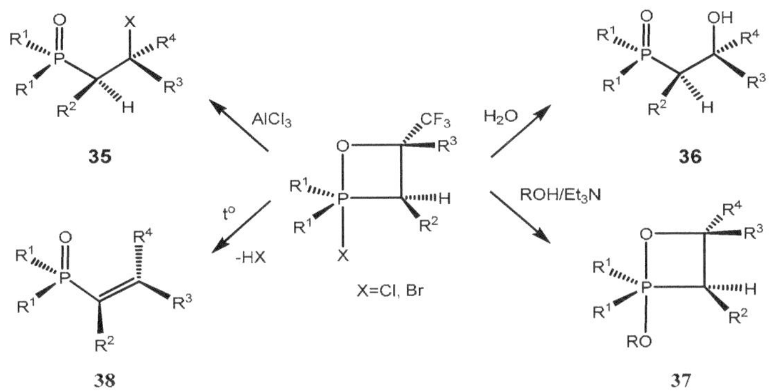

Scheme 23. Chemical reactions of 2-chloro and 2-bromo-1,2λ^5-oxaphosphetanes.

The 2-chloro-1,2λ^5-oxaphosphetanes not containing electron accepting groups at C-3 are unstable and rearrange easily into 2-cloroalkylphosphonates at temperature below +20 °C. The oxaphophetanes bearing at C4 electronegative CF_3 groups convert into vinylphosphonates with elimination of hydrogen chloride at heading up to 150–160 °C. 2-Bromooxaphosphetanes are less stable than 2-chlorooxaphosphetanes and are converted into vinylphosphine oxides at room temperature. At heating (160–190 °C) 2-chloralkylphosphine oxides eliminate HCl to convert into vinylphosphine oxides (Scheme 24) [27–29,42].

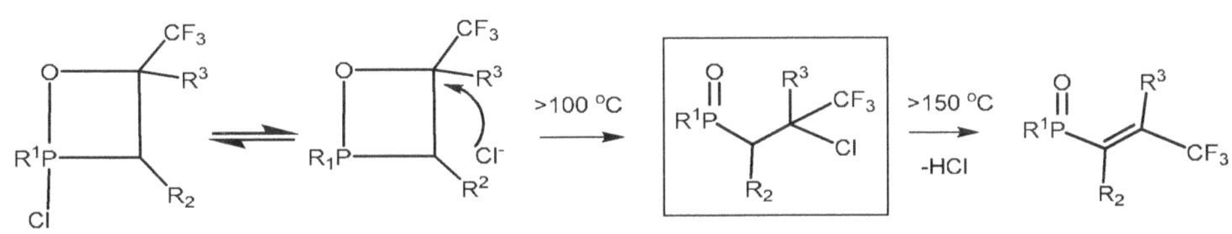

Scheme 24. 2-Chloro-1,2λ^5-oxaphosphetane-2-chloroalkylphosphonate rearrangement.

Table 5. 2-Hydroxyalkylphosphine oxides **23** (Scheme 23) [39,42].

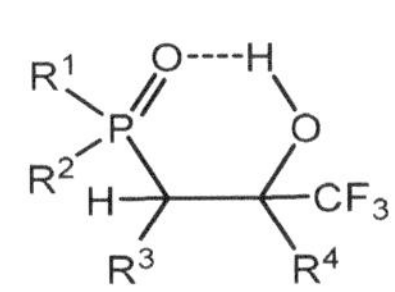

Entry	R^1	R^2	R^3	R^4	Yield, %	m.p., °C	δ_P	δ_F
1	*i*-Pr	*i*-Pr	Ph	Ph	90	131–132	+66.0	+5.63
2	*t*-Bu	*t*-Bu	H	Ph	90	155–156	+68.25	−1.78
3	*t*-Bu	*t*-Bu	H	C_6H_4F-*p*	85	132	+68.7	−2.02; −34.5
4	*t*-Bu	*t*-Bu	H	An-*p*	70	140–141	+68.5	+4.12
5	*t*-Bu	*t*-Bu	Me	Ph	50	152	+73.4	+3.13
6	*t*-Bu	Et_2N	Me	CF_3	50	108	+53.4	+8.1
7	*t*-Bu	Et_2N	H	Ph	65	150	+57.4	−3.06
8	*t*-Bu	Et_2N	H	CF_3	90	46–48	+56.6	+6.1
9	*t*-Bu	*t*-Bu	H	CF_3	90	88-89	+67.5	+6.25

The oxaphosphetanes bearing a trimethylsilyl group at C-3 atom eliminate the trimethylchlorosilane moiety and convert into the phosphorylated alkenes **38**. The conversion of oxaphosphetanes into alkenephosphonates proceeds at room temperature slowly, and at heating faster to give the alkenes **38** in good yields (Scheme 25). The phosphorylated alkenes **38** were purified by distillation under vacuum and isolated as pure compounds. The reaction of ylides **14** with aldehydes is regioselective. For example, the oxaphosphetanes obtained by reaction of silylated *P*-chloroylide **5** with terephthalic aldehyde depending on a ratio of initial reactants led to the formation of 1,4-bis-vinylphosphonobenzene or phosphonovinylbenzaldehyde, that represent interest as reactants for organic synthesis (See Table 6, entries 10, 11) [13,18–20].

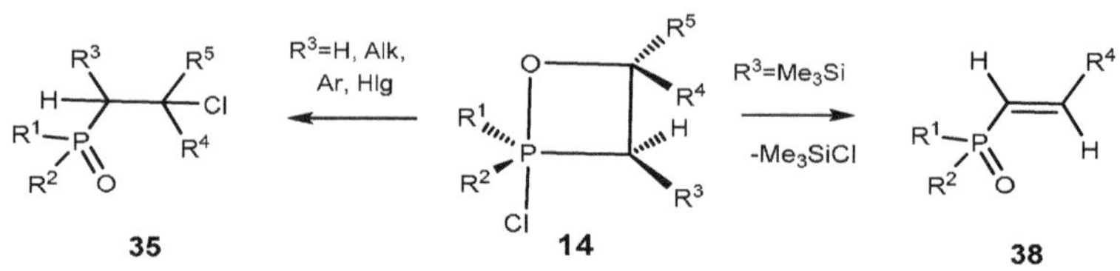

Scheme 25. Conversions of 2-chloro-1,2λ⁵-oxaphosphetanes **14**.

The 2-fluoro-1,2λ⁵-oxaphosphetanes enter readily into a number of interesting organophosphorus compounds proceeding without P—C bond cleavage [26]. Thus, the treatment of 2-fluorooxaphosphetanes **25** with ether solution of HCl led to the formation of 2-chloro-oxaphosphetanes **40**, which was isolated in good yield. At heating the 2-fluorooxaphosphetanes **25**, in contrast to triphenyloxaphosphetanes, convert into phosphorylated alkenes: vinylphosphonates or allylphosphonates. The direction of reaction depended on substituents R^2 and R^3 at C-3 and C-4 of oxaphosphetane cycle. The 2-fluorooxaphosphetanes bearing at C-3 R^2 = H, Alk, Ar, and at C-4 R^3 = Alkyl eliminated HF to convert into the allylphosphonates **42** [39,44,45]. The reaction was catalyzed by boron trifluoride etherate. At the same time the 2-fluorooxaphosphetanes **25** bearing at C-3 substituent R = Me_3Si eliminated Me_3SiF and afforded the vinylphosphonates **41** (Scheme 26, Table 7). This reaction represent a convenient method for the preparation of phosphorylated alkenes that are versatile building blocks for organic synthesis [45–48]. 2-Fluoro-1,2λ⁵-oxaphosphetanes containing two alkyl groups at C3 as well eliminated hydrogen fluoride and afforded the allylphosphonates **42** (Scheme 26, Table 8).

Table 6. Vinylphophonates **38** (Scheme 25).

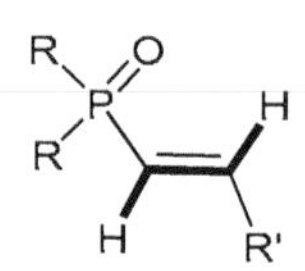

Entry	R	R′	Reaction Conditions			Yields, %	References
			Time, h	t °C	Solvent		
1	*t*-Bu	Ph	0.5	150	a	80	[25,32]
2	EtO	Ph	2	25–100	ether	80	[25,32]
3	EtO	2-BrC_6H_4	6–8	25–100	ether	60	[25,32]
4	*i*-PrO	Ph	2	25–100	ether	60	[25,32]
5	Et_2N	Ph	14	25	ether	80	[25,32]
6	Et_2N	2-FC_6H_4	24	25	ether	60	[25,32]
8	Et_2N	CH=CHPh	18	25	ether	80	[25,32]
9	Et_2N	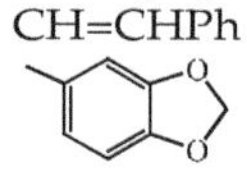	24	25	THF	70	[25,32]
9	EtO	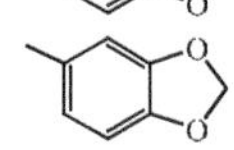	48	35	THF	65	[25,32]
10	Et_2N	C_6H_4CHO-*p*	25	18	THF	50	[25,32]
11	Et_2N	$R_2P(O)CH{=}CHC_6H_4$	25	24	THF	50	[25,32]
12	*i*-PrO	$R_2P(O)CH{=}CHC_6H_4$	25	24	THF	50	[25,32]

a without solvent.

The reaction occurs by the 1,4-elimination as shown in Scheme 26 [25]. Opposite to 2,2,2-triphenyloxaphosphetanes, upon heating, the decyclization of 2-fluorooxaphosphetanes led to the formation of phosphorylated alkenes **41, 42**.

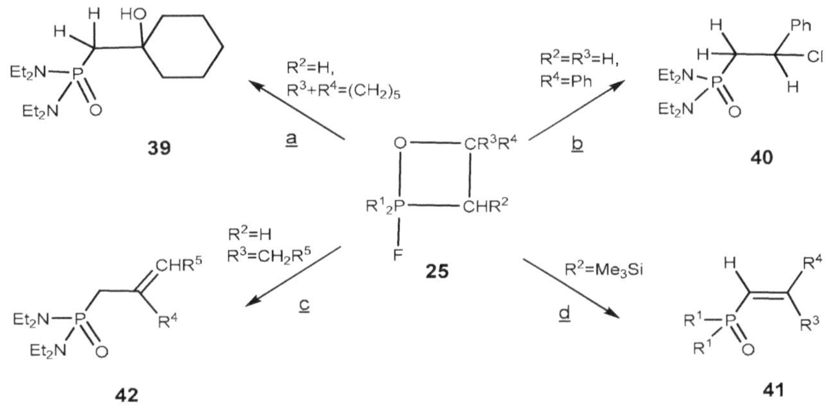

Scheme 26. Chemical properties of 2-fluoro-1,2λ^5-oxaphosphetanes. a = H_2O (-HF); b = HCl, ether (-HF); c = 100–120 °C, ~0.5 h (-HF) or $BF_3 \cdot Et_2O$, +20 °C, 168 h; d = 60–80 °C (-Me_3SiF).

At heating to +100 °C the 2-fluoro-3-silyloxaphosphetanes **25**, bearing CF_3 group at C-4, afforded a mixture of *E*- and *Z*-vinylphosphonates **29, 30** in the ratio 2:1 with elimination of Me_3SiF. However the slow conversion of oxaphosphetane at +20 °C during several days provided almost pure vinylphosphonates *E*-**29**, containing only 2%–3% of *Z*-isomer (Scheme 27) [26]. This effect was explained by formation of carbocation intermediate and rotation of substituents around the C-C bond (Scheme 28).

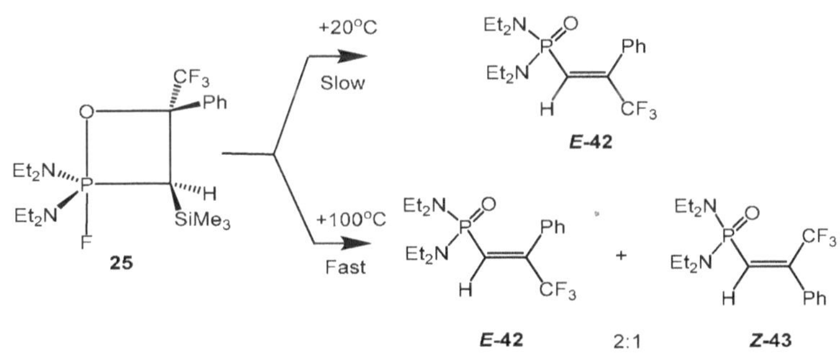

Scheme 27. Effect of temperature on the stereoselectivity of 2-fluoro-1,2λ^5-oxaphosphetane conversion into vinylphosphonates.

The conversion of 2-fluoro-1,2λ^5-oxaphosphetanes bearing alkyl groups at C-3 into allylphosphonates **42** represents an interesting example of 1,4-elimination as Scheme 28 and Table 8 show [45]. The study of the reaction mechanism showed that Lewis and Broensted acids actively catalyze the conversion of 2-fluorooxaphosphetanes into allylphosphine oxides. The reaction is autocatalytic because the evolving hydrogen fluoride catalyzes the transition of 2-fluorooxaphosphetanes into allylphosphine oxides. The decomposition of protonated 2-fluorooxaphosphetanes leads to the formation of oxonium salts and carbocation intermediates under conditions of E_N1 elimination. We suppose that the 2-fluorooxaphosphetanes under condition of acid catalysis (with HF or BF_3) via the formation of an oxonium intermediate F, convert to carbocation intermediate **G** which has a planar configuration [46]. The removal of a proton from the carbocation intermediate **G** depends on electronic effects of the substituents. Alkyl groups possessing the +*I*-effect and the effect of hyperconjugation stabilize the positive charge and reduce the energy of intermediate **G** formation. Therefore, the intermediate **G′** leading to allylphosphonates is energetically more favorable than this one that is converted into vinylphosphonates.

Table 7. Vinylphosphonates **41** (Scheme 26) [26].

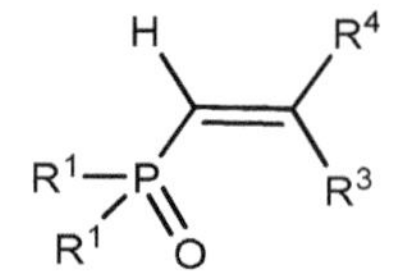

Entry	R^1	R^3	R^4	Yield, % [a]	b.p. °C (*p* mmHg)/ m.p. °C (Solvent)	δ_P, ppm
1	*t*-Bu	H	Ph	50	138 (heptane)	35.0
2	Ph	H	Ph	40	165 (heptane)	22.0
3	Et_2N	H	Ph	85	103.5 (hexane)	24.7
4	Et_2N	H	Me	80	120 (0.05)	24.7
5	Et_2N	H	Pr	72	120 (0.05)	24.3
6	Et_2N	H	Bu	68	120–123 (0.04)	23.95
7	Et_2N	H	C_6H_{13}	80	135 (0.06)	24.1
8	Et_2N	H	C_8H_{17}	79	145–150 (0.05)	24.1
9	Et_2N	H	PhCH=CH-	60	170 (0.06)	23.30
10	Et_2N	H	$Me_2C{=}CHCH_2CH_2C(Me)CH{=}C$-	35	160 (0.08)	25.0
11	Et_2N	CF_3	Ph	74	140 (0.03)	17.0; 18.7 [b]
12	Et_2N	H	$-C_6H_4CHO$-4	50	98 (hexane)	23.7
13	Et_2N	H	$-C_6H_4CH{=}CHP(O)(NEt_2)_2$	50	190 (heptane)	24.8

[a] yield of the isolated product; [b] mixture of *E*/*Z*-diastereomers in 97:3 ratio.

In addition, even in case when the initial 2-fluorooxaphosphetanes **25** exist as a mixture of *threo-* and *erythro*-diastereomers, they converted into pure *E*-vinylphosphonates. Probably, this effect can be explained by rotation of substituents around the C-C bond in carbocation intermediate G. However, the presence of Me_3Si at C-3 in **G″** leads to the elimination of Me_3SiF, which has a high energy of formation, that creates a preference for the formation of vinylphosphonates **45** (Scheme 28) [26,40].

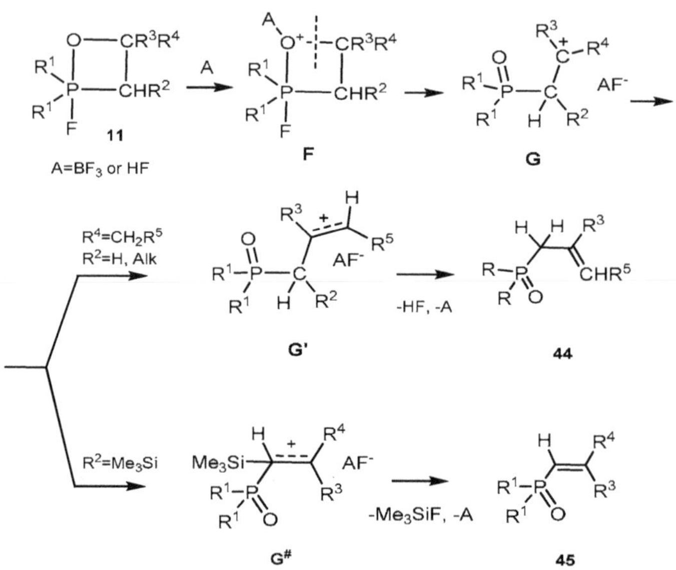

Scheme 28. The mechanism for the 2-fluoro-1,2λ^5-oxaphosphetane conversion into allyl- or vinylphosphonates **44, 45**.

The formation of carbocation intermediate **G** was experimentally confirmed (Scheme 29). The treatment of 2-hydroxyphosphonate **46** with trifluoracetic acid and refluxing for several hours, generated the carbocation intermediate, as a result of acid-catalyzed dehydration of alcohol.

After that the carbocation intermediate is converted into allylphosphonate **47**, which is identical to the one obtained from 2-fluorooxaphosphetane **25** [26,40,45].

Table 8. Allylphosphonates **28**.

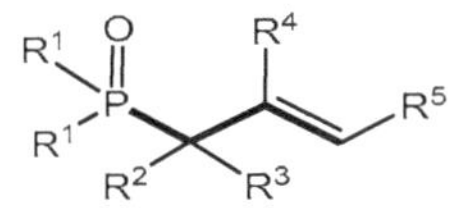

Entry	R^1	R^2	R^3	R^4	R^5	Yield, %	b.p. °C (*P*, mmHg)	References
1	Et_2N	H	H	H	Pr	70	120 (0.06)	[40,45]
2	Et_2N	H	H	H	C_5H_{11}	70	140 (0.08)	[40,45]
3	Et_2N	H	H	Me	H	55	105 (0.08)	[40,45]
4	Et_2N	H	H	Me	Me	60	95 (0.06)	[40,43]
5	Et_2N	H	H	Ph	H	65	145 (0.08)	[40]
6	Et_2N	H	H	$(CH_2)_4$		85	145 (0.08)	[26,45]
7	Et_2N	H	Me	$(CH_2)_4$		75	110 (0.06)	[40,45]
8	Et_2N	Me	Me	$(CH_2)_4$		70	110 (0.06)	[40,45]

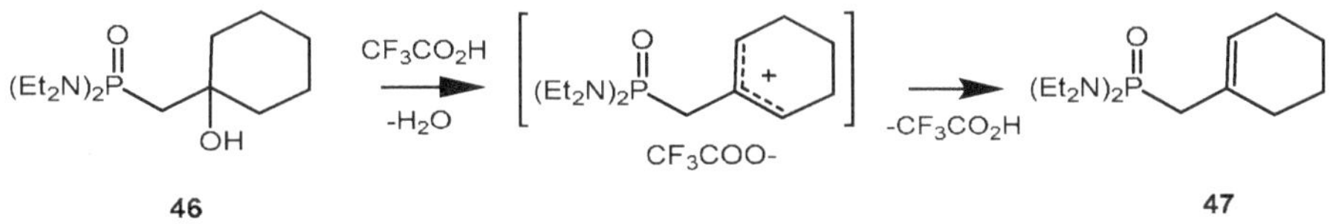

Scheme 29. The conversion of hydroxyphosphonate **46** into the allylphosphonate **47**.

The reaction of *P*-fluoroylides with carbonyl compounds is a convenient method for the synthesis of allylphosphonates having various applications in the synthesis of naturally occurring compounds (Scheme 30) [45–47]. In this case, the carbonyl compounds can be used twice for the constructing of diene structures: first in reaction with *P*-fluoroylide and then in the Wittig reaction with allylphosphonate. This reaction was used for preparing analogs of the juvenile hormone.

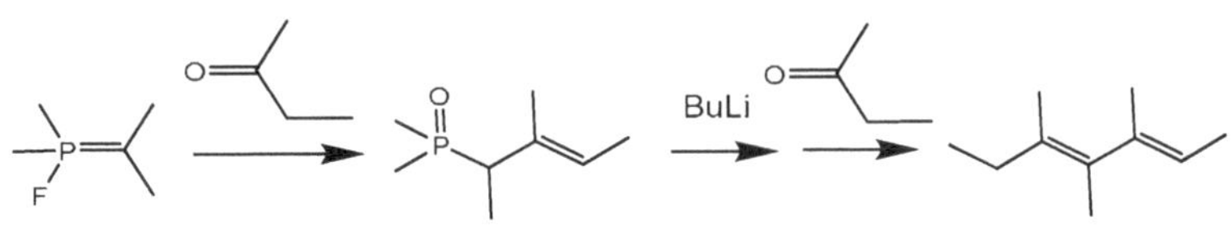

Scheme 30. Synthetic application of allylphosphonates.

The cycloadducts of *P*-fluoroylides with carbon dioxide or with carbon disulfide were isolated as colorless liquids or crystalline substances. Their structure was confirmed by the NMR spectra. Upon gentle heating or at room temperature these cycloadducts **32–35** (Table 2, entries 21, 22) are converted to phosphorylated ketenes or thioketenes **48** (Scheme 31). Stable [2+2]-cycloadducts of 2-fluoroylides with alkyl and aryl isothiocyanates were also synthesized and converted into phosphorylated ketenimines. The reaction of *P*-fluoroylides with carbon disulfide gives unstable cycloadducts that at temperatures above 0 °C converted completely into phosphorylated thioketenes in high yields. Phosphorylated thioketenes are red liquids distillable in vacuum and susceptible for various transformations [29,36,37]. The same phosphorylated ketenes or thioketenes **29** were prepared by reaction of *P*-cloroylides corresponding with CO_2 and CS_2 (See Scheme 31 and Table 9) [37–39].

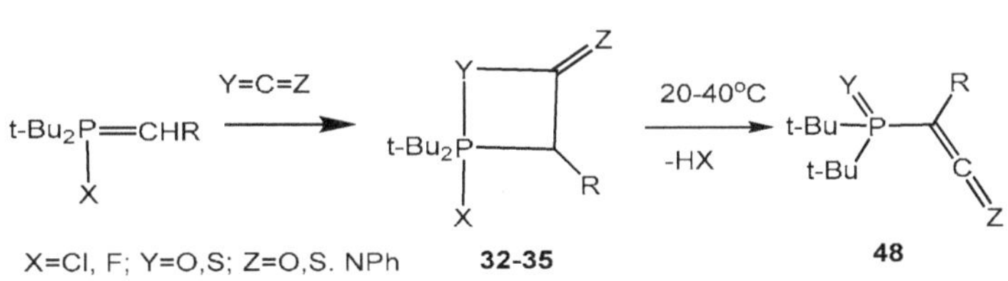

Scheme 31. Preparation of phosphorylated ketenes, thioketenes, ketenimines **48** (Table 9).

Table 9. Preparation of phosphorylated ketenes, thioketenes, ketenimines **48** (Scheme 31).

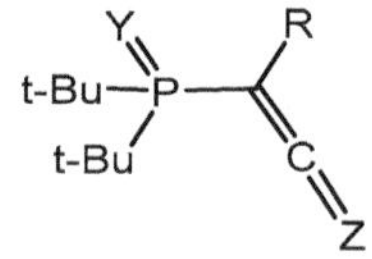

Entry	R	X	Y	Z	Yield, %	b.p. °C (mmHg)	References
1	H	Cl	O	O	80	a	[36,39]
2	Me	F	O	O	70	98 (0.06)	[37]
3	Me	Cl	O	O	85	98 (0.06)	[36,39]
4	Pr	F	O	O	60	102 (0.06)	[37]
5	Pr	Cl	O	O	69	102 (0.06)	[36,39]
6	Me	F	O	NPh	45	150 (0.06)	[37]
7	Pr	F	O	NPh	45	150 (0.06)	[37]
8	Me	Cl	S	S	75	95 (0.06)	[38]
9	Pr	Cl	S	S	80	115 (0.06)	[38]
10	Pr	F	S	S	70	115–120 (0.06)	[37,38]
11	*i*-Pr	Cl	S	S	80	105 (0.06)	[38]

a unstable compound.

3. Conclusions

In conclusion, this review has summarized the achievements in the synthesis and properties of stable four-membered phosphorus heterocycles-2-chloro-, 2-bromo- and 2-fluoro-1,2λ^5-oxaphosphetanes. These interesting compounds were obtained by reaction of *P*-halogenylides with various carbonyl compounds (aldehydes, ketones, isocyanates, carbon dioxide, and others). The 2-chloro and 2-bromo-1,2λ^5-oxaphosphetanes, depending on the halogen nature and substituents at the α-carbon atom, underwent the rearrangement into 2-haloalkylphosphonates or with elimination of hydrogen halides were converted into *trans*-phosphorylated alkenes. Hydrolysis of 2-halo-1,2λ^5-oxaphosphetanes led to the formation of 2-hydroxyalkylphosphonates. The 2-fluoro-oxaphosphetanes, bearing alkyl groups at the C-4 atom at heating as a result of 1,4-elimination of hydrogen fluoride are turned to allylphosphonates. Upon heating 3-silyl-2-fluoro-1,2λ^5-oxaphosphetanes easily eliminate the trimethylsilylfluoride to convert into *E*-vinylphosphonates in high yields. These reactions provide efficient protocols for the preparation of various phosphorylated alkenes (vinylphosphonates, allylphosphonates, phosphorus ketenes, ketenimines, thioketenes) and can be useful for fine organic synthesis.

Acknowledgments: This work was supported by grants of the State Foundation for Basic Research of Ukraine (Projects No 03.07/00047, No. F40.3/034 and No. F53.3/016), grant of the Science and Technology Center in Ukraine, STCU (Project No. 3558).), the research projects of the National Academy of Sciences of Ukraine (No. 0109U001795 and No. 0106U004307).

Author Contributions: O.I.K and A.O.K. conceived and designed the review. O.I.K. wrote the text of review. A.O.K. have prepared reaction schemes and did literature research on relevant articles for this review. Both authors discussed and approved the final version.

References

1. López-Ortiz, F.; López, J.G.; Álvarez Manzaneda, R.; Pérez Álvarez, I.J. Isolable 1,2-Oxaphosphetanes: From curiosities to starting materials for the synthesis of olefins. *Mini-Rev. Org. Chem.* **2004**, *1*, 65–76. [CrossRef]
2. Kawashima, T.; Kato, K.; Okazaki, R. Novel Synthetic Route to Isolable Pentacoordinate 1,2-Oxaphosphetanes and Mechanism of Their Thermolysis, the Second Step of the Wittig Reaction. *J. Am. Chem. Soc.* **1992**, *114*, 4008–2010. [CrossRef]
3. Vedejs, E. Studies in Heteroelement-Based Synthesis. *J. Org. Chem.* **2004**, *69*, 5159–5167. [CrossRef] [PubMed]

4. Vedejs, E.; Marth, C.F. *Phosphorus-31 NMR Spectral Properties in Compound Characterization and Structural Analysis*; Quin, L.D., Verkade, J.G., Eds.; VCH: New York, NY, USA, 1994; Chapter 23; pp. 297–313.
5. Hamaguchi, M.; Iyama, Y.; Mochizuki, E.; Oshima, T. First isolation and characterization of 1,2-oxaphosphetanes with three phenyl groups at the phosphorus atom in typical Wittig reaction using cyclopropylidenetriphenylphosphorane. *Tetrahedron Lett.* **2005**, *46*, 8949–8952. [CrossRef]
6. Robiette, R.; Richardson, J.; Aggarwal, V.K.; Harvey, J.N. Reactivity and selectivity in the Wittig reaction: A computational study. *J. Am. Chem. Soc.* **2006**, *128*, 2394–2409. [CrossRef] [PubMed]
7. Kobayashi, J.; Kawashima, T. Chemistry of pentacoordinated anti-apicophilic phosphorus compounds. *C. R. Chim.* **2010**, *13*, 1249–1259. [CrossRef]
8. Garcia-López, J.; Morán-Ramallal, A.; Gonzalez, J.; Roces, L.; García-Granda, S.; Iglesias, M.J.; Ona-Burgos, P.; López-Ortiz, F. Mechanisms of stereomutation and thermolysis of spiro-1,2-oxaphosphetanes: New insights into the second step of the Wittig reaction. *J. Am. Chem. Soc.* **2012**, *134*, 19504–19507. [CrossRef] [PubMed]
9. Byrne, P.A.; Muldoon, J.; Ortin, Y.; Müller-Bunz, H.; Gilheany, D.G. Investigations on the Operation of Stereochemical Drift in the Wittig Reaction by NMR and Variable-Temperature NMR Spectroscopy of Oxaphosphetane Intermediates and Their Quench Products. *Eur. J. Org. Chem.* **2014**, *2014*, 86–98. [CrossRef]
10. Byrne, P.A.; Gilheany, D.G. The Mechanism of phosphonium ylide alcoholysis and hydrolysis: Concerted addition of the OH bond across the P=C bond. *Chem. Eur. J.* **2016**, *22*, 9140–9154. [CrossRef] [PubMed]
11. Byrne, P.A.; Gilheany, D.G. Unequivocal experimental evidence for a unified lithium salt-free Wittig reaction mechanism for all phosphonium ylide types: Reactions with β-heteroatom-substituted aldehydes are consistently selective for *cis*-oxaphosphetane-derived products. *J. Am. Chem. Soc.* **2012**, *134*, 9225–9239. [CrossRef] [PubMed]
12. Byrne, P.A.; Gilheany, D.G. The modern interpretation of the Wittig reaction mechanism. *Chem. Soc. Rev.* **2013**, *42*, 6670–6696. [CrossRef] [PubMed]
13. Restrepo-Cossio, A.A.; Cano, H.; Mari, F.; Gonzalez, C.A. Theoretical study of the mechanism of the Wittig reaction: Ab initio and MNDO-PM3 treatment of the reaction of unstabilized, semistabilized and stabilized ylides with acetaldehyde. *Heteroatom Chem.* **1997**, *8*, 557–569. [CrossRef]
14. Birum, G.H.; Matthews, C.N. Reactions of Triphenyl-2,2-bis(trifluoromethyl)vinylide-nephosphorane, synthesized from a cyclic ylide-ketone adduct. *J. Org. Chem.* **1967**, *32*, 3554–3559. [CrossRef]
15. Borkenhagen, F.; Neda, I.; Thönnessen, H.; Jones, P.G.; Schmutzler, R. An unusual rearrangement during the oxidative addition of hexafluoroacetone and trifluoroacetophenone to 2-bornanylen (dimethylphosphino)methyl imine-formation of a P=C double-bond. *Z. Anorg. Allg. Chem.* **1998**, *624*, 650–654. [CrossRef]
16. Dieckbreder, U.; Röschenthaler, G.-V.; Kolomeitsev, A.A. P-Bis(trifluoromethyl) ylides: Synthesis and reactions. *Heteroatom Chem.* **2002**, *13*, 650–653. [CrossRef]
17. Kojima, S.; Supino, M.; Matsukawa, S.; Nakamoto, M.; Akiba, K.-Y. First isolation and scheme 4 characterization of an anti-apicophilic spirophosphorane bearing an oxaphosphetane ring: A model for the possible reactive intermediate in the Wittig reaction. *J. Am. Chem. Soc.* **2002**, *124*, 7674–7675. [CrossRef] [PubMed]
18. Kyri, A.W.; Nesterov, V.; Schnakenburg, G.; Streubel, R. Synthesis and Reaction of the First 1,2-Oxaphosphetane Complexes. *Angew. Chem. Int. Ed.* **2014**, *53*, 10809–10812. [CrossRef] [PubMed]
19. Couzijn, E.P.A.; Slootweg, J.C.; Ehlers, A.W.; Lammertsma, K. Stereomutation of pentavalent compounds: Validating the berry pseudorotation. *J. Am. Chem. Soc.* **2010**, *132*, 18127–18140. [CrossRef] [PubMed]
20. Kollar, L.; Berente, Z.; Forintos, H.; Keglevich, G. Detection of the enantiomers of P-stereogenic pentacoordinated phosphorus compounds: ^{31}P-NMR of oxaphosphetes in optically active solvents. *Tetrahedron Asymmetry* **2000**, *11*, 4433–4444. [CrossRef]
21. Appel, M.; Blaurock, S.; Berger, S. A Wittig reaction with 2-furyl substituents at the phosphorus atom: Improved (Z) selectivity and isolation of a stable oxaphosphetane intermediate. *Eur. J. Org. Chem.* **2002**, *2002*, 1143–1148. [CrossRef]
22. Kumara Swamy, K.C.; Satish Kumar, N. New Features in Pentacoordinate Phosphorus Chemistry. *Acc. Chem. Res.* **2006**, *39*, 324–333. [CrossRef] [PubMed]
23. Ugi, I.; Marquarding, D.; Klusacek, H.; Gillespie, P.; Ramirez, F. Berry pseudorotation and turnstile rotation. *Acc. Chem. Res.*, **1971**, *4*, 288–296. [CrossRef]

24. Kolodiazhnyi, O.I.; Schmutzler, R. Synthesis and properties of phosphorus ylides containing fluorine atoms bonded to phosphorus. *Synlett* **2001**, *2001*, 1065–1078. [CrossRef]
25. Kolodiazhna, O.O.; Kolodiazhnyi, O.I. Reaction of *C*-silyl-*P*-chloroalkylidene-phosphoranes with carbonyl compounds. *Phosphorus Sulfur Silicon Relat. Elem.* **2015**, *190*, 322–328.
26. Kolodiazhna, A.O.; Kolodiazhnyi, O.I. Synthesis and properties of four-membered phosphorus heterocycles-2-fluoro-1,2λ^5-oxaphosphetanes. *Phosphorus Sulfur Silicon Relat. Elem.* **2015**, *190*, 2232–2245. [CrossRef]
27. Kolodyazhna, O.O.; Grishkun, E.V.; Sheiko, S.Yu.; Kolodyazhna, A.O.; Kolodyazhnyi, O.I. Synthesis and properties of tert-butylphenylmethylene(chloro)phosphorane. *Russ. J. Gen. Chem.* **2015**, *85*, 1639–1643. [CrossRef]
28. Kolodiazhnyi, O.I. *Phosphorus Ylides. Chemistry and Application in Organic Synthesis*; Wiley-VCH: Weinheim, Germany, 1999; p. 565.
29. Kolodyazhnyi, O.I. Chemistry of P-F ylides. *Russ. J. Gen. Chem.* **2005**, *75*, 1017–1039. [CrossRef]
30. Kolodyazhnyi, O.I.; Kolodyazhnaya, A.O. A new approach towards synthesis of phosphorylated alkenes. *Russ. J. Gen. Chem.* **2015**, *85*, 359–365. [CrossRef]
31. Kolodiazhnyi, O.I. *P*-Halogen-substituted phosphorus ylids. *Russ. Chem. Rev.* **1991**, *60*, 391–409. [CrossRef]
32. Kolodiazhna, A.O.; Kolodiazhnyi, O.I.; Kolodiazhna, O.O. Stereoselective Method for the Synthesis of Phosphorylated Alkenes. U.S. Patent 111025, 10 March 2016.
33. Sotiropoulos, J.-M.; Baceiredo, A.; Bertrand, G. Synthesis and reactivity of diazomethylenephosphoranes ($>P=C=N_2$): New phosphacumulene ylides and first stable pseudounsaturated diazo derivatives. *J. Am. Chem. Soc.* **1987**, *109*, 4711–4712. [CrossRef]
34. Kolodiazhnyi, O.I.; Ustenko, S.N. Synthesis and properties of *P,P*-Difluoroylids. *Phosphorus Sulfur Silicon Relat. Elem.* **1991**, *62*, 111–118. [CrossRef]
35. Ustenko, S.N. Synthesis and Properties of New Types of Phosphorus *P*-halogenylides. Ph.D. Thesis, Institute of Organic Chemistry, National Academy of Sciences of Ukraine, Kiev, Ukraine, 12 October 1995.
36. Golokhov, D.B. *P*-Halogenoylides-Synthesis and Properties. Ph.D. Thesis, Institute of Organic Chemistry, Academy of Sciences of Ukraine, Kiev, Ukraine, 20 September 1990.
37. Kolodyazhnyi, O.I. 2-Halogen-1,2λ^5-oxaphosphetanes. Reaction of *P*-Fluoroylids with carbonyl compounds and carbon disulfide. *Russ. J. Gen. Chem.* **1986**, *56*, 821–827.
38. Kolodiazhnyi, O.I. Thiocetenes phosphores. *Tetrahedron Lett.* **1987**, *28*, 881–884. [CrossRef]
39. Kolodiazhnyi, O.I. Reaction de Wittig "anormale". *Tetrahedron Lett.* **1985**, *26*, 439–442. [CrossRef]
40. Kolodiazhnyi, O.I.; Golokhov, D.B. 2-Halogen-1,2λ^5-oxaphosphetanes. III. About conversion of 2-fluorine-1,2λ^5-oxaphosphetanes to alkenes. *J. Gen. Chem.* **1989**, *59*, 293–306.
41. Burgi, H.B.; Dunitz, I.D.; Lehn, J.M.; Wipff, G. Stereochemistry of reaction paths at carbonyl centres. *Tetrahedron* **1974**, *30*, 1563–1572. [CrossRef]
42. Kolodiazhnyi, O.I. 2-Halogen-1,2λ^5-oxaphosphetanes. *Russ. J. Gen. Chem.* **1986**, *56*, 283–298.
43. Van Bochove, M.A.; Swart, M.; Bickelhaupt, F.M. Nucleophilic substitution at phosphorus centers ($S_N2@P$). *Chem. Phys. Chem.* **2007**, *8*, 2452–2463. [CrossRef] [PubMed]
44. Kolodiazhnyi, O.I.; Kovalenko, A.B. Reaction of P-Halogenylides with carbonyl compounds. *Synthesis of allylphosphonates. Russ. J. Gen. Chem.* **1987**, *57*, 2147–2149.
45. Kolodiazhnyi, O.I. Transformation des 2-fluoro-1,2λ^5-oxaphosphetanes en allylphosphonates. *Tetrahedron Lett.* **1988**, *29*, 3663–3666. [CrossRef]
46. Mayr, H.; Ammer, J.; Baidya, M.; Maji, B.; Nigst, T.A.; Ofial, A.R.; Singer, T. Scales of Lewis basicities toward C-centered Lewis acids (Carbocations). *J. Am. Chem. Soc.* **2015**, *137*, 2580–2599. [CrossRef] [PubMed]
47. Corbridge, D.E.C. *Phosphorus: Chemistry, Biochemistry and Technology*, 6th ed.; CRC Press: Boca Raton, FL, USA, 2013; p. 1473.
48. Robert, A.S., Jr. *Practical Functional Group Synthesis*; John Wiley & Sons, Inc.: Hoboken, NJ, USA, 2016; p. 685.

5

Synergy between Experimental and Theoretical Results of Some Reactions of Annelated 1,3-Azaphospholes

Raj K. Bansal *, Raakhi Gupta and Manjinder Kour

Department of Chemistry, The IIS University, Jaipur 302020, India; raakhi.gupta@iisuniv.ac.in (R.G.); manjinderkour1992@gmail.com (M.K.)
* Correspondence: bansal56@gmail.com

Abstract: Computational calculations have been used successfully to explain the reactivity of the >C=P- functionality in pyrido-annelated 1,3-azaphospholes. Theoretical investigation at the Density Functional Theory (DFT) level shows that the lone pair of the bridgehead nitrogen atoms is involved in extended conjugation, due to which electron density increases considerably in the five-membered azaphosphole ring. The electron density in the azaphosphole is further enhanced by the presence of an ester group at the 3-position making the >C=P- functionality electron-rich. Thus, 1,3-azaphospholo[5,1-*a*]pyridine, i.e., 2-phosphaindolizine having ester group at the 3-position only does not undergo Diels-Alder (DA) reaction with an electron rich diene, such as 2,3-dimethyl-1,3-butadiene (DMB). However, an ester group at 1-position acts as electron-sink, due to which transfer of the electron density to the >C=P- moiety is checked and DA reaction occurs across the >C=P- functionality. The coordination of the Lewis acid to the carbonyl group at the 3-position raises the activation barrier, while it is lowered remarkably when it is coordinated to the P atom. Furthermore, the attack of 1,3-butadiene on the *Si* face of the P-coordinated (*o*-menthoxy)aluminum dichloride complex is a lower activation energy path. Fukui functions could be used to explain relative reactivities of indolizine and 2-phosphaindolizine towards electrophilic substitution reactions.

Keywords: synergy; 1,3-azaphospholes; Diels-Alder reaction; electrophilic substitution; DFT calculations

---

1. Introduction

Almost a century ago, Paul Dirac [1] could foresee the possibility of applying quantum mechanical concepts to solving the experimental problems in chemistry when he remarked, "The underlying physical laws necessary for the mathematical theory of a large part of physics and the whole of chemistry are thus completely known, and the difficulty is only that the exact application of these laws leads to equations much too complicated to be soluble. It therefore becomes desirable that approximate practical methods of applying quantum mechanics should be developed, which can lead to an explanation of the main features of complex atomic systems without too much computation". Dirac's dream inched towards fruition when Kohn, Hohenberg, and Sham [2,3] developed the density functional theory (DFT). It was followed by dramatic progress when many DFT codes were made available commercially, which enabled even less experienced persons with modest computational facilities to do calculations with much accuracy and look into the electronic structures and many other properties of the substances of interest. During the last two decades, the availability of a variety of new and more accurate DFT functionals incorporating specific features has brought DFT computations to the stage where they are used advantageously as a tool not only to explain the experimental results but also to plan and modify experimental conditions for achieving the desired objectives [4–9].

In recent years, there has been a perceptible increase in the number of research papers in chemistry reporting both experimental and theoretical results together. Fristrup and co-workers in a recent article emphasized the synergy between experimental and theoretical methods in the exploration of homogeneous transitional metal catalysis [10]. In another interesting review article titled, "Towards the computational design of solid catalysts", Norskov et al. [11] commented, "*Theoretical methods can be used to describe surface chemical reactions in detail and to understand variations in catalytic activity from one catalyst to another.*" In spite of many challenges in the field, they added "Computational approaches for the discovery and development of catalysts hold great promise for the future." Similar views have been expressed in other reports [12–14].

In this perspective, we shall present mainly our own results highlighting the application of DFT calculations to plan and accomplish successfully some reactions of annelated 1,3-azaphospholes. It will also be illustrated how computational calculations helped to resolve an experimental log jam and paved the way for the use of catalysts, leading to a successful reaction and asymmetric synthesis.

2. Synthesis of Annelated 1,3-Azaphospholes

Before coming to the reactions, it will be appropriate to describe briefly two general methods of synthesis of annelated 1,3-azaphospholes that we developed.

2.1. [4+1] Cyclocondensation Method

The reaction of 2-ethyl-1-phenacylpyridinium bromide (**1**) with PCl_3 in the presence of Et_3N at room temperature (r.t.) afforded 3-benzoyl-1-methyl-1,3-azaphospholo[5,1-*a*]pyridine (**2**) in very good yield (Scheme 1). The general skeleton was named as 2-phosphaindolizine perceiving it to result from a CH/P exchange at the 2-position of indolizine nucleus [15].

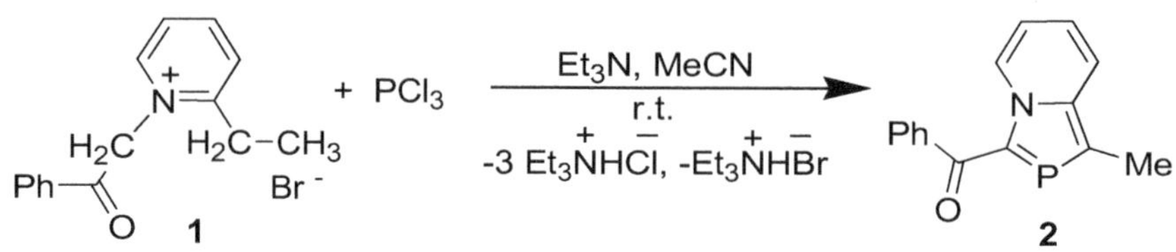

Scheme 1. Synthesis of 2-phosphaindolizine derivative.

Later, this synthetic strategy made accessible not only variously substituted 2-phosphaindolizines [15–17], but also 1-aza-2-phosphaindolizines [18], 3-aza-2-phosphaindolizines [19], and 1,3-diaza-2-phosphaindolizines [20]. Besides, the method could be extended to the synthesis of 1,3-azaphospholes annelated to pyrazine [15], thiazole [21], benzothiazole [21], and oxazole [22] skeletons also (Scheme 2). These reactions have been compiled in a mini-review [23].

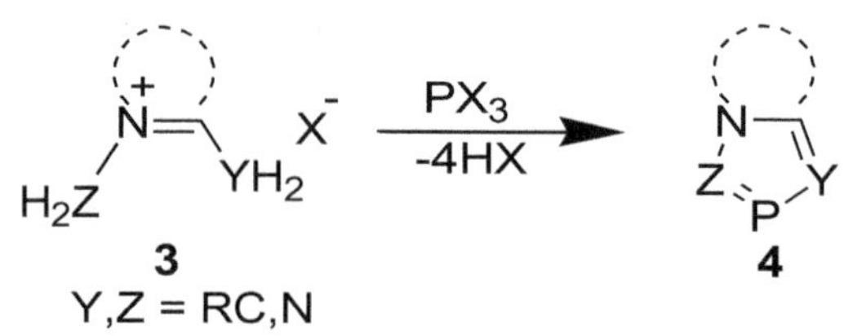

Scheme 2. Synthesis of annelated 1,3-azaphospholes.

2.2. Synthesis Involving Tandem Pyridinium Ylide generation, Disproportionation, and 1,5-Electrocyclization

We obtained 1,3-bis(alkoxycarbonyl)-1,3-azaphospholo[5,1-*a*]pyridines (**9**) serendipitously on reacting (alkoxycarbonyl)methylpyridinium bromide with Et_3N and PCl_3. The reaction involved tandem pyridinium alkoxycarbonyl-dichlorophosphinomethylide (**6**) generation, disproportionation, 1,5-electrocyclization, and 1,2-elimination to furnish the final product (Scheme 3) [24,25].

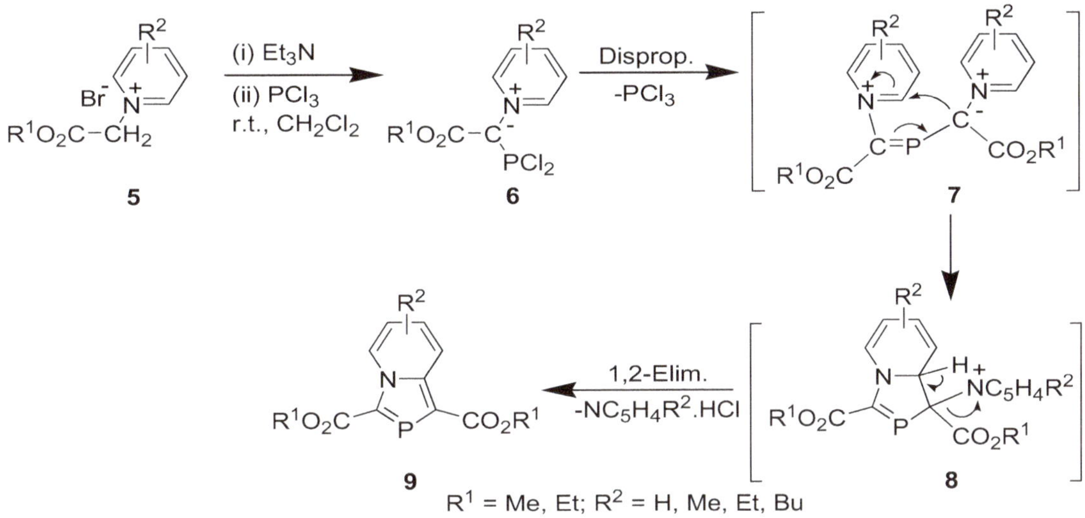

Scheme 3. Synthesis of 1,3-bis(alkoxycarbonyl)-1,3-azaphospholo[5,1-*a*]pyridines.

Subsequently, isoquinoline [26], and phenanthridine [27] analogues of **9** could also be prepared.

3. Reactions Response

3.1. Diels-Alder Reaction

There have been earlier reports about the Diels-Alder (DA) reaction across the >C=P- functionality as dienophile [28,29]. We investigated theoretically the model DA reactions of ethene and phosphaethene with 1,3-butadiene at the DFT(B3LYP/6-31+G**) level [30]. The Frontier Molecular Orbitals (FMO) of the reactants are shown in Figure 1.

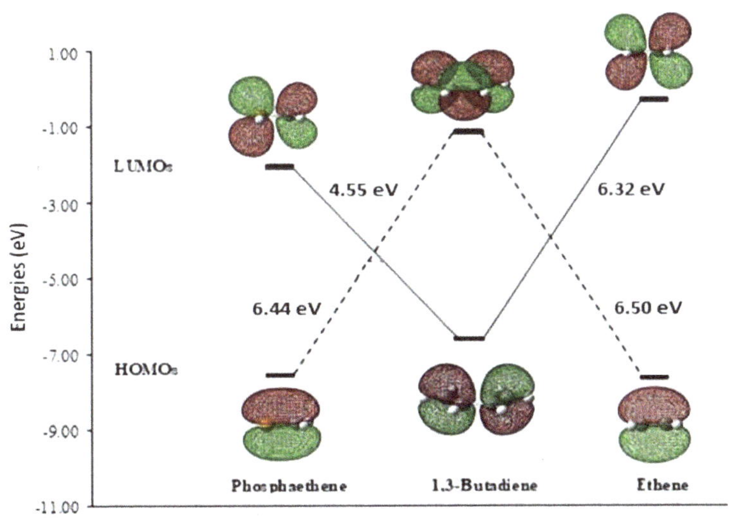

Figure 1. The HOMO-LUMO energy differences between FMOs of phosphaethene, 1,3-butadiene, and ethene calculated at the B3LYP/6-31+G** level.

It may be noted that the $HOMO_{diene}$–$LUMO_{phosphaethene}$ gap (4.55 eV) is much lower than the gap $HOMO_{diene}$–$LUMO_{ethene}$ (6.32 eV). Furthermore, the activation energy barriers (ΔEact.) for the two DA reactions are shown in Scheme 4 [30].

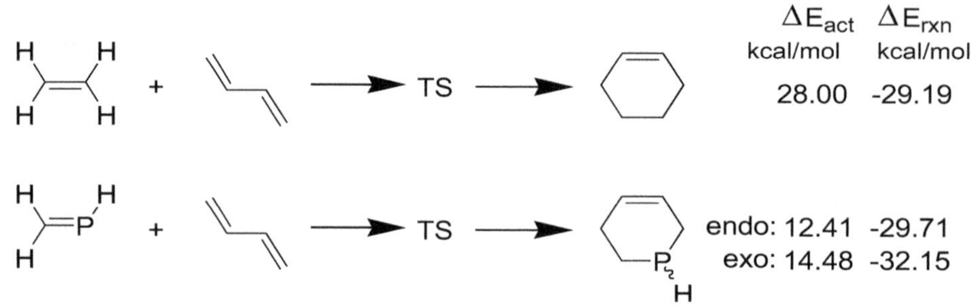

Scheme 4. Model DA reactions of ethene and phosphaethene with 1,3-butadiene computed at B3LYP/6-31+G** in the gas phase.

Encouraged by these theoretical results, we attempted DA reactions of 1,3-azaphospholo[5,1-*a*]pyridines, i.e., 2-phosphaindolizines prepared by two methods. However, to our great surprise, the compounds 3-ethoxycarbonyl-1-methyl-1,3-azaphospholo[5,1-*a*]pyridine **10a** (**10**, Z=Me, R^1=R^2=R^3=H) obtained through [4+1] cyclocondensation and 1,3-bis(ethoxylcarbonyl)-1,3-azaphospholo[5,1-*a*]pyridine **10b** (**10**, Z=CO_2Et, R^1=R^2=R^3=H) showed remarkable difference in their reactivity with 2,3-dimethylbuta-1,3-diene (DMB). It was found that 3-(ethoxycarbonyl)-1-methyl-1,3-azaphospholo[5,1-*a*]pyridine (**10a**) did not undergo DA reaction with DMB, even upon boiling in toluene in the presence of sulfur (used to oxidize phosphorus atom of the initially formed cycloadduct to push the reaction in the forward direction, see later) [31]. But, 1,3-bis(ethoxylcarbonyl)-1,3-azaphospholo[5,1-*a*]pyridine (**10b**) underwent DA reaction with DMB at room temperature (r.t.). However, the reaction at room temperature (ca 25 °C) was very slow and could be completed (δ ^{31}P NMR = 14.1 ppm) after heating at reflux temperature in chloroform for four days. On carrying out the reaction in the presence of sulfur or methyl iodide, it was complete at r.t. in 4 days [31]. Thus, the DA reactions of 1,3-bis(ethoxycarbonyl)-1,3-azaphospholo[5,1-*a*]pyridines and also of an isoquinoline analogue with DMB and isoprene could be accomplished successfully (Scheme 5) [31,32].

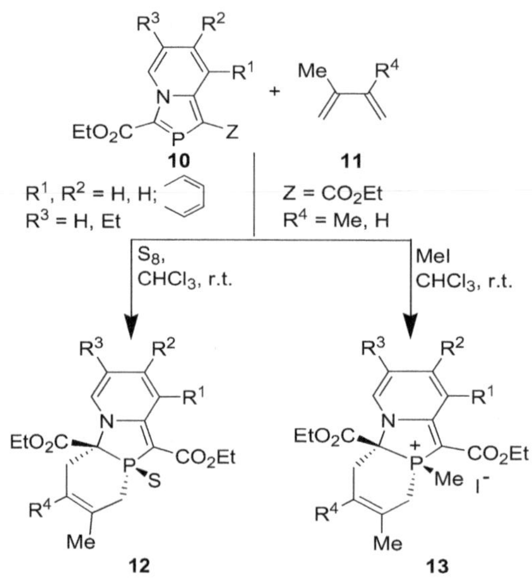

Scheme 5. DA reactions of 1,3-bis-(ethoxycarbonyl)-1,3-azaphospholo[5,1-*a*]pyridines with 2,3-dimethyl-1,3-butadiene (DMB) and isoprene.

All the reactions occurred with complete stereo- and regioselectivity, except the reaction of 1,3-bis(ethoxycarbonyl)-1,3-azaphospholo[5,1-*a*]isoquinoline (**14**) with isoprene in the presence of methyl iodide which afforded two regioisomers, the major (62%) being **15** (Scheme 6) [32].

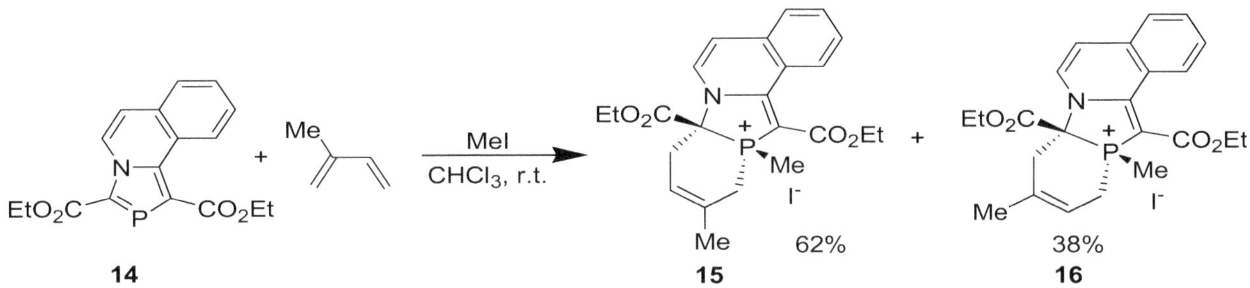

Scheme 6. Regioselectivity in the DA reaction of 1,3-bis(ethoxycarbonyl)-1,3-azaphospholo[5,1-*a*]isoquinoline.

The structure of the cycloadduct **12** (R^1R^2= CH=CH-CH=CH-; R^3=R^4=H) was confirmed by single crystal X-ray diffraction studies [32]. That the [2+4]cycloaddition of annelated 1,3-azaphosphole with DMB precedes the oxidation of the phosphorus atom of the initially formed cycloadduct by sulfur or methyl iodide could be proved unambiguously by carrying out the DA reaction of 1,4,2-diazaphospholo[4-*a*]pyridine derivative **17** in the presence of methyl iodide, when N-methylated product **18** (R=Me) was formed (Scheme 7) [33]. It has been reported that [1,4,2] diazaphospholo[4-*a*]pyridines did not react with methyl iodide [34].

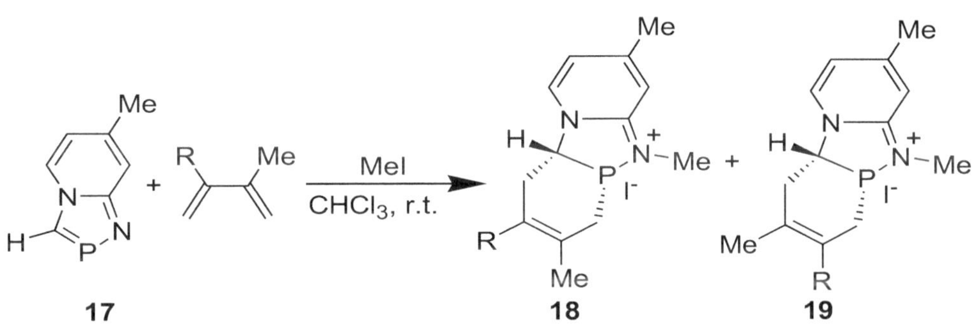

Scheme 7. DA reaction of 7-methyl-1,4,2-diazaphospholo[4,5-*a*]pyridine in the presence of methyl iodide.

The regioselectivity observed in the reaction of 1,3-bis(ethoxycarbonyl)- 1,3-azaphospholo[5,1-*a*] pyridine with isoprene was investigated theoretically at the DFT (B3LYP/6-311++G**// B3LYP/6-31G**) level. Following model reactions were computed (Figure 2). For studying the solvent effect, single point energy was computed using gas phase-optimized geometry at the B3LYP/6-311++G** level with polarizable continuum model (PCM).

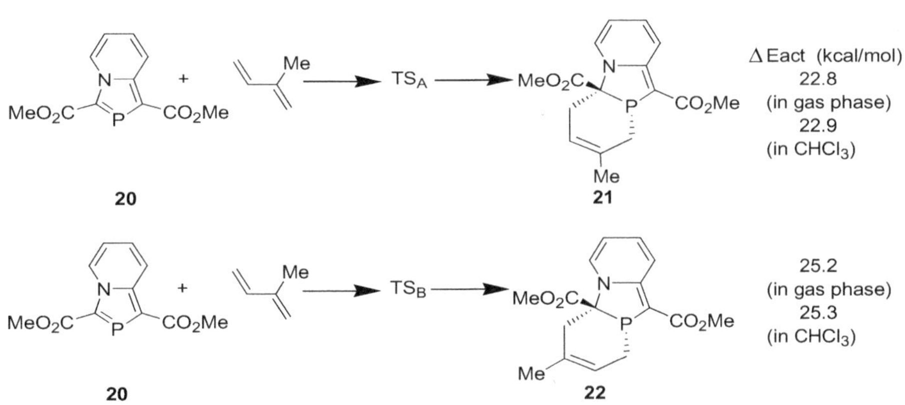

Figure 2. Model DA reactions computed at B3LYP/6-311++G**//B3LYP/6-31G**.

The regioselectivity was calculated by using the formula, $k_1/k_2 = e^{-\Delta E/RT}$ [35]. A difference of 2.4 kcal/mol in the activation energies corresponds to only 2% regioselectivity. It is obvious that the observed regioselectivity of 100% cannot be accounted for at this theoretical level.

We investigated the reason for the difference in the dienophilic reactivities of 3-ethoxycarbonyl -1-methyl-1,3-azaphospholo[5,1-*a*]pyridine (**10a**) and 1,3-bis(ethoxycarbonyl)-1,3-azaphospholo[5,1-*a*] pyridine (**10b**), theoretically. For this purpose, following four model reactions were computed at the DFT (B3LYP/6-31G**) level (Scheme 8) [36].

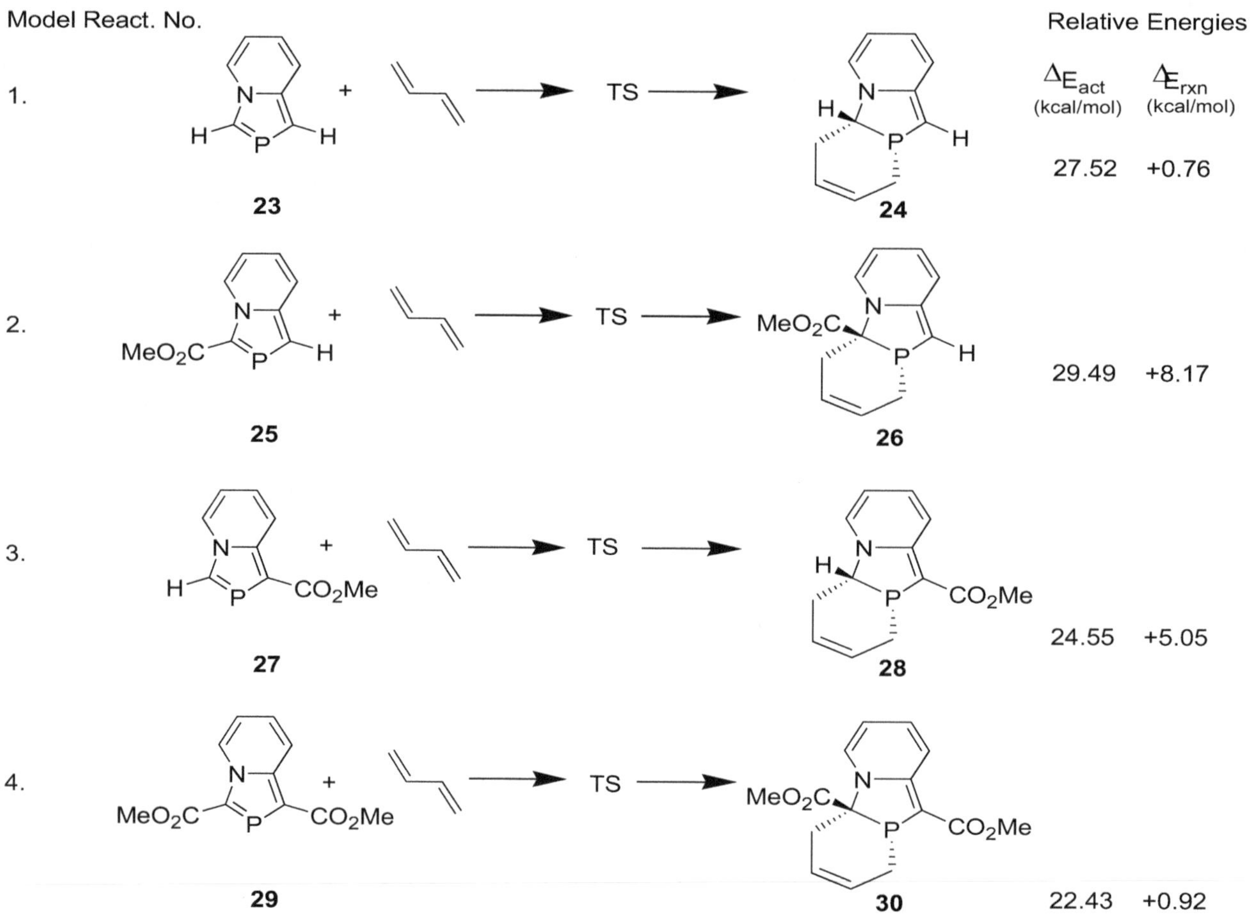

Scheme 8. Model DA reactions computed at the B3LYP/6-31G** level.

As expected, the activation energy barrier for the DA reaction with 1,3-bis(methoxycarbonyl)-1, 3-azaphospholo[5,1-*a*]pyridine (**29**) is lowest. However, it is interesting to find that the activation barrier for the DA reaction with 3-methoxycarbonyl-1,3-azaphospholo[5,1-*a*]pyridine (**25**) is highest (29.49 kcal/mol), higher than even for the DA reaction with the unsubstituted 1,3-azaphospholo[5,1-*a*]pyridine (**23**). It is rather an unexpected effect of an electron-withdrawing group (EWG). To look into the real cause of this unusual effect, we carried out the natural bond orbital (NBO) analysis calculations, which revealed the existence of an interesting phenomenon: due to conjugation of the lone pair of the bridgehead nitrogen in **23**, negative charge density is effectively transferred to the five-membered azaphosphole ring, which is further enhanced by the presence of an EWG at the 3-position. It is shown in Figure 3A. It makes the >C=P- functionality electron-rich, which fails to react with electron-rich diene, DMB. However, a CO_2Me group at the 1-position acts as a trap for the negative charge, the latter being delocalized over to the carbonyl group. It is shown in Figure 3B. Thus in this case, the >C=P- functionality remains unaffected and is able to undergo the DA reaction with DMB.

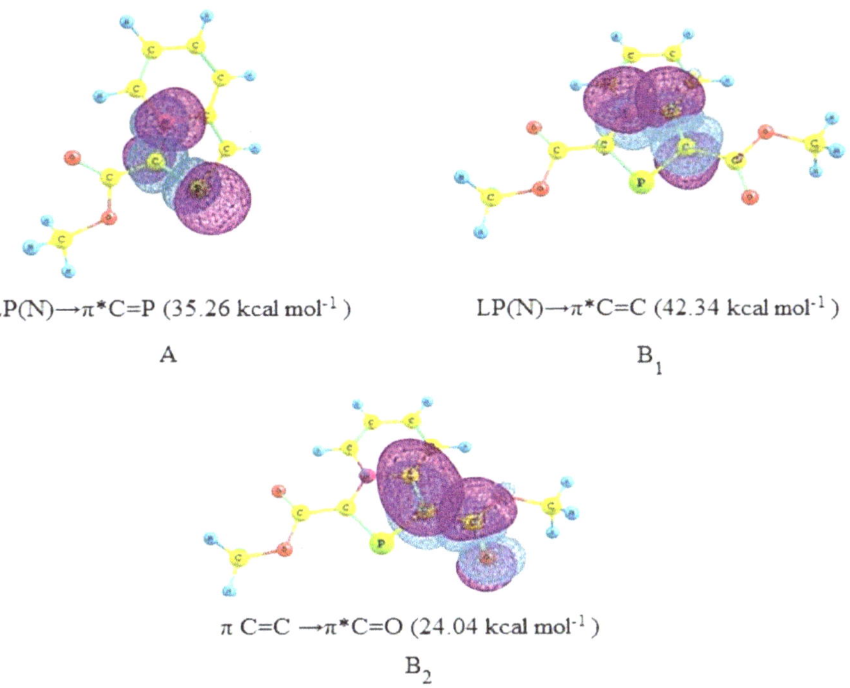

Figure 3. Natural Bond Orbital (NBO) interactions in 3-methoxycarbonyl-1,3-azaphospholo[5,1-*a*]pyridine (**A**) and 1,3-bis(methoxycarbonyl)- 1,3-azaphospholo[5,1-*a*]pyridine (**B**$_1$ and **B**$_2$) with second order perturbation energies (in parenthesis).

3.2. Lewis Acid Catalyzed Diels-Alder Reaction

Then, we explored the possibility of using a Lewis acid as catalyst to accomplish DA reaction with 3-(alkoxycarbonyl)-1-methyl-1,3-azaphospholo[5,1-*a*]pyridines and computed the following model reactions at the DFT (B3LYP/6-31+G**) level (Scheme 9) [37]. For studying the solvent effect, single point energy was computed using gas phase-optimized geometry with PCM solvation model.

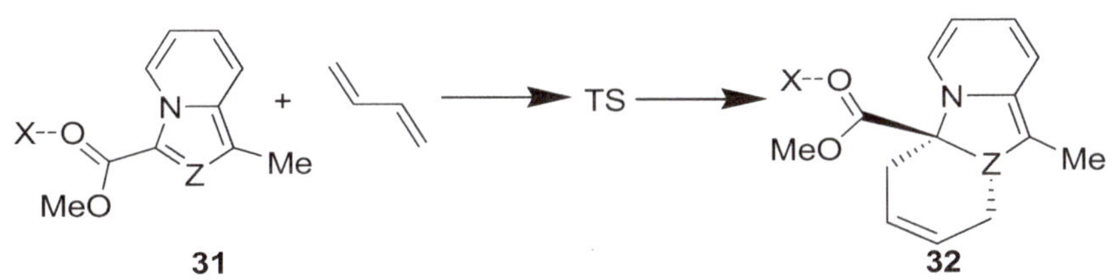

X	Z	ΔE_{act} (kcal/mol) CH_2Cl_2
Nil	CH	45.68
Nil	P	30.80
$AlCl_3$	CH	46.77
$AlCl_3$	P	33.43

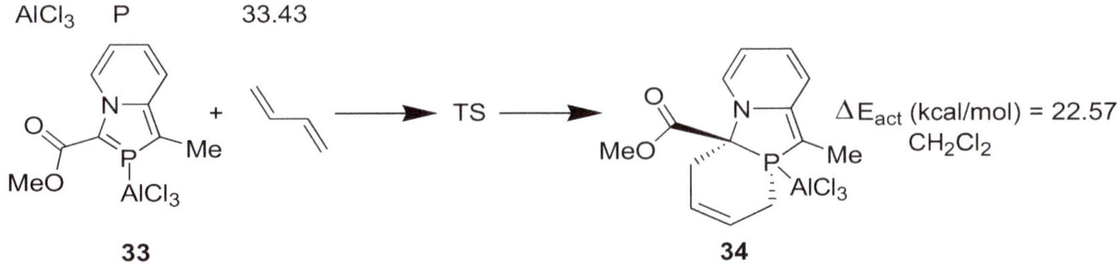

Scheme 9. Model DA reactions computed at the B3LYP/6-31+G** level.

There was another surprise in store for us. When the catalyst, $AlCl_3$, is coordinated to the carbonyl group of indolizine or 1,3-azaphospholo[5,1-*a*]pyridine, the activation energy barrier is increased further as compared with that in the absence of the catalyst. The NBO studies reveal that coordination of $AlCl_3$ with the carbonyl group enhances transfer of the negative charge in these cases, making the >C=C< and >C=P- functionality in indolizine and 1,3-azaphospholo[5,1-*a*]pyridine, respectively, electron-rich. This appears to be the reason why no DA reaction across the >C=C< moiety of indolizine has been reported even in the presence of a catalyst.

In 2-phosphaindolizine, however, there is another site for the coordination of $AlCl_3$: the P atom. It may be noted that on coordination of the catalyst to the P atom, the activation energy barrier is lowered remarkably, to 22.57 kcal/mol (in CH_2Cl_2) only. These results indicated that in contrast to indolizine, DA reactions across the >C=P- moiety of 2-phosphaindolizine having EWG at 3-position only should be possible in the presence of a Lewis acid catalyst. However, in this context a question arises: if coordination of $AlCl_3$ to the carbonyl group is thermodynamically more favorable than its coordination to the P atom by 3.43 kcal/mol, why should it coordinate to the P atom at all? In our opinion, in solution, an equilibrium may exist between the C=O- $\rightarrow$ $AlCl_3$ and P $\rightarrow$ $AlCl_3$ coordinated species, making a small concentration of the latter available that undergoes DA reaction with the diene to produce irreversibly the cycloadduct, thereby shifting the equilibrium to the right (Scheme 10).

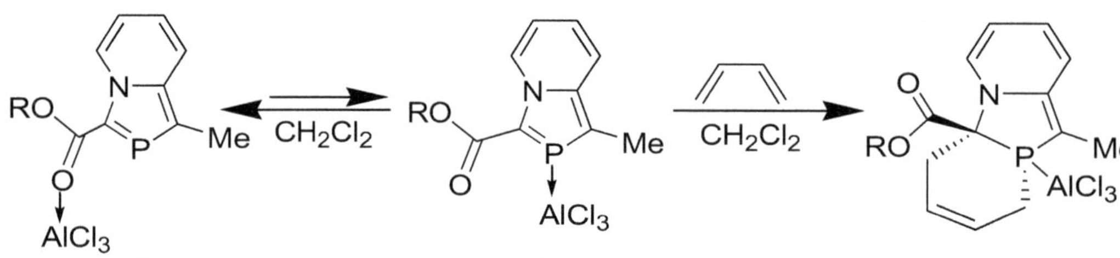

35 36 37

Scheme 10. Equilibrium between C=O $\rightarrow$ and P $\rightarrow$ coordinated species, and DA reaction with the latter.

Guided by the above results, we accomplished successfully the DA reaction of 3-(alkoxycarbonyl/acyl)-1-methyl-2-phosphaindolizines (**38**) with DMB in the presence of ethylaluminium dichloride as a promoter (i.e., using its 1 equiv.) in dichloromethane at r.t. under nitrogen atmosphere (Scheme 11) [38]. The occurrence of a clean reaction was revealed by ^{31}P NMR signal at $\delta = -16.6$ to -5.3 ppm, which is in accordance with the ^{31}P NMR chemical shifts reported for the [2+4] cycloadducts obtained from the DA reaction of P-W(Co)$_5$ complexes of λ^3-phosphinines [39].

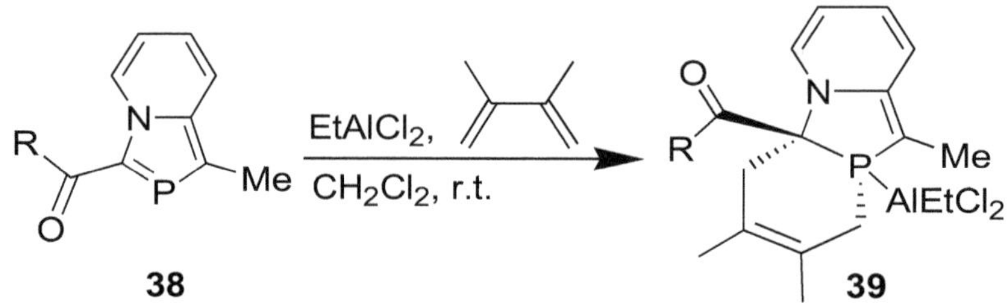

38 39

38, 39	a	b	c	d
R	OMe	OEt	CMe_3	Ph
δ ^{31}P (**39**)	-5.8	-5.3	-16.6	-10.1

Scheme 11. DA reaction of 3-(alkoxycarbonyl/acyl)-1-methyl-2-phosphaindolizines in the presence of $EtAlCl_2$.

3.3. *Asymmetric Diels-Alder Reaction with 2-Phosphaindolizines*

The successful results described above motivated us to investigate asymmetric DA reaction across the >C=P- moiety of 2-phosphaindolizines by using a chiral organoaluminium catalyst, namely, o-menthoxyaluminium dichloride [40].

We computed the following model reactions at the B3LYP/6-31+G* level (Scheme 12).

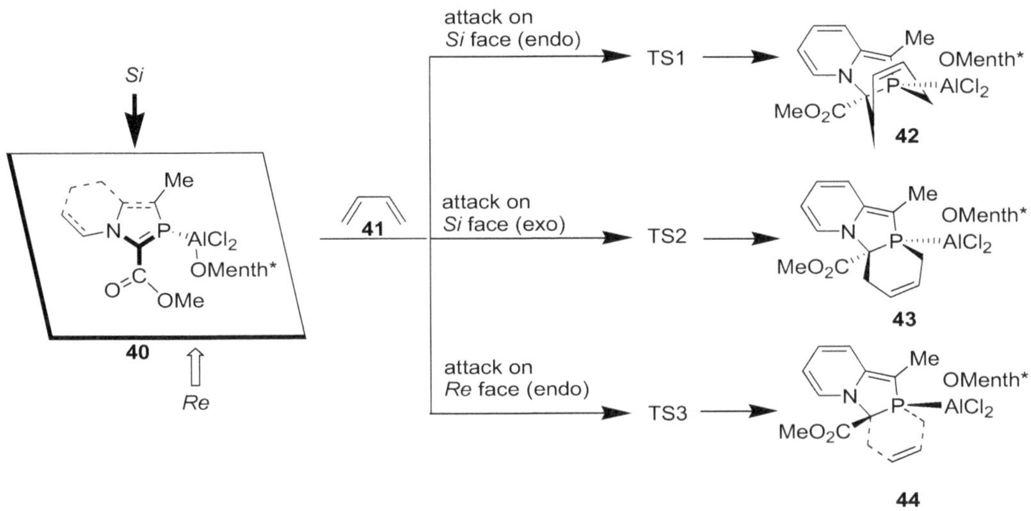

Scheme 12. Attack of 1,3-butadiene on *Si* and *Re* faces of the >C=P- functionality of 2-phosphaindolizine complex computed at the B3LYP/6-31+G* level.

Computational calculations reveal that attack of the diene occurs preferentially from the *Si* face. Furthermore, during attack on the *Si* face, activation energy barrier for the *exo* approach is smaller than for the *endo* approach by ca. 0.3 kcal/mol. Presence of the bulky o-menthoxy group possibly makes the *endo* approach comparatively less accessible. All the reactions are moderately exothermic, but due to decrease in entropy, they are endergonic, Gibbs free energies values being +11.65 to +13.08 kcal/mol.

In accordance with the theoretical results, DA reaction of 2-phosphaindolizines (**47**) could be accomplished with complete diastereoselectivity, as indicated by ^{31}P NMR, with a single product being formed in each case (Scheme 13) [40].

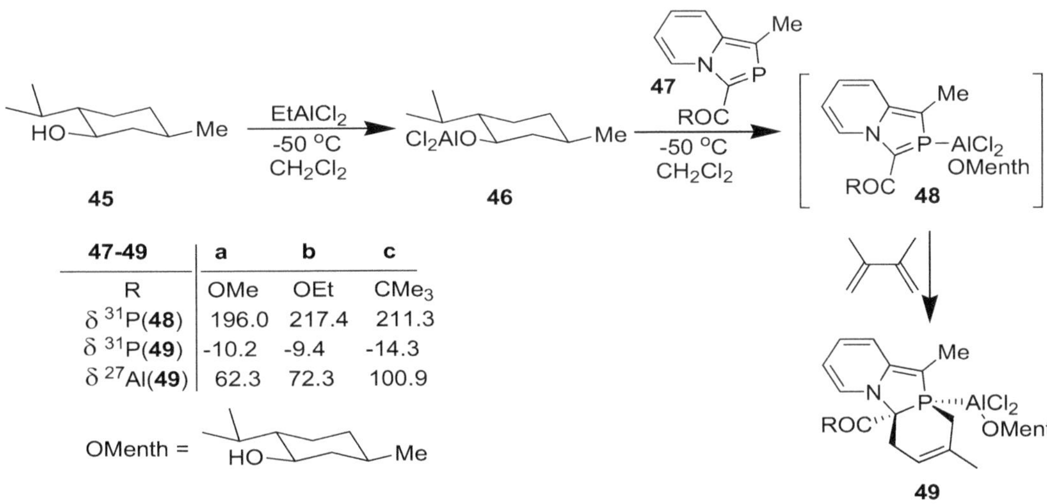

47-49	**a**	**b**	**c**
R	OMe	OEt	CMe$_3$
δ ^{31}P(**48**)	196.0	217.4	211.3
δ ^{31}P(**49**)	-10.2	-9.4	-14.3
δ ^{27}Al(**49**)	62.3	72.3	100.9

Scheme 13. Diels-Alder reaction of 2-phosphaindolizine-n^1-P-aluminium(*o*-menthoxy) dichloride with DMB.

3.4. Electrophilic Substitution

In contrast to indolizines [41], 1-unsubstituted 2-phosphaindolizines failed to react with MeCOCl, PhCOCl, or Me_3SiCl, even on prolonged heating in the presence of Et_3N [42]. However, like indolizines [43], 1-unsubstituted 2-phosphaindolizines reacted with PCl_3 or $PhPCl_2$ (but not with Ph_2PCl) in the presence of Et_3N to afford the phosphynated products (Scheme 14) [44]. It is noteworthy that on reacting with PCl_3, the initially formed dichlorophosphino product (**51**, R=Cl) undergoes dismutation accompanied by the loss of PCl_3 to afford a PCl bridged bis(2-phosphaindolizine) (**52**) as the final product.

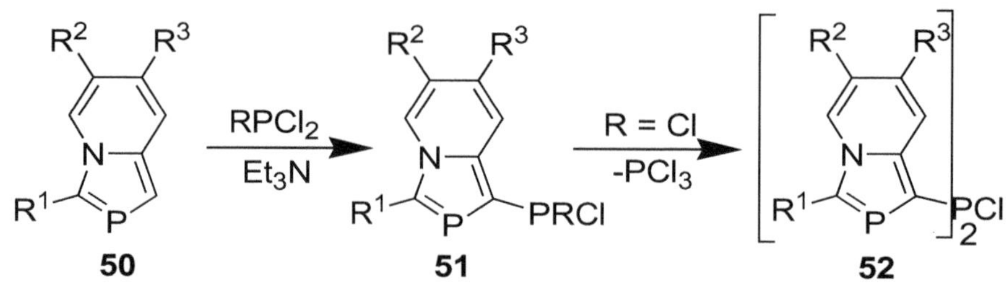

	a	b	c	d	e	f
R^1	COPh	COPh	COPh	CO_2Et	CO_2Et	*p*-$NO_2C_6H_4$
R^2	H	Me	H	H	Me	H
R^3	H	H	^nBu	H	H	H

Scheme 14. Reaction of 1-unsubstituted 2-phosphaindolizines with chlorophosphines.

The concept of hard-soft acid-base (HSAB) theory was extended to explain the global and local reactivity of organic molecules towards nucleophilic and electrophilic reagents [45]. The chemical reactivity at a particular molecular site could be rationalized using a quantitative descriptor, the Fukui function (f(r)). We recently used Fukui function to rationalize 1,2- versus 1,4-addition of amines to maleic anhydride [46]. The difference in the behavior of indolizine and 2-phosphaindolizine towards electrophilic reagents can be rationalized on the basis of the Fukui functions calculated at the B3LYP/6-31+G* level (Figure 4) [47].

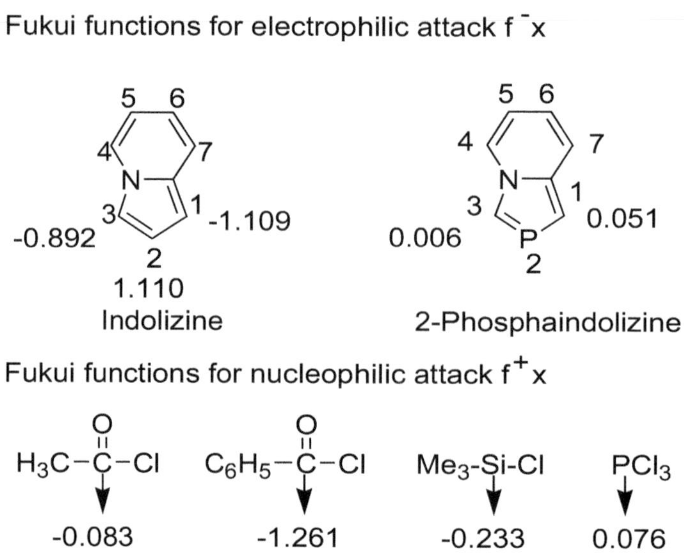

Figure 4. Fukui functions calculated at the B3LYP/6-31+G* level.

Fukui functions reveal that C1 and C3 in indolizine are much harder than C1 and C3 in 2-phosphaindolizine. Thus, hard electrophilic reagents such as acyl chlorides and chlorotrimethylsilane attack at C1 and C3 positions of indolizine, but they fail to react with 2-phosphaindolizine. On the other hand, phosphorus trichloride is comparatively softer and hence reacts with 2-phosphaindolizine at C1 (C3 being substituted in our compounds). Being at the borderline, PCl_3 reacts with indolizine also.

4. Conclusions

The computational calculations could be used as a useful tool to understand and solve problems encountered during reactions of annelated 1,3-azaphospholers. Furthermore, this strategy worked well while planning the reactions such as Lewis acid catalyzed and asymmetric DA reactions across the >C=P- functionality of 2-phosphaindolizines.

Author Contributions: All authors contributed equally in the compilation of this perspective.

Acknowledgments: We are thankful to the authorities of The IIS University, Jaipur (India) for the facilities.

References

1. Dirac, P.A.M. Quantum mechanics of many-electrons systems. *Proc. R. Soc. Lond. Ser. A.* **1929**, *123*, 714–733. [CrossRef]
2. Hohenberg, P.; Kohn, W. Inhomogenous electron gas. *Phys. Rev.* **1964**, *136*, B867–B871. [CrossRef]
3. Kohn, W.; Sham, L.J. Self-consistent equations including exchange and correlation effects. *Phys. Rev.* **1965**, *140*, A1133–A1138. [CrossRef]
4. Zangwill, A. A half century of density functional theory. *Phys. Today* **2015**, *68*, 34–39. [CrossRef]
5. Jain, A.; Shin, Y.; Persson, K.A. Computational predictions of energy materials using density functional theory. *Nat. Rev. Mater.* **2016**, *1*, 1–13. [CrossRef]
6. Cohen, A.J.; Mori-Sánchez, P.; Yang, W. Challenges for density functional theory. *Chem. Rev.* **2012**, *112*, 289–320. [CrossRef] [PubMed]
7. Jones, R.O. Density functional theory: Its origins, rise to prominence, and future. *Rev. Mod. Phys.* **2015**, *87*, 897–923. [CrossRef]
8. Tsipis, A.C. DFT flavor of coordination chemistry. *Coord. Chem. Rev.* **2014**, *272*, 1–29. [CrossRef]
9. Cole, D.J.; Hine, N.D.M. Applications of large-scale density functional theory in biology. *J. Phys. Condens. Matter* **2016**, *28*, 393001. [CrossRef] [PubMed]
10. Lupp, D.; Christensen, N.J.; Fristrup, P. Synergy between experimental and theoretical methods in the exploration of homogeneous transition metal catalysis. *Dalton Trans.* **2014**, *43*, 11093–11105. [CrossRef] [PubMed]
11. Nørskov, J.K.; Bligaard, T.; Rossmeisl, J.; Christensen, C.H. Towards the computational design of solid catalysts. *Nat. Chem.* **2009**, *1*, 37–46. [CrossRef] [PubMed]
12. Lineberger, W.C.; Borden, W.T. The synergy between qualitative theory, quantitative calculations, and direct experiments in understanding, calculating, and measuring the energy differences between the lowest singlet and triplet states of organic diradicals. *Phys. Chem. Chem. Phys.* **2011**, *13*, 11792–11813. [CrossRef] [PubMed]
13. Sierka, M. Synergy between theory and experiment in structure resolution of low-dimensional oxides. *Prog. Surf. Sci.* **2010**, *85*, 398–434. [CrossRef]
14. Mcgrady, J. Guest Editor The themed collection, "The synergy between theory and experiment. *Dalton Trans.* **2009**, 5805–6064. [CrossRef]
15. Bansal, R.K.; Gupta, N.; Karaghiosoff, K.; Schmidpeter, A.; Spindler, C. 2-Phosphaindolizines. *Chem. Ber.* **1991**, *124*, 475–480. [CrossRef]
16. Bansal, R.K.; Kabra, V.; Gupta, N.; Karaghiosoff, K. Synthesis of some new 2-phosphaindolizines. *Indian J. Chem.* **1992**, *31*, 254–256.
17. Gupta, N.; Jain, C.B.; Heinicke, J.; Bharatiya, N.; Bansal, R.K.; Jones, P.G. 2-Phosphaindolizines. *Heteroat. Chem.* **1998**, *9*, 333–339. [CrossRef]
18. Karaghioshoff, K.; Bansal, R.K.; Gupta, N. 1,4,2-Diazaphospholo[4,5-*a*]pyridines. *Z. Naturforsch. B* **1992**, *47b*, 373–378.

19. Bansal, R.K.; Pandey, G.; Karaghiosoff, K.; Schmidpeter, A. 1,2,3-Diazaphospholo[1,5-*a*]pyridines. *Synthesis* **1995**, 173–175. [CrossRef]
20. Bansal, R.K.; Gandhi, N.; Karaghiosoff, K.; Schmidpeter, A. Synthesis of [1,2,4,3]triazaphospholo[1,5-*a*]pyridines. *Z. Naturforsch.* **1995**, *50b*, 558–562. [CrossRef]
21. Bansal, R.K.; Mahnot, R.; Sharma, D.C.; Karaghiosoff, K.; Schmidpeter, A. 1,3-Azaphospholo[5,1-b]thiazolines and benzothiazoles. *Heteroat. Chem.* **1992**, *3*, 351–357. [CrossRef]
22. Bansal, R.K.; Jain, C.B.; Gupta, N.; Kabra, V.; Karaghiosoff, K.; Schmidpeter, A. Synthesis and properties of a 5,6-dihydro-1,3-azaphospholo[5,1-b]oxazole. *Phosphorus Sulfur Silicon* **1994**, *86*, 139–143. [CrossRef]
23. Bansal, R.K.; Karaghiosoff, K.; Gandhi, N.; Schmidpeter, A. 2-Substituted cycloiminium salts in azaphosphole synthesis. *Synthesis* **1995**, 361–369. [CrossRef]
24. Bansal, R.K.; Surana, A.; Gupta, N. 2-Phosphaindolizines via 1,5-electrocyclization. *Tetrahedron Lett.* **1999**, *40*, 1565–1568. [CrossRef]
25. Bansal, R.K.; Gupta, N.; Baweja, M.; Hemrajani, L.; Jain, V.K. Pyridinium dichlorophosphinomethylides. *Heteroat. Chem.* **2001**, *12*, 602–609. [CrossRef]
26. Bansal, R.K.; Hemrajani, L.; Gupta, N. 1,3-Azaphospholo[5,1-*a*]isoquinolines. *Heteroat. Chem.* **1999**, *10*, 598–604. [CrossRef]
27. Singh, D.; Sinha, P.; Gupta, N.; Bansal, R.K. Synthesis of 1,3-azaphospholo[1,5-*f*]phenanthridines through 1,5-electrocyclization. *Phosphorus Sulfur Silicon Relat. Elem.* **2016**, *191*, 488–492. [CrossRef]
28. Appel, R.; Knoll, F.; Ruppert, I. Phospha-alkenes and phospha-alkynes, genesis and properties of the $(p\text{-}p)\pi$-multiple bond. *Angew. Chem. Int. Ed.* **1981**, *20*, 731–744. [CrossRef]
29. Mathey, F.; Mercier, F.; Charrier, C.; Fischer, J.; Mitschler, A. Dicoordinated 2H-phospholes as transient intermediates in the reactions of tervalent phospholes at high temperature. One-step synthesis of 1-phosphabornadienes and phosphorins from phospholes. *J. Am. Chem. Soc.* **1981**, *103*, 4595–4597. [CrossRef]
30. Wannere, C.S.; Bansal, R.K.; Schleyer, P.v.R. Diels-Alder reaction of phosphaethene with 1,3-dienes: An ab-initio study. *J. Org. Chem.* **2002**, *67*, 9162–9174. [CrossRef] [PubMed]
31. Bansal, R.K.; Gupta, N.; Kumawat, S.K.; Dixit, G. Diastereo- and regioselective Diels-Alder reactions of 2-phosphaindolizines. *Tetrahedron* **2008**, *64*, 6395–6401. [CrossRef]
32. Bansal, R.K.; Jain, V.K.; Gupta, N.; Gupta, N.; Hemrajani, L.; Baweja, M.; Jones, P.G. Stereo- and regioselectivity in Diels-Alder reactions of 1,3-azaphospholo[5,1-a]isoquinoline and –[5,1-*a*]pyridine. *Tetrahedron* **2002**, *58*, 1573–1579. [CrossRef]
33. Bansal, R.K.; Karaghiosoff, K.; Gupta, N.; Gandhi, N.; Kumawat, S.K. Diastereo- and regioselectivity in Diels-Alder reaction of [1,4,2]diazaphospholo[4,5-*a*]pyridines. *Tetrahedron* **2005**, *61*, 10521–10528. [CrossRef]
34. Karaghiosoff, K.; Cleve, C.; Schmidpeter, A. Chloromethyl-dichlorophosphane: A useful reagent for the synthesis of new heterocycles with dicoordinated phosphorus. *Phosphorus Sulfur Silicon Relat. Elem.* **1986**, *28*, 289–296. [CrossRef]
35. Pellegriner, S.C.; Silva, M.A.; Goodman, J.M. Theoretical evaluation of the origin of the regio- and stereoselectivity in the Diels-Alder reactions of dialkylvinylboranes: Studies on the reactions of vinylborane, dimethylvinylborane, and vinyl-9-BBN with *trans*-piperylene and isoprene. *J. Am. Chem. Soc.* **2001**, *123*, 8832–8837. [CrossRef]
36. Bansal, R.K.; Gupta, N.; Dixit, G.; Kumawat, S.K. Theoretical investigation of an unusual substituent effect on the dienophilicity of >C=P– Functionality in 2-phosphaindolizines. *J. Phys. Org. Chem.* **2009**, *22*, 125–129. [CrossRef]
37. Gupta, N.; Jangid, R.K.; Bansal, R.K.; Hopffgarten, M.V. Catalytic effect of organoaluminium chloride reagents on the dienophilic reactivity of indolizine and 2-phosphaindolizine towards [2+4] cycloaddition: A DFT investigation. *Curr. Cat.* **2012**, *1*, 93–99. [CrossRef]
38. Jangid, R.K.; Gupta, N.; Bansal, R.K.; Hopffgarten, M.V.; Frenking, G. Diels-Alder reaction of 2-phosphaindolizines catalyzed by organoaluminium reagent: Theoretical and experimental results. *Tetrahedron Lett.* **2011**, *52*, 1721–1724. [CrossRef]
39. Alcaraz, J.P.; Mathey, F. Accroissement de la reactivite des phosphorines en tant que dienes et philodienes par complexation du phosphore. *Tetrahedron Lett.* **1984**, *25*, 207–210. [CrossRef]
40. Jangid, R.K.; Sogani, N.; Gupta, N.; Bansal, R.K.; Hopffgarten, M.V.; Frenking, G. Asymmetric Diels-Alder reaction with >C=P– functionality of the 2-phosphaindolizine-η^1-P-aluminium(*o*-menthoxy) dichloride complex: Experimental and theoretical results. *Beilstein J. Org. Chem.* **2013**, *9*, 392–400. [CrossRef] [PubMed]

41. Swinbourne, F.J.; Hunt, J.H.; Klinkert, G. *Advances in Indolizine Chemistry in the Series: Advances in Heterocyclic Chemistry*; Katritzky, A.R., Boulton, A.J., Eds.; Academic Press: New York, NY, USA, 1979; Volume 23, pp. 103–170.
42. Bansal, R.K.; Heinicke, J. Anellated heterophospholes and phospholides and analogies with related non-phosphorus systems. *Chem. Rev.* **2001**, *101*, 3549–3578. [CrossRef] [PubMed]
43. Tolmachev, A.A.; Yurchenko, A.A.; Kozlov, E.S.; Shulezhko, V.A.; Pinchuk, A.M. Phosphorylated indolizines. *Heteroat. Chem.* **1993**, *4*, 343–360. [CrossRef]
44. Bansal, R.K.; Gupta, N.; Kabra, V.; Spindler, C.; Karaghiosoff, K.; Schmidpeter, A. Substitution of 2-phosphaindolizines by bromine and by chlorophosphines. *Heteroat. Chem.* **1992**, *3*, 359–366. [CrossRef]
45. Mendez, F.; Gázquez, J.L. Chemical reactivity of enolate ions: The local hard and soft-acids and bases principle viewpoint. *J. Am. Chem. Soc.* **1994**, *116*, 9298–9301. [CrossRef]
46. Kour, M.; Gupta, R.; Bansal, R.K. Experimental and theoretical investigation of the reaction of secondary amines to maleic anhydride. *Aust. J. Chem.* **2017**, *70*, 1247–1253. [CrossRef]
47. Gupta, R.; Paul, B.; Bansal, R.K. Application of Fukui functions for comparing reactivities of indolizine and 2-phosphaindolizine towards electrophilic substitution. 2018, unpublished work.

6

Protolysis and Complex Formation of Organophosphorus Compounds—Characterization by NMR-Controlled Titrations

Gerhard Hägele

Institute of Inorganic Chemistry and Structural Chemistry, Heinrich-Heine-University Düsseldorf, Universitätsstraße 1, D-40225 Düsseldorf, Germany; haegele@uni-duesseldorf.de

Academic Editors: György Keglevich and Kamyar Afarinkia

Abstract: Phosphonic acids, aminophosphonic acids, and phosphonocarboxylic acids are characterized by an advanced hyphenated technique, combining potentiometric titration with NMR spectroscopy. Automated measurements involving ^{13}C, ^{19}F and ^{31}P nuclei lead to "pseudo 2D NMR" spectra, where chemical shifts or coupling constants are correlated with analytical parameters. Dissociation constants, stability constants, dynamic and specific chemical shifts are determined. Macroscopic and microscopic dissociation equilibria are discussed.

Keywords: phosphonic acids; aminophosphonic acids; phosphonocarboxylic acids; NMR-controlled titration; dissociation constants; stability constants; dynamic and specific NMR parameters

---

1. Introduction

NMR-controlled titration, also known as NMR titration, a useful tool combining NMR and analytical aspects, is based on fundamental observations in the dawn of NMR spectroscopy: "Early phosphorus NMR studies of condensed phosphates showed that raising the acidity of phosphate solutions increased the shielding of the phosphorus nucleus, causing a shift of the ^{31}P resonances to higher fields by several ppm" [1]. "Later studies on the short chain condensed phosphates exhibited that the pH variations of the chemical shifts and spin-coupling constants where, when measured to sufficient precision, sensitive functions of the molecular structure and the bonding". A first titration curve of H_3PO_4 shown as δ_P vs. pH was derived in this paper [2]. ^{13}C-NMR measurements on linear aliphatic acids revealed that COOH groups in $C_nH_{2n+1}COOH$ (n = 0 to 4) exhibit higher chemical shifts δ_C than COO^- groups of corresponding anions $C_nH_{2n+1}COO^-$. A characteristic downfield shift of δ_C ranging from 5.1 to 4.7 ppm was observed for deprotonation by addition of tetramethylammonium hydroxide to carboxylic acids [3].

In subsequent years, those phenomena attracted the attention of numerous studies dealing with inorganic and organic phosphorus chemistry. A higher level of sophistication was achieved by combining the analytical theory of protolysis and complex formation for acids and bases with advanced NMR technologies and expanding the range of sensor nuclei to ^{1}H, ^{13}C, ^{15}N, ^{19}F, ^{31}P and spin active metal nuclei.

The first NMR titration curves for phosphonoacetic acid $HOOC\text{-}CH_2\text{-}PO_3H_2$ using ^{13}C and ^{31}P NMR were reported as δ_C vs. pH and δ_P vs. pH functions. For the first time, a characteristic deprotonation sequence was established: $HOOC\text{-}CH_2\text{-}PO_3H_2 \rightarrow HOOC\text{-}CH_2\text{-}PO_3H^- \rightarrow {}^-OOC\text{-}CH_2\text{-}PO_3H^- \rightarrow {}^-OOC\text{-}CH_2\text{-}PO_3{}^{2-}$. In addition, $^{1}J_{PC}$ was related to the s-electron density around the central C-atom [4].

Several decades of creative work followed those early observations. Induced by synthetic, analytical, biological, or technical aspects, interests were concentrated on several classes of organophosphorus compounds. Particular attention was drawn towards analogues of amino acids, e.g., aminophosphonic acids and strong complexing agents like NTMP ($N(CH_2PO_3H_2)_3$) and EDTMP ($(H_2O_3PCH_2)_2NCH_2CH_2N(CH_2PO_3H_2)_2$), which are phospha analogues of the classical complexones NTA ($N(CH_2COOH)_3$) and EDTA ($(HOOCCH_2)_2NCH_2CH_2N(CH_2COOH)_2$). Dissociation constants, stability constants for protonation and metal complex formation were studied as quoted with a few selected key papers [5–16].

Further interests concentrated on phospha analogues of carboxylic acids, e.g., phosphonocarboxylic acids and geminal bisphosphonic acids. ^{31}P- and ^{13}C-NMR spectra of cyclohexyl- and phenylphosphonic acid showed that chemical shifts δ_P and δ_C including coupling constants $^nJ_{PC}$ (n = 1–4) of cyclohexanephosphonic acid and benzenephosphonic acid proved to be pH-dependent [17].

A key paper in understanding the NMR titration of geminal bisphosphonate structures described three asymmetric esters of chlodronic acid $(HO)_2(O)P\text{-}CCl_2\text{-}P(O)(OiPr)OH$, $(HO)_2(O)P\text{-}CCl_2\text{-}P(O)(OiPr)_2$, and $HO(iPrO)(O)P\text{-}CCl_2\text{-}P(O)(OiPr)_2$. Proton coupled ^{31}P-NMR titration spectra revealed the coupling constants $^2J_{PP}$ in a range between 15.6 and 17.9 Hz. This significant parameter is not accessible for the symmetric ester $HO(iPrO)(O)P\text{-}CCl_2\text{-}P(O)(OiPr)OH$ since this compound gives rise to a dynamic deceptively simple spectrum ranging from singlet to triplet as a result of the parent symmetric $[AM_6X]_2$ spin system [18]. ^{31}P-NMR measurements at 202.5 MHz showed that the chemical shift δ_P of $CH_3C(OH)[P(O)(OH)_2]_2$ (HEDP) is sensitive towards pH and the concentration of $[(CH_3)_4N]^+$ when $[(CH_3)_4N]Cl$ was used as an ion buffer [19].

The determination of high pK values (pK > 13) and low pK values (pK < 1) required specific, advanced techniques for NMR titration. Comments on measurements at high and low pH were reported [19,20]. $^1H/^{31}P$ NMR pH indicator series were used to eliminate the glass electrode in NMR spectroscopic pK determinations, leading to "electrodeless titrations" [21]. Comprehensive guidelines for NMR measurements for the determination of high and low pK values were given in a IUPAC Technical Report. Those sophisticated and detailed instructions should be followed for accurate analytical and NMR measurements, data evaluation and subsequent publications [22].

1.1. *Developing Technical Setups for Automated NMR Titrations*

In general, NMR titrations for various nuclei were performed in single sample techniques, which proved to be rather laborious and time consuming. For practical reasons, the number of data points were limited in those early titration curves. Hence, attempts were made to develop the technology of automated NMR titrations.

An innovative set up was constructed, which permitted the acquisition of spectra from spinning 20 mm NMR tubes, adding a solution of base under efficient mixing while monitoring the pH. This apparatus worked together with the wide-bore magnet of a Bruker CXP-300 spectrometer, yielding approximately 80 titration points within a couple of hours. This technology was successfully used to titrate H_3PO_4 vs. KOH and provided a smooth NMR titration curve [23].

Further progress for automated NMR titrations inside spinning 10 mm NMR tubes was described and the novel installation applied to monitor the complex formation between $Tl(I)^+$ and Cl^- in aqueous solutions. A Bruker CXP-100 spectrometer was used, operating at 51.9 MHz for ^{205}Tl [24].

Very recently, an elegant low-cost construction was developed for a gravity-driven pH adjustment inside a 5 mm NMR tube. No hardware modifications of the NMR spectrometer were requested. This technique was applied to site-specific protein pK measurements [25]. It might be useful for future studies of organophosphorus compounds.

A different route to automated NMR titrations was chosen by the Düsseldorf group. We were intrigued by the technology of 2D NMR spectra and the graphical representation of such spectra by standard spectrometer software. Hence, a hyphenated technique was envisaged, replacing the f2 axis of 2D NMR spectra by analytical parameters.

Bypass constructions were developed in several generations of increasing accuracy. A 10 mm NMR tube was attached to a special homemade insert and used with a Bruker AM 200 SY NMR spectrometer operating at 81 MHz for ^{31}P NMR. This insert acted as bypass to a precision titration equipment. A series of pH dependent 1D NMR spectra were recorded and processed (using standard spectrometer software) to yield instructive "pseudo 2D NMR" spectra (e.g., in analogy to COSY spectra). Chemical shift δ_P data were correlated with analytical data like pH or the degree of titration τ. The technical setup and two examples are shown in [26]. Phosphaalanine was used as an example where deprotonation and complex formation with Zn^{2+} cations were observed by titrations vs. tetramethylammonium hydroxide (TMAOH) [26].

This equipment was used to characterize a series of aminomethylphosphine oxides $(CH_3)_{3-n}(CH_2NH_2)_nPO$ (n = 1–3), adding n equivalents of HCl and back-titrating vs. NaOH. Ion-specific chemical shifts δ_P and pK data were obtained for those aminomethylphosphine oxide bases. In addition, technical details of NMR, analytics, software concepts and programs used were described [27].

A brief overview of "^{31}P NMR controlled titrations of Phosphorus-Containing Acids and Bases in Protolysis and Complex Formation" reported about the 81 MHz $^{31}P\{^1H\}$ NMR titration of phosphonoacetic acid [28]. The hardware and software concepts were shown.

Typical "pseudo 2D NMR" spectra correlating the chemical shift δ_P vs. the degree of titration τ were obtained for the pair of isomers 1- and 2-aminoethanephosphonic acids CH_3-$CH(NH_2)$-PO_3H_2 and NH_2-CH_2-CH_2-PO_3H_2 (α-Ala-P and β-Ala-P) and for diphosphaasparaginic acid H_2O_3P-$CH(NH_2)$-CH_2-PO_3H_2 (Asp-P_2). Macroscopic dissociation constants and ion-specific chemical shifts are reported. The *p*-aminophenylene-substituted phosphonic acid *p*-NH_2-C_6H_4-PO_3H_2 was compared [29]. Hardware and software concepts used in NMR titration were demonstrated. A subsequent UV-controlled titration revealed the microscopic dissociation scheme of *p*-NH_2-C_6H_4-PO_3H_2. Corresponding deprotonation patterns were discussed [30].

In practice, ^{13}C-NMR titrations in single sample techniques proved to be very time consuming. Hence, it seemed advisable to use the technology described above for automated 50.29 MHz $^{13}C\{^1H\}$ or ^{13}C-NMR measurements. As practical examples, the pair of isomers 1- and 2-aminoethanephosphonic acids were titrated vs. NaOH. Within this context, CH_3-$CH(NH_2)$-PO_3H_2 and the fluorinated analogue CF_3-$CH(NH_2)$-PO_3H_2 were compared using $^{31}P\{^1H\}$ and $^{19}F\{^1H\}$ NMR titrations. Replacing the CH_3 by a CF_3 group reduces the basicity of the NH_2 function, which is reflected in δ_P vs. τ and in the δ_F vs. τ correlations [31].

The experimental set up described above requested individual titrations for each nucleus wanted. Hence, multinuclear studies (e.g., ^1H and ^{13}C and ^{31}P) demanded high spectrometer times.

At this stage, special probe heads were developed by Bruker for another hyphenated technique combining liquid chromatography with HR NMR. In our laboratory, a Bruker LC probe head LC-TXO-NMR was successfully introduced to a DRX 500 NMR spectrometer and used for advanced NMR titrations. It became routine to run consecutively $^{31}P\{^1H\}$, ^{31}P, and ^1H-NMR spectra for each titration step, thus saving time, gaining higher sensitivity and reducing the necessary concentrations (and amounts) of titrands. Excellent spectra resulted with a high S/N ratio and high digital resolution in the chemical shift or frequency axis.

In addition, a special ^{19}F-LC probe head was available, combining ^{19}F and ^1H-NMR techniques. The high field stability of the supercon magnet allowed measurements in H_2O solutions (without D_2O), thus avoiding the problems with "mixed" stability and dissociation constants resulting from D_2O/H_2O mixed solvents.

A comprehensive report about the technical designs of NMR and analytical components, software, data evaluation, error calculations and applications was written in 2002 and incorporated into the Bruker NMR Guide collection, freely accessible for Bruker spectrometer users [32] only. This detailed review is now open for free downloads to all interested readers: (a) https://www.theresonance.com/nmr-controlled-titration-download-the-paper/, (b) https://www.bruker.com/fileadmin/user_upload/8-PDF-Docs/MagneticResonance/NMR/NMR_controlled_titration.pdf.

NMR titrations using 200 MHz and 500 MHz spectrometers were described using geminal bisphosphonic acids, e.g., HEDP and Pamidronic acid, as model systems. The TXO-HPLC probe head improved the signal-to-noise ratio of "pseudo 2D NMR" spectra and reduced the concentration of titrand required by this procedure: The following concentrations for sensor nuclei are recommended: ^{1}H: 0.25–0.01 mol/L, ^{13}C: 0.50–0.005 mol/L, ^{19}F: 0.01–0.005 mol/L, ^{31}P: 0.01–0.001 mol/L, and ^{113}Cd: 0.25–0.1 mol/L. ^{113}Cd NMR was used when studying protolytic and complex formation equilibria of $(H_2O_3P\text{-}CH_2)_2NCH_2CH_2N(CH_2PO_3H_2)_2$ (EDTMP).

Within this context, a special computer program MultipleNMRGraphics was developed which is able to generate four characteristic "pseudo 2D NMR" plots, e.g., δ_P vs. pH or δ_P vs. τ either as contour or as stacked plots, in black-and-white or color design [33]. Those graphics have a lower storage demand than the previously used "pseudo 2D NMR" spectra generated by the routine Bruker spectrometer software.

Some examples relevant to phosphorous chemistry and organic chemistry dealt with in [32] are listed in Table 1:

Table 1. Some examples for NMR-controlled titrations of phosphonic acids, phosphinic acids, and carboxylic acids as discussed in [32]. 1) Retro titration; 2) Micro dissociation; 3) Second-order ^{1}H-NMR spin systems.

Examples	NMR	Remarks
Phosphonic acids		
$CH_3P(O)(OH)_2$	$^{31}P\{^{1}H\}$	
$LiOOC\text{-}CH_2\text{-}P(O)(OLi)_2$	$^{31}P\{^{1}H\}$	1)
$(HO)_2(O)P\text{-}CH_2\text{-}CH_2\text{-}P(O)(OH)_2$	$^{13}C\{^{1}H\}$	
$CH_3\text{-}C(OH)[P(O)(OH)_2]_2$, HEDP, etidronic acid	$^{31}P\{^{1}H\}$	
$NH_2\text{-}CH_2\text{-}CH_2\text{-}C(OH)(P(O)(OH)_2)_2$, pamidronic acid	$^{31}P\{^{1}H\}$	
$HOOC\text{-}CH_2\text{-}CH(COOH)\text{-}CH(COOH)\text{-}P(O)(OH)_2$, PPTC	$^{31}P\{^{1}H\}$	
Phosphinic acids		
$(CH_3)_2P(O)OH$	$^{31}P\{^{1}H\}$	
$HOOC\text{-}CH_2\text{-}CH_2\text{-}P(CH_3)(O)OH$	$^{13}C\{^{1}H\}$, ^{1}H	2)
$HO(O)(CH_3)P\text{-}CH_2\text{-}CH_2\text{-}P(CH_3)(O)OH$	$^{13}C\{^{1}H\}$	
$HO(O)(CH_3)P\text{-}CH_2\text{-}CH_2\text{-}C(H)(NH_2)COOH$	^{1}H	2, 3)
Carboxylic acids		
CH_3COOH	$^{13}C\{^{1}H\}$	
$CH(CH_3)_2\text{-}CH_2\text{-}CH(NH_2)\text{-}C(O)\text{-}NH\text{-}CH(CH_3)\text{-}COOH$, peptide Leu-Ala	$^{13}C\{^{1}H\}$	
$CH_2{=}CF\text{-}CH_2\text{-}C(CH_3)(NH_2)\text{-}COOH$	^{19}F	

A modification of our design for automated NMR titrations shown above was adjusted to the local conditions of a Bruker 250 MHz spectrometer and applied to study the complexation of Zn^{2+}, Cd^{2+} and Pb^{2+} with diazacrown ethers substituted by phosphonate groups [34].

Particular attention was drawn towards microscopic dissociation constants going back to early studies on $NH_2\text{-}CH_2\text{-}CH_2\text{-}NH\text{-}CH_2\text{-}COOH$. 60 MHz and 100 MHz ^{1}H-NMR titrations evaluated the pH dependence of a singlet for the methylene group $NH\text{-}CH_2\text{-}COOH$, while the ethylene function $N\text{-}CH_2\text{-}CH_2\text{-}N$ appeared with the spectral character, changing from a deceptively simple singlet towards an AA′BB′ ($[AB]_2$) system. The analytical formalism and microscopic dissociation constants were derived [35]. For deeper reading, an up-to-date and comprehensive survey on the theory and practice of proton microspeciation based on NMR-pH titrations is recommended [36].

As an example, *S*-2-amino-4-(methylphosphinoyl)butyric acid (*S*-phosphinothricine, GLUFOSINATE) $HOOC\text{-}CH(NH_2)\text{-}CH_2\text{-}CH_2\text{-}P(CH_3)(O)OH$ was characterized by $^{31}P\{^{1}H\}$- and ^{1}H-NMR titrations. Microscopic dissociation and intramolecular rotational equilibria were discussed [32,37]. Within this context, a program LAOTIT was developed, which is able to simulate series of pH-dependent second-order NMR spectra. A practical example for $AFGMNQ_3X$ spin systems of GLUFOSINATE in a pH range from 1 to 6 was shown in [37].

The ring-chain tautomerism and protolytic equilibria of an effectively three-basic 3-hydroxy-3-phosphonoisobenzofuranone was studied by ^{1}H-, ^{13}C{^{1}H}- and ^{31}P{^{1}H}-NMR-controlled titrations. A complex pattern of macroscopic and microscopic deprotonation steps leading from the starting H_3L to the final L^{3-} (Scheme 1) was discussed.

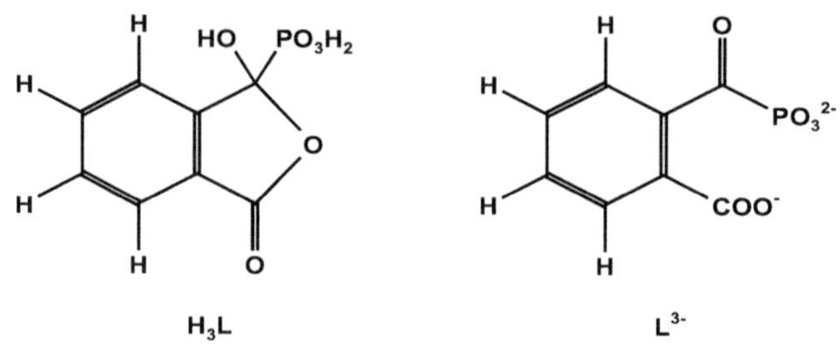

Scheme 1. Showing 3-hydroxy-3-phosphonoisobenzofuranone H_3L and its terminal anion L^{3-}.

The OPIUM program enabled the simultaneous evaluation of potentiometric, ^{31}P{^{1}H}- and ^{1}H-NMR titrations using the four individual ^{1}H signals from the ABCD system [38]. Macroscopic dissociation constants, $pK_1 = 0.445 \pm 0.008$, $pK_2 = 5.792 \pm 0.003$, $pK_3 = 6.486 \pm 0.002$. δ_H, δ_C, and δ_P of H_3L (0.2510 mol/L in D_2O) and L^{3-} (0.1919 mol/L in 1 mol/L KOD), were determined. For details of the complex equilibrium system, see [39,40].

NMR-controlled titration was successfully used to analyze the mixture of diastereomers from 1-phosphonopropane-1,2,3-tricarboxylic acid, HOOC-CH_2-CH(COOH)-CH(COOH)-PO_3H_2 (PPTC). The genuine product from synthesis consisted of 64% of the RS/SR and 36% of the RR/SS forms. ^{31}P{^{1}H}-NMR-controlled titration revealed two diastereospecific titration curves which were individually identified by additional 1D and 2D NMR studies using ^{1}H, ^{31}P and ^{13}C nuclei. Dissociation constants and ion-specific chemical shifts δ_P were calculated for the pair of diastereomers [41–43]. It seems evident to use automated NMR titration for production control in research and industrial chemistry.

1.2. Some Comments on Macroscopic Protolytic Equilibria—Dissociation and Stability Constants

Organophosphorus compounds studied by potentiometric or NMR-controlled titrations may be described by two numerical indices: a, the number of acidic functions (e.g., P(O)OH, C(O)OH, etc.) and b, the number of basic functions (e.g., NH_2, NHR, NR_2, etc.). The minimal protonated species corresponds to the n-valent base L^{-a} having a anionic centers and b neutral base centers in $(^{0}N)_b$-R-$(O^{-})_a$. Total protonation leads to the n-valent acid H_nL^{b+} (n = a + b) having a neutral centers and b cationic acid centers in $(^{+}HN)_b$-R-$(OH^{0})_a$.

Protonation equilibria of the n-valent base are described by Equation (1):

$$iH^{+} + L^{-a} \rightleftharpoons H_iL^{i-a} \quad i = 1 \text{ to } n \tag{1}$$

and by brutto-stability constants following Equation (2):

$$\beta_i = \frac{[H_iL^{i-a}]}{[H^{+}]^{i}[L^{-a}]} \quad i = 1 \text{ to } n \quad \beta_0 = 1 \tag{2}$$

Stepwise dissociation equilibria of the n-valent acid are described by Equation (3):

$$H_{n+1-i}L^{b+1-i} \rightleftharpoons H^{+} + H_{n-i}L^{b-i} \quad i = 1 \text{ to } n \tag{3}$$

while corresponding dissociation constants K_i are given by Equation (4):

$$K_i = \frac{[H^+][H_{n-i}L^{b-i}]}{[H_{n+1-i}L^{b+1-i}]} \quad i = 1 \text{ to } n \tag{4}$$

Stoichiometric stability constants and stoichiometric dissociation constants are connected by Equations (5) and (6):

$$pK_i = \lg\beta_{n+1-i} - \lg\beta_{n-i} \quad i = 1 \text{ to } n \tag{5}$$

$$\lg\beta_i = \sum_{j=1}^{i} pK_{n+1-j} \quad i = 1 \text{ to } n \tag{6}$$

This paper will use stoichiometric variables (containing concentrations instead of activities) in abbreviated forms: pK_i—macroscopic acid dissociation constant; pk_i—microscopic acid dissociation constant; pK_w—ion product of water. pH stands for the concentration-based pH = $-\lg(c_H)$. Glass electrodes were calibrated by blank titration. The more complex situation of activities and activity-based parameters exceeds the scope of this paper and hence will not be discussed at this stage.

The molar fractions x_i of protolytic species H_iL^{i-a} are derived from Equation (7):

$$x_i = \frac{10^{\lg\beta_i - i\cdot pH}}{\sum\limits_{j=0}^{n} 10^{\lg\beta_j - j\cdot pH}} \quad i = 0 \text{ to } n \tag{7}$$

Each protolytic species H_iL^{i-a} present in the equilibrium contributes specific NMR parameters $\delta(H_iL^{i-a})$ in an exchange reaction, which is rapid on the NMR timescale. Effectively, only one signal is observed when monitoring NMR during the course of titrations. A dynamically averaged chemical shift δ follows Equation (8):

$$\delta = \sum_{i=0}^{n} x_i \cdot \delta_{H_iL^{i-a}} \quad i = 0 \text{ to } n \tag{8}$$

A gradient called the deprotonation shift Δ_i [ppm] is given by Equation (9):

$$\Delta_i = \delta_{H_{n-i}L^{b-i}} - \delta_{H_{n+1-i}L^{b+1-i}} \quad i = 1 \text{ to } n \tag{9}$$

This gradient defines the change of chemical shift for each deprotonation step. Signs and magnitudes of gradients are used to elucidate the deprotonation and protonation pathways of multifunctional acids, bases and ligands as shown in examples below.

As deduced above, the dynamically averaged chemical shift δ is a function of pH. Experimentally, the pH of solutions may be varied by titration with a strong univalent base or a strong univalent acid. While the experiment directly provides the well-known titration curve pH = $f(V_{Titrator})$, it is more convenient to calculate the inverse function $V_{Titrator}$ = f(pH) with suitable computer programs. Within this paper, a reduced parameter τ, commonly called degree of titration, will be used to describe the status of a titration process. τ is a ratio defined by Equation (10):

$$\tau = \frac{n_{Titrator}}{n_{Titrand}} \tag{10}$$

The sign of τ is positive if $n_{Titrator}$ corresponds to the molar amount of a strong monovalent base (e.g., NaOH, KOH, TMAOH), but is negative for a strong monovalent acid (HCl, HNO_3, $HClO_4$). $n_{Titrand}$ corresponds to the molar amount of a n-basic acid H_nL.

Details of the basic principles and experimental equipment with hardware and software are described in [31,32] and in references given herein.

2. Results and Discussion

In the following sections, a few examples will be shown for automated NMR-controlled titrations using hardware and software concepts described above. Chemical shifts δ_C [ppm] quoted below were referenced vs. $(CH_3)_3Si\text{-}CH_2\text{-}CH_2\text{-}SO_3Na$, while δ_P [ppm] was virtually referenced towards external H_3PO_4. Coupling constants $^nJ_{XY}$ are given in [Hz].

2.1. Phosphonic Acids

Methanephosphonic acid **1** and phenylphosphonic acid **2** shown in Scheme 2 were chosen from [31,44], which will be presented below:

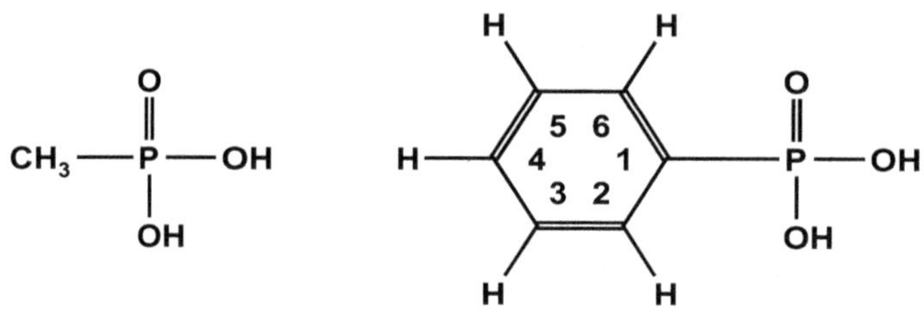

1 H_2L **2** H_2L

Scheme 2. Phosphonic acids: methanephosphonic **1** and phenylphosphonic acid **2**.

2.1.1. Methanephosphonic Acid 1

The results from a proton-coupled ^{31}P-NMR-controlled titration of methanephosphonic acid **2** vs. NaOH are shown as a contour plot in Figure 1. A quartet structure from the parent A_3X spin system of the P-CH_3 fragment is recognized. Numerical results are given in Table 2. The deprotonation of both P-OH functions induces a decrease in chemical shifts δ_P and a decrease in the absolute values of $^2J_{PH}$.

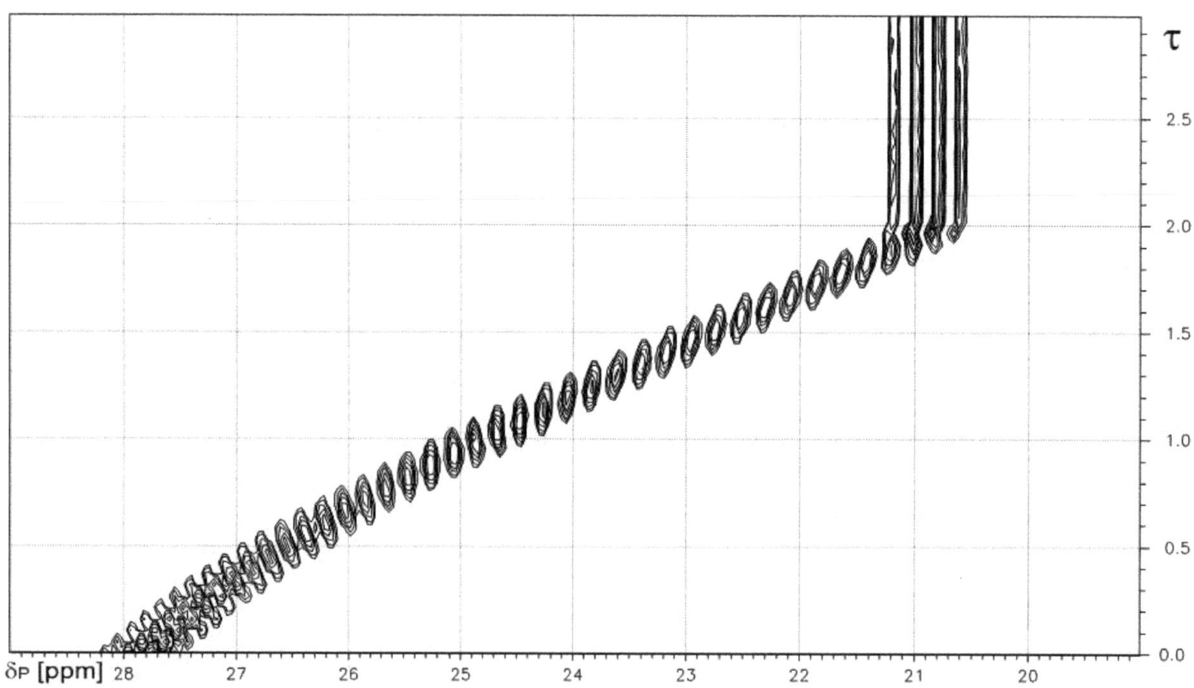

Figure 1. ^{1}H-coupled ^{31}P-NMR-controlled titration of methanephosphonic acid **1** vs. NaOH. Note: quartet fine structure from X-part of AX_3 system of P-CH_3 fragment. *X*-axis: δ_P [ppm]. *Y*-axis: degree of titration τ.

Table 2. Specific chemical shifts δ_C, δ_P and coupling constants $^1J_{PC}$ and $^2J_{PH}$ of methanephosphonic acid **1** were obtained by $^{13}C\{^1H\}$-, $^{31}P\{^1H\}$-, and ^{31}P-NMR-controlled titrations in H_2O. $\Delta_i = \delta(H_{n-i}L) - \delta(H_{n+1-i}L)$ or $\Delta_i = {}^nJ_{PC}(H_{n-i}L) - {}^nJ_{PC}(H_{n+1-i}L)$, respectively. i = 1 to n. n = 2. Experimental data: $C_{Titrand}$: a) 0.269 mol/L. b) 0.01220 mol. c) 0.0095 mol/L. $C_{Titrator}$: a) 4.82 mol/L KOH. b) 0.0971 mol/L NaOH. c) 0.0971 mol/L NaOH. d) Early data from results from titration vs. KOH [45].

			1 in H_2O			
Method	$^{13}C\{^1H\}$		$^{31}P\{^1H\}$	^{31}P		$^{31}P\{^1H\}$
Exp.	a)		b)	c)		d)
Species	δ_C	$^1J_{PC}$	δ_P	δ_P	$^2J_{PH}$	δ_P
	[ppm]	[Hz]	[ppm]	[ppm]	[Hz]	[ppm]
H_2L	14.27	135.92	33.03	33.03	−17.65	31.76
HL^-	15.53	133.82	24.79	24.79	−16.52	24.94
L^{2-}	16.61	129.95	21.08	21.08	−15.52	20.94
Gradients	δ_C	$^1J_{PC}$	δ_P	δ_P	$^2J_{PH}$	δ_P
Δ_1	1.26	−2.10	−8.24	−8.24	1.13	
Δ_2	1.08	−3.87	−3.71	−3.71	1.00	
pK_i						
pK_1	2.27		2.06	2.00		2.33
pK_2	7.85		7.66	7.68		7.78

pK_i values found are consistent with results from potentiometric titrations of $CH_3P(O)(OH)_2$ [46].

2.1.2. Phenylphosphonic Acid 2

Chemical shifts δ_P for protolytic species H_2L, HL^-, and L^{2-} of phenylphosphonic acid **2** together with dissociation constants pK_1 and pK_2 are listed in Table 3. The deprotonation of both P-OH groups leads to characteristic high field shifts for δ_P as indicated by negative gradients Δ_1 and Δ_2.

Table 3. Specific chemical shifts δ_P [ppm], gradients [ppm] and dissociation constants pK_i from $^{31}P\{^1H\}$-NMR-controlled titration of phenylphosphonic acid **2** vs. TMAOH and NaOH. Exp.: $C_{Titrator}$: a) 0.09894 mol/L TMAOH. b) 0.09925 mol/L NaOH. $C_{Titrand}$: a) 0.02 mol/L **3**. b) 0.008 mol/L **2**. Shifts and gradients given in ppm.

	Phenylphosphonic Acid 2	
	vs. TMAOH	vs. NaOH
	a)	b)
$\delta_P(H_2L)$	18.39	17.77
$\delta_P(HL^-)$	13.77	13.75
$\delta_P(L^{2-})$	11.69	11.72
Δ_1	−4.62	−4.02
Δ_2	−2.08	−2.03
pK_1	1.74	1.86
pK_2	7.28	7.16

Higher concentrations are required for $^{13}C\{^1H\}$-NMR-controlled titrations as shown for the titration of phenylphosphonic acid **2** vs. KOH in Figure 2:

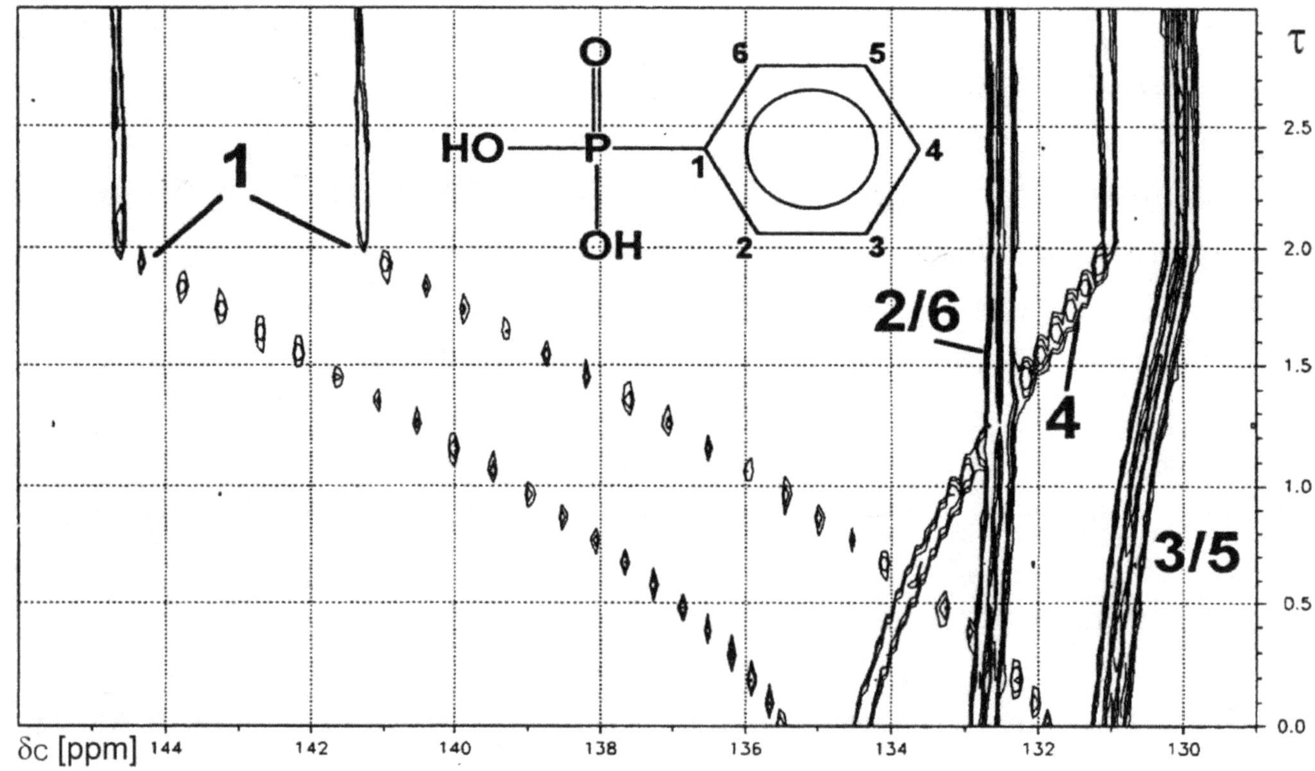

Figure 2. $^{13}C\{^{1}H\}$-NMR-controlled titration of phenylphosphonic acid **2** vs. KOH yielded a contour plot for chemical shift δ_C as a function of the degree of titration τ. *X*-axis: δ_C [ppm]. *Y*-axis: degree of titration τ.

The deprotonation of each of the two P-OH functions led to a strong low field shift for the *ipso*-C1 carbon. For the remaining carbons high field shifts are observed, an effect decreasing in the order *para*-C4 > *meta*-C3/5 > *ortho*-C2/5.

Semi-empirical calculations with VAMP 4.4 using parameter set AM1 showed that the electron density at C1 increases with deprotonation in the order $PhPO_3H_2$ < $PhPO_3H^-$ < Ph-PO_3^{2-}, while the electron density of C4 decreases in this order [44,47]. In addition, the deprotonation of P-OH led to a decrease for all $^{n}J_{PC}$ (n = 1 to 4). Particularly indicative is $^{1}J_{PC}$ from the *ipso*-carbon C1, which reaches a minimum at total deprotonation. Numerical results for compound 2 are listed in Table 4:

Table 4. Specific chemical shifts δ_C [ppm] and coupling constants $^{n}J_{PC}$ (n = 1 to 4) [Hz] for the $^{13}C\{^{1}H\}$-NMR-controlled titration of phenylphosphonic acid **2** vs. KOH. pK_1 = 1.75. pK_2 = 6.92. Experimental data: $C_{Titrand}$: a) 0.237 mol/L **2**. $C_{Titrator}$: 4.53 mol/L KOH.

	δ_C and $^{n}J_{PC}$ for Species			Gradients	
Parameters	**H_2L**	**HL^-**	**L^{2-}**	**Δ_1**	**Δ_2**
δ_C C1	133.66	138.15	143.83	+4.48	+5.69
δ_C C2/6	133.68	133.45	133.43	−0.24	−0.02
δ_C C3/5	131.99	131.59	130.95	−0.40	−0.64
δ_C C4	135.66	134.03	131.98	−1.58	−2.06
$^{1}J_{PC}$	183.48	177.02	167.32	−6.46	−9.70
$^{2}J_{PC}$	10.51	9.72	8.79	−0.79	−0.93
$^{3}J_{PC}$	14.84	13.90	12.65	−0.94	−1.25
$^{4}J_{PC}$	3.08	2.91	2.74	−0.17	−0.17

Laborious and time-consuming single sample $^{31}P\{^{1}H\}$- and $^{13}C\{^{1}H\}$-NMR studies on phenylphosphonic acid **2** were performed, where δ_P, δ_C, and $^{n}J_{PC}$ data are consistent with findings derived from the automated titrations presented in this paper [17].

2.2. *Comparison of Aliphatic and Aromatic Aminophosphonic Acids*

α-Aminoethylphosphonic acid (α-Ala-P) **3** [29,44], β-aminoethylphosphonic acid (β-Ala-P, CILIATIN) **4** [29], and *p*-aminophenylphosphonic acid **5** [29,30,43] shown in Scheme 3 will be studied as examples in the following section.

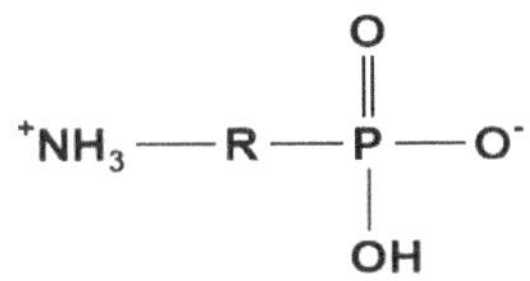

3: R = -CH(CH$_3$)- ; 4: R = -CH$_2$-CH$_2$- ; 5: R = -p-C$_6$H$_4$- .

Scheme 3. Aminophosphonic acids **3** to **5**.

2.2.1. Aliphatic Aminophosphonic Acids

Aminophosphonic acids NH_2-R-PO_3H_2, such as examples **3** to **5**, exist as betainic forms $^{+}NH_3$-R-PO_3H^{-} in solid and in solution state. Protolytic equilibria of aminophosphonic acids are described by macroscopic and microscopic formalisms as shown in Table 5 below:

Table 5. Macroscopic and microscopic dissociation species of aminophosphonic acids.

Dissociation Species		
Macroscopic	**Microscopic**	
H_3L^+	$^{+}NH_3$-R-PO_3H_2	
H_2L	$^{+}NH_3$-R-PO_3H^-	NH_2-R-PO_3H_2
HL^-	$^{+}NH_3$-R-PO_3^{2-}	NH_2-R-PO_3H^-
L^{2-}		NH_2-R-PO_3^{2-}

If R is aliphatic (e.g., in **3** and **4**), the deprotonation takes place following route a): $^{+}NH_3$-R-PO_3H_2 → $^{+}NH_3$-R-PO_3H^- → $^{+}NH_3$-R-PO_3^{2-} → NH_2-R-PO_3^{2-}. But if R is aromatic (e.g., in **5**. R = *p*-C_6H_4-), the deprotonation dominantly will follow b): $^{+}NH_3$-R-PO_3H_2 → $^{+}NH_3$-R-PO_3H^- → NH_2-R-PO_3H^- → NH_2-R-PO_3^{2-}. Consistent conclusions are supported by a combination of potentiometric titrations and $^{31}P\{^{1}H\}$-NMR-controlled titrations as shown for examples **3** to **5**, and in addition by $^{13}C\{^{1}H\}$-NMR-controlled titrations for examples **3** and **4**. Owing to its low solubility, **5** was not suitable for $^{13}C\{^{1}H\}$-NMR-controlled titrations.

Macroscopic dissociation constants pK_i of **3** and **4** are listed in Table 6. pK_i data of compounds **4** and **5** were discussed in [44,45,48–50].

Specific chemical shifts δ_P and gradients Δ for compounds **3** to **5** obtained by $^{31}P\{^{1}H\}$-NMR-controlled titrations are given in Table 7:

The deprotonation of the P-OH groups led to high field shifts for δ_P connected with negative gradients. The final deprotonation of the NH_3^+ group gave rise to a low field shift for δ_P. This effect is stronger in α-aminophosphonic acid **4** than in β-aminophosphonic acid **5**. Earlier results for chemical shifts δ_P of H_2L, HL^- and L^{2-} species of **3** and **4** were mentioned in [5,45]. In addition, δ_P of H_3L^+ was accessible for **4** but not for **3**.

Table 6. Macroscopic dissociation constants pK_i of compounds α-Ala-P **3** and β-Ala-P **4** obtained by $^{13}C\{^1H\}$- and $^{31}P\{^1H\}$-NMR-controlled titrations and by potentiometric titrations [4a]. Note: $pK_3 - pK_2 > 3$ and $pK_2 - pK_1 > 3$ for compounds **3** and **4**. Exp.: **3:** $C_{Titrand}$: a) 0.0867 mol/L **3** + 0.0834 mol/L HNO_3. b) and c) 0.005 mol/L **3** + 0.00476 HNO_3 + 0.0917 mol/L $NaNO_3$. $C_{Titrator}$: a) 3.98 mol/L NaOH. b) and c): 0.100 mol/L NaOH. **4:** $C_{Titrand}$: d) 0.139 mol/L **4** + 0.139 mol/L HNO_3. e) 0.010 mol/L **4** + 0.010 mol/L HNO_3. f) 0.010 mol/L **4** + 9.747 mmol/L HNO_3. g) 0.005 mol/L **4** + 0.005 mol/L HNO_3 + 0.100 mol/L TMACl. $C_{Titrator}$: d) 0.98 mol/L NaOH. e): 0.0991 mol/L TMAOH. f): 0.0993 mol/L NaOH. g): 0.099 mol/L TMAOH.

	3			**4**			
	$^{13}C\{^1H\}$ **a**	**$^{31}P\{^1H\}$** **b**	**Pot.** c	**$^{13}C\{^1H\}$** **d**	**$^{31}P\{^1H\}$** **e**	**$^{31}P\{^1H\}$** **f**	**Pot.** **g**
	NaOH	NaOH	NaOH	NaOH	TMAOH	NaOH	TMAOH
pK_1	0.70	0.31	0.3	1.02	1.22	1.26	1.14
pK_2	5.72	5.63	5.58	6.38	6.23	6.24	6.34
pK_3	10.64	10.21	10.28	11.50	11.06	11.08	11.06

Table 7. Specific chemical shift δ_P [ppm] and gradients Δ [ppm] for compounds **3** to **5** [44]. *) Not iterated. Experimental details for **3** and **4** are given in Table 6. Titrator: a) NaOH; b) TMAOH; c) for **5** were used: $C_{Titrand}$ = 1.6953 mol/L **5** and 3.6935 mol/L TMAOH. Titrator = 0.09993 mol/L HCl.

	3	**4**		**5**
	$^{31}P\{^1H\}$ **a**	**$^{31}P\{^1H\}$** **b**	**$^{31}P\{^1H\}$** **a**	**$^{31}P\{^1H\}$** c
Species	**δ_P**	**δ_P**	**δ_P**	**δ_P**
H_3L^+	15 *	22.9	23.4	13 *
H_2L	14.92	19.29	19.36	12.16
HL^-	13.08	16.80	16.81	15.29
L^{2-}	22.25	19.39	19.72	12.70
Gradients				
Δ_1	−0.08	−3.61	−4.04	−0.84
Δ_2	−1.84	−2.49	−2.55	+3.13
Δ_3	+9.17	+2.59	+3.91	−2.59

$^{13}C\{^1H\}$-NMR-controlled titrations of compounds **4** and **5** led to specific chemical shifts δ_C, coupling constants $^1J_{PC}$, and gradients Δ as listed in Table 8:

Table 8. Specific chemical shifts δ_C [ppm], coupling constants $^1J_{PC}$ [Hz], and corresponding gradients Δ for α-aminoethanephosphonic acid (α-Ala-P) **3** and β-aminoethanephosphonic acid (β-Ala-P, CILIATIN) **4**. Spin enumerations: **3**: C2-C1(N)-P; **4**: (N)C2-C1-P. $^1J_{PC}$ shows a minimum for species HL^- of **3** and **4**. $^2J_{PC}$ was not resolved for compounds **3** and **4**.

	3 in H_2O			**4 in H_2O**		
Species	**δ_C(C1)**	**$^1J_{PC}$**	**δ_C(C2)**	**δ_C(C1)**	**$^1J_{PC}$**	**δ_C(C2)**
H_3L^+	46.80	151.5	16.0	27.50	137.4	37.09
H_2L	47.70	143.8	16.43	28.73	131.4	38.22
HL^-	49.07	134.5	17.23	29.47	124.8	39.28
L^{2-}	48.15	138.0	19.79	35.45	126.5	39.76
Gradients						
Δ_1	+0.9	−8.7	+0.43	+1.23	−6.0	+1.13
Δ_2	+1.37	+9.3	+0.80	+0.74	−6.6	+1.04
Δ_3	−0.92	+3.5	+2.55	+5.98	+1.3	+0.47

NMR was used to monitor the complex formation of aminophosphonic acids with biorelevant cations in homogeneous solutions. An instructive example is the $^{31}P\{^1H\}$-NMR-controlled titration of CILIATIN/Mg^{2+} vs. NaOH where the formation of [MgL] and $[MgHL]^+$ was monitored [44].

2.2.2. Aromatic *p*-Aminophenylphosphonic Acid **5**

The deprotonation of PO_3H^- in aliphatic aminophosphonic acids **3** and **4** is affiliated with a high field shift (gradients Δ_2 are negative), while the deprotonation of the ammonium function $^+NH_3$ leads to a low field shift (gradients Δ_3 are positive).

The aromatic *p*-aminophenylphosphonic acid **5** exhibits a different pattern: while gradient Δ_2 is positive, Δ_3 is negative (Scheme 4).

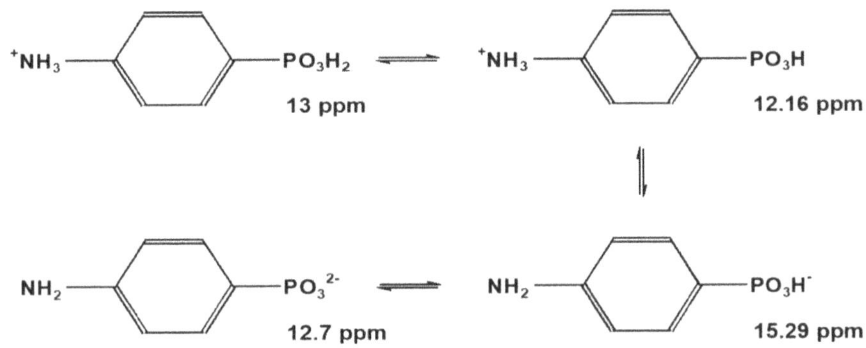

Scheme 4. Specific chemical shifts δ_P [ppm] derived from the $^{31}P\{^1H\}$-NMR-controlled retro titration of *p*-aminophenylphosphonic acid **5**. Δ_2 = +3.13 ppm. Δ_3 = −2.59 ppm.

But is it sufficient to assume a simple first-order macroscopic dissociation scheme for compound **5**? Deeper insight might be obtained from the microscopic dissociation scheme. In principle $^{13}C\{^1H\}$-NMR-controlled titration should lead to specific chemical shifts and coupling constants $^nJ_{PC}$ indicative for microscopic dissociations species of **5**. But *p*-aminophenylphosphonic acid **5** is less soluble in water than the aliphatic aminophosphonic acids **3** and **4**. The S/N-ratio of $^{13}C\{^1H\}$-NMR spectra of **5** is not sufficient to perform evaluable $^{13}C\{^1H\}$-NMR-controlled titrations. In this situation, UV/VIS-controlled titration, which allows for lower concentrations suitable for conclusive measurements, will help to study both the macroscopic and the microscopic dissociation equilibrium of **5** **[30]**. In addition, the parent compounds $C_6H_5PO_3H_2$ **2** and $C_6H_5NH_2{*}HCl$ **6** were compared. The following macroscopic pK_i data were found by potentiometric titration and listed in Table 9:

Table 9. Dissociation constants of compounds *p*-aminophenylphosphonic acid **5**, phenylphosphonic acid **2**, and anilinium hydrochloride **6**.

	5	2	6
pK_1	0.44	1.88	4.68
pK_2	3.95	7.15	
pK_3	7.56		

Those data point towards a dominating deprotonation sequence for **5** following $^+NH_3$-R-PO_3H_2 → $^+NH_3$-R-PO_3H^- → NH_2-R-PO_3H^- → NH_2-R-$PO_3{}^{2-}$. But is it justified to exclude the alternative route $^+NH_3$-R-PO_3H_2 → $^+NH_3$-R-PO_3H^- → $^+NH_3$-R-$PO_3{}^{2-}$ → NH_2-R-$PO_3{}^{2-}$? Evaluating the macroscopic dissociation constants of **5** shows that between pH = 1.5 and pH = 10, only three macroscopic species exist: H_2L, HL^-, and L^{2-}. The UV/VIS-controlled titration of **5** [30] showed that the maximum concentration for macroscopic HL^- is reached at pH = 5.75, consisting of two microdissociation species NH_2-R-PO_3H^- and $^+NH_3$-R-$PO_3{}^{2-}$ in a ratio of 9:1. Thus, the results from the UV/VIS-controlled titration of **5** [30] confirm the dominance of NH_2-R-PO_3H^- as previously assumed for the macroscopic deprotonation sequence derived from the $^{31}P\{^1H\}$-NMR-controlled titration of **5** [44].

2.3. *Phosphonocarboxylic Acids HOOC-(CH$_2$)$_n$-PO$_3$H$_2$ **7a** to **7d** (n = 0 to 3)*

Phosphonocarboxylic acids HOOC-$(CH_2)_n$-PO_3H_2 gave rise to potentiometrically [4,44] and NMR-controlled titrations [4,28,31,44]. Those neutral acids of type H_3L deprotonate dominantly in a sequence: HOOC-$(CH_2)_n$-PO_3H_2 → $^-$OOC-$(CH_2)_n$-PO_3H^- → $^-$OOC-$(CH_2)_n$-PO_3^{2-}. Corresponding dissociations constants for compounds shown in Scheme 5 are listed in Table 10 below:

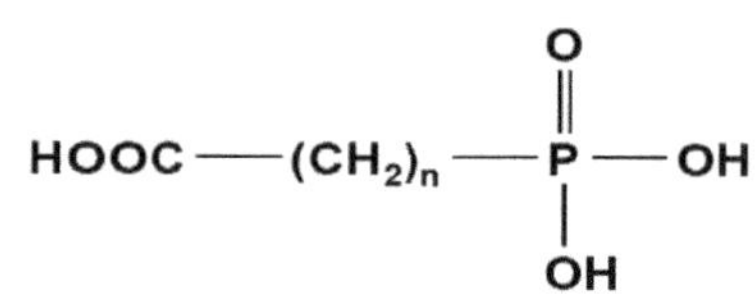

7a: n = 0 ; **7b**: n = 1 ; **7c**: n = 2 ; **7d**: n = 3 .

Scheme 5. Phosphonocarboxylic acids **7a** to **7d**.

Table 10. Dissociation constants of phosphonocarboxylic acids HOOC-$(CH_2)_n$-PO_3H_2 (n = 0 to 3) **7a** to **7d** [8,44]. Experimental data: a) $C_{Titrand}$: 0.153 mol/L FOSCARNET (trisodium phosphonoformate hexahydrate), $C_{Titrator}$: 2.002 mol/L HNO_3, retro titration; b) $C_{Titrand}$: 0.220 mol/L **7b**, $C_{Titrator}$: 0.980 mol/L NaOH. c) $C_{Titrand}$: 0.200 mol/L **7c**, $C_{Titrator}$: 3.986 mol/L NaOH. d) $C_{Titrand}$: 0160 mol/L **7d**, $C_{Titrator}$: 3.986 mol/L NaOH.

	HOOC-$(CH_2)_n$-PO_3H_2					
	7a	7a [4]	7b	7c	7c [4]	7d
	n = 0 [a]	n = 0	n = 1 [b]	n = 2 [b]	n = 2	n = 3 [b]
pK_1	0.78	1.7 ± 0.1	1.22 ± 0.166	2.58 ± 0.013	2.26 ± 0.04	2.276 ± 0.006
pK_2	3.60	3.59 ± 0.02	4.942 ± 0.004	4.633 ± 0.004	4.63 ± 0.02	4.776 ± 0.004
pK_3	7.57	7.56 ± 0.02	8.099 ± 0.003	7.738 ± 0.003	7.75 ± 0.02	7.969 ± 0.003

$^{13}C\{^1H\}$-NMR-controlled titrations yielded the specific chemical shift δ_C and coupling constants $^nJ_{PC}$ (n = 1 to 3) of phosphonocarboxylic acids HOOC-$(CH_2)_n$-PO_3H_2 (n = 0 to 3) **7a** to **7d** as listed in Table 11a. Gradients are given in Table 11b. Note: The deprotonation of P-OH groups and of C-OH led to a low field shift for all carbon atoms. Some chemical shifts and coupling constants $^1J_{PC}$ of **7a** and **7c** were obtained and discussed in an early key paper [4].

2.3.1. Compound **7c**: $^{13}C\{^1H\}$-NMR-Controlled Titration of 3-Phosphonopropionic Acid HOOC-CH_2-CH_2-PO_3H_2 7c.

3-Phosphonopropionic acid **7c** was chosen as an example to show practical results from $^{13}C\{^1H\}$-NMR-controlled titrations (see Figure 3a,b below). The deprotonation of C-OH and of both P-OH functions induces low field shifts for C1, C2, and C3. Hence, the corresponding gradients are negative. Lorentzian deconvolution of $^{13}C\{^1H\}$ signals yielded $^nJ_{PC}$, where absolute values follow the sequence: $^1J_{PC}$ >> $^3J_{PC}$ > $^2J_{PC}$. See Table 11a,b above.

Table 11. (**a**). Specific chemical shifts δ_C [ppm] and coupling constants ${}^nJ_{PC}$ [Hz] of phosphonocarboxylic acids HOOC-$(CH_2)_n$-PO_3H_2 (n = 0 to 3) **7a** to **7d**. For experimental data, see preceding Table 10. Remarks: n. r. = not resolved; (**b**). Gradients Δ_i of specific chemical shifts δ_C [ppm] and coupling constants ${}^nJ_{PC}$ [Hz] of phosphonocarboxylic acids HOOC-$(CH_2)_n$-PO_3H_2 (n = 0 to 3) **7a** to **7d**. For experimental data, see preceding Table 10.

(a)

7	n	Species	δ_C(C1)	δ_C(C2)	δ_C(C3)	δ_C(C4)	${}^1J_{PC}$	${}^2J_{PC}$	${}^3J_{PC}$	${}^4J_{PC}$
a	0	H_3L	176.8				246.6			
		H_2L^-	178.7				236.7			
		HL^{2-}	181.8				231.8			
		L^{3-}	187.3				220.0			
b	1	H_3L	172.91	37.68			128.6	n. r.		
		H_2L^-	175.44	39.30			117.8	n. r.		
		HL^{2-}	178.85	41.64			119.2	n. r.		
		L^{3-}	181.74	43.50			112.6	n. r.		
c	2	H_3L	179.25	30.09	24.51		138.5	3.6	17.3	
		H_2L^-	180.68	31.40	25.91		135.1	3.2	17.8	
		HL^{2-}	185.01	34.27	27.50		133.0	4.1	18.7	
		L^{3-}	186.71	35.61	28.99		130.3	3.6	19.8	
d	3	H_3L	180.38	36.67	20.44	28.19	135.2	4.0	17.4	n. r.
		H_2L^-	181.12	37.46	21.58	29.74	133.5	3.8	17.2	n. r.
		HL^{2-}	185.86	41.45	23.13	30.46	132.5	3.9	17.7	n. r.
		L^{3-}	186.41	42.01	24.08	31.92	130.1	3.4	17.9	n. r.

(b)

7	n	Gradients	δ_C(C1)	δ_C(C2)	δ_C(C3)	δ_C(C4)	${}^1J_{PC}$	${}^2J_{PC}$	${}^3J_{PC}$
a	0	Δ_1	+1.9				−9.9		
		Δ_2	+3.1				−4.9		
		Δ_3	+5.5				−11.8		
b	1	Δ_1	+2.53	+1.62			−10.8		
		Δ_2	+3.41	+2.34			+1.4		
		Δ_3	+2.89	+1.86			−6.6		
c	2	Δ_1	+1.43	+1.31	+1.40		−3.4	−0.3	0.5
		Δ_2	+4.33	+2.87	+1.59		−2.1	0.9	0.9
		Δ_3	+1.70	+1.34	+1.49		−2.7	−0.5	1.1
d	3	Δ_1	+0.74	+0.79	+1.14	1.56	−1.7	−0.2	−0.2
		Δ_2	+4.74	+3.99	+1.55	0.72	−1.0	0.1	0.5
		Δ_3	+0.55	+0.56	+0.95	1.46	−2.4	−0.5	0.2

The 81 MHz ${}^{31}P\{{}^1H\}$-NMR-controlled titration of 3-phosphonopropionic acid **7c** vs. NaOH yielded Figure 3c. The deprotonation sequence reported in [4] corresponds to: HOOC-CH_2-CH_2-PO_3H_2 → HOOC-CH_2-CH_2-PO_3H^- → ${}^-$OOC-CH_2-CH_2-PO_3H^- → ${}^-$OOC-CH_2-CH_2-$PO_3{}^{2-}$. Deprotonation at PO_3H_2 or PO_3H^- is affiliated with high field shifts of δ_P, while deprotonation at HOOC induces a low field shift for δ_P.

Specific chemical shifts δ_P for **7c** and corresponding anions together with gradients are listed in Table 12.

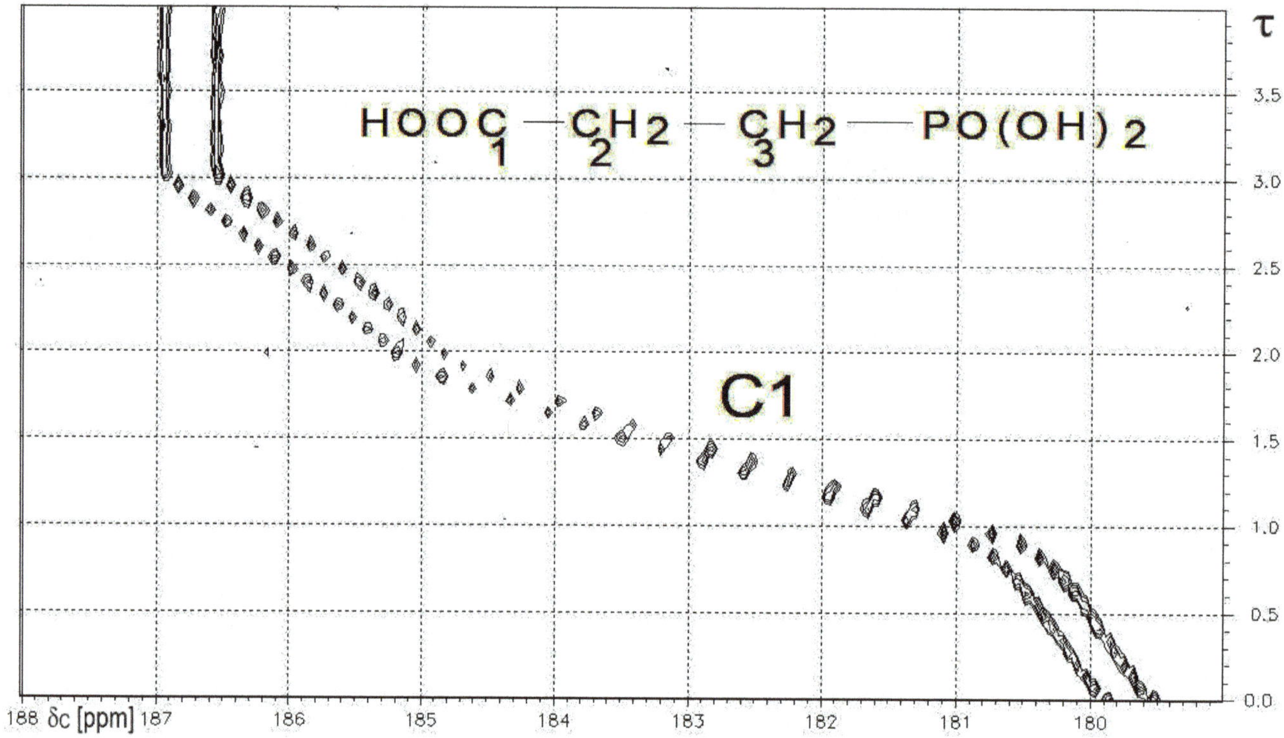

a

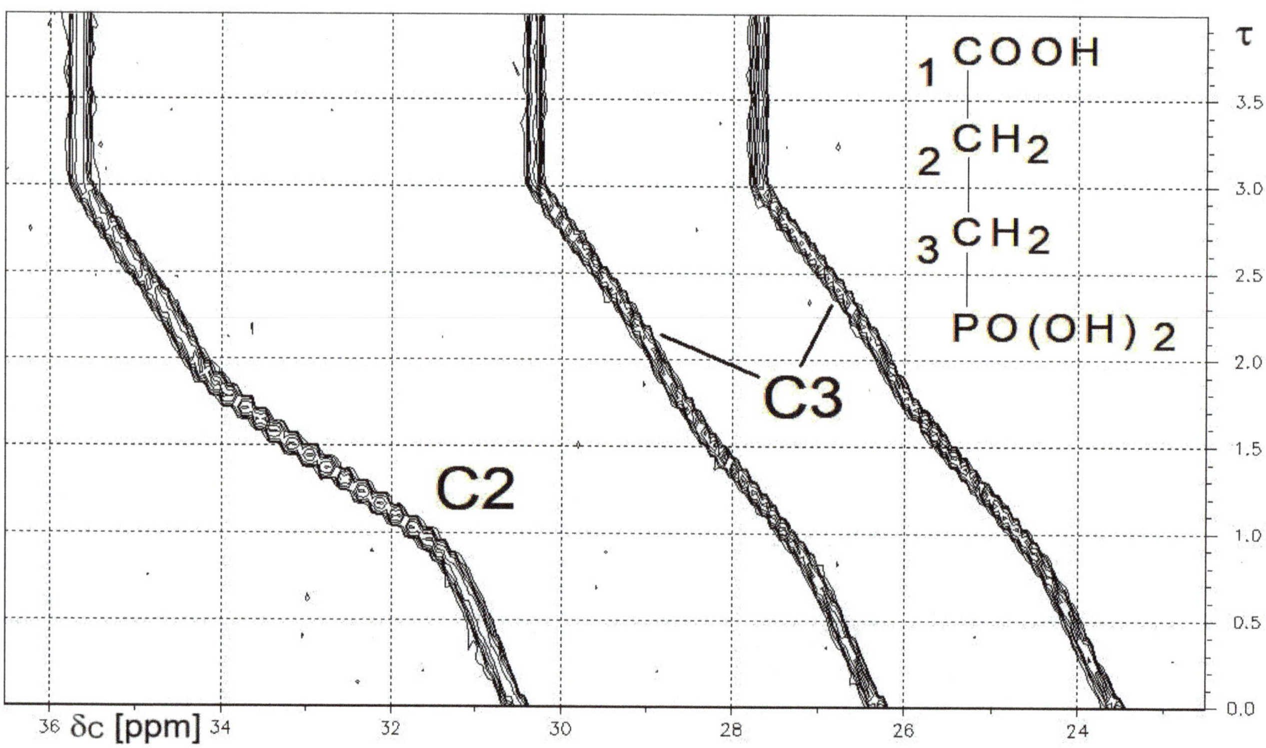

b

Figure 3. *Cont.*

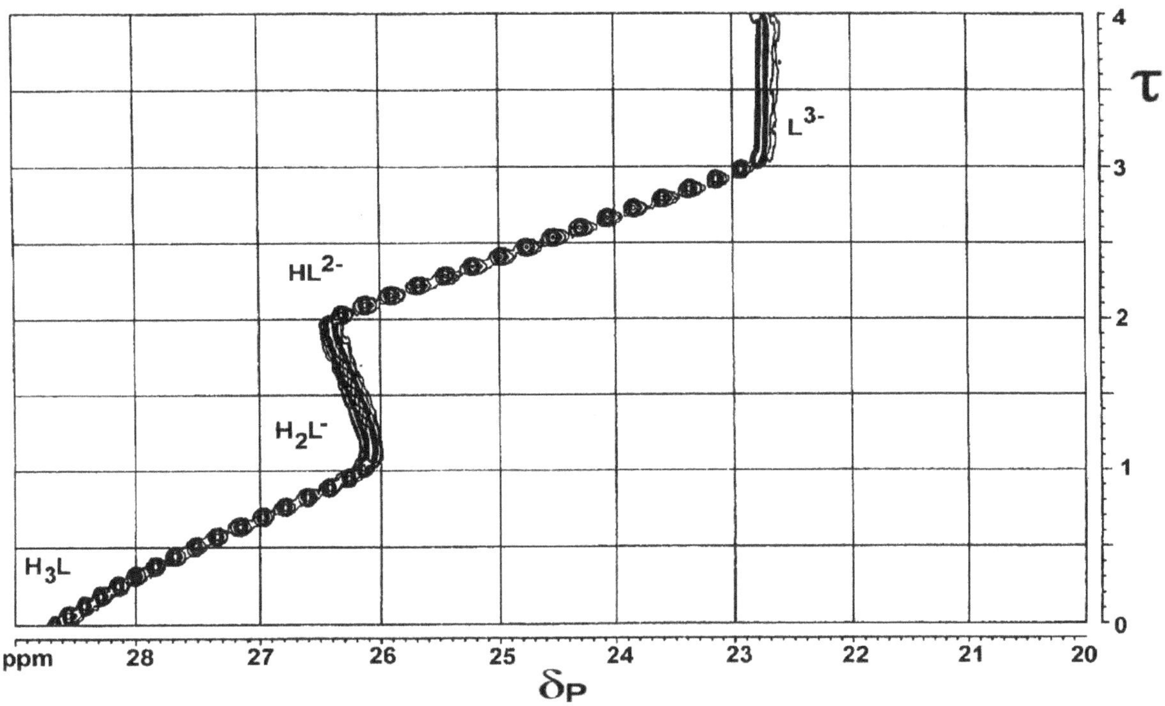

c

Figure 3. (**a**) $^{13}C\{^1H\}$-NMR-controlled titration of 3-phosphonopropionic acid **7c** vs. NaOH. Contour plot showing the range of the carboxylic carbon C1. For experimental data, see Table 10. *X*-axis: δ_C [ppm]. *Y*-axis: degree of titration τ; (**b**) $^{13}C\{^1H\}$-NMR-controlled titration of 3-phosphonopropionic acid **7c** vs. NaOH. Contour plot showing the range of the aliphatic carbons C2 and C3. For experimental data, see Table 10. *X*-axis: δ_C [ppm]. *Y*-axis: degree of titration τ; (**c**) 81 MHz $^{31}P\{^1H\}$-NMR-controlled titration of 3-phosphonopropionic acid **7c** vs. NaOH *X*-axis: δ_P [ppm]. *Y*-axis: degree of titration τ. $C_{Titrand}$: 0.010 mol/L of **7c**. $C_{Titrator}$: 0.10 mol/L TMAOH. Digital resolution: 0.6 Hz/point in *x*-axis. [51].

Table 12. Gradients Δ_i of specific chemical shifts δ_P [ppm] of **7c**. Experimental data: $C_{Titrand}$: 0.010 mol/L of **7c**. $C_{Titrator}$: 0.10 mol/L TMAOH.

Shifts	δ_P	Error
H_3L	29.93	±0.25
H_2L^-	24.56	±0.02
HL^{2-}	25.88	±0.01
L^{3-}	22.06	±0.01
Gradients		
Δ_1	−5.37	
Δ_2	+1.32	
Δ_3	−3.82	

2.3.2. ^{19}· F-NMR-Controlled Retro Titrations of Lithium Salts $LiOOC\text{-}CH_{2\text{-}n}F_n\text{-}PO_3Li_2$ **8a** and **8b**

The trilithium salts $LiOOC\text{-}CH_{2\text{-}n}F_n\text{-}PO_3Li_2$ (**8a** and **8b**; n = 1 and 2, Scheme 6) were used for retro titrations vs. HNO_3, since the parent mono- and difluorophosphonoacetic acids **8c** and **8d** were not available for ^{19}F-NMR- and $^{31}P\{^1H\}$-NMR titrations. Corresponding dissociation constants pK_i of **8c** and **8d** were calculated as listed in Table 13, while chemical shifts δ_F and δ_P and coupling constants $^2J_{PF}$ are given in Table 14. As expected, the introduction of fluorine into the skeleton of the parent phosphonoacetic acid led to lower pK_1 and pK_2 values. The deprotonation of P-OH groups induces a low field shift for δ_F in fluorinated phosphonic acids **8c** and **8d**.

8a: n = 1 ; **8b**: n = 2 . **8c**: n = 1 ; **8d**: n = 2 .

Scheme 6. Trilithium salts $LiOOC\text{-}CH_{2\text{-}n}F_n\text{-}PO_3Li_2$ **8a** and **8b** and free acids $HOOC\text{-}CH_{2\text{-}n}F_n\text{-}PO_3H_2$ **8c** and **8d**.

Table 13. Dissociation constants of fluorinated phosphonocarboxylic acids $HOOC\text{-}CH_{2\text{-}n}F_n\text{-}PO_3H_2$ (**8c** and **8d**; n = 1 and 2) obtained by the retro titration of $LiOOC\text{-}CH_{2\text{-}n}F_n\text{-}PO_3Li_2$ (**8a** and **8b**; n = 1 and 2) vs. $HClO_4$. Experimental data: $C_{Titrand}$: 0.85 mol/L **8a** or **8b** resp. $C_{Titrator}$: 0.3928 mol/L $HClO_4$. No ion buffer.

	$HOOC\text{-}CH_{2\text{-}n}F_n\text{-}PO_3H_2$		
	8c n = 1	8d n = 2	
	[44]	[44]	[52]
pK_1	1.05	0.52	1.30
pK_2	3.43	2.22	1.95
pK_3	7.08	6.36	6.16

Table 14. Chemical shifts δ_F [ppm], coupling constants ${}^2J_{PF}$ [Hz] and corresponding gradients of fluorinated phosphonocarboxylic acids $HOOC\text{-}CH_{2\text{-}n}F_n\text{-}PO_3H_2$ (**8c** and **8d**, n = 1 and 2) obtained by ${}^{19}F$-NMR-controlled titrations vs. HNO_3 of $LiOOC\text{-}CH_{2\text{-}n}F_n\text{-}PO_3Li_2$ (**8c** and **8d**, n = 1 and 2). For experimental data, see Table 13. δ_F is virtually referenced to $\delta_F(CF^{35}Cl_2{}^{37}Cl) = 0$ ppm.

	$HOOC\text{-}CH_{2\text{-}n}F_n\text{-}PO_3H_2$			
	8c	8d	8c	8d
	n = 1	n = 2	n = 1	n = 2
	^{19}F		^{19}F	
Species	δ_F	δ_F	${}^2J_{PF}$	${}^2J_{PF}$
H_3L	−38.27	44.24	67.8	76.5
H_2L^-	−38.72	50.24	65.5	88.6
HL^{2-}	−29.31	53.05	70.3	92.8
L^{3-}	−27.15	55.24	63.4	82.0
Gradients				
Δ_1	−0.45	+6.00	−2.3	+12.1
Δ_2	−9.41	+2.81	+4.8	+4.2
Δ_3	−8.16	+2.19	−6.9	−10.8

2.3.3. 2,4-Diphosphonobutane-1,2-Dicarboxylic Acid (DPBDC) 9

Strong practical interests focused on polyfunctional phosphonocarboxylic acids, e.g., phosphonosuccinic acid (PBS), 1-phosphonopropane-1,2,3-tricarboxylic acid (PPTC), and 2-phosphonobutane-1,2,4-tricarboxylic acid (PBTC), which gave rise to analytical and NMR studies of protolytic and complex formation equilibria [42–44,53].

The following section will deal with 2,4-diphosphonobutane-1,2-dicarboxylic acid (DPBDC) **9** to demonstrate the potential of automated NMR titrations. Dissociation constants of this 6-valent acid DPBDC **9** were obtained from precision potentiometric [53] and by ${}^{13}C\{{}^1H\}$-NMR-controlled titrations vs. NaOH [44].

${}^{13}C\{{}^1H\}$-technique yielded Figure 4a–d. The spin enumeration used in subsequent tables and figures is given in Scheme 7:

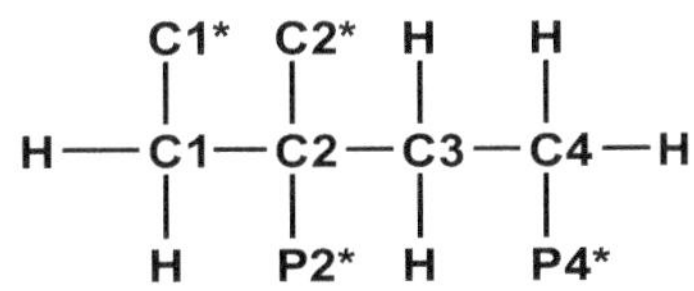

Scheme 7. Spin enumeration in DPBDC 9 used for $^{13}C\{^{1}H\}$-NMR C1* and C2* = COOH, COO^{-}. P2* and P4* = PO_3H_2, PO_3H^{-}, PO_3^{2-}.

Results for those six carbon atoms C1*, C2*, and C1 to C4 will be presented as (δ,τ))-contour plots in four separate spectral ranges shown in Figure 4a–d. Numerical results including specific chemical shifts δ_C and coupling constants $^{n}J_{PC}$ of DPBDC are listed in Tables 15 and 16.

Table 15. Dissociation constants of 2,4-diphosphonobutane-1,2-dicarboxylic acid (DPBDC) as obtained from $^{13}C\{^{1}H\}$-NMR-controlled and potentiometric titrations vs. NaOH. * Concentrations given in mol/L.

	$^{13}C\{^{1}H\}$ NMR [44]	Potentiometric [44]	Potentiometric [53]
pK_1	1.07	0.6	1.806 ± 0.066
pK_2	2.73	2.42	2.250 ± 0.021
pK_3	4.82	4.32	4.078 ± 0.005
pK_4	7.05	6.46	6.562 ± 0.004
pK_5	8.95	8.18	8.664 ± 0.006
pK_6	11.62	10.75	12.839 ± 0.007
$C_{Titrand}$	0.262 (DPBDC) *	0.0050 (DPBDC) *	0.01119 (DPBDC) *
$C_{Titrator}$	3.986 (NaOH) *	0.0975 (NaOH) *	0.09863 (TMAOH) *
$C_{Ion\ buffer}$	0	0.1 (NaCl) *	0.09863 ($TMANO_3$) *

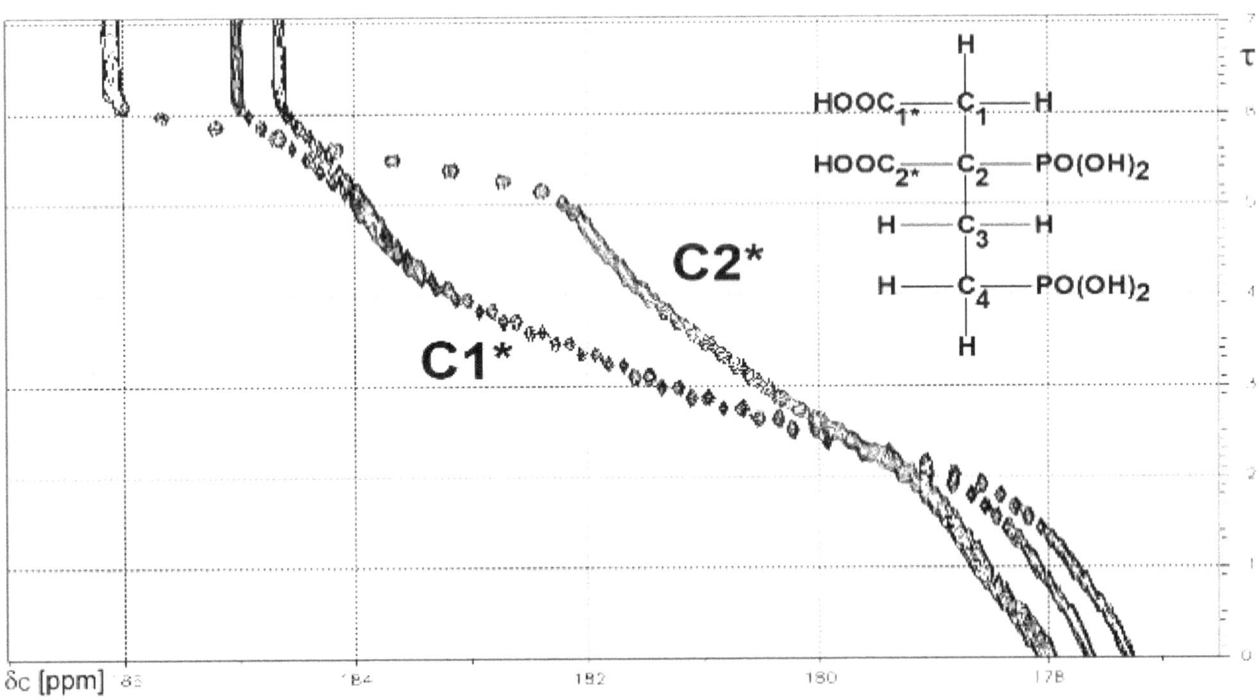

(a)

Figure 4. *Cont.*

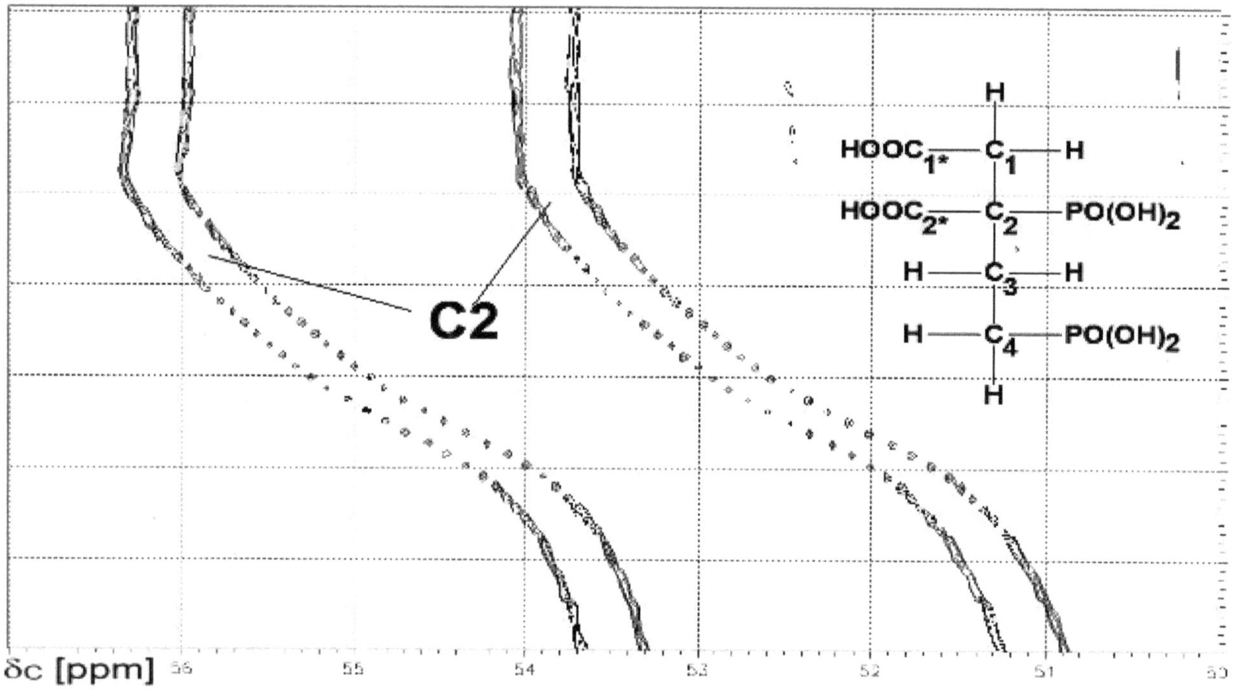

(**b**)

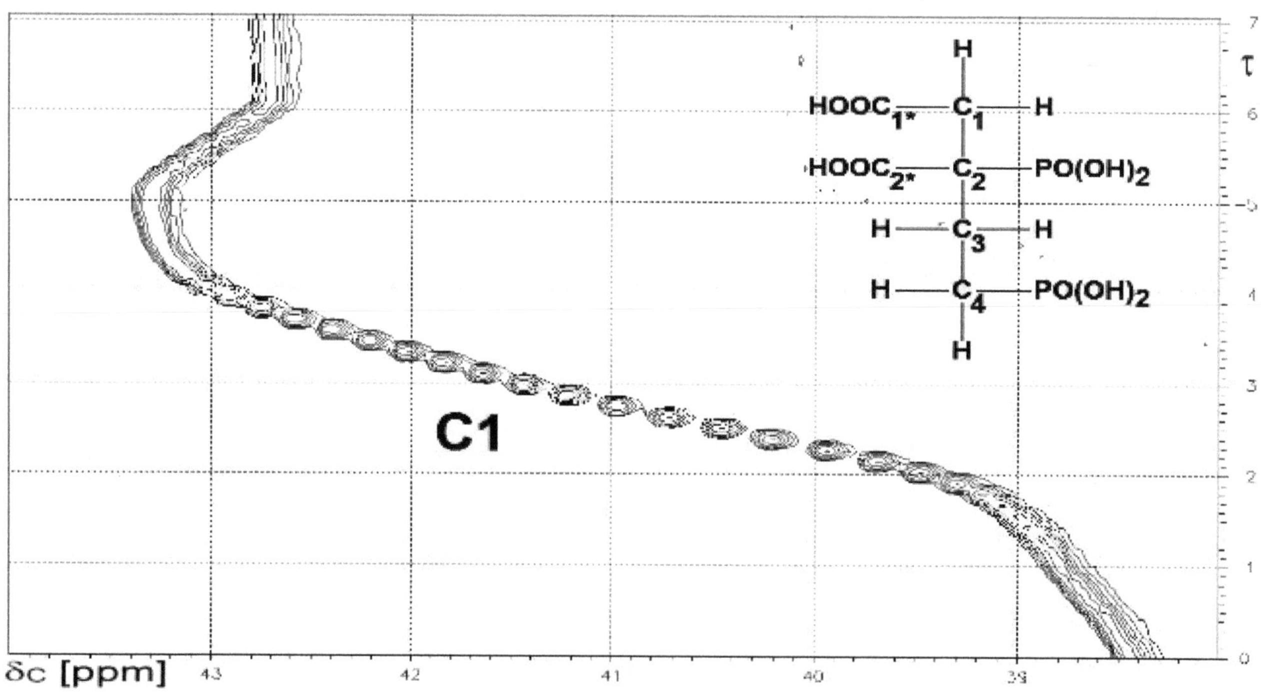

(**c**)

Figure 4. *Cont.*

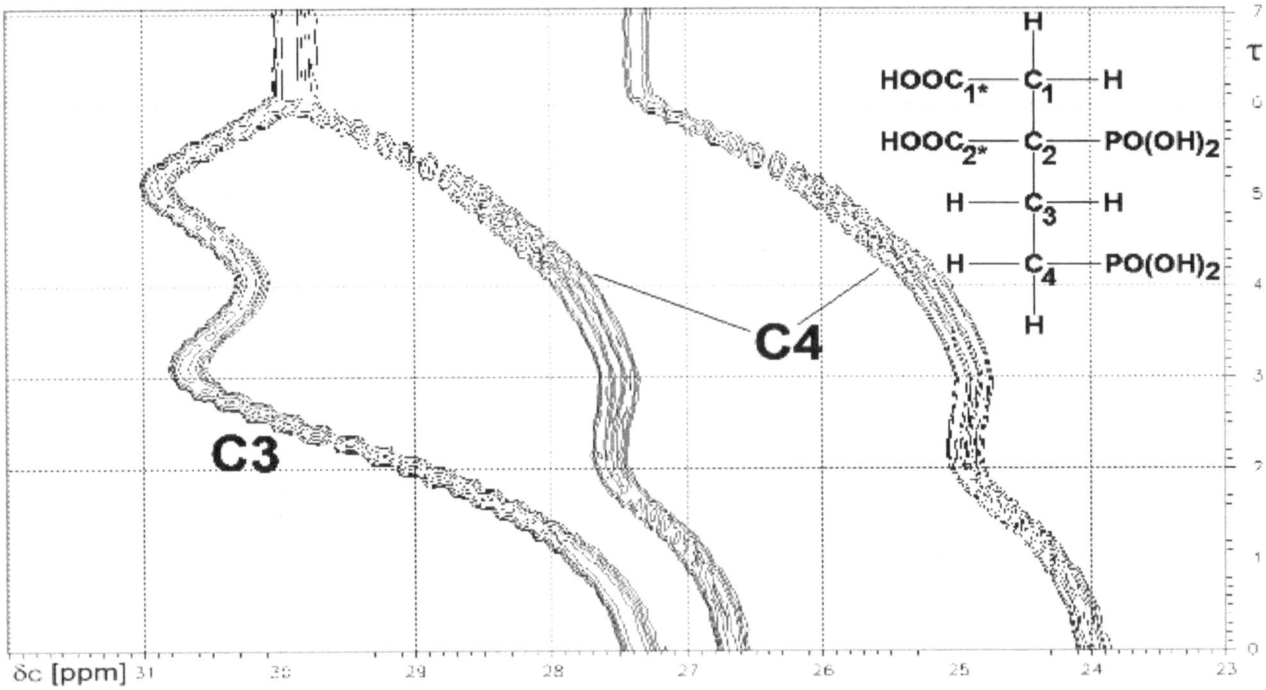

(**d**)

Figure 4. (**a**) The carboxylate functions C1* (note: $^3J_{PC}$) and C2*. δ_C = 186 to 177 ppm; (**b**) The skeleton carbon C2 (note: $^1J_{PC}$ and $^3J_{PC}$). δ_C = 56 to 50 ppm.; (**c**) The skeleton carbon C1. δ_C = 44 to 38 ppm; (**d**) The skeleton carbons C3 and C4 (note: $^1J_{PC}$). δ_C = 32 to 23 ppm.

Table 16. Specific chemical shifts δ_C [ppm], coupling constants $^nJ_{PC}$ [Hz] and gradients Δ_i [ppm] of 2,4-diphosphonobutane-1,2-dicarboxylic acid DPBDC **9**. For experimental data, see Table 15.

	C1	C2			C3	C4			C1*		C2*
Species	δ_C	δ_C	$^1J_{PC}$	$^3J_{PC}$	δ_C	δ_C	$^1J_{PC}$	$^3J_{PC}$	δ_C	$^3J_{PC}$	δ_C
H_6L	38.29	52.19	124.2	18.4	27.26	25.11	134.0	4.7	177.23	17.5	177.53
H_5L^-	38.70	52.42	116.8	18.2	27.51	25.56	132.4	5.1	177.82	15.9	178.66
H_4L^{2-}	39.38	52.98	116.9	17.3	29.09	26.34	131.3	6.3	178.85	12.5	179.18
H_3L^{3-}	41.78	54.06	116.7	16.8	30.95	26.15	131.3	9.0	181.69	6.4	180.70
H_2L^{4-}	43.28	54.66	115.9	16.6	29.98	26.54	130.3	9.9	183.46	4.3	181.53
HL^{5-}	43.26	55.04	115.8	15.9	30.99	27.51	128.3	9.4	184.08	5.3	182.53
L^{6-}	42.70	55.01	111.7	16.3	29.87	28.65	127.7	-	184.82	18.9	186.15
Gradients											
Δ_1	+0.41	+0.23	−7.4	−0.2	+0.25	+0.45	−1.6	+0.4	+0.59	−1.6	+1.13
Δ_2	+0.68	+0.56	+0.1	−0.9	+1.58	+0.78	−1.1	+1.2	+1.03	−3.4	+0.52
Δ_3	+1.40	+1.08	−0.2	−0.5	+1.86	−0.19	0.0	+2.7	+2.84	−6.1	+1.52
Δ_4	+1.50	+0.60	−0.8	−0.2	−0.97	+0.39	−1.0	+0.9	+1.77	−2.1	+0.83
Δ_5	−0.02	+0.38	−0.1	−0.7	+1.01	+0.97	−2.0	−0.5	+0.62	+1.0	+1.00
Δ_6	−0.56	−0.03	−4.1	+0.4	−1.12	+1.14	−0.6	-	+0.74	+13.6	+3.62

Some Comments on DPBDC 9

A complicated example for NMR-controlled titration which needs some discussion is 2,4-diphosphonobutane-1,2-dicarboxylic acid DPBDC **9**. Measurements and data evaluation were performed according to the state of technique. But it is not possible to explain all the parameters found for compound **9** by comparison with data from analogous structural elements of HOOC-(CH_2)n-PO_3H_2 (n = 1 to3) **7b** to **7d**, H_2O_3P-$(CH_2)_3$-PO_3H_2, and phosphonopolycarboxylic acids **10** to **12** shown in Scheme 8:

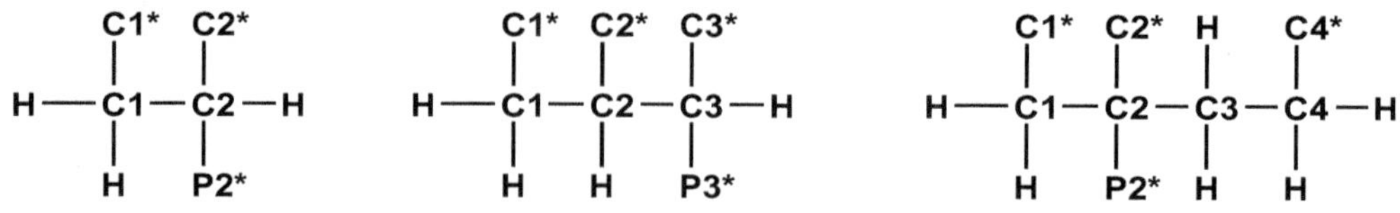

PBC **10** PPTC **11** PBTC **12**

Scheme 8. Phosphonopolycarboxylic acids **10** to **12** used for comparative $^{13}C\{^{1}H\}$-NMR-controlled titrations.

In a starting phase, 1D and 2D NMR techniques involving ^{1}H-, $^{1}H\{^{31}P\}$-, ^{13}C-, $^{13}C\{^{1}H\}$-, and C,H-COSY spectra were combined to assign the carbons C1*, C2*, C1 to C4 and phosphonate functions P2* and P4*.

For $^{13}C\{^{1}H\}$-NMR-controlled titration, the deprotonation steps may be divided into three sections (see Table 15 and Figure 4a,d). For $\tau = 0$ to 2 deprotonation $PO_3H_2 \rightarrow PO_3H^-$ takes place, first at P2* and then at P4*. In the second section for $\tau = 2$ to 4, the carboxylic units C1* and C2* are deprotonated. Finally for $\tau = 4$ to 6 the deprotonation $PO_3H^- \rightarrow PO_3^{2-}$ takes place at P2* and P4*.

(1) Comments on Chemical Shifts δ_C of Carbon Atoms in DPBDC **9**

The deprotonation of PO_3H_2, PO_3H^- and COOH functions in DPBDC **9** leads to a monotonous down field shift for δ_C C1* and C2* (see Figure 4a), while carbons C1 to C4 exhibit specific non-monotonous trends (see Figure 4b,d).

Since gradient Δ_6 for δ_C (C1*) > Δ_5 for δ_C (C1*), the final sixth deprotonation steps is affiliated with P2*. This conclusion is confirmed by Δ_6 for δ_C (C2*) >> Δ_5 for δ_C (C1*). Hence, the fifth deprotonation step of **9** is due to $PO_3H^- \rightarrow PO_3^{2-}$ of P4*. Dynamic chemicals shifts δ_C of C1* span a range from 177.5 to 184.61 ppm, while δ_C of C2* is found from 178 to 185.1, as shown in Figure 4a.

A tentative explanation may be found using Δ_3 for δ_C (C1*) > Δ_4 for δ_C (C1*) and Δ_3 for δ_C (C2*) < Δ_4 for δ_C (C1*). These findings imply that the carboxylic function C1* is more acidic than C*2.

Similar arguments for the relative acidity of C1* and C2* may be derived from the chemical shift δ_C of the skeleton carbon C2 (see Figure 4b). δ_C (C2) of H_6L corresponds to 52.2 ppm, while the totally deprotonated form L^{6-} is found at 55 ppm. Deprotonation at C1* and C2* is characterized again by Δ_3 of δ_C (C2) > Δ_4 of δ_C (C2).

Chemical shifts δ_C of C3 span a range of 38.4 to 43.3 ppm. Surprisingly, the final deprotonation $HL^{5-} \rightarrow L^{6-}$, due to $PO_3H^- \rightarrow PO_3^{2-}$ of P2* reduces δ_C (C3) from 43.26 to 42.70 ppm. This is the first observation (within this context) of a negative gradient ($\Delta_6 = -0.56$ Hz) connected to deprotonation at a PO_3H^- unit.

The situation is even more complex for the chemical shift δ_C of C2 covering a range from 27.4 to 30.9 ppm (Figure 4d). Two negative gradients are observed: $\Delta_4 = -0.97$ ppm for $H_3L^{3-} \rightarrow H_2L^{4-}$ due to deprotonation at C2* and $\Delta_6 = -1.12$ ppm for $HL^{5-} \rightarrow L^{6-}$ induced by deprotonation at P2*. For simpler compounds CH_3-$(CH_2)_n$-COOH (n = 0 to 3) and HOOC-$(CH_2)_n$-COOH (n = 1 to 3), solely positive gradients were described [54].

In addition, we did not observe negative gradients for compounds **7b** to **7d** and **10** (PBC), but in **11** (PPTC) and in **12** (PBTC) [43].

The major RR/SS diastereomer of PPTC **11** exhibited a negative gradient Δ_5 (C1) = −0.44 ppm for the final deprotonation step $PO_3H^- \rightarrow PO_3^{2-}$ at P3*. An upfield shift occurred, since δc (C1) of HL^{4-} = 42.17 ppm and δc (C1) of L^{5-} = 41.73 ppm. This effect might be due to opening of hydrogen bridges and conformational changes. In contrast here to is the minor RR/SS diastereomer of PPTC, it does not show a negative gradient Δ_5 (C1) [42,44].

Weaker negative gradients $\Delta_5 = -0.22$ ppm are observed for chemical shifts δ_C of both carbons C1 and C3 in PBTC **12.** The final deprotonation $PO_3H^- \rightarrow PO_3^{2-}$ at P2* is affiliated with following data:

δc (C1) of HL^{4-} = 43.54 ppm, δc (C1) of L^{5-} = 43.32 ppm, and δc (C3) of HL^{4-} = 32.60 ppm, δc (C3) of L^{5-} = 32.60 ppm. In contrast hereto carbons C2 and C4 in PBTC **12** exhibit positive gradients Δ_5.

Those unexpected observations for chemical shifts δc in **9** and conformational aspects will be mentioned in the following section on coupling constants $^{n}J_{PC}$ as well.

(2) Comments on Coupling Constants nJPC (n = 1 to 3) of DPBDC **9**

The vicinal coupling $^{3}J_{PC}$ (P2*C1*) is remarkably sensitive towards the protonation state (see Figure 5):

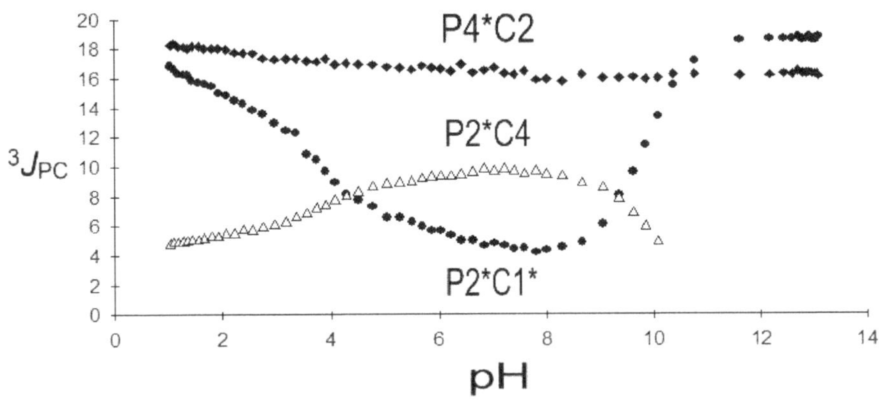

Figure 5. Three coupling constants $^{3}J_{PC}$ (P2*C1*), $^{3}J_{PC}$ (P2*C4), and $^{3}J_{PC}$ (P4*C12) in DPBDC **9**.

For the protolytic species H_6L to H_2L^{4-} of **9,** a decrease in $^{3}J_{PC}$ (P2*C1*) from 17.5 Hz down to a minimum of 4.3 Hz is observed, followed by an increase from 5.3 Hz to 18.6 Hz due to HL^{5-} and finally L^{6-}. Between pH = 7 and 8, a maximum of the protolytic species H_2L^{4-} is expected, while HL^{5-} dominates around pH = 9. Those observations indicate changes of the dihedral angle of P2*-C2-C1-C1* possibly involving hydrogen bridges as indicated by Scheme 9 below:

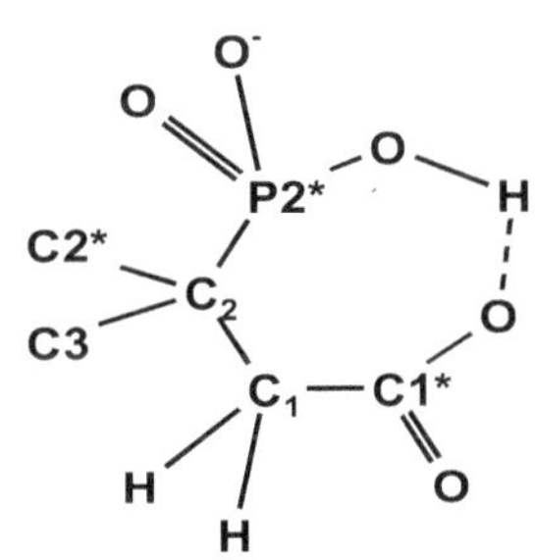

Scheme 9. Tentative hydrogen bridges for protolytic species H_3L^{3-} to HL^{5-}.

A corresponding bridge -C1-P-O—H—O-P-C2- was discussed for the HL^{3-} species of ethane-1,2-bisphosphonic acid [44].

$^{2}J_{PC}$ (P2*C1), $^{2}J_{PC}$ (P2*C3), $^{2}J_{PC}$ (P2*C2*), and $^{2}J_{PC}$ (P4*C3) were not resolved in $^{13}C\{^{1}H\}$-NMR spectra obtained by NMR-controlled titrations.

$^{3}J_{PC}$ (P2*C4) shows a monotonous increase from 4.7 Hz to a maximum of 9.9 Hz for the sequence H_6L to H_2L^{4-} followed by a decrease to 9.4 Hz for HL^{5-}. This observation points towards an increase in the dihedral angle in P2*-C2-C3-C4.

$^{3}J_{PC}$ (P4*C2) is less sensitive towards deprotonation but larger than $^{3}J_{PC}$ (P2*C4) and found in a range from 18.4 to 16.3 Hz possibly indicating a tendency towards trans-conformation of the fragment P4*-C4-C3-C2. For comparison, $^{3}J_{PC}$ in HOOC-$(CH_2)_3$-PO_3H_2 **7d** appeared in a corresponding range from 17.2 to 17.9 Hz.

${}^1J_{PC}$ (P2*C2), ranging from 124.2 to 111.7 Hz, is markedly smaller than ${}^1J_{PC}$ (P4*C4), which is observed from 134.0 to 127.7 Hz. A ${}^1J_{PC}$, pH diagram is given in Figure 6 below:

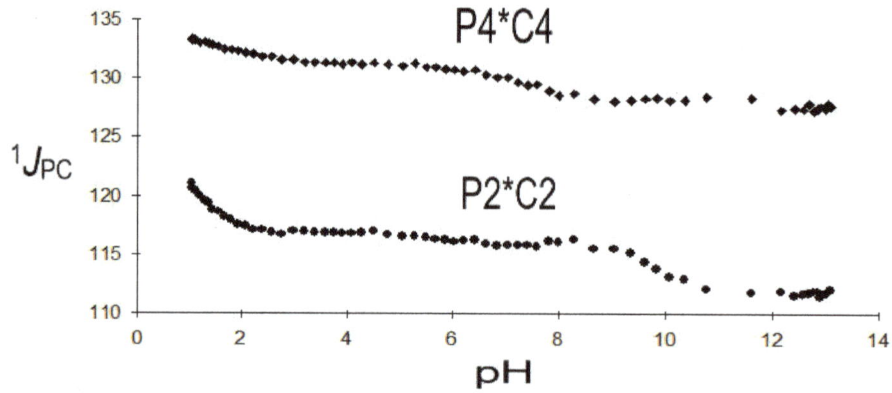

Figure 6. ${}^1J_{PC}$ (P2*C2) and ${}^1J_{PC}$ (P4*C4) of DPBDC **9**.

${}^1J_{PC}$ (P2*C2) is very indicative and selective for the deprotonation of the phosphonic functions $PO_3H_2 \rightarrow PO_3H^-$ and $PO_3H^- \rightarrow PO_3^{2-}$. It indicates that the first deprotonation ($pK_1 = 1.07$) of DPBDC **9** takes place at P2* with a strong gradient Δ_1 ${}^1J_{PC}$ (P2*C2) = −7.4 Hz. The final deprotonation ($pK_6 = 11.62$) is affiliated with P2* as well as indicated by Δ_6 ${}^1J_{PC}$ (P2*C2) = −4.1 Hz. This assignment leaves $pK_2 = 2.73$ and $pK_5 = 8.95$ to the deprotonation of P4*. Deprotonation at the carboxylic groups ($pK_3 = 4.62$) and $pK_4 = 7.05$) does not significantly affect ${}^1J_{PC}$ (P2*C2) and ${}^1J_{PC}$ (P4*C4).

Results on ${}^1J_{PC}$ (P2*C2) of **9** are consistent with observations on phosphonosuccinic acid **10**, where ${}^1J_{PC}$ (P2*C2) is found in a range from 133.7 to 112.1 Hz. Strong gradients $\Delta_1 = -17.1$ Hz and $\Delta_4 = -4.5$ Hz are affiliated with the deprotonation of P2*, while deprotonations of C1* and C2* do not significantly influence ${}^1J_{PC}$ (P2*C2).

3. Conclusions

Automated NMR-controlled titrations efficiently combine ^{1}H, ^{13}C-, ^{19}F-, and ^{31}P-NMR spectroscopy with potentiometric titrations to determine dissociation constants, specific chemicals shifts and coupling constants. Results are presented in two-dimensional plots, where NMR parameters (chemical shifts, coupling constants) are correlated with analytical parameters (pH, degree of titration τ). High digital resolution and high S/N are achieved in time- and material-saving measurements. These hyphenated techniques are powerful instruments to identify the structure and purity of research and industrial compounds. Limitations of accuracy due to the nature of glass electrodes occur at very low and very high pH values, obscuring the lower and higher pK_i values. In those situations, single sample NMR methods are recommended. ^{13}C{^{1}H}-NMR-controlled titrations may be used for conformational analysis. For more complicated structures, additional studies using pH-dependent high-resolution ^{1}H- and ^{1}H{^{31}P}-NMR spectra, X-ray diffraction of selected salts and molecular modelling of acids and anions are needed to solve details of conformational problems. The latter topics are laborious and beyond the scope of this paper.

4. Experimental

Details of the hardware and software used in NMR-controlled titration together with references for this field are described in [31,32]. Most model compounds were obtained from external sources listed under the acknowledgements.

Funding: Our research was supported by Fonds der Chemischen Industrie e. V. (VCI), Frankfurt am Main, Germany, and by the German-Israeli Foundation (GIF), Jerusalem, Israel, Research Grant I-316-186-05/93. Material and technical help was obtained from BRUKER Spectrospin, Rheinstetten, Germany, and from SCHOTT Geräte GmbH, Hofheim, Germany.

Acknowledgments: α-Ala-P **1** and β-Ala-P **2** were obtained from R. Tyka and P. Mastalerz, Wroclaw, Poland. E. Breuer, Jerusalem, Israel, provided $LiOOC\text{-}CH_{2\text{-}n}F_n\text{-}PO_3Li_2$ (n = 0 to 2). Technical compounds were obtained from BAYER AG, Leverkusen, Germany, (PBS, PPTC, PBTC, DPBDC, $CH_3\text{-}CH_{2\text{-}n}F_n\text{-}PO(O)(OMe)_2$), HENKEL AG, Düsseldorf, Germany (PPTC), BOZZETTO, Bergamo, Italy (NTMP, EDTMP, DETPMP), and via K. Kellner, Halle, Germany (FOSCARNET). T. Clark, Erlangen, Germany, kindly provided VAMP 4.4 for CONVEX 220 (1989). Special thanks are due to members of the Düsseldorf team, M. Grzonka, M. Batz, J. Peters, A. Bier, H. J. Majer, J. Ollig, I. Reimann, C. Arendt, S. Hermens, and Z. Szakács, for dedicated cooperation and discussions concerning preparative, analytical and NMR studies including development of novel hardware and software.

References and Notes

1. Van Wazer, J.R.; Callis, C.F.; Shoolery, J.N.; Jones, R.C. Principles of Phosphorus Chemistry. II. Nuclear Magnetic Resonance Measurements. *J. Am. Chem. Soc.* **1956**, *78*, 5715–5726. [CrossRef]
2. Crutchfield, M.M.; Callis, C.F.; Irani, R.R.; Roth, G.C. Phosphorus Nuclear Magnetic Resonance Studies of Ortho and Condensed Phosphate. *Inorg. Chem.* **1962**, *1*, 813–817. [CrossRef]
3. Hagen, R.; Roberts, J.D. Nuclear Magnetic Resonance Spectroscopy. ^{13}C Spectra of Aliphatic Carboxylic Acids and Carboxylate Anions. *J. Am. Chem. Soc.* **1969**, *91*, 4504–4514. [CrossRef]
4. Heubel, P.-H.; Popov, A.I.J. Acid Properties of Some Phosphonocarboxylic Acids. *Sol. Chem.* **1979**, *8*, 615–625. [CrossRef]
5. Appleton, T.G.; Hall, R.J.; Harris, A.D.; Kimlin, H.A.; McMahon, I.J.N.M.R. Study of Acid-Base Equilibria of Aminoalkylphosphonic Acids, $^{+}NH_3(CH_2)_nPO_3H^{-}$ (n = 1, 2, 3); Evidence for Cyclization in Solution. *Austr. J. Chem.* **1984**, *37*, 1833–1840. [CrossRef]
6. Appleton, T.G.; Hall, R.J.; McMahon, I.J. NMR Spectra of Iminobis(methylenebisphosphonic acid), $HN(CH_2PO_3H_2)_2$ and Related Ligands and of Their Complexes with Platinum(II). *Inorg. Chem.* **1985**, *25*, 726–734. [CrossRef]
7. Sawada, K.; Araki, T.; Suzuki, T. Complex Formation of Amino Polyphosphonates. 1. Potentiometric and Nuclear Magnetic Resonance Studies of Nitrilotris(methylenephosphonate) Complexes of the Alkaline Earth-Metal Ions. *Inorg. Chem.* **1987**, *26*, 1199–1204. [CrossRef]
8. Sawada, K.; Araki, T.; Suzuki, T.; Doi, K. Complex Formation of Amino Polyphosphonates. 1. Stability and Structure of Nitrilotris(methylenephosphonate) Complexes of the Divalent Transitions-Metal Ions in Aqueous Solution. *Inorg. Chem.* **1989**, *28*, 2687–2688. [CrossRef]
9. Sawada, K.; Kanda, T.; Naganuma, Y.; Suzuki, T. Formation and Protonation of Aminopolyphosphonate Complexes of Alkaline-earth and Divalent Transition-metal Ions in Aqueous Solution. *J. Chem. Soc. Dalton Trans.* **1993**, *17*, 2558–2562. [CrossRef]
10. Sawada, K.; Myagawa, T.; Sakaguchi, T.; Doi, K. Structure and Thermodynamic Properties of Aminopolyphosphonate Complexes of the Alkaline-earth Metal Ions. *J. Chem. Soc. Dalton Trans.* **1993**, *24*, 3777–3784. [CrossRef]
11. Sawada, K.; Ichikawa, T.; Uehara, K. Eight-membered chelate-ring complexes of cobalt(III)-polyamine complexes of aminopolyphosphonates in aqueous solution. *J. Chem. Soc. Dalton Trans.* **1996**, *14*, 3077–3085. [CrossRef]
12. Matczak-Jon, E.; Kurzak, B.; Kamecka, A.; Sawka-Dobrowolska, W.; Kafarski, P.; Lejjczak, B. Zink(II) complexes of phosphonic acid analogues of glutamic acid. *J. Chem. Soc. Dalton Trans.* **1996**, 3455–3464. [CrossRef]
13. Matczak-Jon, E.; Kurzak, B.; Kamecka, A.; Sawka-Dobrowolska, W.; Kafarski, P. Interactions of zinc(II), magnesium(II) and calcium(II) with iminodimethylenediphosphonic acids in aqueous solutions. *J. Chem. Soc. Dalton Trans.* **1999**, *20*, 3627–3637. [CrossRef]
14. Rohovec, J.; Kývala, M.; Vojtišek, P.; Hermann, P.; Lukeš, I. Synthesis, Crystal Structures, and Solution Properties of N-Methylene(phenyl)phosphinic Acid Derivatives of Cyclen and Cyclam. *J. Inorg. Chem.* **2000**, 195–203. [CrossRef]

15. Popov, K.; Popov, A.; Rönkkömäki, H.; Lajunen, L.H.J.; Hannu-Kuure, M.; Vendilo, A.; Tsirul'nikova, N. ^{31}P, ^{23}Na and ^{133}Cs NMR equilibrium study of iminobis(methylenephosphonic acid) complexes with alkali metals. *Inorg. Chimica Acta* **2002**, *344*, 1–6. [CrossRef]
16. Popov, A.; Rönkkömäki, H.; Popov, K.; Lajunen, L.H.J.; Vendilo, A. ^{31}P NMR protonation equilibria study of iminobis(methylenphosphonic) acid and its derivatives at high pH. *Inorg. Chimica Acta* **2003**, *353*, 1–7. [CrossRef]
17. Ohms, G.; Grossmann, G. Über die pH-Abhängigkeit der ^{31}P- und ^{13}C-NMR-Spektren von Cyclohexan-, Cyclohexen- und Benzenphosphonsäuren. *Z. Anorg. Allg. Chem.* **1987**, *544*, 232–240. [CrossRef]
18. Rönkömäki, H.; Jokisaari, H.; Lajunen, L.H. ^{31}P NMR and Potentiometric Studies on the Protonation of Isopropyl Esters of Chlodronic Acid. *Acta Chem. Scand.* **1993**, *47*, 331–337. [CrossRef]
19. Popov, K.; Niskanen, E.; Rönkkömäki, H.; Lajunen, L.H.J. ^{31}P NMR Study of organophosphonate protonation equilibrium at high pH. *New J. Chem.* **1999**, *23*, 1209–1213. [CrossRef]
20. Szakács, Z.; Hägele, G. Accurate determination of low pK values by ^{1}H NMR titration. *Talanta* **2004**, *62*, 819–825. [CrossRef]
21. Szakács, Z.; Hägele, G.; Tyka, R. ^{1}H/^{31}P NMR pH indicator series to eliminate the glass electrode in NMR spectroscopic pK_a determinations. *Anal. Chim. Acta* **2004**, *522*, 247–258. [CrossRef]
22. Popov, K.; Rönkkömäki, H.; Lajunen, L.H. Guidelines for The NMR Measurements for Determination of High And Low pK_a Values. *Pure Appl. Chem.* **2006**, *78*, 663–675. [CrossRef]
23. Yesinowski, J.P.; Sunberg, R.J.; Benedict, J.J. pH Control and Rapid Mixing in Spinning NMR Samples. *J. Magn. Res.* **1982**, *47*, 85–90. [CrossRef]
24. Glaser, J.; Henriksson, U.; Klason, T. A 205 TL NMR Titration Study of the Complex Formation between Tl(I) and Cl- in Aqueous Solution. *Acta Chem. Scand.* **1986**, *A40*, 344–349. [CrossRef]
25. Li, W. Gravity-driven pH adjustment for site-specific protein pK_a measurement by solution-state NMR. *Meas. Sci. Technol.* **2017**, *28*. [CrossRef]
26. Hägele, G.; Grzonka, M.; Kropp, H.-W.; Ollig, J.; Spiegl, H. Phosphonic and phosphinic acids: Monitoring protolytic and complex formation equilibria by titration dependent stopped-flow-NMR-techniques. *Phosphorus Sulfur Silicon* **1993**, *77*, 85–88. [CrossRef]
27. Hägele, G.; Varbanov, S.; Ollig, J.; Kropp, H.-W. Aminomethylphosphine Oxides: Synthesis, Dissociation and Stability Constants, ^{31}P{^{1}H}-NMR-controlled Titration. *Z. Anorg. Allg. Chemie* **1994**, *620*, 914–920. [CrossRef]
28. Hägele, G. ^{31}P NMR controlled titrations of Phosphorus-Containing Acids and Bases in Protolysis and Complex Formation. In *Phosphorus-31 NMR Spectral Properties in Compound Characterization and Structural Analysis*; Quin, L.D., Verkade, J.G., Eds.; VCH Publishers: New York, NJ, USA, 1994; Chapter 30; pp. 395–409.
29. Hägele, G.; Ollig, J. NMR-controlled titrations of phosphorus containing acids and bases. *Comput. Chem.* **1995**, *19*, 287–294.
30. Ahrendt, C.; Hägele, G. The Photo-T-Concept: Hard- and software combination for the determination of macroscopic and microscopic dissociation constants. *Comput. Chem.* **1995**, *19*, 263–268. [CrossRef]
31. Hägele, G.; Szakács, Z.; Ollig, J.; Hermens, S.; Pfaff, C. NMR-Controlled Titrations: Characterizing Aminophosphonates and Related Structures. *Heteroat. Chem.* **2000**, *11*, 562–582. [CrossRef]
32. Hägele, G.; Grzonka, M.; Peters, J.; Spiegl, H.; Kropp, H.W.; Ollig, J.; Hermens, S.; Augner, S.; Uhlemann, C.; Pfaff, C.; et al. NMR-Controlled Titrations—Principles and Progress: Monitoring Protonation and Complex Formation Equilibria in Aqueous Solutions. Available online: https://www.theresonance.com/nmr-controlled-titration-download-the-paper/ (accessed on 2 September 2019).
33. Pfaff, C.G. Softwareentwicklung zur Auswertung und Visualisierung Kernresonanzspektroskopisch Kontrollierter Titrationen. Ph.D. Thesis, Heinrich-Heine-Universität, Düsseldorf, Germany, 2002.
34. Peters, M.; Siegfried, L.; Kaden, T.A. pH-metric and NMR studies of complexation of Zn^{2+}, Cd^{2+} and Pb^{2+} with diazacrown ethers having dangling phosphonate groups. *J. Chem. Soc. Dalton Trans.* **2000**, *24*, 4664–4668. [CrossRef]
35. Rabenstein, D.L.; Sayer, T.L. Determination of Microscopic Acid Dissociation Constants by Nuclear Magnetic Resonance Spectrometry. *Anal. Chem.* **1976**, *48*, 1141–1146. [CrossRef]
36. Szakacs, Z.; Kraszni, M.; Noszal, B. Determination of microscopic acid-base parameters from NMR-pH titrations. *Anal. Bioanal. Chem.* **2004**, *378*, 1428–1448. [CrossRef]

37. Szakacs, Z. NMR-Titrationen und neue Auswertekonzepte zur Aufklärung der Mikroskopischen Dissoziation und der pH-abhängigen Konformation von Biorelevanten Phophinsäuren und Carnosinderivaten. Ph.D. Thesis, Heinrich-Heine-Universität, Düsseldorf, Germany, 2002.
38. Kývala, M.; Lukeš, I. Chemometrics 95. In *Book of Abstracts, Proceedings of the 4th International Chemometrics Conference of the Czech Chemical Society, Pardubice, Czech Republic, 3–7 July 1995*; University of Pardubice: Pardubice, Czech Republic, 1995; p. 63.
39. Augner, S.; Kehler, J.; Szakács, Z.; Breuer, E.; Hägele, G. Ring-chain tautomerism and protolytic equilibria of 3-hydroxy-3-phosphonoisobenzo-furanone studied by ^{1}H-, ^{13}C-, and ^{31}P-NMR-controlled titrations. *New J. Chem.* **2008**, *32*, 1608–1616. [CrossRef]
40. Augner, S. Neue Entwicklungen zur Durchführung automatisierter NMR-Kontrollierter Titrationen von Phosphon- und Phosphinsäuren. Ph.D. Thesis, Heinrich-Heine-Universität, Düsseldorf, Germany, 2002.
41. Kropp, H.-W.; Hägele, G. Unpublished results from my student Kropp. Interested readers may contact G. Hägele. Unpublished work. 1994.
42. Lindner, A. 1-Phosphonopropan-1,2,3-tricarbonsäure—NMR- und Konformationsanalytische Untersuchungen. Ph.D. Thesis, Heinrich-Heine-Universität, Düsseldorf, Germany, 2000.
43. Hägele, G. NMR-controlled titrations characterizing organophosphorus compounds. *Phosphorus Sulfur Silicon Relat. Elem.* **2019**, *194*, 361–363.
44. Ollig, J. Untersuchungen zur Titrationsabhängigen Kernresonanzspektroskopie. Ph.D. Thesis, Heinrich-Heine-Universität, Düsseldorf, Germany, 1996.
45. Robitaille, P.-M.L.; Robitaille, P.A.; Brown, G.G., Jr.; Brown, G.G. An Analysis of the pH-Dependent Chemical-Shift Behavior of Phosphorus-Containing Metabolites. *J. Magn. Res.* **1991**, *91*, 73–84. [CrossRef]
46. Popov, K.; Rönkkömäki, H.; Lajunen, L.J.J. Critical evaluation of stability constants of phosphonic acids. *Pure Appl. Chem.* **2001**, *73*, 1641–1677. [CrossRef]
47. Clark, T.; Alex, A.; Beck, B.; Chandrasekhar, J.; Gedeck, P.; Horn, A.H.C.; Hutter, M.; Martin, B.; Rauhut, G.; Sauer, W.; et al. *VAMP 4.4*; Universität Erlangen-Nürnberg: Erlangen, Germany, 1990.
48. Wozniak, M.; Nowogrocky, G. Acidites et complexes des acides (alkyl- et aminoalkyl-) phosphoniques—I: Determination potentiometrique des constantes d'acidite par affinement multiparametrique: Prise en compte de l'impurete carbonate. *Talanta* **1979**, *26*, 1135–1141. [CrossRef]
49. Mohan, M.S.; Abbott, E.H. Metal complexes of biologically occurring aminophosphonic acids. *J. Coord. Chem.* **1978**, *8*, 175–182. [CrossRef]
50. Kiss, T.; Balla, J.; Nagy, G.; Kozzlowski, H.; Kowalik, J. Complexes of aminophosphonates. I. Transition metal complexes of aminophosphonic acid analogues of α-alanine, β-alanine, phenylalanine and tyrosine. *Inorg. Chim. Acta* **1987**, *138*, 25–30. [CrossRef]
51. Kropp, H.-W.; Hägele, G. Analytische und NMR-spektroskopische Untersuchungen an Organophosphorsäuren. Unpublished work.
52. Burton, D.J.; Sprague, L.G.; Pietrzyk, D.J.; Edelmuth, S.H. A Facile Synthesis of Difluorophosphonoacetic Acid. *J. Org. Chem.* **1984**, *49*, 3437–3438. [CrossRef]
53. Bier, A. Computereinsatz in der Analytischen Chemie zur Untersuchung von Protolyse- und Komplexbildungs-Gleichgewichten am Beispiel der Phosphonocarbonsäuren. Ph.D. Thesis, Heinrich-Heine-Universität, Düsseldorf, Germany, 1993.
54. Rabenstein, D.L.; Sayer, T.L. Carbon-13 Chemical Shift parameters for Amines, Carboxylic Acids, and Amino Acids. *J. Magn. Res.* **1978**, *24*, 27–39. [CrossRef]

7

Chemistry of Phosphorylated Formaldehyde Derivatives

Vasily P. Morgalyuk

Laboratory of Organophosphorus Compounds, Nesmeyanov Institute of Organoelement Compounds, Russian Academy of Sciences, Vavilova str., 28, Moscow 119991, Russia; morgaliuk@mail.ru

Abstract: The underinvestigated derivatives of unstable phosphorylated formaldehyde acetals and some of the structurally related compounds, such as thioacetals, aminonitriles, aminomethylphosphinoyl compounds, are considered. Separately considered are halogen aminals of phosphorylated formaldehyde, acetals of phosphorylated formaldehyde of H-phosphinate-type and a phosphorylated gem-diol of formaldehyde. Synthetic methods, chemical properties and examples of practical applications are given.

Keywords: phosphorylated acetals; -thioacetals; -aminonitriles; -aminomethylphosphinoyl compounds; -chloro(or bromo)aminals; -gem-dioles; -ketene acetale; H-phosphinate

---

1. Introduction

Among organophosphorus compounds, α-phosphorylated carbonyl compounds stand out by the capacity of cleavage of phosphorus–carbon bond under mild conditions when reacted with nucleophiles [1–5]. The cleavage of phosphorus–carbon bond may proceed spontaneously as well. At the same time, α-oxoalkylphosphinoyl compounds also retain properties inherent in carbonyl compounds, for example, they undergo cross aldol condensation.

The least stable compounds among them are phosphorylated formaldehyde derivatives, dialkyl formylphosphonates (**1**) [6–11] and *N,N,N′,N′*-tetraalkyl formylphosphondiamides (**2**) [12] (Figure 1), whose existence was even disputed in the first half of the 1970s [13,14].

1 **2**

Figure 1. Structures of dialkyl formylphosphonates (**1**) and *N,N,N′,N′*-tetraalkyl formylphosphondiamides (**2**).

However, the first synthesis of **1** by reaction of sodium derivatives of dialkyl phosphites with acetic formic anhydride was described in 1974 [15]. Compounds **1** were shown to be unstable and prone to spontaneous degradation [9,10,16,17]; thus, diethyl formylphosphonate (**3**) prepared by the reaction of triethyl phosphite (**4**) with acetic formic anhydride (**5**) at low temperature begins to undergo decarbonylation even at −10 °C [9,10] to give diethyl phosphite (**6**) (Scheme 1).

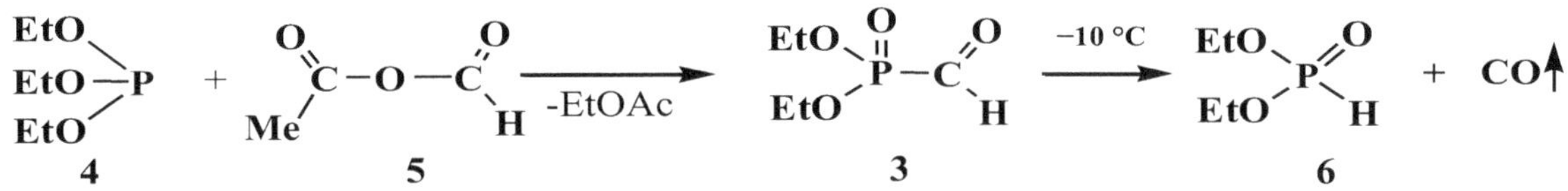

Scheme 1. Syntheses and destruction of diethyl formylphosphonate (**3**).

N,N,N′,N′-Tetraisopropylformylphosphondiamide (**7**) obtained by saponification of *N,N,N′,N′*-tetraisopropyl[(*N″,N″*-diisopropylamino)methylydeniminium]phosphondiamide dichlorophosphate (**8**) with potassium hydroxide in tetrahydrofuran at 20 °C proved to be slightly more stable; it undergoes decarbonylation only above 40 °C [12] to give *N,N,N′,N′*-tetraisopropylphosphonicdiamide (**9**) (Scheme 2).

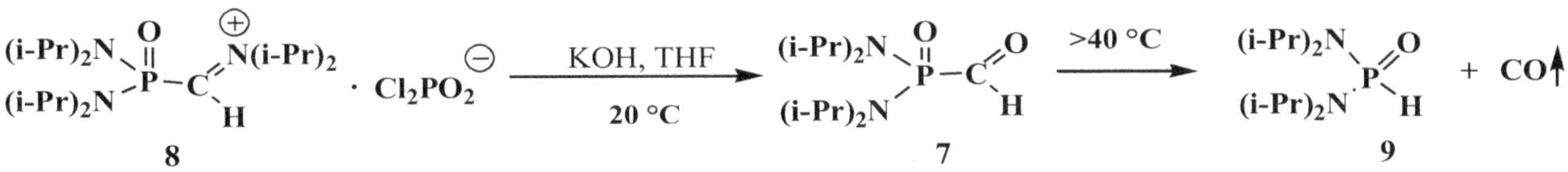

Scheme 2. Syntheses and destruction of *N,N,N′,N′*-tetraisopropyl formylphosphondiamide (**7**).

The stability of the formylphosphondiamide **7** allowed the study some of its chemical properties. It was shown that in its reactions with methylene(triphenyl)phosphorane ($Ph_3P{=}CH_2$) and 2,4-dinitrophenylhydrazine (NH_2NH-DNP) **7** behaves as a typical aldehyde, yielding the corresponding α-phosphorylated olefin **10** and hydrazone **11** [12] (Scheme 3).

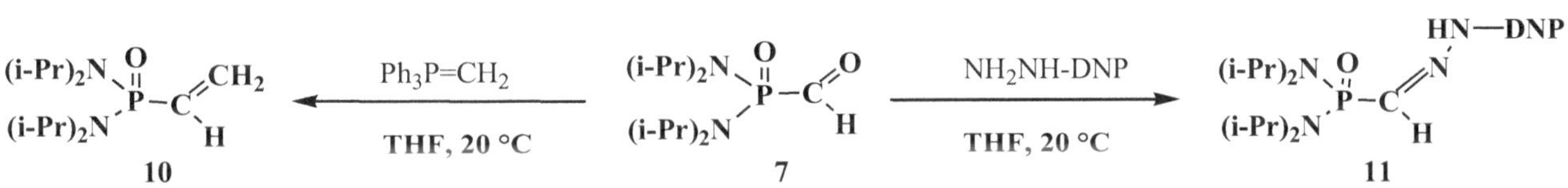

Scheme 3. Reactions *N,N,N′,N′*-tetraisopropylformylphosphondiamide (**7**) as an aldehyde (DNP means a 2,4-dinitrophenyl moiety).

However, many authors showed that formylphosphinoyl compounds are also stable in the form of formylphosphonic acid (**12**) [12,14,15], its disodium salt [15], aldehyde group derivatives such as acetals **13** [18–20] and structurally related compounds. These compounds are thioacetals **14** [21–23], aminonitriles **15** [24–26], diphosphinoyl (*N,N*-dialkylaminomethyl)methanes **16** [27–29], chloro(or bromo)aminals **17** [30–34], mixed S,O-thioacetals **18** [35–37], aminals **19** [30,38], aminoacetals **20** [31,32,38,39], aminothioacetals **21** [32], chloroacetals **22** [40,41], chloro- (or bromo-) thioacetals **23** [40,42,43], α-chlorosulfinyl derivatives **24** [44,45], α-alkoxynitriles **25** [46–48], α-thionitriles **26** [46], α-dihalo derivatives **27** [49–51], α-alkoxydiphosphoryl compounds **28** [41], α-mercaptodiphosphoryl compounds **29** [40,52,53], α-alkoxysilyl derivatives **30** [54], α-aminosilyl derivatives **31** [29,55], or α-mercaptosilyl derivatives **32** [56] (Figure 2):

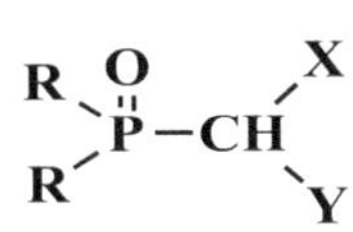

13-32

Figure 2. Molecular structure of phosphorylated formaldehyde acetals (**13**) and related compounds (**14–32**), where R = OAlk, Alk_2N, Ph; X, Y are the combinations of OAlk, Alk_2N, CN, $(AlkO)_2P(O)$, S(Alk, Ar), S(O)(Alk, Ar), N(H)C(O)Alk, C(O)Ar, $N(H)S(O)_2Alk$, $S(O)_2Ar$, Hal, Me_3Si [18–56].

Among these compounds, phosphorylated acetals **13**, thioacetals **14**, α-dimethylaminonitriles **15**, aminodiphosphinoyl compounds **16** and chloro- (or bromo-) aminals **17** are used in contemporary organic synthesis. Nonetheless, the phosphorylated acetals of formaldehyde and structurally related compounds remain poorly studied types of organophosphorus compounds until now. The chemical properties of this type of compounds were most studied on an example of dialkyl(dialkoxymethyl)phosphonates **33**. The properties of other compounds **18–32** have been studied in much less detail, and the majority of them are examined in one-sided manner, only as precursors for the synthesis of ketene acetals and similar compounds by the Horner reaction.

To date, it is known that—except for formylphosphonic acid (**12**), phosphorylated acetals **13**, and similar compounds **14–32**—the phosphorylated formaldehyde derivatives are stable in the form of H-phosphinate acetals: alkyl (diethoxymethyl)phosphinate **34** [57–59], and geminal dioles, phosphorylated formaldehyde hydrates (hydrates of phosphorylated formaldehyde) **35** [60–62] (Figure 3).

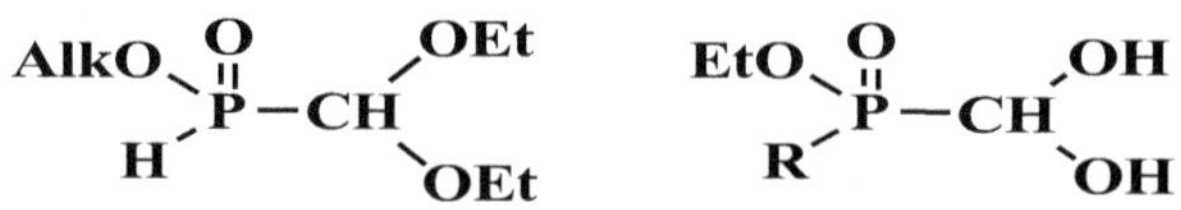

34

35

R = OEt, Ph

Figure 3. Molecular structure of alkyl (diethoxymethyl)phosphinate **34** and phosphorylated formaldehyde hydrates **35**.

Noteworthy are nitrogen-containing analogs of phosphorylated formaldehyde: *N*-substituted imines (**36**) [16,63,64], *N*-alkylnitrones **37** [8,65,66], phosphorylated diazomethane (**38** [67–69] and O-alkylated oximes of diethyl formylphosphonates **39** [7,70], used in contemporary organic synthesis (Figure 4).

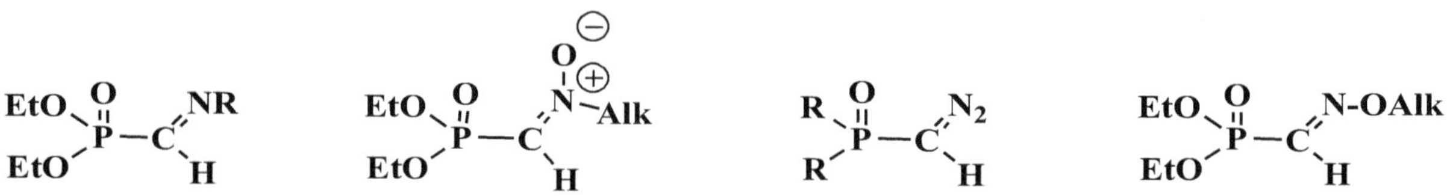

36 **37** **38** **39**

R=Alk, AlkC(O), Ar, ArC(O)

R = OAlk, Ph

Figure 4. Molecular structures of formylphosphonates of *N*-substituted imines **36**, N-alkylnitrones **37**, phosphorylated diazomethane **38** and alkylated oximes **39**.

However, the properties of compounds **36–39** differ significantly from those of compounds **13–32**. Therefore, compounds **36–39** will be considered in a separate publication.

2. Chemistry of Phosphorylated Formaldehyde Derivatives

2.1. *Syntheses and Chemical Properties of Formylphosphonic Acid* (**12**)

Formylphosphonic acid (**12**), which shows distinct antiviral activity [71] was obtained for the first time in 1974 [14] as a byproduct of the electrolysis of an aqueous solution of nitrilotrimethylphosphonic acid (**40**) (Scheme 4). Later it was shown that **12** might be also obtained in high yields by the catalytic oxidation of nitrilotrimethylphosphonic acid (**40**) with hydrogen peroxide in the presence of vanadyl sulfate or potassium tungstate (88–93% yield) [17] or activated carbon (up to 82% yield) [72]. Compound **12** can be also prepared by the oxidation of hydroxymethylphosphonic acid (**41**) in aqueous solution with air oxygen or hydrogen peroxide in the presence of Raney copper in a yield up to 65% [73] (Scheme 4).

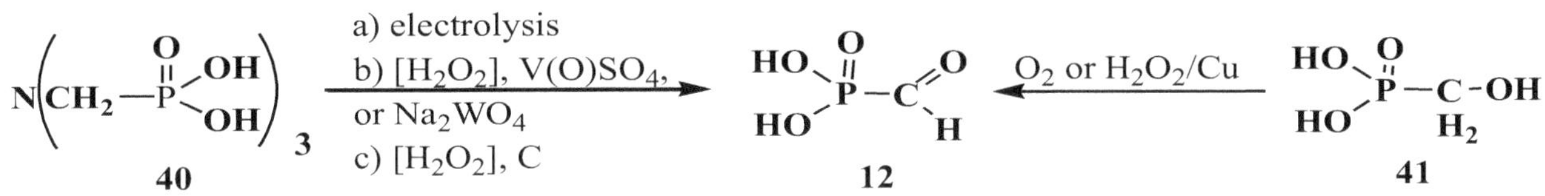

Scheme 4. Possible routes of syntheses of formylphosphonic acid (**12**).

In a third method of synthesis, **12** was obtained via a two-stage synthesis starting from dimethyl (dimethoxymethyl)phosphonate (**42**), which was converted by treatment with bromotrimethylsilane (**43**) into bis(trimethylsilyl)(dimethoxymethyl)phosphonate (**44**). Hydrolysis of the latter resulted in formylphosphonic acid (**12**) [74] (Scheme 5). See also Section 3.2.—Synthesis of formacetalphosphonic acids.

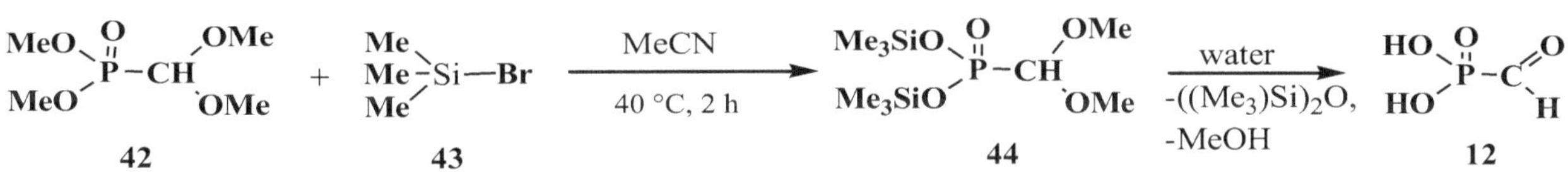

Scheme 5. Synthesis of formylphosphonic acid (**12**) from dimethyl (dimethoxymethyl)phosphonate (**42**).

However, the chemical properties of formylphosphonic acid (**12**) have been poorly studied until now. It is known that heating **12** in acidic medium leads to the cleavage of the P–C bond to form phosphoric and formic acids **45** and **46**. In the presence of bases **12** undergoes disproportionation (Cannizzaro reaction) to give hydroxymethylphosphonic **41** and carboxyphosphonic acids **47**. Acid **47** undergoes rapid decarboxylation on acidification of the reaction medium [14] (Scheme 6).

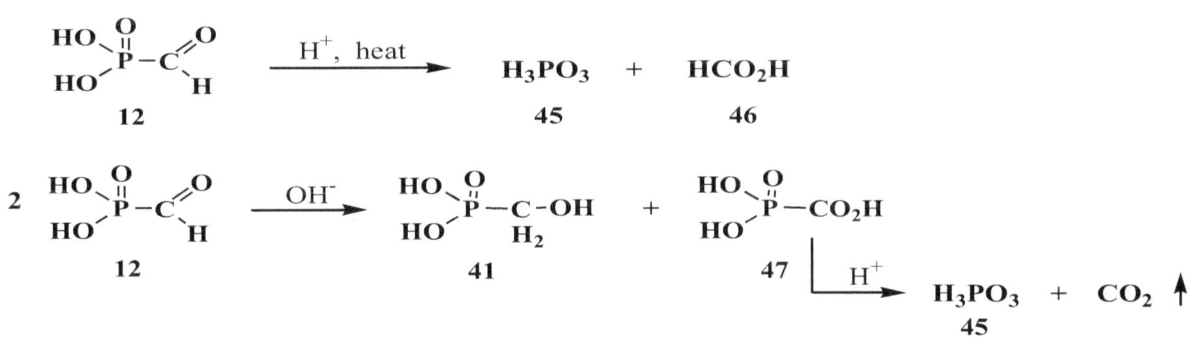

Scheme 6. Degradation of formylphosphonic acid (**12**) to form phosphoric **45** and formic acids **46** and disproportionation of **12** in aqueous solutions.

2.2. *Chemistry of Phosphorylated Formaldehyde Acetals* 13

Chemistry of phosphorylated formaldehyde acetals **13** started as a chemistry of dialkyl (dialkoxymethyl)phosphonates **33** due to their more ready availability as compared with the analogs—*N,N,N′,N′*-tetraalkyl(dialkoxymethyl)phosphondiamides **48**, dialkyl (dialkoxymethyl)-phosphine oxides **49** or diphenyl(dialkoxymethyl)phosphine oxide **50**. Many types of acetals are known to date, but their chemical properties are still insufficiently studied, although they are more studied than the other derivatives of phosphorylated formaldehyde.

2.2.1. Methods of Synthesis of Phosphorylated Formaldehyde Acetals **13**

First phosphorylated formaldehyde acetals **13** were obtained by the reaction of hydrophosphinoyl compounds **51–53** with orthoformate esters **54** on heating. The reaction with dialkyl phosphites **51** [19,75] is conducted by heating to 182 °C [76,77] or at 60 °C in the presence of $BF_3{\cdot}Et_2O$ as catalyst (without the catalyst the yields decrease from 69–90% to 25% [75]). In the case of *sec*-phosphine oxides—dialkylphosphine oxides **52** [77] and diphenylphosphine oxide **53** [77,78], the reaction proceeds at 100 °C [19,77]. This method provides high yields of compounds **33, 49** and **50** (Scheme 7).

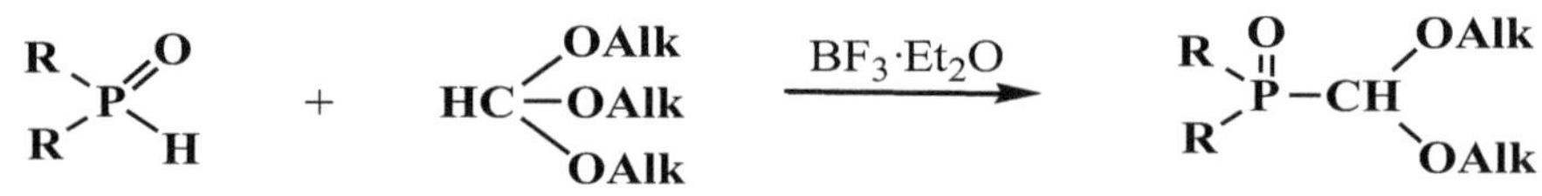

51,52,53 **54** **33, R=OAlk'; 49, R=Alk"; 50, R=Ph**

Scheme 7. Synthesis of phosphorylated formaldehyde acetals **33, 49, 50** from dialkyl phosphites **51** or *sec*-phosphine oxides **52, 53** by means of orthoformate esters **54**.

A variant of this method consists in the reaction of phosphoric acid (H_3PO_3, **45**) or phosphoric anhydride (**55**) with triethyl orthoformate (**56**) in 1:3 ratio that results in diethyl (diethoxymethyl)phosphonate (**57**) with minimal effort [79] (Scheme 8).

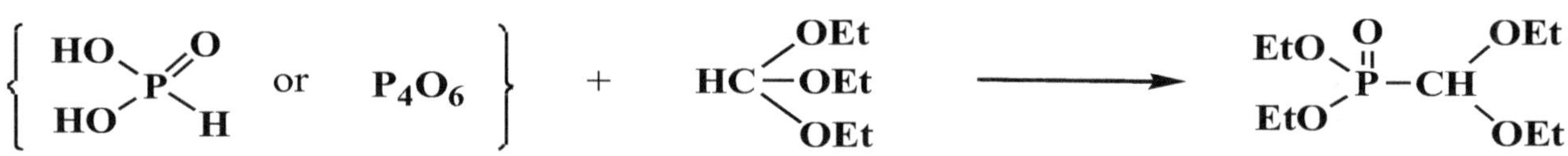

45 **55** **56** **57**

Scheme 8. Synthesis of diethyl (diethoxymethyl)phosphonate (**57**) from phosphoric acid (**45**) or phosphorous anhydride (**55**).

Compounds **33** can be also prepared by the reaction of trialkyl phosphites **58** with orthoformate esters **54**. It was shown that phosphites **58** do not react directly with **54** even on heating [19]. However, as in the case of dialkyl phosphites **51**, orthoformates **54** react with trialkyl phosphites **58** and their analogs $(AlkO)_2POX$ **59**, where X = Me_3Si, $(AlkO)_2P$, $(AlkO_2)P(O)$ on heating in the presence of catalytic amounts of boron trifluoride etherate $BF_3{\cdot}Et_2O$ [19]. The reaction of **58** with orthoformate esters **54** is also possible in the presence of phosphorus trichloride (PCl_3, **60**) or diethoxychloromethane $(EtO)_2C(H)Cl$ [80,81] (Scheme 9).

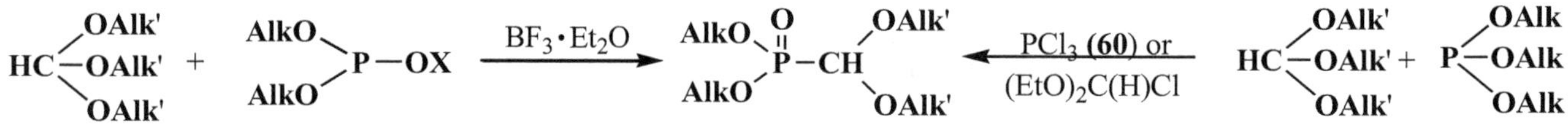

54 **59** **33** **54** **58**

Scheme 9. Synthesis of dialkyl (dialkoxymethyl)phosphonates **33** from orthoformate esters **54**, trialkyl phosphites **58** and their analogs $(AlkO)_2POX$ **59**.

It was reported that the reaction of trialkyl phosphites **58** with acetoxy(diethoxy)methane (**61**), which is more reactive derivative than orthoformate esters **54** [19,82], results in dialkyl (diethoxymethyl)phosphonates **62** (Scheme 10).

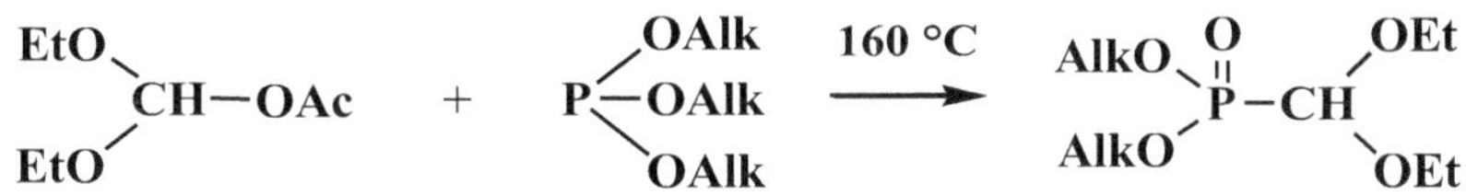

61 58 62

Scheme 10. Synthesis of dialkyl (diethoxymethyl)phosphonates **62** from orthoformate esters **54** and acetoxy(diethoxy)methane (**61**).

The most general method of synthesis of phosphorylated formaldehyde acetals **13** is the reaction of phosphorus trichloride derivatives **60** with orthoformate esters **54**. Excess of **54** reacts on heating with **60** [79,81,83–85] and mono- and dialkyl $AlkPCl_2$ **63**, Alk_2PCl **64**, alkoxy $(AlkO)PCl_2$ **65**, $(AlkO)_2PCl$ **66**, and phenyl $PhPCl_2$ **67**, Ph_2PCl **68** [80,84,86–89] derivatives of **60** (Scheme 11), to form symmetrical and unsymmetrical acetals **33**, **49**, **50** and **69**, **70**, respectively. It was noted that the reaction of 2-chloro-1,2,3-dioxaphospholanes **71** with **54** leads to the opening of the dioxaphospholane ring to form ethyl (β-chloroethyl) ([1,3]-dioxolan-2-yl)phosphonates **72** [90].

33, R=OAlk' or OAlk"; 49, R=Alk; 50, R=Ph

54

69, R=Alk; 70, R=Ph

Scheme 11. Syntheses of phosphorylated formaldehyde acetals **33**, **49**, **50**, **69**, **70** from phosphorus trichloride derivatives **63**–**68** with orthoformate esters **54**. See the text above.

The reaction of *N,N,N′,N′*-tetraalkyl(chloro)phosphindiamides **73** with orthoformates **54** proceeds with a partial exchange of dialkylamino groups at phosphorus atom of **73** to give both *N,N,N′,N′*-tetraalkyl (dialkoxymethyl)phosphondiamides **48** and mixed phosphonamidates **74** [19,91,92]. In the case of *N,N*-dialkyl(dichloro)phosphinamides **75**, a total exchange of dialkylamino groups at the phosphorus atom for alkoxy groups takes place to yield dialkyl (dialkoxymethyl)phosphonates **33** [85] (Scheme 12).

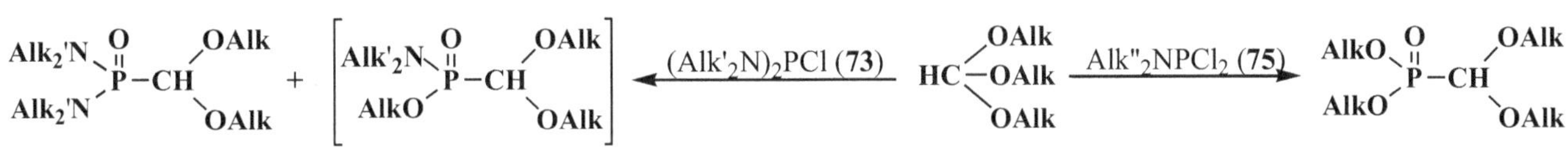

48 74 54 33

Scheme 12. Interaction of *N,N,N′,N′*-tetraalkyl(chloro)phosphindiamides **73** and *N,N*-dialkyl(dichloro)phosphinamides **75** with orthoformates **54**.

Quaternary ammonium salts of dimethylformamide acetals can be used instead of orthoformates **54** [18,93]. The method only allows preparation of cyclic phosphorylated formaldehyde dialkyl acetals **76** since the quaternary ammonium salts of dialkylformamide acetals of linear structure are unstable and undergo fast degradation [93] (Scheme 13).

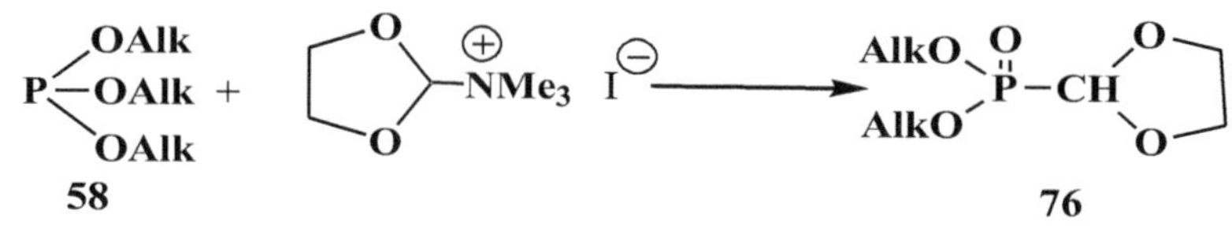

Scheme 13. Synthesis of cyclic acetals **76** by interaction of trialkyl phosphites **58** with the quaternary ammonium salts of dimethylformamide acetals.

Nonetheless, linear compounds **33** can be obtained by this method in low yield (15–25%) using a one-pot method from methyl iodide, dialkylformamide acetals **77**, and trialkyl phosphites **58** [18].

The reaction of phosphorus trichloride (**60**) with dimethylformamide dimethylacetal (**78**) also leads to the formation of dimethyl (dimethoxymethyl)phosphonate (**42**) (along with a certain amount of tetramethyl (*N,N*-dimethylaminomethyl)diphosphonate (**79**)) [83] (Scheme 14).

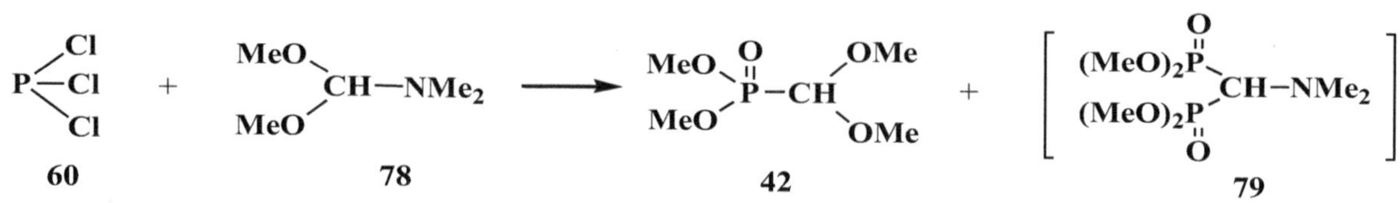

Scheme 14. Syntheses of dimethyl (dimethoxymethyl)phosphonate (**42**) from phosphorus trichloride (**60**) and dimethylformamide dimethylacetal (**78**).

Dialkyl acetals **33** can be also prepared by the reaction of alcohols with tetraalkyl bis(dialkoxymethyl)pyrophosphonates **80**. The method provides a possibility to obtain formaldehyde dialkyl acetals with different alkoxy substituents at phosphorus atom **81** [89] that are difficult to prepare from trivalent phosphorus derivatives and orthoformates **54** or their derivatives because of competitive exchange of substituents at phosphorus atom [19,89] (Scheme 15). Alkyl (dialkoxymethyl)phosphonic acid **82** is also formed at the same time. See also Schemes 22 and 23.

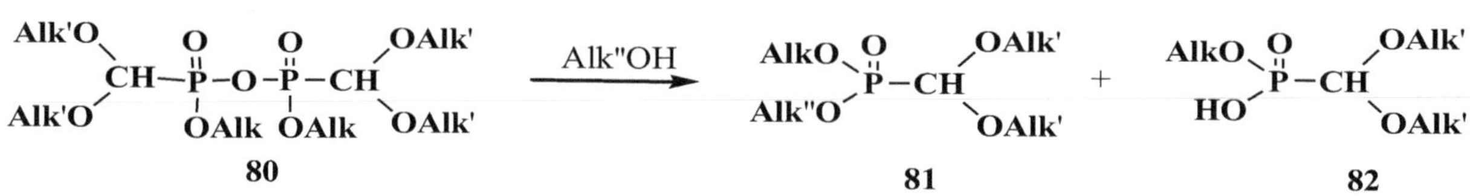

Scheme 15. Preparation of acetals **81** with different alkoxy substituents at phosphorus atom by means of alcoholysis of tetraalkyl bis(dialkoxymethyl)pyrophosphonates **80**.

Alkyl alkyl(dialkoxymethyl)phosphinates **83** unsymmetrically substituted at the phosphorus atom were synthesized by the Arbuzov reaction of dialkyl (dialkoxymethyl)phosphonites **84** with alkyl iodides [82,84], for example see Scheme 16.

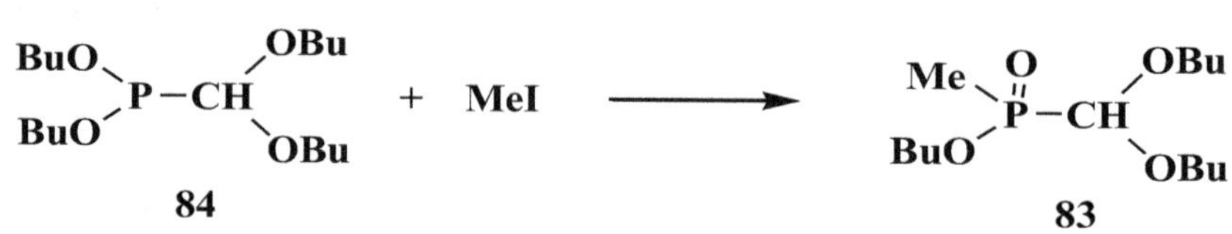

Scheme 16. Example synthesis of acetals with different substitutients at phosphorus atom by the Arbuzov reaction. See also Scheme 11, Scheme 15, Scheme 17, Schemes 21–23 and 111–114.

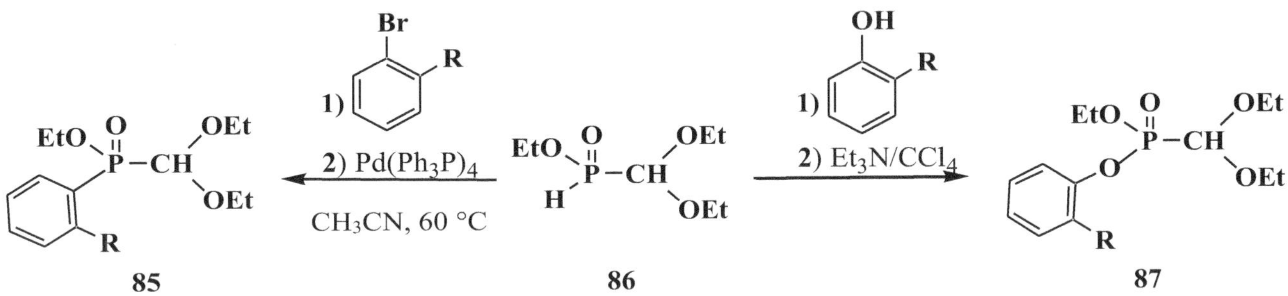

Scheme 17. Synthesis of ethyl aryl(diethoxymethyl)phosphinates **85** and ethyl aryl (diethoxymethyl)phosphonates **87** from ethyl (diethoxymethyl)phosphinate (**86**), R are Alk or Hal.

The catalytic synthesis of ethyl aryl(diethoxymethyl)phosphinates **85** by the reaction of ethyl (diethoxymethyl)phosphinate (**86**) with *ortho*-substituted aryl bromides in the presence of tetrakis(triphenylphosphine)palladium(0) $Pd(Ph_3P)_4$ in 75–90% yields was reported in 1995 [94]. The Todd-Atherton reaction of **86** with *ortho*-substituted phenols in the presence of triethylamine at 0–23 °C leads to ethyl aryl (diethoxymethyl)phosphonates **87** [94] (Scheme 17).

The attempted preparations of novel acetales **33** by the transesterification of phosphorus ester groups failed because they gave rise to intractable mixtures of compounds [95]. Ethoxyphosphoryl groups in acetal **57** were replaced by butoxyphosphoryl groups only under cathode electrolysis conditions. As a result, butyl ethyl (diethoxymethyl)phosphonate (**88**) and dibutyl (diethoxymethyl)phosphonate (**89**) were obtained in low yields—11% and 5%, respectively [96] (Scheme 18).

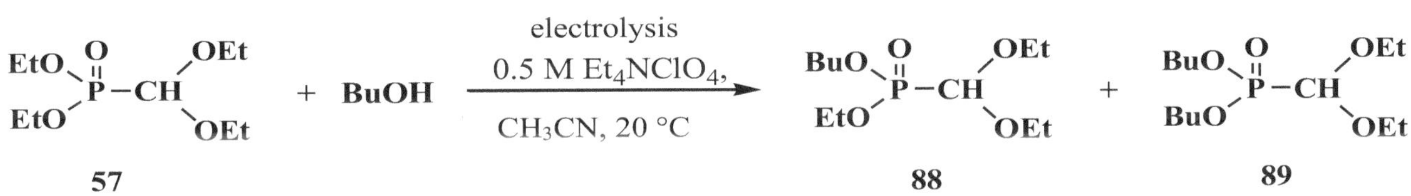

Scheme 18. Synthesis of acetales **88** and **89** by transesterification under the conditions of electrolysis.

However, the heating of **57** with 1,3-dimethylpropanediols in the presence of a catalytic amount of benzenesulfonic acid ($PhSO_3H$) results in the replacement of ethoxy groups of the acetal fragment to give cyclic diethyl (5-dimethyl-[1,3]-dioxan-2-yl)phosphonates **90** [97], for example, see Scheme 19.

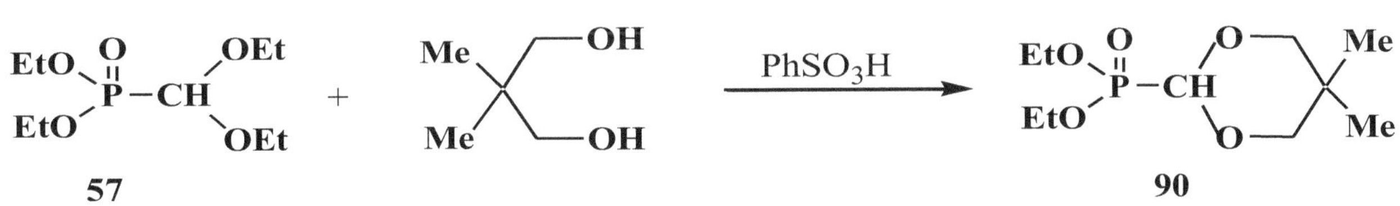

Scheme 19. Preparation of cyclic diethyl (5-dimethyl-[1,3]-dioxan-2-yl)phosphonates **90** by the transesterification under acid catalysis.

2.2.2. Chemical Properties of Phosphorylated Formaldehyde Acetals 13

The chemistry of phosphorylated formaldehyde acetals **13** was initially developed for the most part as a chemistry of available dialkyl (dialkoxymethyl)phosphonates **33**. Therefore the properties of acetals as a separate type of organophosphorus compounds were studied mainly by the examples of compounds **33** whose reactivity is affected by the presence of both phosphorus ester and acetal groups.

Hydrolysis of acetals **33** was studied by the example of compound **57** and a compound with aromatic substituents in the acetal group, diethyl (5,6-dichloro-1,3-benzodioxomethyl)phosphonate

(**91**). However, the attempted acid hydrolysis of acetal **57** on heating lead to the cleavage of phosphorus–carbon bond [77,95,98]. Compound **91** underwent acid hydrolysis on heating to give (5,6-dichloro-1,3-benzodioxomethyl)phosphonic acid (**92**) (Scheme 20). See also the section "Cleavage of Phosphorus–Carbon Bond under the Action of Acids and Acidic Reagents", Scheme 54.

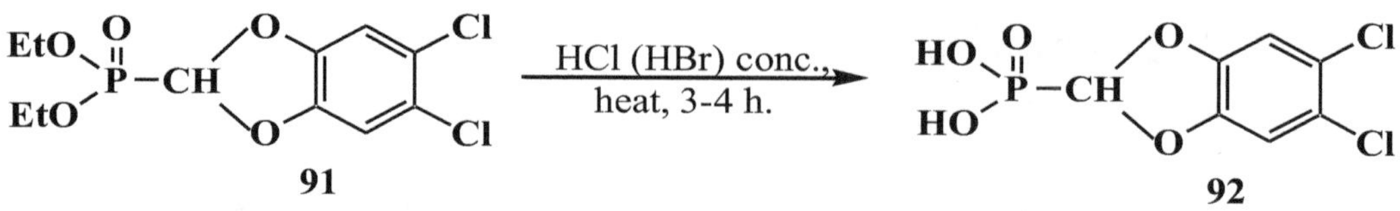

Scheme 20. Acid hydrolysis of diethyl (5,6-dichloro-1,3-benzodioxomethyl)phosphonate (**91**).

The heating of a solution of diethyl (dialkoxymethyl)phosphonate **93** in absolute ethanol with sodium ethoxide (NaOEt) leads to dealkylation of one of the ethoxy groups by phosphorus atom to form ethyl sodium (dialkoxymethyl)phosphonate **94**, which produces the free acid **95** on acidification [95]. Heating of **93** with sodium iodide NaI leads to the same result [93]. The reaction of ethyl sodium (diethoxymethyl)phosphonate (**96**) with electrophilic reagents brings about the formation of phosphonates **97** with different substituents at the phosphorus atom [95] (Scheme 21).

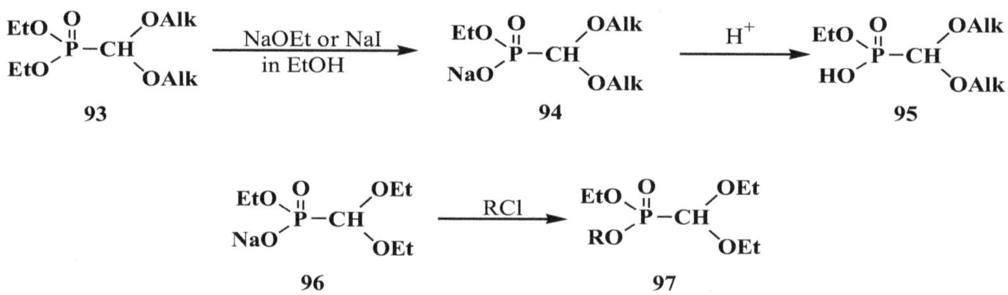

Scheme 21. Acetals **93** dealkylation at their interaction with sodium ethoxide or sodium iodide. Synthesis of acetals **97** with different substituents at phosphorus atom, where R = Alk, Ac, Me_3Si, $MeOCH_2$, $ArOCH_2$.

Acids **95** show typical properties of hydroxy compounds. Using as example ethyl (diethoxymethyl)phosphonic acid (**98**) it is shown that they react with diazomethane and thionyl chloride. The reaction products are ethyl methyl (diethoxymethyl)phosphonate (**99**) and ethyl (diethoxymethyl)phosphonic chloride (**100**) [95], which reacts with phenylmagnesium bromide in tetrahydrofuran to produce ethyl phenyl(diethoxymethyl)phosphinate (**101**) (Scheme 22).

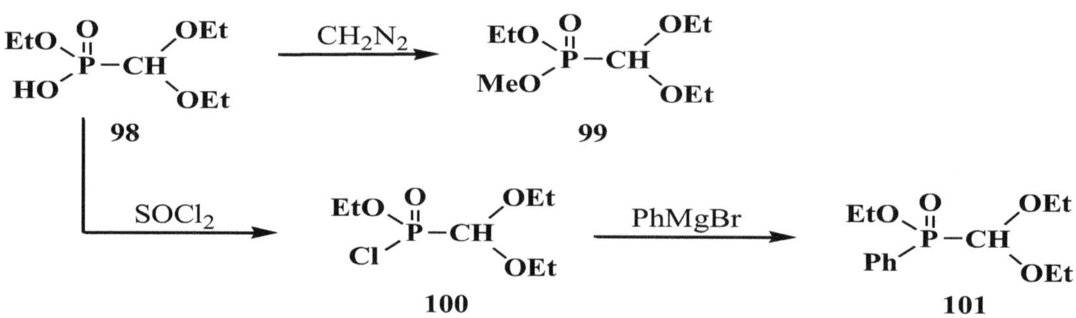

Scheme 22. Ethyl (diethoxymethyl)phosphonic acid (**98**) transformations.

Unsymmetrical ethyl aryl (diethoxymethyl)phosphonates **87** containing aryloxy substituents at the phosphorus atom [94] undergo rearrangement in the presence of equimolar amount of lithium diisopropylamide LDA in tetrahydrofuran at –70 °C to yield ethyl (2-hydroxyaryl)-(diethoxymethyl)phosphinates **102** (Scheme 23).

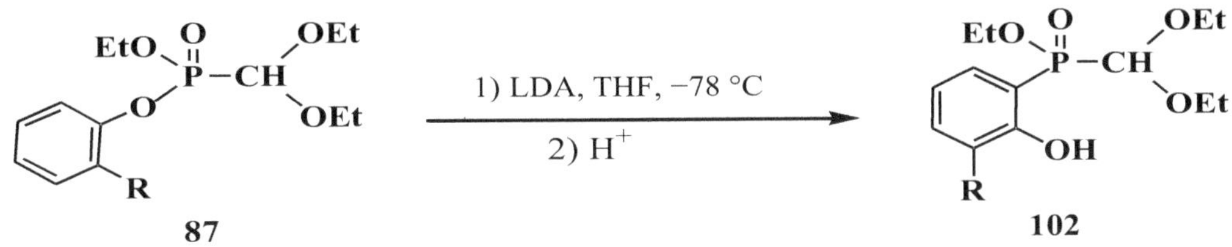

Scheme 23. Rearrangement of compounds **87**, R = Alk, Hal.

Heating of diethyl (diethoxymethyl)phosphonate (**57**) with a catalytic amount of $BF_3{\cdot}Et_2O$ gives rise to formation of tetraethyl (ethoxymethyl)diphosphonate (**103**) in a 14% yield [19] (Scheme 24).

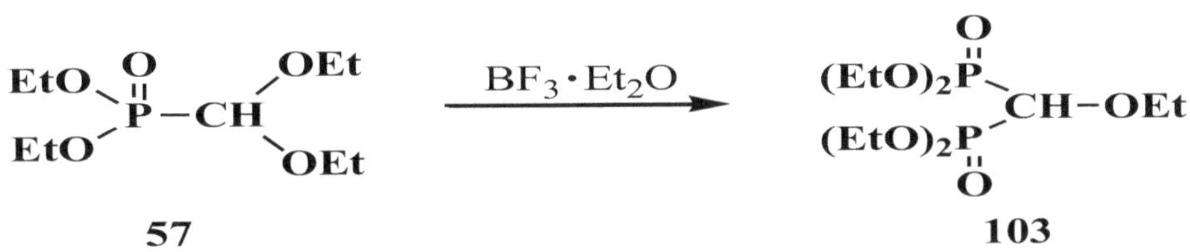

Scheme 24. Formation of tetraethyl (ethoxymethyl)diphosphonate (**103**) from acetal **57** under a catalysis by $BF_3{\cdot}Et_2O$.

The acetal group of compounds **13** is rather stable to the action of co-reactants. Nonetheless, a series of transformations of phosphorylated formaldehyde acetals (**33**) that involves dialkoxyacetal group is described.

The halogenation of compounds **33** with *N*-bromosuccinimide leads to alcoxycarbonylphosphonates **104**. Five-membered cyclic acetals **76** by halogenation with *N*-chlorosuccinimide and azodiisobutyro-nitrile mix produce β-haloethoxycarbonylphosphonates **105** [99]. Reactions proceed via a radical mechanism (Scheme 25).

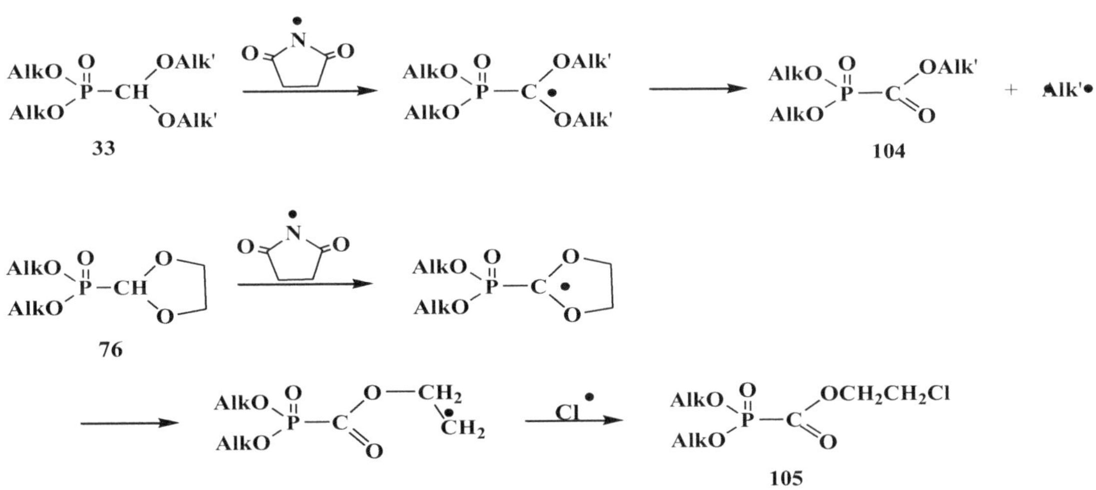

Scheme 25. Interaction of compounds **33** and **76** with N-bromosuccinimide or N-chlorosuccinimide.

The reaction of **57** with o-aminophenol in oxygen flow at 160 °C results in diethyl (2-benzoxazolyl)phosphonate (**106**) [100] (Scheme 26).

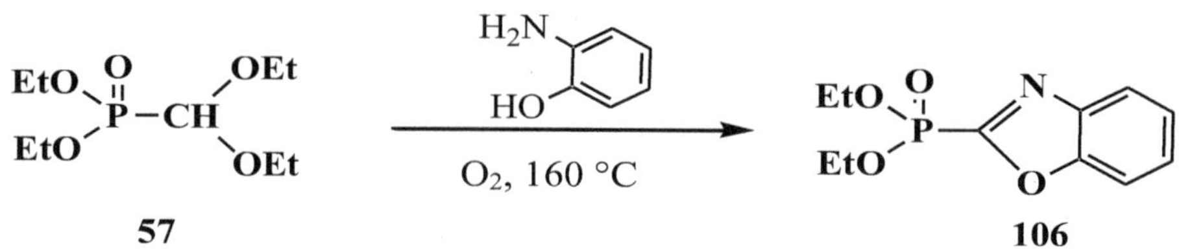

Scheme 26. The reaction of diethyl (diethoxymethyl)phosphonate (**57**) with o-aminophenol.

When compound **33** reacted with thiols in a 1:1 mixture of acetic and hydrochloric acid at 0 °C, dialkyl (dialkylthiomethyl)phosphonates **107** were obtained in 61–64% yield [21] (Scheme 27).

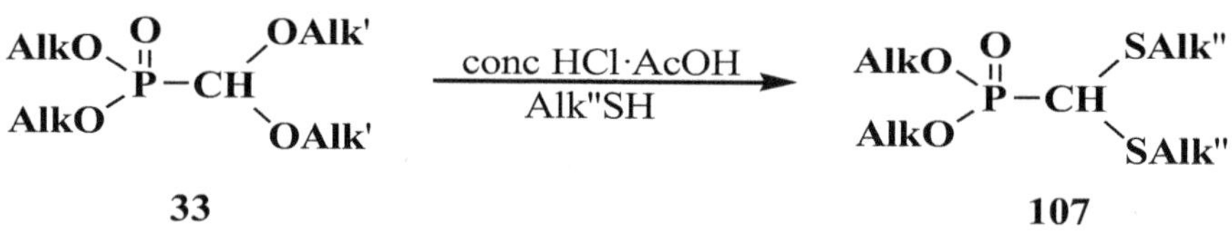

Scheme 27. Transformation of compounds **33** into dialkyl (dialkylthiomethyl)phosphonates **107**.

Diethyl (diethoxymethyl)phosphonate (**57**) reacts with titanium tetrachloride $TiCl_4$ or tetrabromide $TiBr_4$ in diethyl ether to give diethyl [(ethoxy)chloromethyl]phosphonate (**108**) or diethyl [(ethoxy)bromomethyl]phosphonate (**109**) [61] (Scheme 28).

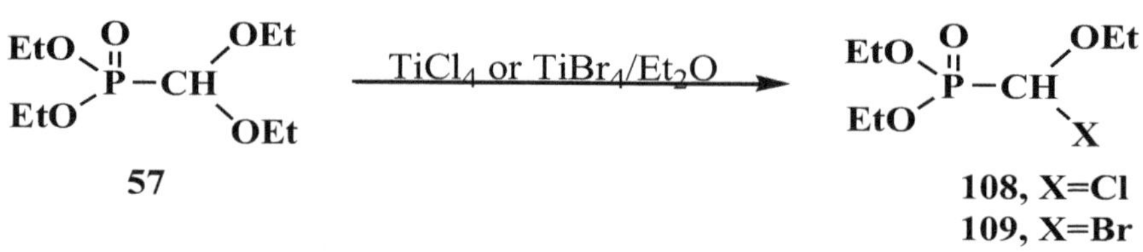

Scheme 28. Syntheses of phosphorylated halogenacetals **108**, **109** from diethyl (diethoxymethyl)phosphonate (**57**).

Cyclic diisopropyl [(4-bromomethyl-[1,3]-dioxolan)-2-yl]phosphonates **110** were used in the synthesis of analogs of natural purine and pyrimidine nucleotides [101]. Guanine analog **111** was prepared by the scheme (Scheme 29).

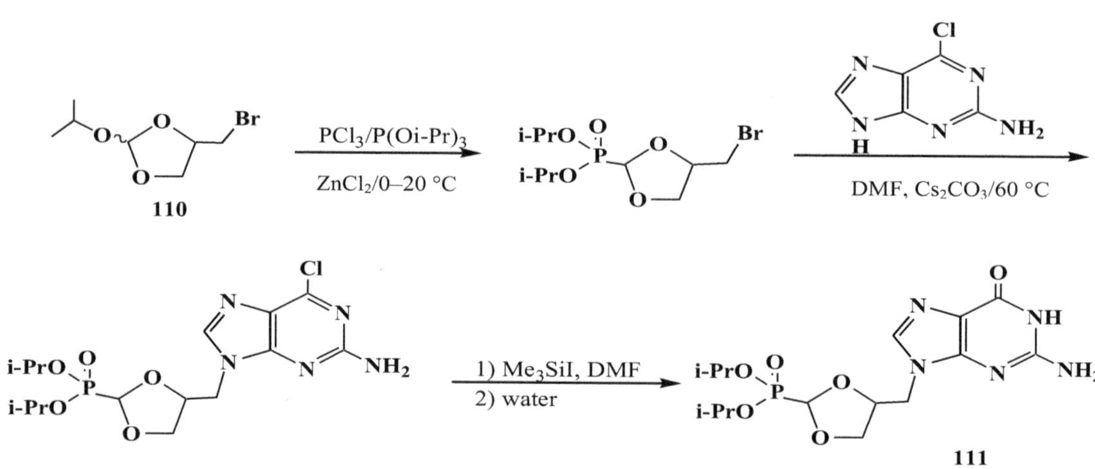

Scheme 29. Synthesis of [1,3]-dioxolane analog guanine nucleotide **111**.

The corresponding uracil analog was obtained in a similar manner. Both compounds showed moderate activity in vitro toward human cytomegalovirus [101]. See also Scheme 16, Scheme 18, Scheme 19, Schemes 54, 57, 58, 60, 62, 86–88 and 90–92.

2.3. *Phosphorylated Formaldehyde Thioacetals* **14**

The first syntheses of compounds **14**, linear dialkyl (dialkylthiomethyl)phosphonates **107** or dialkyl (diphenylthiomethyl)phosphonates **112** [21,22], cyclic dialkyl ([1,3]-dithiolan-2-yl)phosphonates **113**, dialkyl ([1,3]-dithian-2-yl)phosphonates **114** [102] and dialkyl (1,3-benzodithiolylmethyl)phosphonates **115** [103] (Figure 5) were reported almost simultaneously in 1976–1977.

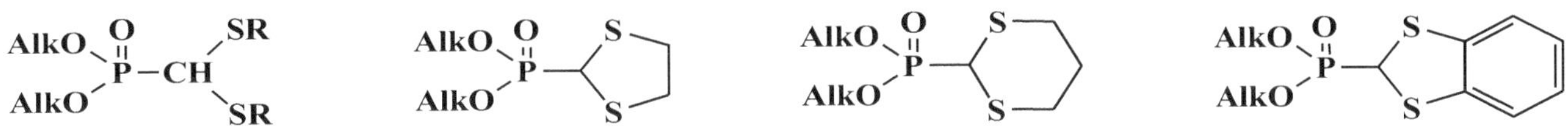

107 (R=Alk), **112** (R=Ph) **113** **114** **115**

Figure 5. First representatives of the phosphorylated formaldehyde thioacetals **107**, **112–115**.

2.3.1. Methods of Synthesis of Phosphorylated Formaldehyde Thioacetals 14

The first syntheses of linear and cyclic dialkyl (dialkylthiomethyl)phosphonates **107**, **112–114** were performed by the analogy with the syntheses of dialkyl (dialkoxymethyl)phosphonates **33** (Schemes 9 and 13), namely, by the Arbuzov reaction of trialkyl phosphites **58** with linear (dialkylthio)chloromethanes **116** or (diphenylthio)chloromethane (**117**) [21,22], and cyclic 2-chloro-1,3-dithiolane [22,102] (**118**) or 2-chloro-1,3-dithiane (**119**) with 93–95% yields (or quaternary ammonium salts of dimethylformamide thioacetal (only for **113**), 50–85%) [21,22] (Scheme 30).

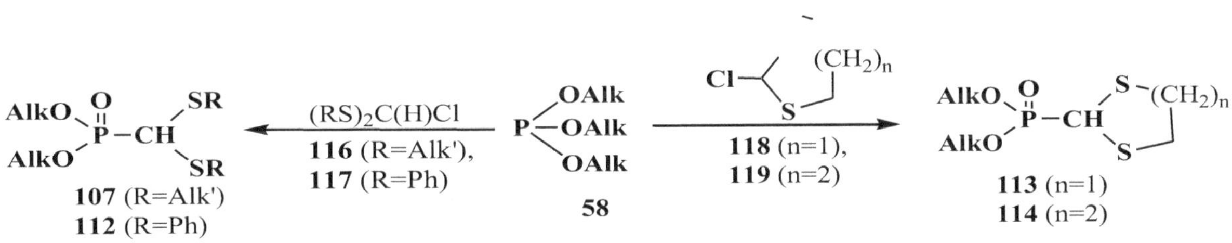

Scheme 30. The first syntheses of linear **107**, **112** and cyclic **113**, **114** phosphorylated formaldehyde thioacetals.

At present, the method is used for the synthesis of cyclic dialkyl (dialkylthiomethyl)phosphonates **113**, **114**, because initial 2-chloro-1,3-dithiolane (**118**) and 2-chloro-1,3-dithiane (**119**) are readily prepared by the reaction of 1,3-dithiolane and 1,3-dithiane with *N*-chlorosuccinimide [102,104–106]. The method allows one to obtain in good yields both cyclic dialkyl (dialkylthiomethyl)phosphonates **113**, **114** (55–96%), [102,104,106]) and diphenyl(dialkylthiomethyl)phosphine oxides **121**, **122** (57–85%), [56,104, 105]) (Scheme 31).

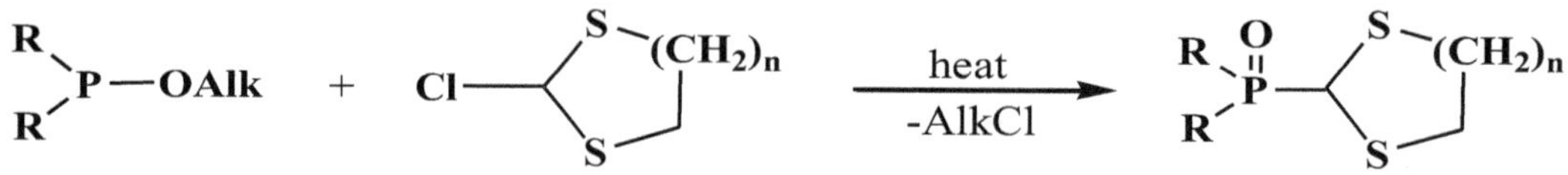

58 (R=OAlk)
121 (R=Ph)

118 (n=1),
119 (n=2)

113, n=1; R=OAlk;
114, n=2; R=OAlk;
121, n=1; R=Ph;
122, n=2; R=Ph

Scheme 31. Syntheses of cyclic dialkyl (dialkylthiomethyl)phosphonates **113**, **114** and diphenyl(dialkylthiomethyl)phosphine oxides **121**, **122**.

However, the poor availability of linear dialkylthiochloromethanes [21] confined the use of this method of synthesis of phosphorylated thioacetals **14** and stimulated the search for alternative methods for their synthesis.

In 1979, a new two-step method was proposed for the synthesis of unsymmetrical diethyl (dialkylthiomethyl)phosphonates **123** starting from diethyl (methylthiomethyl)phosphonate (**124**). After one-pot treatment with butyllithium and elemental sulfur (sulfenylation) followed by aqueous treatment, **124** was converted into diethyl [(methylthio)mercaptomethyl]phosphonate (**125**), which under phase transfer catalysis conditions was further alkylated to give final unsymmetrical thioacetals **123** [107] (Scheme 32). In further work, alkyl halides (AlkHal) were introduced in the reaction medium immediately after sulfenylation, which enabled the preparation of thioacetals **123** by a one-pot method [108].

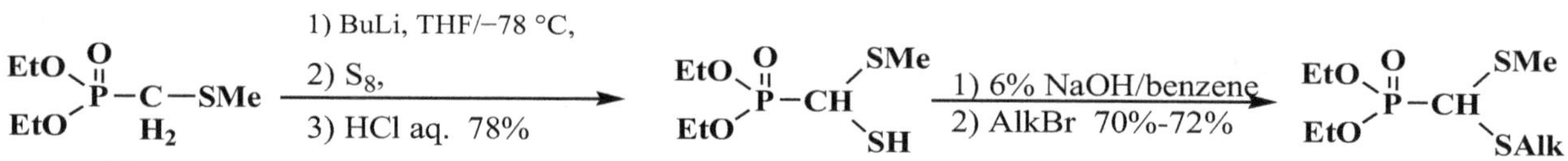

124 125 123

Scheme 32. The synthesis of unsymmetrical diethyl (dialkylthiomethyl)phosphonates (**123**).

Diphenyl(diphenylthiomethyl)phosphine oxide (**126**) [109] was previously obtained in similar manner in 75% yield from diphenyl(phenylthiomethyl)phosphine oxide (**127**) by its interaction with diphenyldisulfide (Scheme 33).

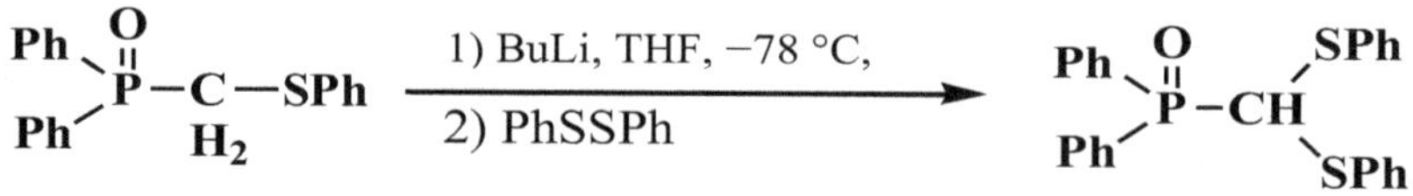

127 126

Scheme 33. Syntheses of diphenyl(diphenylthiomethyl)phosphine oxide (**126**) from diphenyl(phenylthiomethyl)phosphine oxide (**127**).

It was shown later that diethyl (diphenylthiomethyl)phosphonate (**128**) can be prepared from dialkyl methylphosphonate (**129**) [108] by a one-pot technique in 84% yield (Scheme 34).

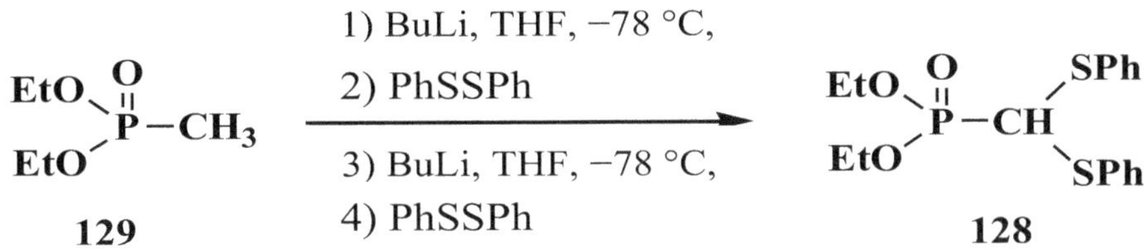

Scheme 34. Syntheses of diethyl (diphenylthiomethyl)phosphonate (**128**) from dialkyl methylphosphonate (**129**) by a one-pot technique.

Diethyl [(methylthio)(trimethylsilyl)methyl]phosphonate (**130**) undergoes a similar transformation with diethyldisulfide to form diethyl (diphenylthiomethyl)phosphonate (**131**) (yield 62–75%) [110] (Scheme 35).

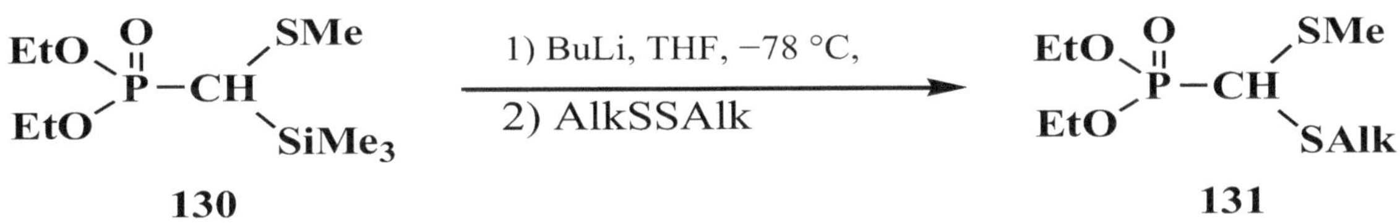

Scheme 35. Interaction of diethyl [(methylthio)(trimethylsilyl)methyl]phosphonate (**130**) with diethyldisulfide.

This method [107] and its variations [108–110] provide a possibility to synthesize linear dialkyl (dialkylthiomethyl)phosphonates **107** containing different substituents in the thioacetal group.

The reduction of phosphonodithioformates **132** with sodium borohydride ($NaBH_4$) or borane–dimethyl sulfide adduct $BH_3{\cdot}SMe_2$ [111] followed by alkylation of the resulted dialkyl [(methylthio)mercaptomethyl]phosphonates **133** finally results in unsymmetrical dialkyl [(methylthio)(alkylthio)methyl]phosphonates **134** (Scheme 36).

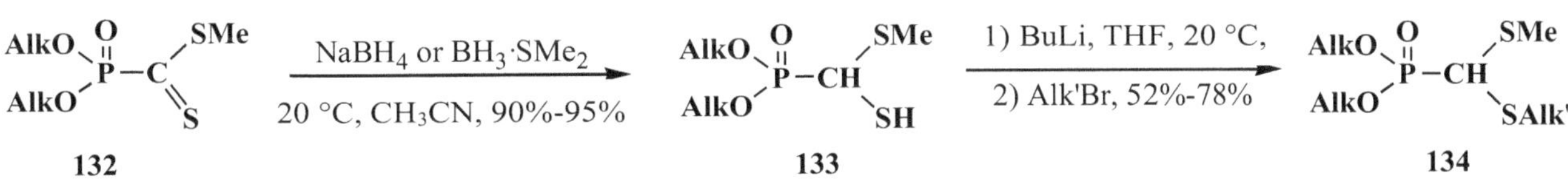

Scheme 36. Transformation of phosphonodithioformates **132** into dialkyl [(methylthio)(alkylthio)methyl]phosphonates **134**.

It was shown later that dialkyl and diaryl disulfides in the presence of catalysts (Cat.) like $BF_3{\cdot}Et_2O$, rhodium(II) tetraacetate $Rh_2(OAc)_4$, or copper(II) sulfate $CuSO_4$, can react with diazomethanephosphonates **38**, where R=OAlk. Phosphorylated carbenes produced in the reaction undergo insertion into S–S bond to afford dialkyl (dialkylthiomethyl)phosphonates **107** or dialkyl (diarylthiomethyl)phosphonates **135** in 42–93% yields [56,112]. Reactions proceed by the carbene mechanism (Scheme 37) [112].

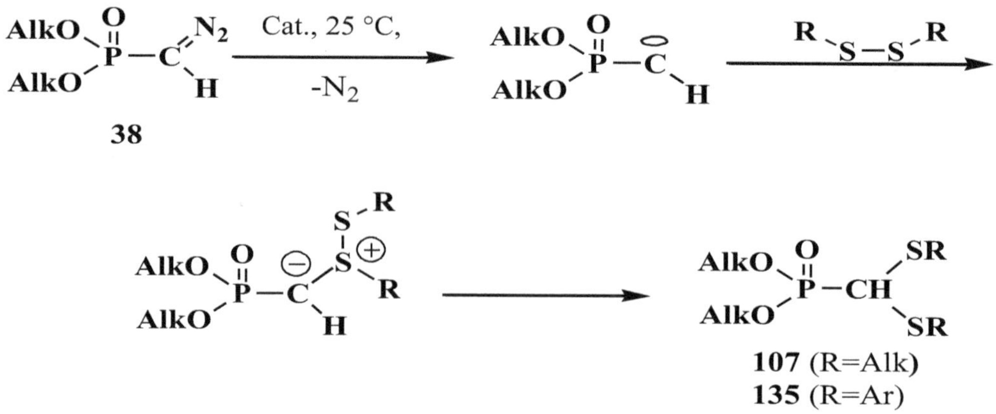

Scheme 37. Diazomethanephosphonates 38 reaction with organic disulfides with formation of dialkyl (dialkylthiomethyl)phosphonates 107 or dialkyl (diarylthiomethyl)-phosphonates 135.

A method of synthesis of substituted diphenyl([1,3]-dithian-2-yl)phosphine oxides **136** (yields 34–85%) starting from substituted 1,3-dithianes and chlorodiphenylphosphine (**68**) was also suggested [104,113,114]. Initially formed diphenyl([1,3]-dithian-2-yl)phosphines **137** undergo further oxidation with molecular oxygen to final **136** (Scheme 38). Diphenyl([1,3]-dithian-2-yl)phosphine sulfides **138** or selenides **139** can be obtained by this method when elemental sulfur or selenium are used as oxidants for **137** [104,113].

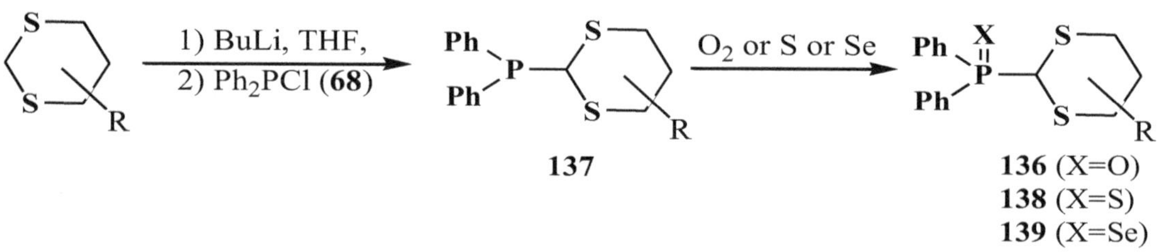

Scheme 38. Obtaining of substituted diphenyl([1,3]-dithian-2-yl)phosphine oxides **136** and corresponding phosphine sulfides **138** and phosphine selenides **139** from dithianes.

The attempted preparation of **107** and **114** in one step from dialkylthiomethanes or 1,3-dithianes and dialkyl chlorophosphate in the presence of strong bases failed, as evidenced by the negligible yield of final products [21,115]. The syntheses of **107** by the interaction of thioles with dialkyl [(*N*,*N*-dimethylamino)alkoxymethyl]phosphonates **140** or dialkyl [(alkylthio)chloromethyl]phosphonates **141** also failed [21].

Hovewer a method of synthesis of linear phosphorylated formaldehyde thioacetals **14** by the reaction of dialkyl [(arylthio)chloromethyl]phosphonates **142** with thiols at 0 °C in the presence of equimolar amount of tin(IV) tetrachloride $SnCl_4$ was succesful. This route provided a preparation of unsymmetrical dialkyl [(alkylthio)(arylthio)methyl]phosphonates **143** in 73–84% yields [43] (Scheme 39).

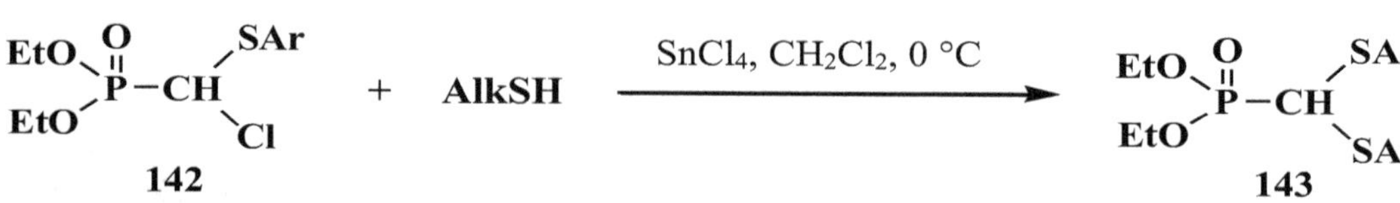

Scheme 39. Obtaining of dialkyl [(alkylthio)(arylthio)methyl]phosphonates **143** by the reaction of dialkyl [(arylthio)chloromethyl]phosphonates **142** with thiols.

Dialkyl (dialkylthiomethyl)phosphonates **107** were also obtained by the reaction of dialkyl (dialkoxymethyl)phosphonates **33** with thiols in the presence of acids [21] (see Scheme 27).

Dialkyl (1,3-benzodithiolylmethyl)phosphonates **115** were prepared for the first time by the reaction of 1,3-benzodithiolyl tetrafluoroborate with **58** in the presence of NaI [103]. The method is used at present without changes [116] (Scheme 40).

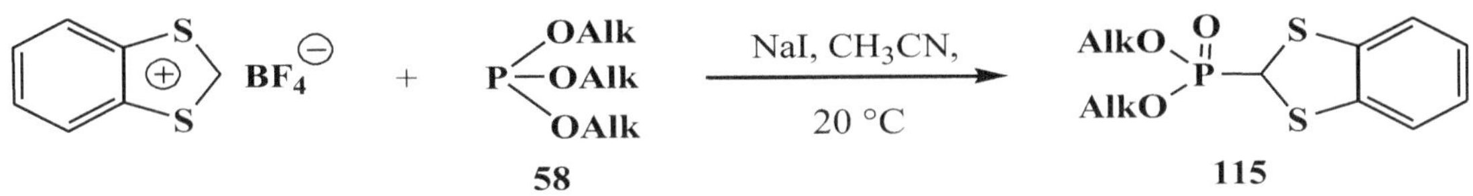

Scheme 40. The synthesis of dialkyl (1,3-benzodithiolylmethyl)phosphonates **115** from 1,3-benzodithiolyl tetrafluoroborate.

2.3.2. Chemical Properties of Phosphorylated Formaldehyde Thioacetals 14

The chemical properties of phosphorylated formaldehyde thioacetals **14** have been studied in much less detail compared with the corresponding acetals **13**. Their properties are determined by the presence of sulfur atoms and a disubstituted phosphoryl group. Both sulfur atoms are oxidized when diethyl ([1,3]-dithian-2-yl)phosphonate (**144**) is treated with sodium periodate $NaIO_4$ in aqueous methanol at 20 °C [106] to yield the dioxo form **145** (Scheme 41).

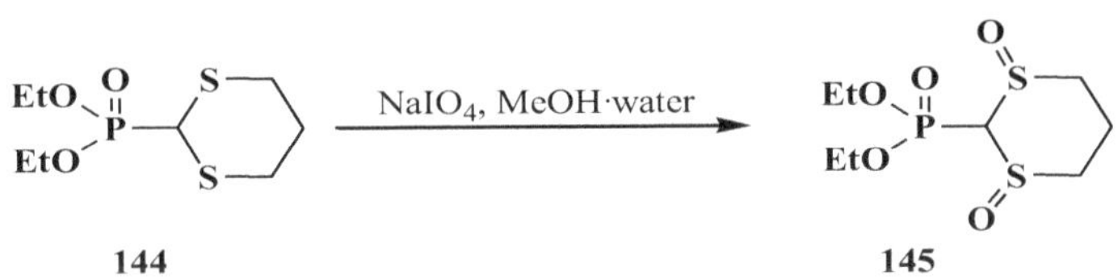

Scheme 41. Diethyl ([1,3]-dithian-2-yl)phosphonate (**144**) oxidation by periodate.

The heating of diphenyl([1,3]-dithiolan-2-yl)phosphine oxides **121** [115] and diphenyl([1, 3]-dithian-2-yl)phosphine oxides **122** [56,104] with phosphorus pentasulfide P_2S_5 in benzene or *N,N*-diethylaniline $PhNMe_2$ leads to the replacement of phosphinoyl oxygen by sulfur to yield diphenyl([1,3]-dithiolan-2-yl)phosphine sulfide (**146**) and diphenyl([1,3]-dithian-2-yl)phosphine sulfide, respectively (**147**) (Scheme 42). See also Scheme 38.

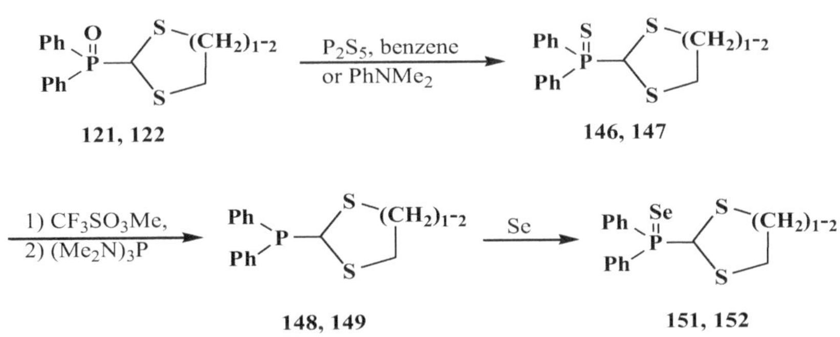

Scheme 42. Transformation of phosphorylated cyclic thioacetals **122, 123** into corresponding phosphines **148, 149**, phosphine sulfides **146, 147** and phosphine selenides **151, 152**.

The sequential treatment of the prepared phosphine sulfides **146** and **147** with trifluoromethylsulfonate CF_3SO_3Me and tris(dimethylamino)phosphite $(Me_2N)_3P$ finally affords diphenyl([1,3]-dithiolan-2-yl)phosphine (**148**) [105] and diphenyl([1,3]-dithian-2-yl)phosphines **149** [56, 104]. The reaction of diphenyl(5-*tert*-butyl-[1,3]-dithian-2-yl)phosphine sulfide (**150**) with trichlorosilane leads to the same result [56]. The obtained phosphines **146, 147** combine with elemental selenium to yield diphenyl([1,3]-dithiolan-2-yl)phosphine selenide (**151**) [105] and diphenyl([1,3]-dithian-2-yl)phosphine selenide (**152**) [56,104]. See also Schemes 59, 60, 64, 65, 74–81.

2.4. *(N,N-dialkylamino)cyanomethyl Derivatives of Phosphorylated Formaldehyde (α-dialkylamino-nitriles)* **15**

For 1982 till now, only two derivatives of this type of organophosphorus compounds of phosphonate series were obtained, diethyl [(*N,N*-dimethylamino)cyanomethyl]phosphonate (**153**) [24] and diethyl [(*N*-morpholino)cyanomethyl]phosphonate (**154**) [26].

2.4.1. Methods of Synthesis of Diethyl [(N,N-Dialkylamino)cyanomethyl]phosphonates **15**

Diethyl [(*N,N*-dimethylamino)cyanomethyl]phosphonate (**153**) was obtained for the first time by a reaction similar to the preparation of phosphorylated acetals **13** (Scheme 7) from diethyl phosphite (**6**) and (*N,N*-dimethylamino)(methoxy)cyanomethane [24,25] (Scheme 43).

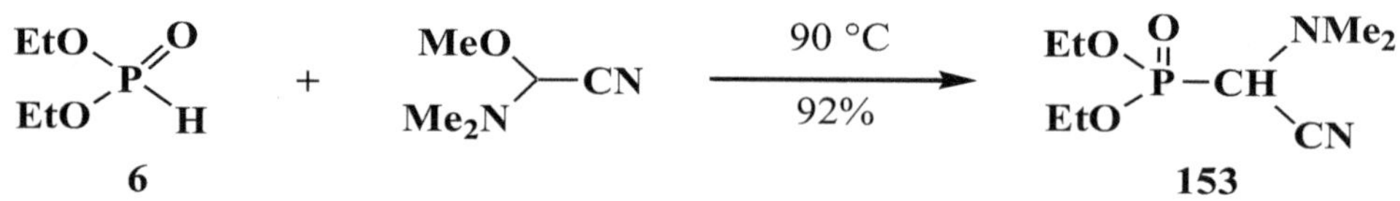

Scheme 43. Syntheses of diethyl [(*N,N*-dimethylamino)cyanomethyl]phosphonate (**153**).

Diethyl [(*N*-morpholino)cyanomethyl]phosphonate (**154**) was prepared by the second method from diethyl chlorophosphate (**155**) and (*N*-cyanomethyl)morpholine [26] (Scheme 44) and was used in a subsequent reaction without isolation.

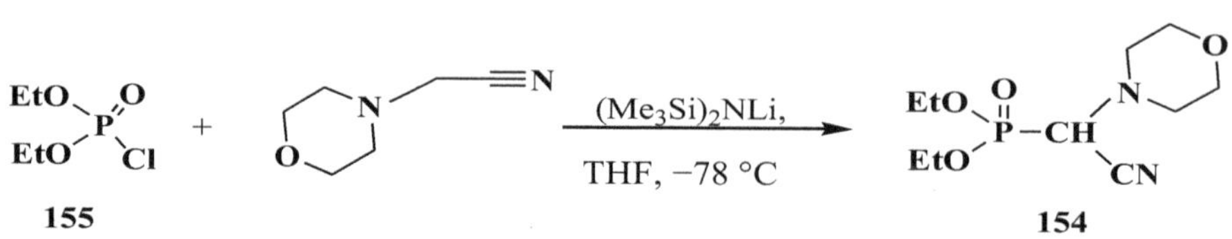

Scheme 44. Syntheses of diethyl [(*N*-morpholino)cyanomethyl]phosphonate (**154**) from diethyl chlorophosphate (**155**).

2.4.2. Chemical Properties of Diethyl [(N,N-Dialkylamino)cyanomethyl]phosphonates **15**

The chemical properties of this type of compounds have been studied almost completely using the example of diethyl [(*N,N*-dimethylamino)cyanomethyl]phosphonate (**153**). The treatment of **153** with methyl iodide (MeI) or dimethyl sulfate Me_2SO_4 [117,118] gives rise to methylation of nitrogen atom (quaternization) of the dimethylamino group and formation of diethyl [(*N,N,N*-trimethylammonio)cyanomethyl]phosphonate cation (**156**), whose α proton shows enhanced acidity and breaks off under the action of aqueous potassium carbonate solution K_2CO_3. The resultant diethoxy [(*N,N,N*-trimethylammonio)cyanomethylidium]phosphonate (**157**) can react with electrophiles—MeI and molecular chlorine Cl_2 [117], to give diethyl [(1-(*N,N,N*-trimethylammonio))(1-cyano)ethan-1-yl]phosphonate cation (**158**) and diethyl [chloro(1-*N,N,N*-trimethylammonio)cyanomethyl]phosphonate cation (**159**), respectively (Scheme 45).

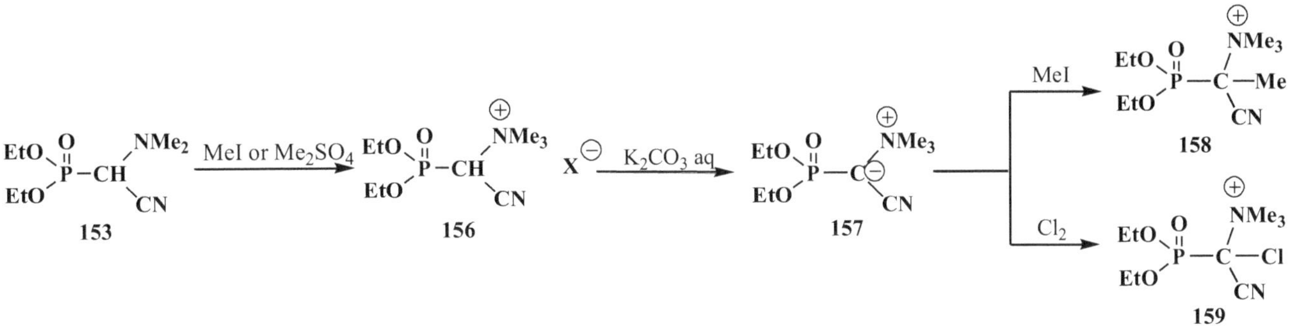

Scheme 45. Transformation of diethyl [(*N*,*N*-dimethylamino)cyanomethyl]phosphonate (**153**).

Under phase transfer catalysis conditions—50% KOH solution, $[Et_3NCH_2Ph]^+Cl^-$ (TEBAC) [24] phosphonate **153** combines with nitrosobenzene PhN=O similarly to the Horner reaction to afford [(*N*,*N*-dimethyl)(*N*′-phenyl)amidinoyl]oxalnitrile (**160**) in 58% yield (Scheme 46). See also Schemes 55, 60, 66, 67 and 82–85.

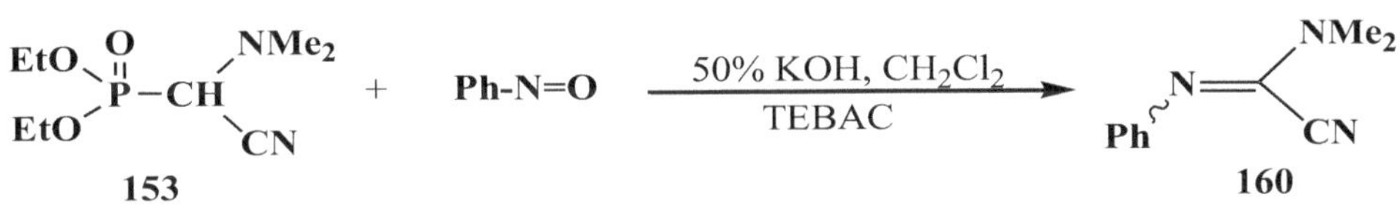

Scheme 46. Horner—analog reaction of phosphonate **153** with nitrosobenzene.

2.5. *Diphosphinoyl N,N-Dialkylaminomethanes* 16

2.5.1. Methods of Synthesis of Diphosphinoyl N,N-Dialkylaminomethanes 16

The first report on the synthesis of tetraethyl (*N*,*N*-dimethylaminomethyl)diphosphonate (**161**) was published in 1968. Compound **161** was obtained in 62% yield by heating a 2:1 mixture of diethyl phosphite (**6**) and dimethylformamide dimethylacetal (**78**) [27]. Dialkyl phosphites **51**, dialkyl- and diphenylphosphine oxides **52**, **53** can be also involved in this reaction [31,32,119]. Intermediate dialkyl [(*N*,*N*-dimethylamino)methoxymethyl]phosphonates (**162**) or dialkyl[(*N*,*N*-dimethylamino)-methoxymethyl]phosphine oxides **163** may be isolated in many cases [32,119]. This allows preparation of symmetrical **164** and unsymmetrical diphosphinoyl *N*,*N*-dimethylaminomethanes **165** [31,32,119,120] (Scheme 47).

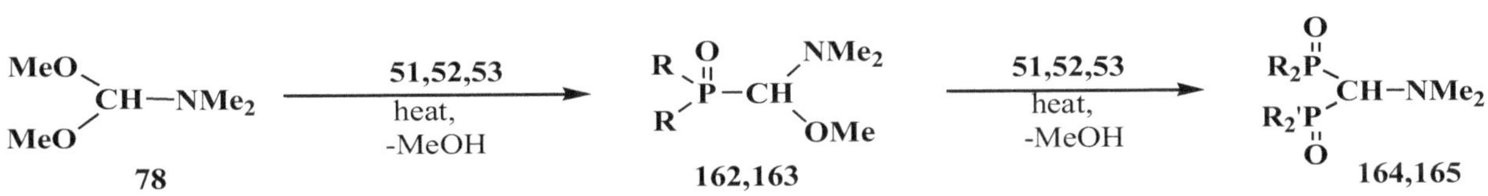

Scheme 47. Synthesis of diphosphinoyl *N*,*N*-dimethylaminomethanes **164** and **165** from dimethylformamide dimethylacetal (**78**), where R, R′ = OAlk, Alk, Ph. See text above.

Other methods of synthesis of symmetrical tetraalkyl (*N*,*N*-dimethylaminomethyl)diphosphonates **166** were proposed later. Since trialkyl phosphites **58** do not react with dimethylformamide dimethylacetal (**78**) [18], mixed dialkyl trimethylsilyl phosphites $(EtO)_2PSiMe_3$ were successfully employed in the reaction. The reaction proceeds spontaneously at 20 °C [121] in 36–66% yield or upon heating in the presence of zinc chloride in 72–77% yield [122,123] (Scheme 48).

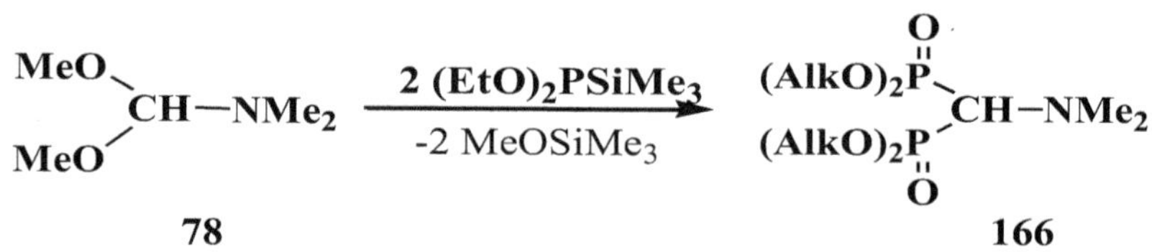

Scheme 48. Use mixed dialkyl trimethylsilyl phosphites for the syntheses of tetraalkyl diphosphonates **166** from dimethylformamide dimethylacetal (**78**).

A convenient method of synthesis of unsymmetrical tetraalkyl (*N,N*-dialkylaminomethyl)-diphosphonates **167** by the reaction of trialkyl phosphites (**58**) with *N,N*-dialkylhalo-methylideneiminium halides $[HalC(H)=NAlk_2]^+Hal^-$, where Hal = Cl (compounds **168**) [28,31,32,120] or Br (compounds **169**) [29] in 2:1 ratio was proposed in 1969.

It was shown that the reaction of trialkyl phosphites **58** with *N,N*-dimethylchloro-methylideneiminium chloride (**170**) proceeds via intermediate formation of dialkyl [(*N,N*-dimethylamino)chloromethyl]phosphonate **171** [31,32], specially prepared dimethyl [(*N,N*-dimethylamino)chloromethyl]phosphonate (**172**) and diethyl [(*N,N*-dimethylamino)chloromethyl]- phosphonate (**173**) may be also involved in the reaction [30,31,124]. This method enables preparation of symmetrical and unsymmetrical tetraalkyl phosphonates **174** as well (Scheme 49).

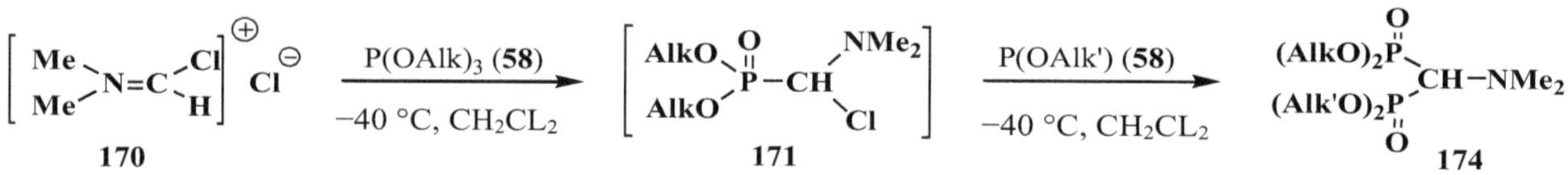

Scheme 49. Method of synthesis of symmetrical and unsymmetrical dialkyl (*N,N*-dimethylaminochloromethyl)phosphonates **174** from *N,N*-dimethylchloromethylideneiminium chloride (**170**) and trialkyl phosphites (**58**).

Dialkyl phosphites **51**, dialkylphosphine oxides **52** and diphenylphosphine oxide (**53**) also react with *N,N*-dialkylchloromethylideneiminium chlorides **168** [31,120] and their analogs **175** obtained from *N,N*-dialkylformamides and phosphorus oxychloride [28] (Scheme 50). The reaction also proceeds via intermediate compounds **176**, where X=Cl or Cl_2PO_2. This allows also the preparation of symmetrical or unsymmetrical diphosphinoyl compounds **177** and **178**, respectively.

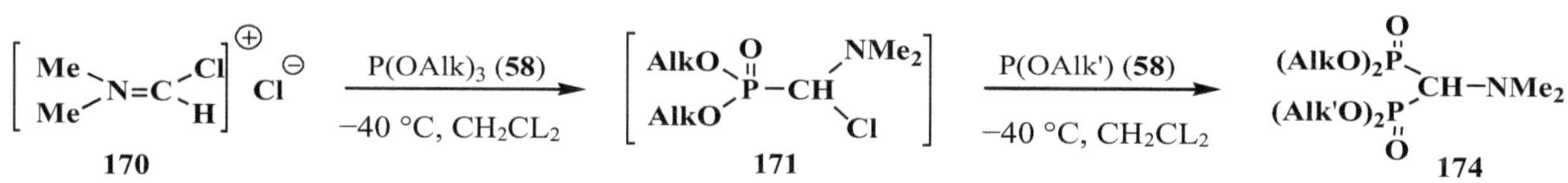

Scheme 50. Synthesis of diphosphinoyl *N,N*-dialkylaminomethanes **177** and **178** from *N,N*-dimethylchloromethylideneiminium chloride (**170**) and hydrophosphorylic compounds **51**, **52**, **53**. See text above.

See also Scheme 14.

2.5.2. Chemical Properties of Diphosphinoyl N,N-Dialkylaminomethanes **16**

The chemical properties of this type of organophosphorus compounds are studied insufficiently and almost exclusively by the example of tetraethyl (*N,N*-dimethylaminomethyl)diphosphonate (**161**). Their properties are attributable to the presence of both an amino group and disubstituted phosphoryl groups.

Hydrolysis of tetraalkyl (*N*,*N*-dimethylaminomethyl)diphosphonates **174** was studied using the example of compound **161**. Boiling diphosphonate **161** with concentrated hydrochloric acid leads to (*N*,*N*-dimethylaminomethyl)diphosphonic acid (**179**) [98,120] in almost quantitative yield (Scheme 51).

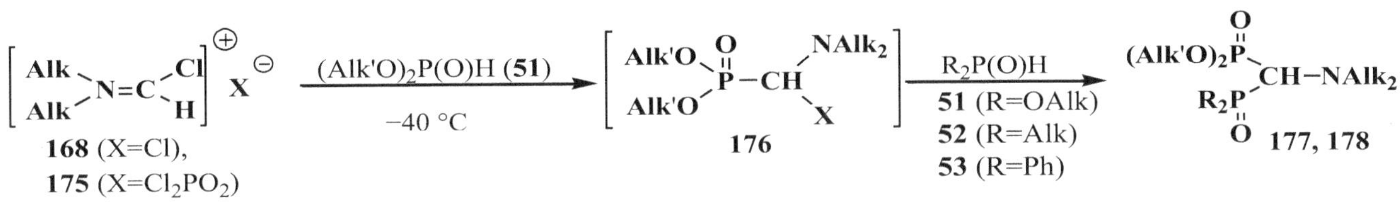

Scheme 51. Acid hydrolysis of tetraethyl (*N*,*N*-dimethylaminomethyl)diphosphonate (**161**).

The amino group of diphosphonate **161** undergoes methylation (quaternization) when reacted with methyl iodide (MeI) or dimethyl sulfate (Me_2SO_4) to give tetraethyl (*N*,*N*,*N*-trimethylammoniomethyl)diphosphonate cation (**180**) in 76 and 85% yields, respectively [118, 125]. The α-proton of the latter shows enhanced acidity and undergoes elimination under the action of aqueous solution of potassium carbonate (K_2CO_3) to afford tetraethyl [(*N*,*N*,*N*-trimethylammonio)methylidium]diphosphonate (**181**) [120,125] (Scheme 52).

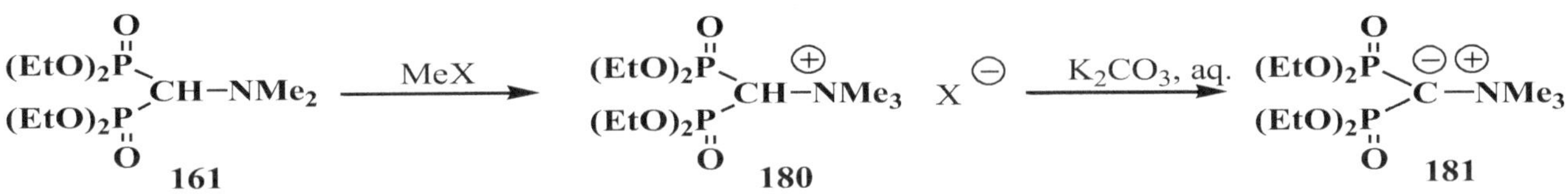

Scheme 52. Transformation of tetraethyl (*N*,*N*-dimethylaminomethyl)diphosphonate (**161**) into methylidiumdiphosphonate **181**, where X = I, $MeSO_3$.

Ylide **181** exhibits an enhanced stability: it can be stored for a long time in air, it is thermally stable and does not react with methyl iodide [126]. However, ylide **181**, like cyanomethylide **154**, reacts with molecular chlorine to yield tetraethyl [(*N*,*N*,*N*-trimethylammonio)chloromethyl)diphosphonate chloride (**182**) that undergoes fast dealkylation on storage to afford the corresponding betaine **183** in 85% yield [120] (Scheme 53).

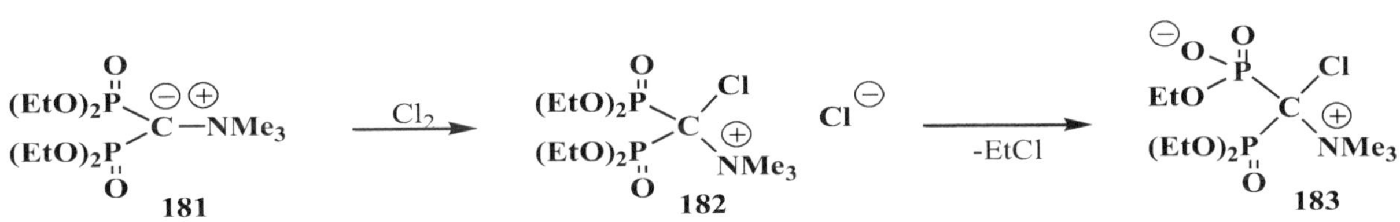

Scheme 53. Interaction of methylidiumdiphosphonate **181** with chlorine with the subsequent formation of betaine **183**.

See also Schemes 60, 68 and 72–74.

3. General Chemical Properties of Phosphorylated Formaldehyde Acetals 13 and Structurally Related Compounds 14–16

3.1. *Phosphorus–Carbon Bond Cleavage*

3.1.1. Cleavage of Phosphorus–Carbon Bond under the Action of Acids and Acidic Reagents

The general property of phosphorylated formaldehyde acetals **13** and structurally related compounds **14–16** is the possibility of phosphorus–carbon bond cleavage under the action of acids and acidic reagents [24,77,95,98], organic reagents [19,119], and bases [37,127,128].

The cleavage of phosphorus–carbon bond under acidic conditions compounds **33** is studied on examples of diethyl (diethoxymethyl)phosphonate (**57**). Phosphonate **57** was shown to undergo cleavage of the phosphorus–carbon bond on heating with 5% hydrochloric acid [77,98] to form diethyl phosphites (**6**), ethyl formate (**184**) and ethanol. Similar degradation of compound **57** occurs in a flow of hydrogen chloride at 20 °C [95] (ethyl chloride is formed as a byproduct) (Scheme 54). Unlike **57**, interaction of diethyl (diphenoxymethyl)phosphonate (**185**) with dry hydrogen chloride does not lead to the cleavage of phosphorus–carbon bond, but leads to the dealkylation of one ethoxy substituent at the phosphorus atom with obtaining ethyl (diphenoxymethyl)phosphonic acid (**186**) (Scheme 54).

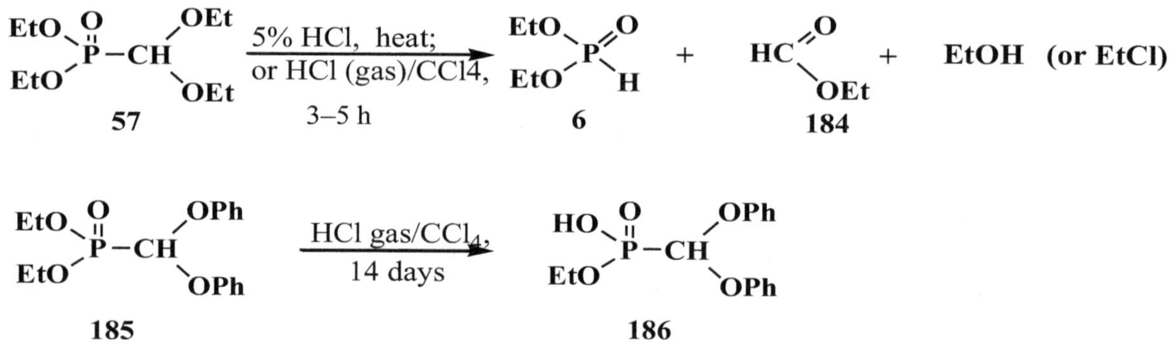

Scheme 54. Interaction of acetals **57** and **185** with hydrochloric acid (only for **57**) and hydrogen chloride.

It was shown that acid hydrolysis of diethyl (diethylthiomethyl)phosphonates (**187**) is accompanied by a partial cleavage of the phosphorus–carbon bond [98]. Prolonged refluxing of diethyl [(*N,N*-dimethylamino)cyanomethyl]phosphonate (**153**) in concentrated hydrochloric acid leads to *N,N*-dimethylaminoacetic acid (**188**) in 87% yield [24] (Scheme 55).

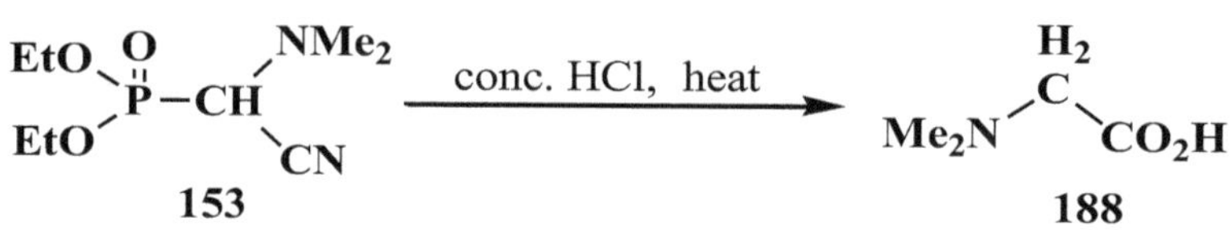

Scheme 55. Obtaining *N,N*-dimethylaminoacetic acid (**188**) by acid hydrolysis of cyanomethylphosphonate **153**.

It was shown in [95] that the attempted transesterification of ethoxy substituents at the phosphorus atom of phosphonate **57** by catechol residue also leads to the cleavage of the phosphorus–carbon bond (Scheme 56).

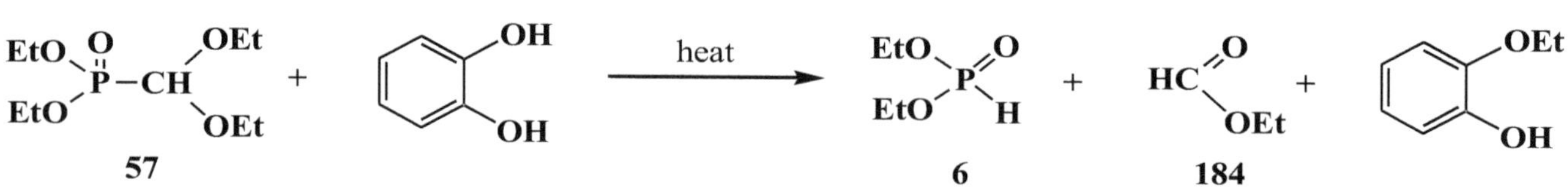

Scheme 56. Destruction of acetales **57** at its interaction with catechol.

The cleavage of phosphorus–carbon bond may also occur when diethyl (dialkoxymethyl)phosphonates **93** are exposed to phosphorus pentachloride [95].

3.1.2. The Cleavage of Phosphorus–Carbon Bond in Reactions with Organic Coreactants

The reactions of **57** with dichloromethoxymethane $Cl_2C(H)OMe$ or acetyl bromide MeC(O)Br was shown to be accompanied by the cleavage of phosphorus–carbon bond [19] to form diethoxychlorophosphine (**189**) and diethyl acethylphosphonate (**190**) (Scheme 57).

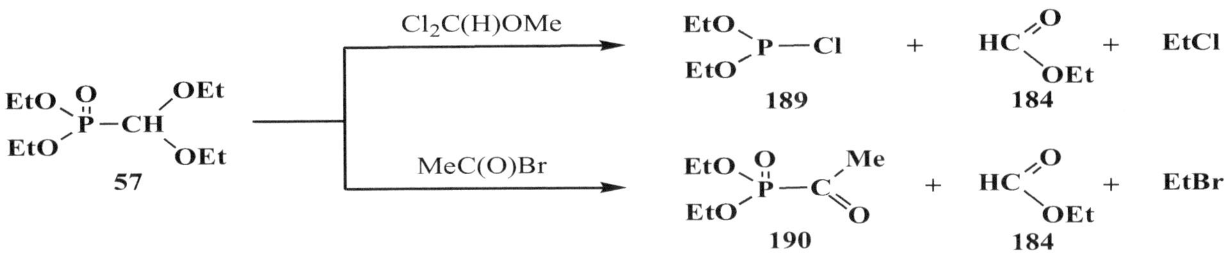

Scheme 57. Acetales **57** interaction with $Cl_2C(H)OMe$ and MeC(O)Br leading to the cleavage of phosphorus–carbon bond.

The reaction of diethyl phosphite (**6**) with dipropyl [(*N,N*-dimethylamino)ethoxymethyl]phosphonate (**191**) is also accompanied by the partial cleavage of the phosphorus–carbon bond [119].

3.1.3. Phosphorus–Carbon Bond Cleavage under the Action of Bases

The reaction of the lithium derivatives of diphenyl(dialkoxymethyl)phosphine oxides **50** with *n*-octanal and *p*-isopropylbenzaldehyde leads to α-phosphorylated alcohols **192**, the products of addition of diphenylphosphinite anion (**193**) to the carbonyl group (Abramov reaction) [127,128], rather than ketene *O,O*-acetals **194** as expected products of the Horner reaction [9,10] (Scheme 58). The reason of this course is the instability of diphenyl (dialkoxymethyl)phosphine oxide anion (**195**) that compound **50** decomposes after long exposition under the reaction conditions to form anion **193**, which further undergoes addition to the aldehyde carbonyl group [127,128].

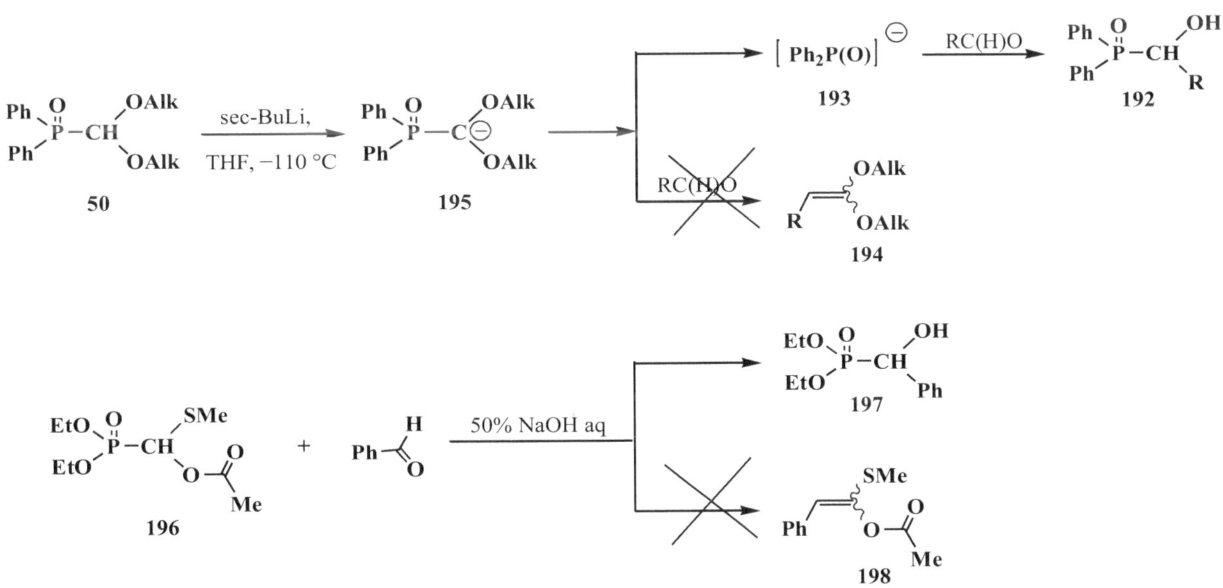

Scheme 58. Cleavage of phosphorus–carbon bond of acetal **57** and O,S-acetal **196** leading to the formation of α-phosphorylated alcohols **192** and **198**.

Like phosphine oxides **50**, the reaction of diethyl [(acetoxy)methylthiomethyl]phosphonate (**196**) with benzaldehyde PhC(H)O under phase transfer catalysis conditions gives rise to diethyl [(hydroxy)phenylmethyl]phosphonate (**197**), the Abramov reaction product, instead of the expected ketene O,S-acetal **198** resulting from the Horner reaction [37] (Scheme 58).

Lithiated anions **199** of diethyl [(methylthio)(alkylthio)methyl]phosphonates **123** react quickly with molecular oxygen with the cleavage of phosphorus–carbon bond and formation of dialkyl dithiocarbonates **200** (yields 71–72%) [129] (Scheme 59).

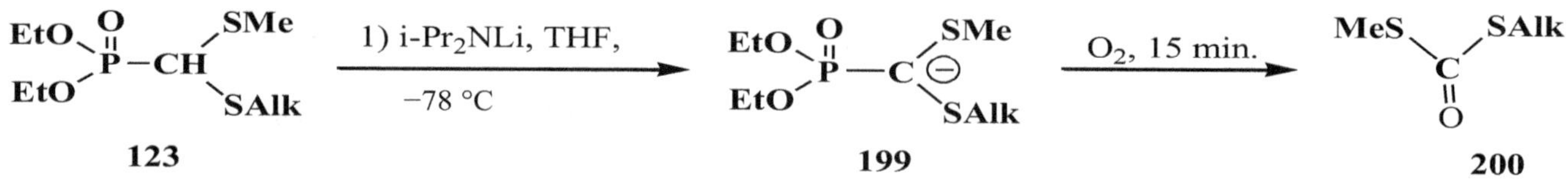

Scheme 59. Destruction of thioacetal **123** in the presence of oxygen, leading to formation of dithiocarbonates **200**.

See also Schemes 63, 96, 103, 104, 113–117 and 120.

3.2. Synthesis of Formacetalphosphonic Acids

Attempted preparations of formacetalphosphonic acids **201**, **202** and **203** from the corresponding phosphorylated acetal **57**, thioacetal **187** and aminonitrile **153** by acid hydrolysis lead to cleavage of the phosphorus–carbon bond, but in the case of compound **91** with a aromatic substituent in the acetal group, it undergoes acid hydrolysis to give the expected phosphonic acid **92** (Scheme 20). See also Section 3.1.1. "Cleavage of phosphorus–carbon bond under the action of acids and acidic reagents".

However, the target formacetalphosphonic acids **201–203** may be successfully prepared by the reaction of acetals **33**, diethyl (diethylthiomethyl)phosphonate (**187**), diethyl [(*N,N*-dimethyl-amino)cyanomethyl]phosphonate (**153**), and tetraethyl(*N,N*-dimethylaminomethyl)diphosphonate (**161**) with trimethylsilyl bromide (Me_3SiBr) in acetonitrile. Alkyl groups at the phosphorus atom are eliminated as alkyl halides to afford the corresponding intermediate bis(trimethylsilyl) phosphonates **204** (or tetrakis(trimethylsilyl) phosphonate (**205**) from diphosphonate **161**), which are further readily hydrolyzed by water treatment to give:

- (dialkoxymethyl)phosphonic acids **201** in 93–100% yield from **33** [19,98]. It was shown that the reaction **33** with an equimolar mixture of trimethylsilyl chloride Me_3SiCl and NaBr or LiBr [130] in acetonitrile of trimethylsilyl chloride and NaI in methylene chloride [131] leads to the same result,
- (diethylthiomethyl)phosphonic acid (**202**) from **187** in 58% yield [98],
- (1-dimethylamino-1-cyanomethyl)phosphonic acid (**203**) from **153** in 55% yield [24],
- (*N,N*-dimethylaminomethyl)diphosphonic acid (**179**) from **161** in 65% yield [98] (Scheme 60).

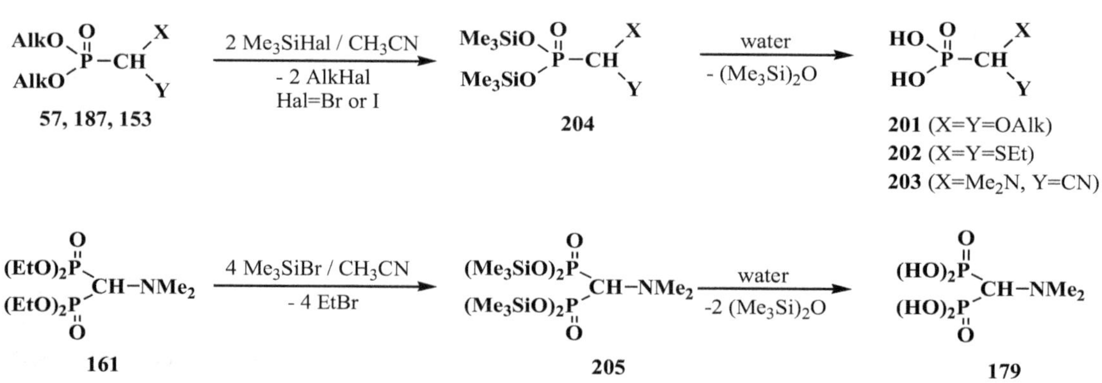

Scheme 60. Syntheses of formacetalphosphonic acids **201–203** and **179** from compounds **57, 187, 153, 161** by means of Me_3SiBr or Me_3SiI. See text above.

See also Schemes 15, 20–22, 29 and 54.

3.3. Alkylation (Acylation) of the Formacetal Carbon Atom

Dialkyl (dialkoxymethyl)phosphonates **33** produce no stable phosphorylated carbanion **206** when reacted with bases (no metallation occurs, even under the action of *tert*-butyllithium (*t*-BuLi), which provides no possibility for further alkylation and acylation of the formacetal group [23,132] (Scheme 61).

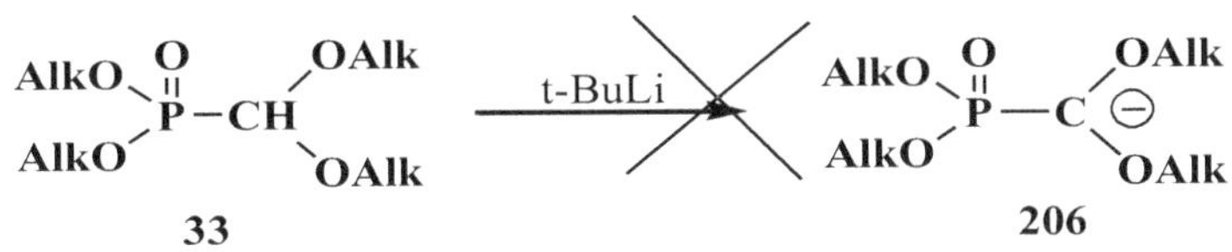

Scheme 61. Phosphorylated acetals **33** do not produce carbanions **206**.

This fact was explained by insufficient stabilization of the negative charge of carbanion on the two oxygen atoms in the α-position [23]. However, it was shown in 1983 [133] that, in contrast to phosphonates **33**, diphenyl(dialkoxymethyl)phosphine oxides **50** produce phosphorylated anions **195** at −110 °C that undergo metallation. The reason for the stability of the lithium derivatives of phosphine oxides **50** is the ability of diphenylphosphinoyl group to delocalize the negative charge of carbanion **195** [20,134] (Scheme 58). By the example of anion **207** of diphenyl(dimethoxy-methyl)phosphine oxide (**208**), it was shown that it is rather stable to subsequent alkylation with alkyl halides and acylation with benzoyl chloride [20]. The reactions afford diphenyl[(dimethoxy)alkylmethyl]phosphine oxides **209**, in 30–94% yields, and diphenyl-[(dimethoxy)benzoylmethyl]phosphine oxide (**210**) (60%) (Scheme 62).

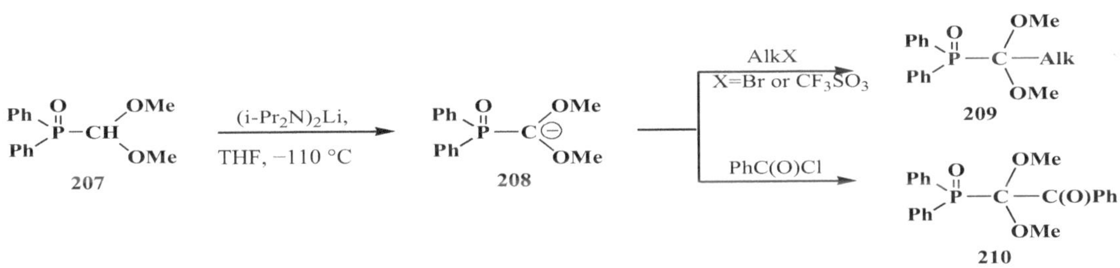

Scheme 62. Alkylation and acylation of carbanion **207**.

In acidic medium at 20 °C, phosphine oxides **209** are readily decomposed with cleavage of the phosphorus–carbon bond. The resultant methyl carboxylates **211** are homologous to the initial alkyl halides—carbon chain elongation by one atom (Scheme 63).

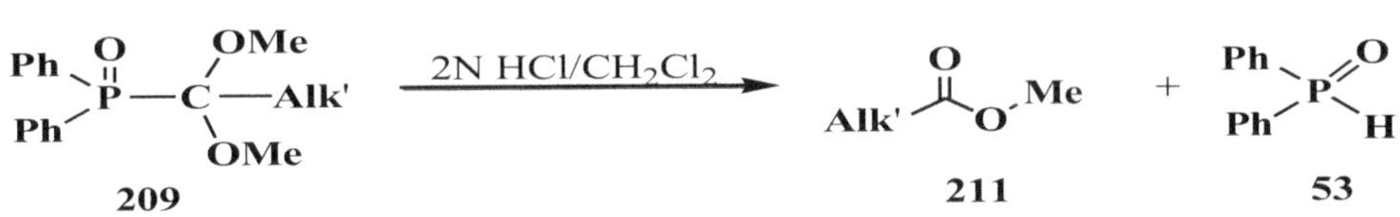

Scheme 63. Hydrolysis of phosphine oxide **209** leads to phosphorus–carbon bond cleavage.

Similarly, according to ^{1}H-NMR spectroscopy, the methanolysis of diphenyl[(1,1-dialkoxy)nonan-1-yl)phosphine oxide (**212**) in the presence of trifluoroacetic acid leads to 1,1,1-trimethoxynonane (**213**) in 65% yield [20].

Nonetheless, the storage of a solution of lithiated anion **207** for two hours even at −110 °C causes the cleavage of phosphorus–carbon bond (see Scheme 58).

Distinct from acetals **33** [23,132], the two sulfur atoms of thioacetals **113**, **114** stabilize well the neighboring carbanion [23], therefore the α-hydrogen atom in the thioacetal group is readily removed under the action of strong bases [23,132] in both dialkyl (dialkythiomethyl)phosphonates **113**, **114** [135] and diphenyl(dialkythiomethyl)phosphine oxides **121**, **122** [113]. Further, the carbanions are readily alkylated with alkyl halides [113,135]. For example, dialkyl ([1,3]-dithian-2-yl)phosphonates **114** in this reaction produce dialkyl [(2-alkyl-[1,3]-dithian)-2-yl]phosphonates **214**, and their oxidative decomposition may result in α-phosphorylated carbonyl compounds **215** (Scheme 64).

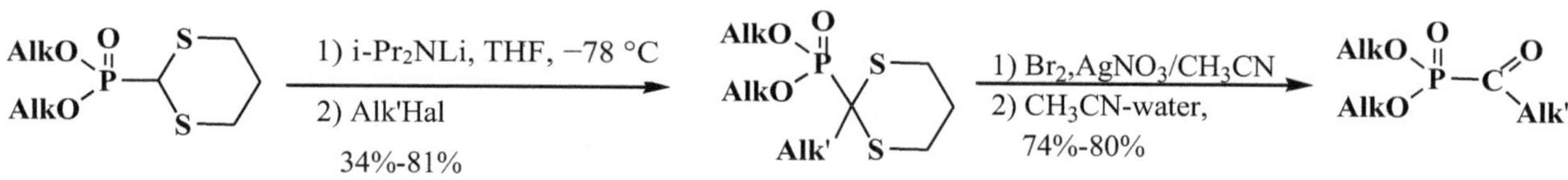

114 **214** **215**

Scheme 64. Thioacetals **114** alkylation with the subsequent transformation of thioketals **214** into α-phosphorylated carbonyl compounds **215**.

The possibility to alkylate diethyl ([1,3]-dithian-2-yl)phosphonate (**144**) was used for the elongation of the hydrocarbon chain in the synthesis of 3-deoxy-D-manno-octulosonic acid (**216**), through compound **217** as alkylated form **144** [136] (Scheme 65).

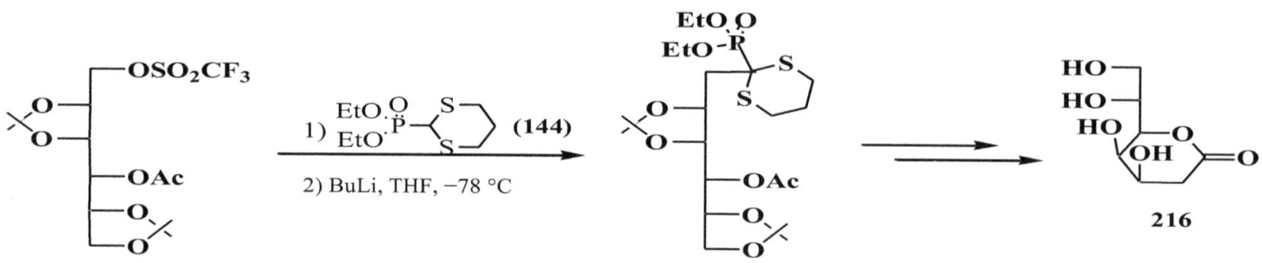

216 **217**

Scheme 65. Synthesis of 3-deoxy-D-manno-octulosonic acid (**216**) by the alkylation of thioacetal **144**.

In contrast to dialkyl (dialkoxymethyl)phosphonates **33** and similarly to phosphorylated formaldehyde thioacetals **14**, diethyl [(*N*,*N*-dimethylamino)cyanomethyl]phosphonate (**153**) is readily deprotonated under the action of sodium hydride in dioxane or dimethyl sulfoxide or 50% KOH solution under phase transfer catalysis conditions [24] as well as with butyllithium BuLi in THF [25]. Lithium derivative **218** of diethyl [(*N*,*N*-dimethylamino)cyanomethyl]phosphonate (**153**) proved to be so stable that it could be stored without decomposition for several months [25] (Scheme 66).

153 **218**

Scheme 66. Synthesis of stable lithium derivative **218** of diethyl [(*N*,*N*-dimethylamino)-cyanomethyl]phosphonate (**153**).

The anion of **218** undergoes alkylation at the carbon atom when treated with methyl iodide MeI [24,25,117], dimethyl sulfate Me_2SO_4 [117] or benzyl chloride $PhCH_2Cl$ [24] to give α-alkylated derivatives **219** and **220**. Benzyl derivative **220** prepared by this method eliminates hydrogen cyanide on heating to give α-phosphorylated enamine **221** [24] (Scheme 67).

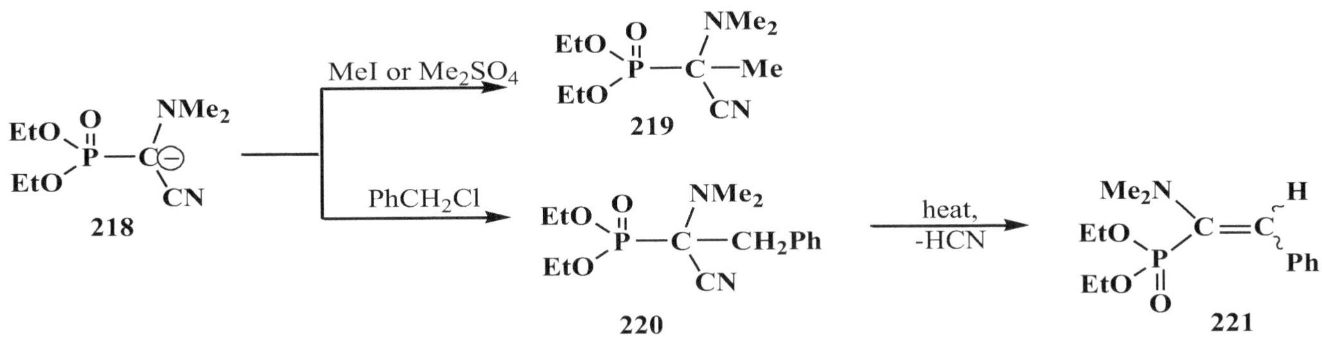

Scheme 67. Anion **218** alkylation.

Similarly to diphenyl(dialkoxymethyl)phosphine oxides (**50**), cyclic dialkyl (dialkylthiomethyl)-phosphonates **113**, **114**, cyclic diphenyl(dialkylthiomethyl)phosphine oxides **121**, **122**, and diethyl [(*N*,*N*-dimethylamino)cyanomethyl]phosphonate (**153**), and tetraethyl (*N*,*N*-dimethylaminomethyl)-diphosphonate (**161**) under the action of strong bases readily eliminate a proton from the acetal carbon atom to give anion **222**. This provides an opportunity for its further alkylation (compounds **223**) [28], which has been used in the synthesis of pesticides (Scheme 68).

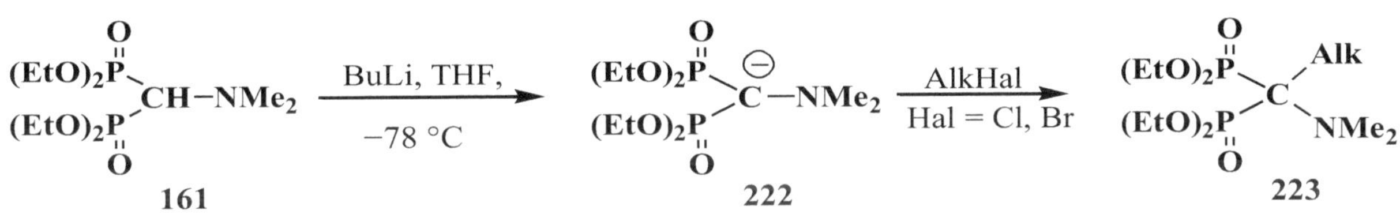

Scheme 68. Alkylation of methyldiphosphonate **161**, (Hal = Cl, Br).

3.4. Horner Reaction

In 1958 and 1959 L. Horner and co-authors reported their discovery of a new reaction [137, 138] that they named as "P=O-activated olefination". The authors showed that the reaction of alkyl(diphenyl)phosphine oxides **224** and dialkyl alkylphosphonates **225** with aldehydes and ketones in the presence of strong bases produce olefins **226** (Scheme 69). Key reaction intermediates—lithium derivatives of carbanion of the initial phosphinoyl compounds **227** and β-phosphorylated hydroxy derivatives **228** were identified on the example reaction of benzyl(diphenyl)phosphine oxide (**229**) with benzaldehyde in the presence of phenyllithium [138].

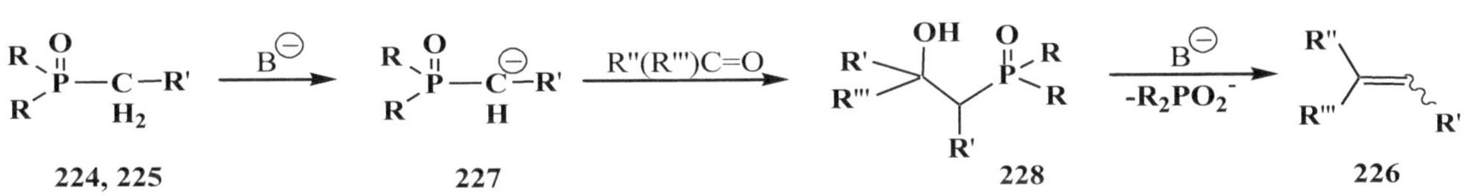

Scheme 69. Syntheses of olefins **226** from aldehydes and ketones by means of Horner's reaction, R = OAlk, Ph; R′ = Alk; R″, R‴ = H, Alk, Ar.

The reaction has a number of advantages in comparison with similar reaction of phosphorus ylides previously described by L. Wittig [139] where ketones are difficult to react, whereas both aldehydes and ketones undergo the Horner reaction. It was further shown that Horner reaction has a larger synthetic potential and is applicable for the synthesis of other types of organic compounds, for example, allenes, cyclopropanes, terminal [140] and disubstituted alkynes [132]. The involvement of phosphonates functionalized at the α-position with dialkylamino, alkoxy or alkylthio groups in the reaction leads to enamines, vinyl ethers [132,141–143] and vinyl thioethers [141,143]. Their subsequent hydrolysis affords aldehydes and ketones with elongated hydrocarbon chain in high yields (homologation).

The Horner reaction also provides the possibility to prepare carboxylic acids homologized by one carbon atom via the shortest route starting from phosphinoyl compounds functionalized at the α-position with two heteroatoms, namely, phosphorylated formaldehyde acetals and structurally related compounds [132,141]. In this case, carbanions **230** of phosphorylated formaldehyde acetals and structurally related compounds **13–16, 18–32** behave as a masked form of triply functionalized carbanions **231** that may be considered as a synthetic equivalent or carrier of reversed-polarity formate carbanion $[O{=}C{-}OH]^-$ **232** [20,141] (Scheme 70).

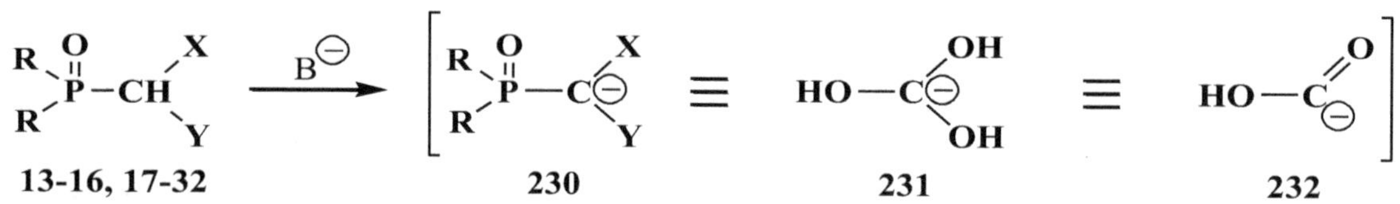

Scheme 70. Compounds **13–16, 18–32** as hidden form of reversed-polarity formate carbanion **232**, where R = OAlk, Ph; X, Y = AlkO, AlkS, Alk_2N, CN, $R_2P(O)$, Hal, AlkS(O).

Carbanions **230**, prepared by the deprotonation of the initial phosphoryl compound, react with carbonyl compounds to afford β-phosphorylated alcohols **233**, which can be isolated. The subsequent treatment of alcohols **233** with strong bases, usually potassium *tert*-butoxide, leads to ketene acetals and structurally related compounds **234** that are valuable precursors in the synthesis of organic compounds of different kinds [23,24,132,133,143]. Further acid hydrolysis of compounds **234** produces carboxylic acids **235** (Scheme 71) or their derivatives, for example esters **236**, or thioesters **237**, depending on the conditions.

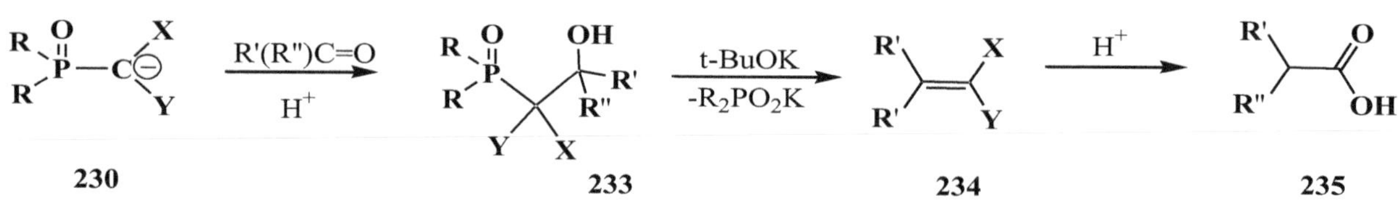

Scheme 71. Syntheses of carboxylic acids **235** by means of Horner's reaction, R′,R″ = H, Alk, Ar.

However, the simplest and most available phosphorylated formaldehyde acetals, dialkyl (dialkoxymethyl)phosphonates **33**, do not form stable carbanions [23,132], therefore the attempted synthesis of carboxylic acids and their derivatives by Horner reaction failed for a long time. Among acetals of phosphonate type compounds, only diethyl (5,6-dichloro-1,3-benzodioxomethyl)phosphonate (**91**) participated in the reaction with ketones at 90 °C in dioxane in the presence of sodium hydride NaH to give ketene acetals in 19–32% yields [23]. See also Section 3.3 "Alkylation of formacetal carbon atom".

Carboxylic acids were obtained for the first time by Horner reaction in 75–90% yields in 1968 by reacting tetraethyl (*N,N*-dimethylaminomethyl)diphosphonate (**161**) with aliphatic and aromatic aldehydes [27]. After formation of the carbanion **222**, the reaction proceeds through the sequential formation of 1-dimethylaminoalkenylphosphonates—α-phosphorylated enamines **236**, then α-phosphinoylacyl derivatives **237**, and finally yields free linear acids **238** (Scheme 72).

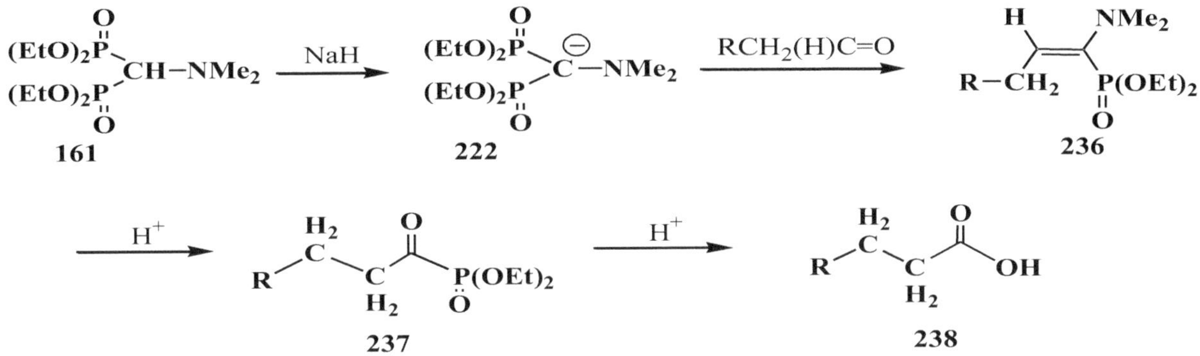

Scheme 72. Synthesis of carboxylic acids **238** by reacting tetraethyl (*N*,*N*-dimethylaminomethyl)diphosphonate (**161**) with aliphatic and aromatic aldehydes.

Since phosphorylated enamines **236** (synthesized from aliphatic aldehydes only) contain an anion-stabilizing diethoxyphosphinoyl group in the α-position, the methylene group in the γ-position is readily deprotonated under the action of strong bases. The resulting phosphorylated aminoallyl anions **239** react with alkyl halides and finally form carboxylic acids **240** branched at the β-position via the phosphorylated enamines **241**. The reaction of anions **239** with aldehydes gives rise to hydroxy compounds (**242**) and then to β,γ-disubstituted γ-butyrolactones **243** over unisolated γ-hydroxy carboxylic acids that undergo fast cyclization under the reaction conditions [144] (Scheme 73).

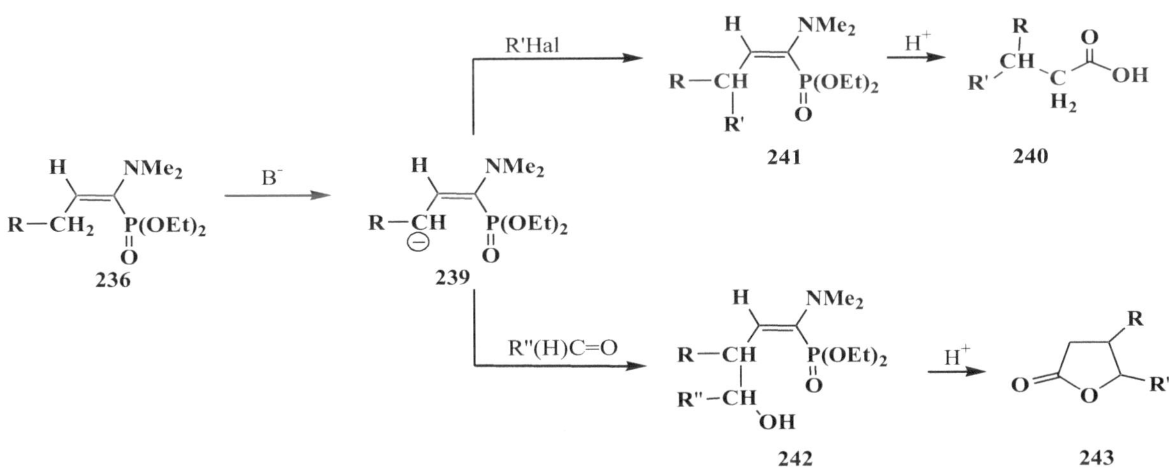

Scheme 73. Syntheses of carboxylic acids **240** and γ-butyrolactones **243** from phosphorylated enamines **236**.

Because only aldehydes react with compound **161** [27,132], this method of synthesis of carboxylic acids is not widely used. However, compound **161** is employed for the preparation of substituted acetic acids as intermediate stages in the synthesis of potential pharmaceuticals [124,145] and pesticides [28], for example, acid **244** (Scheme 74).

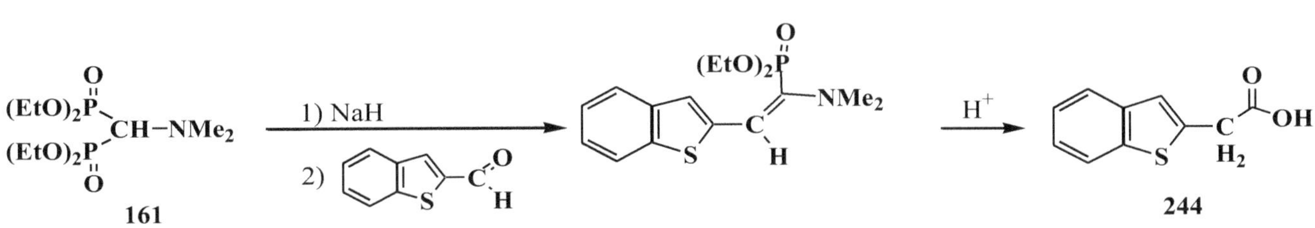

Scheme 74. Synthesis of 2-benzothienylacetic acid **244** by Horner's reaction.

The successful homologation of aldehydes with the use of diphosphonate **161** stimulated further search for the synthetic equivalents of formate carbanion **232** among organophosphorus compounds. In 1976–1977, linear (**107**) and cyclic dialkyl (dialkythiomethyl)phosphonates **113, 114** were proposed [21, 146]. These compounds can form a stable carbanion **245**, and react with both aldehydes and ketones to form ketene thioacetals **246** and further under subsequent hydrolysis (over thioesters **237**) produce homologous carboxylic acids **235** (Scheme 75), see also section "Alkylation of formacetal carbon atom". Ketene thioacetals **246** were obtained from ketones and aldehydes [102,146], including those unsaturated, in 66–82% and 80–96% yields, respectively, as mixtures of *Z*/*E* isomers [23].

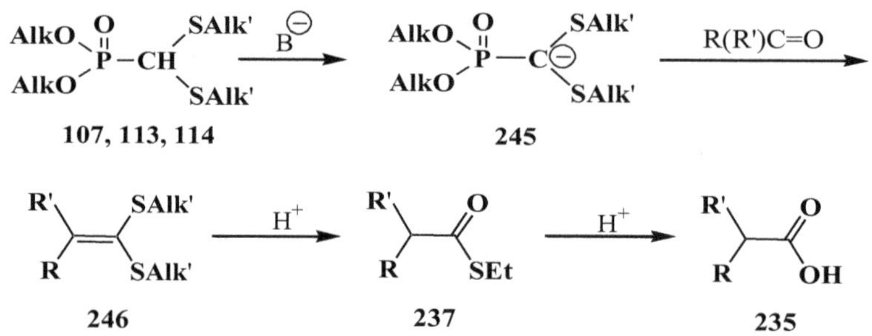

Scheme 75. Syntheses of carboxylic acids **235** by reacting thiomethylphosphonates **107, 113, 114** with aldehydes and ketones.

The reaction is observed also for diphenyl([1,3]-dithian-2-yl)phosphine oxide (**122**) [123,124] and diethyl (1,3-benzodithiolylmethyl)phosphonate (**247**) [113,147]. Phosphonate **247** produces benzo-analogs of ketene thioacetals, 1,4-benzodithiafulvenes **248**, in 92–98% yields when reacted with carbonyl compounds (Scheme 76).

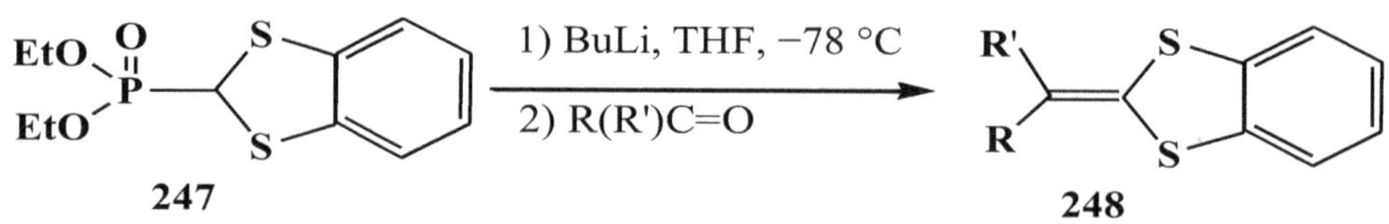

Scheme 76. Syntheses 1,4-benzodithiafulvenes **248** from diphenyl([1,3]-dithian-2-yl)phosphine oxide (**122**) by the reaction with carbonyl compounds.

Phosphorylated thioacetals **107, 113, 114** and thioacetals produced from its ketene **246, 248** have a synthetic importance because, along with carboxylic acid synthesis, they undergo numerous reactions to afford various products [23]. For example, aldehydes **250**, branched out in α-position are formed in the reduction of 1,4-benzodithiafulvenes **248** followed by hydrolysis (over the stage of reduced compounds **249**) [148] (Scheme 77).

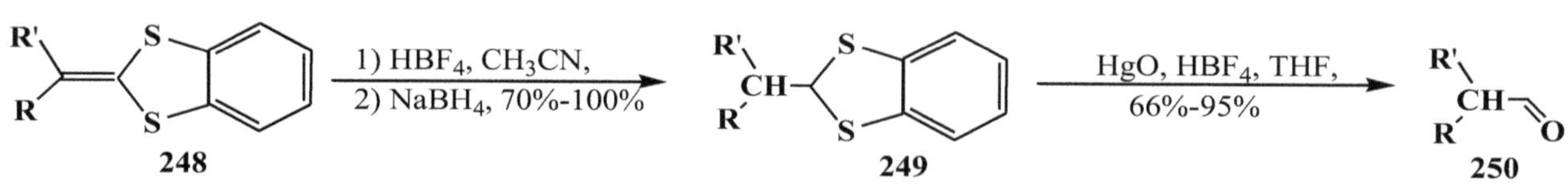

Scheme 77. Obtaining branched aldehydes **250** from 1,4-benzodithiafulvenes **248**.

Methyl esters **251** result from thioacetal **252** methanolysis [149] (Scheme 78).

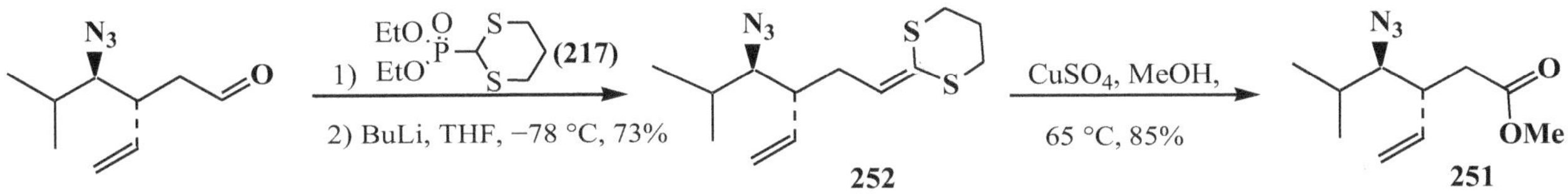

Scheme 78. Synthesis of methyl esters **251** by methanolysis of thioacetal **252**.

Dimethyl (dimethylthiomethyl)phosphonate **253** is used in the practice of contemporary organic chemistry, for example, in the intermediate stages of synthesis of biologically active dipeptide mimetics **254** [149] (Scheme 79), the antibiotic thienamycin (**255**) [150] (Scheme 80), and "organic metals" **256** [116,151–154], for example (Scheme 81).

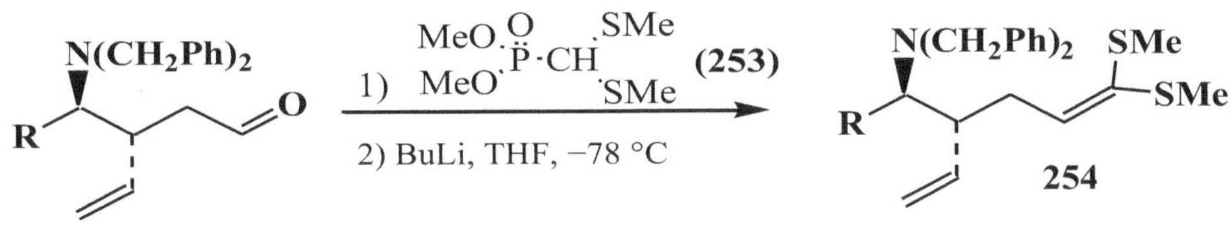

Scheme 79. Synthesis of compound **254**—an intermediate stage of synthesis of dipeptide mimetics.

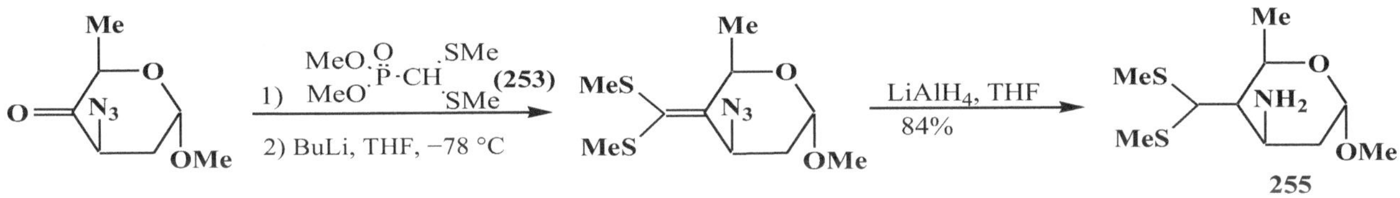

Scheme 80. Synthesis of the antibiotic thienamycin (**255**) by means of thiomethylphosphonate **253**.

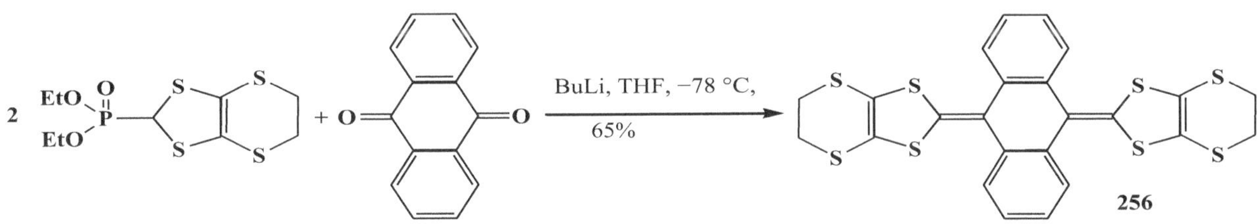

Scheme 81. Example of synthesis of "organic metals" **256**.

However, the long duration of the two-stage conversion of ketene thioacetals **246** into acids **235**, often in the presence of mercury Hg^{2+} [141] or copper Cu^{2+} salts [149] and the necessity of working with mercaptans [23,141,154] limits the application of phosphorylated formaldehyde thioacetals **14** in the Horner reaction.

Therefore, the search for efficient precursors for the synthesis of carboxylic acids **235** from carbonyl compounds by the Horner reaction has continued. From the mid-1970s to the early 1980s, many acetal-like derivatives of diethyl formylphosphonates **13–32** [18,19,21–56], where the negative charge of the carbanion was stabilized by two heteroatoms of the "acetal" group, were obtained [54]. See also Figure 2, where R = OEt. In the case of X = Me_3Si, Peterson olefination prevails over the Horner reaction [54–56,110] and trimethylsyloxy fragment is a leaving group.

Because the majority of compounds **18–32** have no substantial advantages over the phosphorylated formaldehyde thioacetals **14**, the study of Horner reaction with their participation was confined mainly to academic interest. More detailed studies of reactivity of the majority of these compounds were not conducted.

Among the compounds synthesized over this period, diethyl [(*N,N*-dialkylamino)cyanomethyl]phosphonate (**153**) and diphenyl(dialkoxymethyl)phosphine oxides **50** were involved in the practice of organic synthesis.

Compound **153** proposed in 1982 [24] reacts like dialkyl (dialkylthiomethyl)phosphonates **107** with aldehydes, in 50–69% yields, and acetophenone as ketone example, in 24% yield [24,132]. The products of Horner reaction in this case are cyanoenamines **257**, whose acid hydrolysis produces linear carboxylic acids **238** homologous to the initial carbonyl compounds [24,25,132] (Scheme 82). See also Scheme 72.

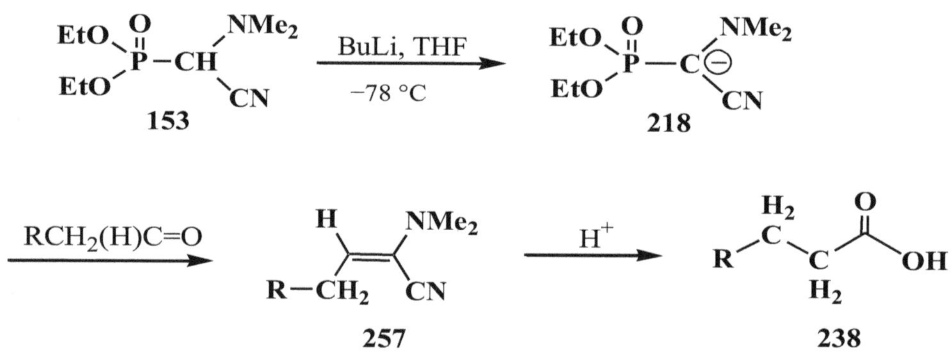

Scheme 82. Carboxylic acids **238** synthesis by means of diethyl [(*N,N*-dialkylamino)cyanomethyl]phosphonates **153**, where R = Alk, Ph.

Like phosphorylated enamines **236** (Scheme 73), compound **257** contains an anion-stabilizing CN-group in the α-position. Resulting cyanoaminoallyl anions **258** combine with alkyl halides to form carboxylic acids **240** through cyanoenamines **259**, while the reaction with aldehydes leads to γ-hydroxy acids that undergo cyclization to give β,γ-disubstituted γ-butyrolactones **243** through cyano aminoallylalcohols **260** [24,25,132,144] (Scheme 83).

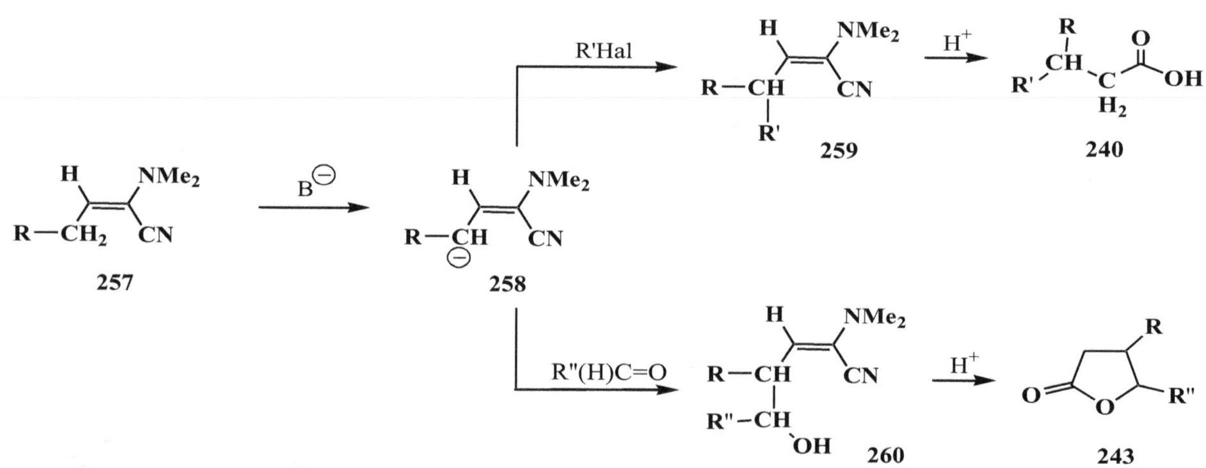

Scheme 83. Enamines **257** alkylation by means of alkyl halides or aldehydes leads to obtaining branched acids **240** or butyrolactones **243**.

Diethyl [(*N*-morpholino)cyanomethyl]phosphonate (**154**) was used in the synthesis of the colerofragarone fragment—terminator of fungi *Collerotrichum fragariac* [155] (Scheme 84, compound **261**).

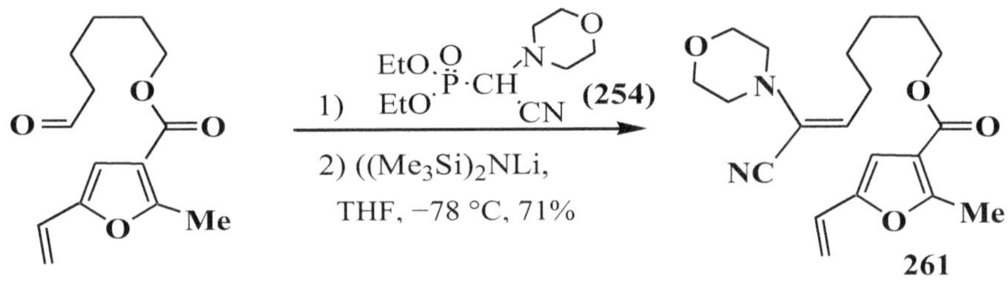

Scheme 84. Synthesis of colerofragarone fragment **262** using diethyl [(*N*-morpholino)cyanomethyl]phosphonate (**154**).

It was shown that compound **153** can react also with cyclic semiacetal **262**, which was used in one of the stages of synthesis of prostaglandin analog cloprosterol PGF_2 [155,156] (Scheme 85, compound **263**).

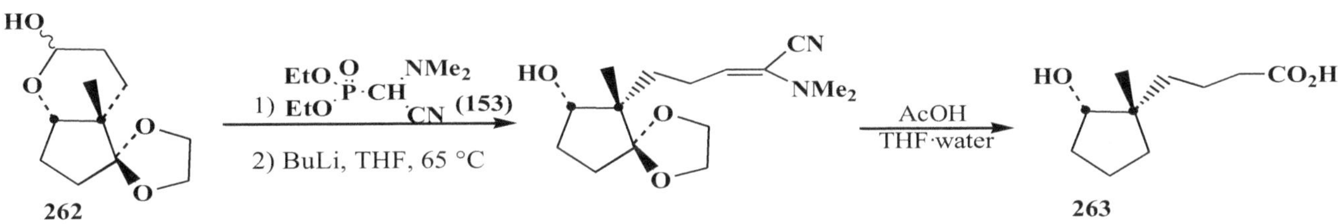

Scheme 85. Synthesis of **263**, a semi-product of synthesis of the analog of cloprosterol PGF_2 from cyclic semiacetal **262** by means of phosphonate **153**.

However, compound **153** was not widely used because of the low yields in its reactions with ketones [25] and formation of hydrogen cyanide on hydrolysis of cyanoenamines. Almost simultaneously with **153**, a successful synthesis of ketene acetal at −110 °C starting from diphenyl(dialkoxymethyl)phosphine oxides **50**, over carbanion **195** and both aldehydes and ketones was reported in 1983 [133]. Intermediate β-hydroxydiphenylphosphinoyl derivatives **264** were isolated in almost quantitative yield that further react with potassium *tert*-butoxide (*t*-BuOK) to give ketene acetals **265** in 45–85% yields [133] (Scheme 86).

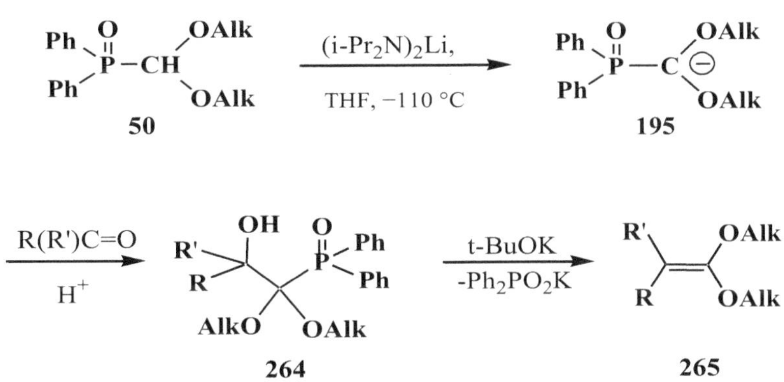

Scheme 86. Synthesis of ketene acetals **265** from diphenyl(dialkoxymethyl)phosphine oxides **50** through conpounds **264** as intermediates.

This publication attracted no attention for a long time, although in 1993 it was shown that the reaction of diphenyl(diethoxymethyl)phosphine oxide (**266**) with substituted cyclohexadienone in the presence of lithium diisopropylamide leads to the formation of carboxylic acid **267** through the stages of ketene acetal **268** and ethyl ester **269** formation [20,157] (Scheme 87).

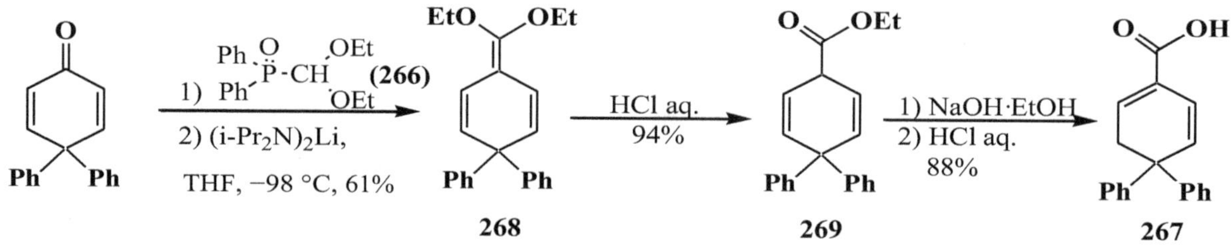

Scheme 87. Synthesis of carboxylic acid **267** from (diethoxymethyl)phosphine oxide (**266**) through ketene acetal **268** and ethyl ester **269**.

It was found in 2002, however, that the reaction of diphenyl(dimethoxymethyl)phosphine oxide (**209**) with aliphatic aldehydes, in contrast to other phosphorylated equivalents of formate anion—compounds **107**, **112–115**, **153** (**154**), **161** (**166**), may also result in derivatives of α-hydroxycarboxylic acids **270** [140,158,159]. Initially formed β-hydroxydiphenylphosphinoyl derivatives **271** in wet acidified dichloromethane readily decompose with the cleavage of the P–C bond to afford methyl esters of α-hydroxycarboxylic acids **272** in 41–89% yield and diphenylphosphine oxide (**53**) (Scheme 88).

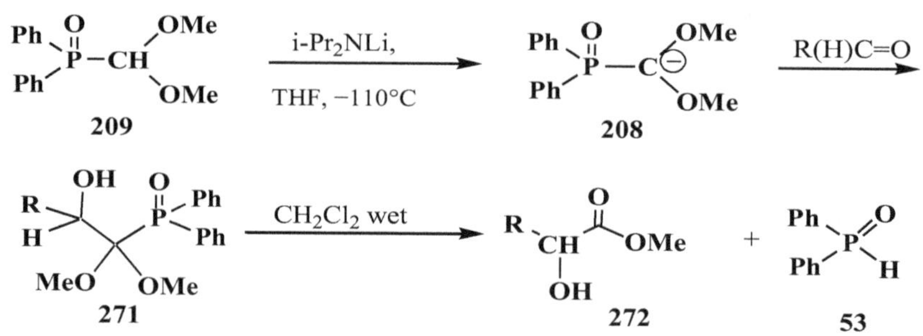

Scheme 88. Synthesis and decomposition of β-hydroxydiphenylphosphinoyl compounds **271** lead to methyl esters of α-hydroxycarboxylic acids **272**.

It was further shown that ketene acetals **273** resulting from aldehydes and phosphine oxide **266** may be successfully oxidized [20,128,160] in the Sharpless asymmetric dihydroxylation—AD reaction [161]. As a result, the esters of chiral α-hydroxycarboxylic acids **274**, **275** were obtained in 49–94% yields and enantiomeric excess up to 98% (Scheme 89).

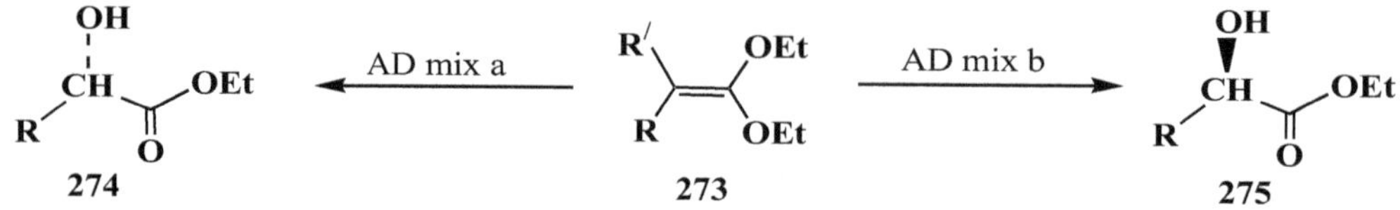

Scheme 89. Using of Sharpless asymmetric dihydroxylation (AD reaction) for synthesis of chiral α-hydroxycarboxylic acids **274**, **275** from ketene acetals **273**.

It should be noted that ketene thioacetals **246** do not undergo the AD reaction. Mixed *O,S*-acetal **18** reacts but yields and enantiomeric excess of α-hydroxycarboxylic acids do not exceed 7–37% and 80%, correspondingly [127,128]. The combination of Horner and AD reactions was successfully used on one of stages of synthesis of the diterpenoid tonantzitlolone (**276**) [162] (Scheme 90).

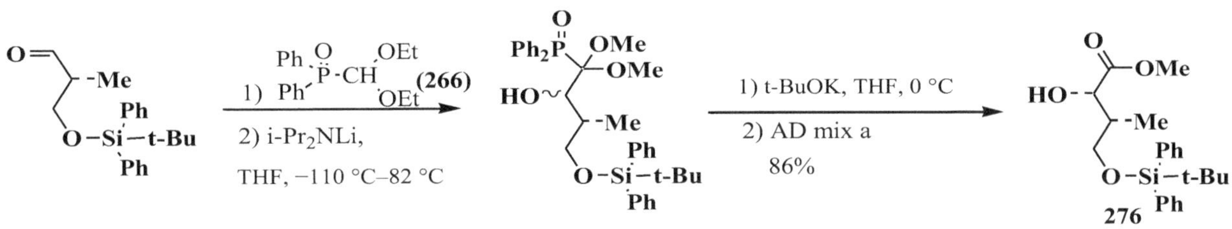

Scheme 90. Using of combination of Horner and AD reactions to obtain compound **276**—intermediate stage of synthesis of the diterpenoid tonantzitlolone.

The involvement of chiral α-alkylaminoaldehydes in the Horner reaction, correspondingly, leads to the synthesis of esters of chiral β-amino acids **277** [20] (Scheme 91).

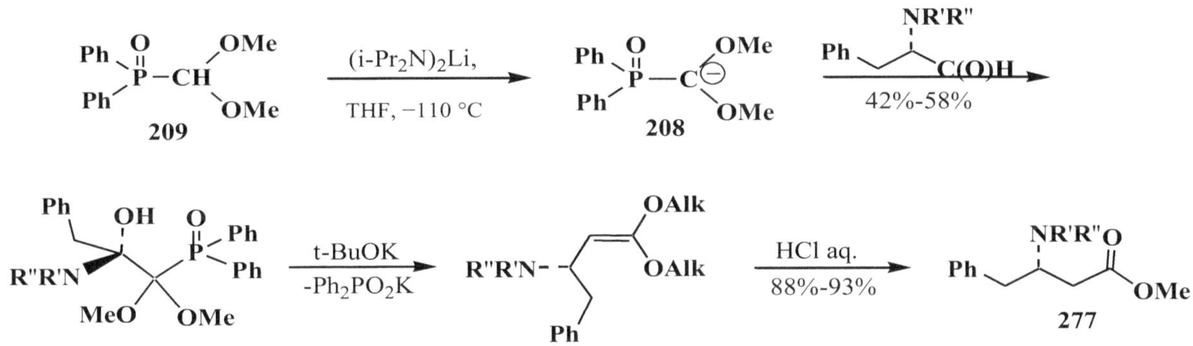

Scheme 91. Synthesis of chiral β-amino acids **277** by the Horner reaction.

Sharpless asymmetric dihydroxylation followed of Horner reaction also provides an opportunity to synthesize α-hydroxy-β-amino acid diastereomers **278**, **279** [20,128] (Scheme 92).

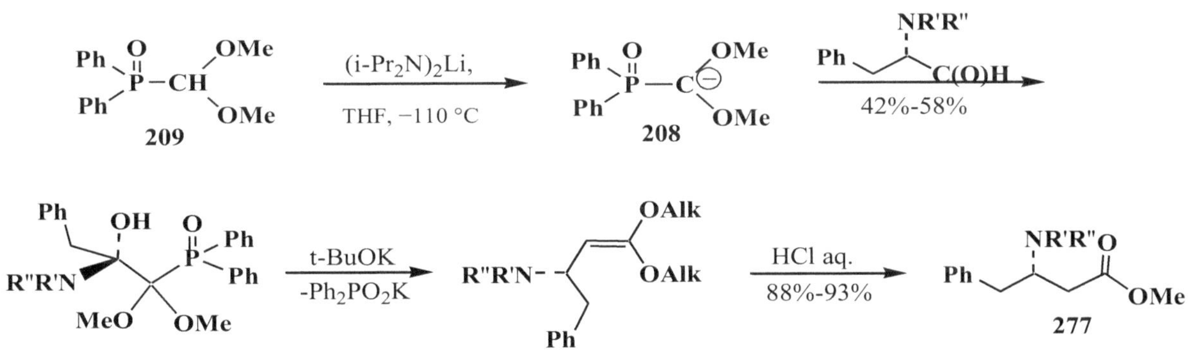

Scheme 92. Using of combination of Horner and AD reactions to form α-hydroxy-β-amino acid diastereomers **278**, **279** (R = *t*-BuOC(O)).

4. Phosphorylated Formaldehyde Halogenoaminals (Phosphorylated Vilsmeier–Haak Reagents) 17

At present, four types of phosphorus compounds are known that can be related to phosphorylated Vilsmeier–Haak reagents: *N*,*N*,*N'*,*N'*-tetraisopropyl [(*N''*,*N''*-diisopropylamino)methylydeniminium] phosphondiamide dichlorophosphate (**8**) [12], dialkyl [(*N*,*N*-dimethylamino)chloromethyl]phosphonates **171** [31,32], diphenyl[(*N*,*N*-dialkylamino)chloromethyl]phosphine oxides **280** [163], and structurally related to the phosphorylated Vilsmeier–Haak reagents, diethyl [(*N*-acylamino)bromomethyl]phosphonates **281** [164] (Figure 6).

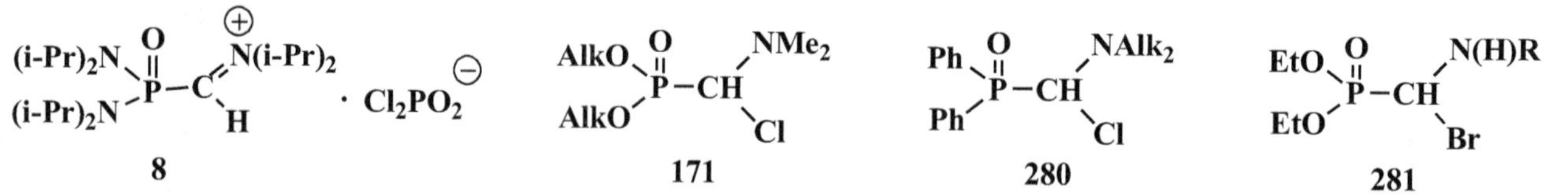

Figure 6. Representatives of phosphorylated Vilsmeier–Haak reagents **8**, **171**, **280**, **281** synthesized to date.

No general methods of synthesis have been found for the compounds of this type **8**, **171**, **280**, **281**. Their chemical properties also differ significantly from each other.

4.1. Synthesis of N,N,N′,N′-Tetraisopropyl [(N″,N″-Diisopropylamino)methylydeniminium] Phosphondiamide Dichlorophosphate (**8**)

A sole paper was published in 1999 [12] on the synthesis of *N,N,N′,N′*-tetraisopropyl [(*N″,N″*-diisopropylamino)methylydeniminium]phosphondiamide dichlorophosphate (**8**) by the oxidation of *N,N,N′,N′*-tetraisopropyl [(*N″,N″*-diisopropylamino)methylydeniminium]phosphindiamide dichlorophosphate (**282**) with dimethyl sulfoxide DMSO (Scheme 93). However, the chemical properties of **8** are insufficiently studied (for some chemical properties of **8** and its derivatives, see Section 1. Introduction).

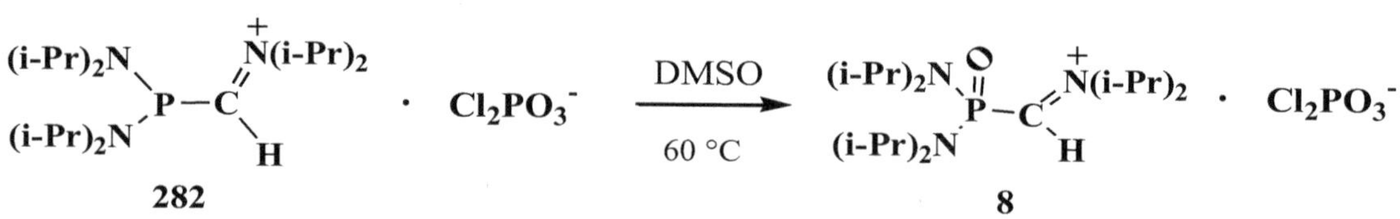

Scheme 93. Phosphindiamide dichlorophosphate **282** oxidation by means of DMSO leads to the formation of phosphondiamide dichlorophosphate **8**.

4.2. Synthesis and Chemical Properties of Dialkyl [(N,N-Dimethylamino)chloromethyl]phosphonates **171**

The first compounds of this type, dimethyl [(*N,N*-dimethylamino)chloromethyl] phosphonate (**172**) and diethyl [(*N,N*-dimethylamino)chloromethyl]phosphonate (**173**) were synthesized in 1969 [31,32]. For the first time, these compounds were obtained as intermediates in the synthesis of tetraalkyl (*N,N*-dimethylaminomethyl)diphosphonates **174** by reaction of N,N-dimethyl(chloromethylideneiminium) chloride (**171**) with trialkyl phosphites **58** [32]—see Scheme 49, section "Diphosphinoyl *N,N*-dialkylaminomethanes".

Preparative amounts of these compounds were prepared by a two-step scheme [30,31]. First, the reaction of dimethylformamide dimethylacetal (**78**) with dialkyl phosphites **51** resulted in dialkyl [(*N,N*-dimethylamino)methoxymethyl]phosphonates **162**, which were further reacted with thionyl chloride ($SOCl_2$) at 0 °C to yield the final products **172**, **173** (Scheme 94). See also Schemes 49 and 50.

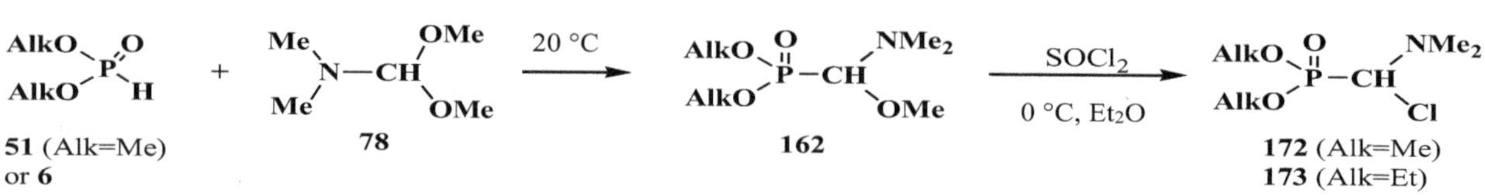

Scheme 94. Reaction dialkyl phosphites **51** and dimethylformamide dimethylacetal (**78**) leads to phosphorilated halogenoaminals **172** (Alk = Me), **173** (Alk = Et).

This scheme is currently used [33,165]; phosphorus trichloride can be used instead of thionyl chloride [83]. Dimethyl [(*N,N*-dimethylamino)chloromethyl]phosphonate (**172**) was found to undergo spontaneous dealkylation on storage [31], therefore only diethyl [(*N*,*N*-dimethylamino)chloromethyl]phosphonate (**173**) is used in organic synthesis. Phosphonate **173**, as a phosphorylated Vilsmeier–Haak reagent where one chlorine atom is substituted by a diethoxyphosphoryl group, show electrophilic properties [166] inherent in compounds of such kind and readily reacts with nucleophiles. Thus, compound **173** reacts vigorously with secondary amines, carbazole, alcohols, thiols, and hydrophosphinoyl compounds **51**, **53** in the presence of equimolar amount of triethylamine. The reaction of α-phosphono-α-aminomethylation leads to the corresponding diethoxyphosphinoylformaldehyde derivatives: asymmetric aminals **283**, **284**, aminoacetals **285**, aminothioacetals **286**, and diphosphorylated *N*,*N*-dimethylaminomethanes **164** [30], including unsymmetrical **165** [165] (Scheme 95). The reaction with aromatic and aliphatic amides $RC(O)NH_2$, urea and thiourea $(NH_2)_2C{=}X$ [30] proceeds in a similar manner. Products of reactions are phosphorylated amidoaminals **287**, **288** and symmetrically disubstituted derivatives of urea and thiourea **289**, **290** (Scheme 95). See also Scheme 53, section "Diphosphinoyl *N*,*N*-dialkylaminomethanes".

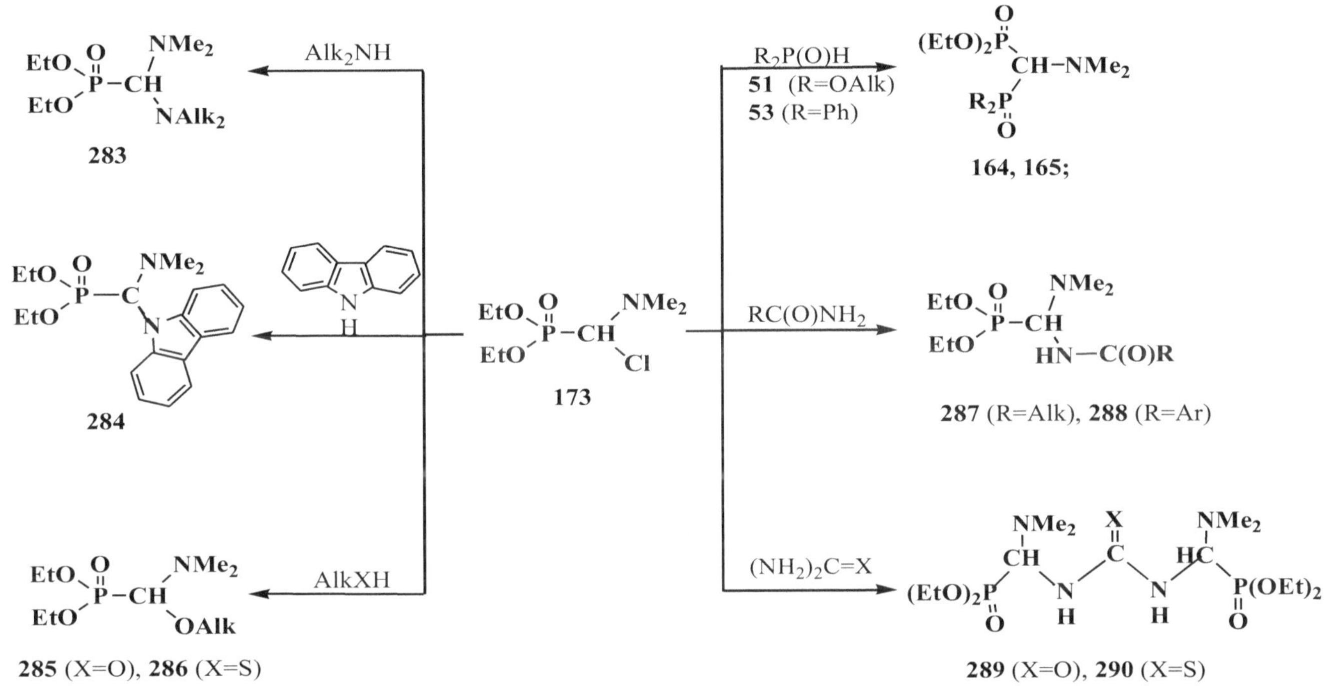

Scheme 95. Reactions of diethyl [(*N*,*N*-dimethylamino)chloromethyl]phosphonate (**173**) with nucleophilic co-reagents.

Diethyl [(*N*,*N*-dimethylamino)chloromethyl]phosphonate (**173**) reacts with arylsulfonamides to lead to the cleavage of the phosphorus–carbon bond and give *N*-sulfonyl-substituted derivatives of formamidines **291** and diethyl phosphite (**6**) [30] (Scheme 96). See also Section 3.1. Cleavage of phosphorus–carbon bond.

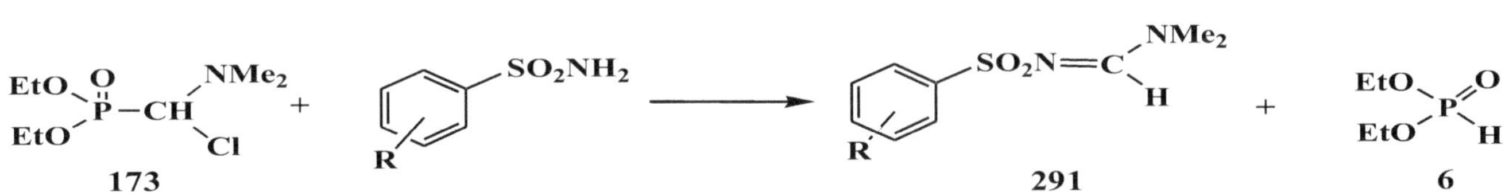

Scheme 96. Cleavage of phosphorus–carbon bond by means of the reaction of compound **173** with arylsulfonamides.

Compound **173** reacts with methyl aryl ketones [30] and dialkyl or aryl alkyl ketones [33] to give diethyl [(*N*,*N*-dimethylamino)(aroyl(or alkanoyl)alkylmethyl)methyl]phosphonates **292** and **293** [30,33], phosphorylated analogs of natural α-amino acids potentially possessing biological activity [33] (Scheme 97). However, reaction stereoselectivity is low, and the *anti/syn* ratio is 2:1.

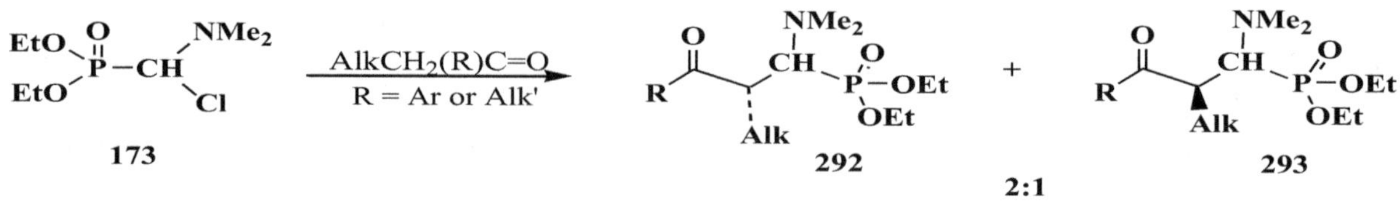

Scheme 97. Reaction of compound **173** with ketones leads to diethyl [(*N,N*-dimethylamino)(aroyl(or alkanoyl)alkylmethyl)methyl]phosphonates **292** and **293**.

Like ketones, the reaction of **173** proceeds also with aldehydes branched at the α-position to the carbonyl group to give phosphorylated aminoaldehydes **294** and **295** (Scheme 98), but the reaction is not stereoselective: the stereoisomer ratio of the resulting α-formyl-α-methylaminophosphonates is 1:1 [33].

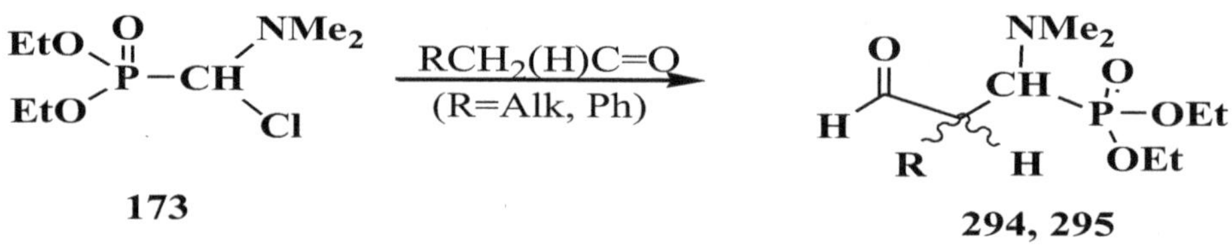

Scheme 98. Reaction of compound **173** with aldehydes leads to phosphorylated aminoaldehydes **294** and **295**.

The involvement of enamines instead of ketones as their synthetic equivalents in the reaction with phosphonate **173** provides stereoselective synthesis of diethyl (α-methylamino-α-aroyl(or alkanoyl))methylphosphonates **292**; the reaction leads to the products with only *anti* configuration [33] (Scheme 99).

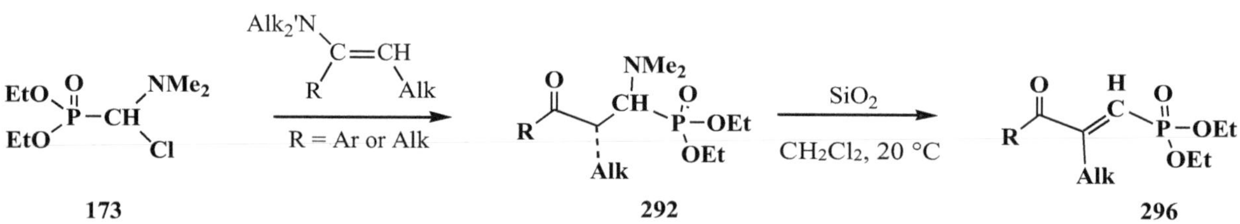

Scheme 99. Synthesis of diethyl (α-methylamino-α-aroyl(or alkanoyl))methylphosphonates **292** and phosphorylated vinyl ketones **296**.

Diethyl [(*N,N*-dimethylamino)(aroyl(or alkanoyl)alkylmethyl)methyl]phosphonates **292, 293** (Schemes 101–103) are obtained as hydrochlorides in the above-stated syntheses. They are unstable and on storage undergo spontaneous β-elimination of dimethylamino group as dimethylammonium chloride—with conversion up to 40% over 24 h [30] to form the corresponding phosphorylated vinyl ketones **296**. The same result was achieved on the heating of hydrochlorides of compounds **292** and **293** to 100 °C over 45 min [30] or stirring their solutions in methylene chloride with silica gel at 20 °C for 15 h [33]. The reaction of amino group elimination with the use of silica gel is also stereoselective: the anti stereoisomers **292** produce only *E* isomers of vinyl compounds **297** [33] (Scheme 99).

The reaction of β-enaminonitriles **298** with compound **173** was used in the synthesis of substituted nitriles of nicotinic acid **299** [167] (Scheme 100).

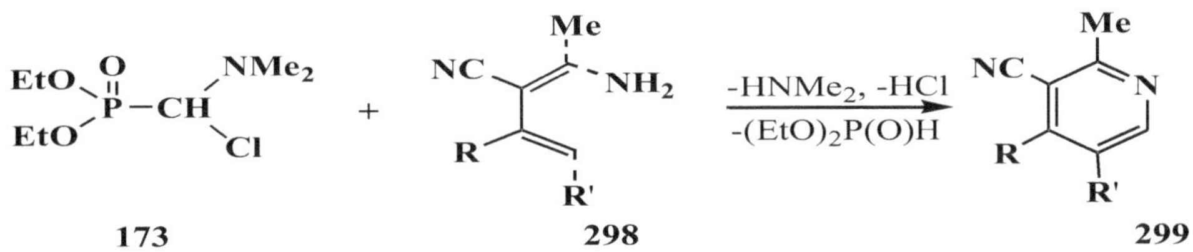

Scheme 100. The reaction of compound **173** with β-enaminonitriles **298** leads to nitriles of nicotinic acid **299**.

As a phosphorylated Vilsmeier–Haak reagent, phosphonate **173** undergoes typical reactions with activated aromatic compounds: *N*,*N*-dimethylaniline, triethylammonium salts of β-naphthol and *p*-cresol, and heteroaromatic compounds: *N*-methylindole, *N*-methylpyrrole, α-methylfuran. The synthesis of aromatic α-aminomethylphosphonates **300** is regiospecific: only one addition product forms in each case [33] (Scheme 101).

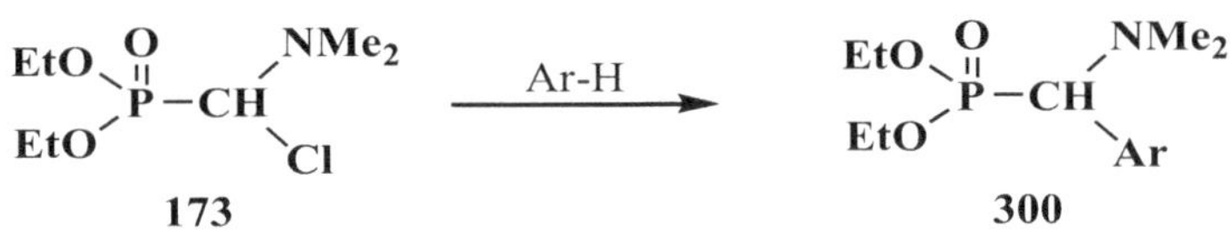

Scheme 101. Interactions of phosphonate **173** with aromatic compounds (Ar = 4-dimethylaminophenyl; 2-hydroxynaphthyl-1; 2-hydroxy(5-methyl)phenyl; 1-methylaminoindol-3-yl; 1,5-dimethylaminopyrrol-2-yl; 5-methylfur-2-yl).

4.3. *Syntheses and Chemical Properties of Diphenyl[(N,N-dialkylamino)chloromethyl]phosphine Oxides* **280**

Diphenyl[(*N*,*N*-dialkylamino)chloromethyl]phosphine oxides **280** are the diphenylphosphinoyl analogs of dialkyl [((*N*,*N*-dialkylamino)chloromethyl]phosphonates. They are obtained by the reaction of chlorodiphenylphosphine (**68**) with *N*,*N*-dialkylformamides in the presence of 10–20% of *N*,*N*-dialkylchloromethyleniminium chloride $[Alk_2N{=}C(H)Cl]^+Cl^-$ (**168**)—Vilsmeier–Haak reagent [168,169] according to Scheme 102.

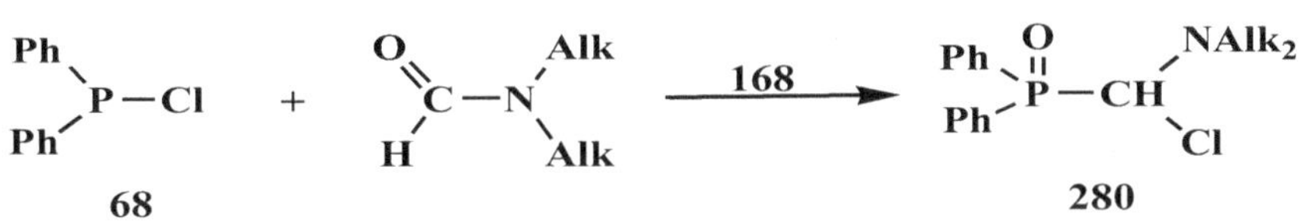

Scheme 102. Synthesis of diphenyl[(*N*,*N*-dialkylamino)chloromethyl]phosphine oxides **280**.

Although phosphine oxides **280** are the formal analogs of dialkyl [(*N*,*N*-dialkylamino)chloromethyl]phosphonates **171**, they differ substantially in reactivity. It was shown by the example of diphenyl[(*N*,*N*-dimethylamino)chloromethyl]phosphine oxide (**301**) that compounds **280** can dissociate in solutions with the cleavage of C–Cl and C–P bonds. See also section "Cleavage of phosphorus–carbon bond". The dissociation in solution results in the formation of both diphenyl[(*N*,*N*-dimethylamino)chloromethyliden]phosphine oxide cation (**302**) and diphenylphosphinite anion (**193**) [170] (Scheme 103).

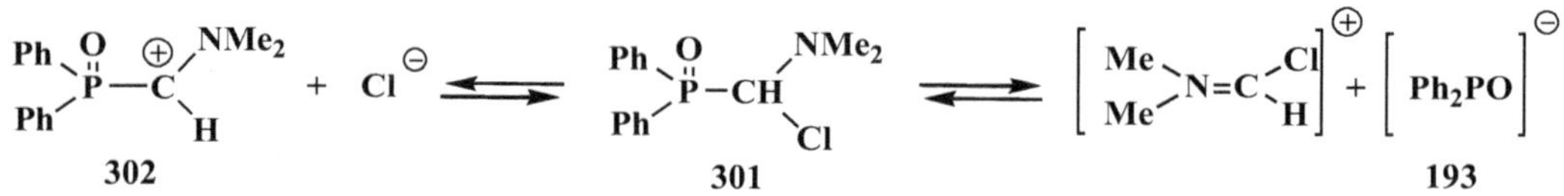

Scheme 103. Diphenyl[(*N,N*-dimethylamino)chloromethyl]phosphine oxide (**301**) dissociation in solutions with a cleavage as bond **C-Cl** and bond **C-P**.

Therefore there are two kinds of reactivity for **301** (and accordingly **280**) [163,170]: electrophilic diphenyl[(*N,N*-dimethylamino)chloromethylidene]phosphine oxide cation (**302**) reacts with nucleophiles such as hydrophosphinoyl compounds **6**, **53**, mixed diethyl trimethylsilyl phosphite $(EtO)_2POSiMe_3$ and *N,N*-dimethylamino aniline $PhNMe_2$, to give symmetrical **303**, R=Ph and unsymmetrical **304**, R=OEt diphosphorylated *N,N*-dimethylaminomethanes and diphenyl[(bis(4-*N,N*-dimethylamino)phenyl)methyl] phosphine oxide (**305**), the product of substitution of both chloro and amino groups at the carbon atom of **301**. However, nucleophilic diphenylphosphinite anion (**193**) reacts with electrophiles: acetic aldehyde MeC(O)H, acetone $Me_2C(O)$, phenyl isocyanate PhNCO, triethyl orthoformate $(EtO)_3CH$, dimethylformamide dimethylacetal (**79**), and bis(diethylamino)methane $(Et_2N)_2CH_2$ to form addition or substitution products, α-phosphorylated compounds: alcohols **306** and **307**, carbamoyl compound **308**, acetal **309**, aminoacetal **310** and amine **311** (Scheme 104).

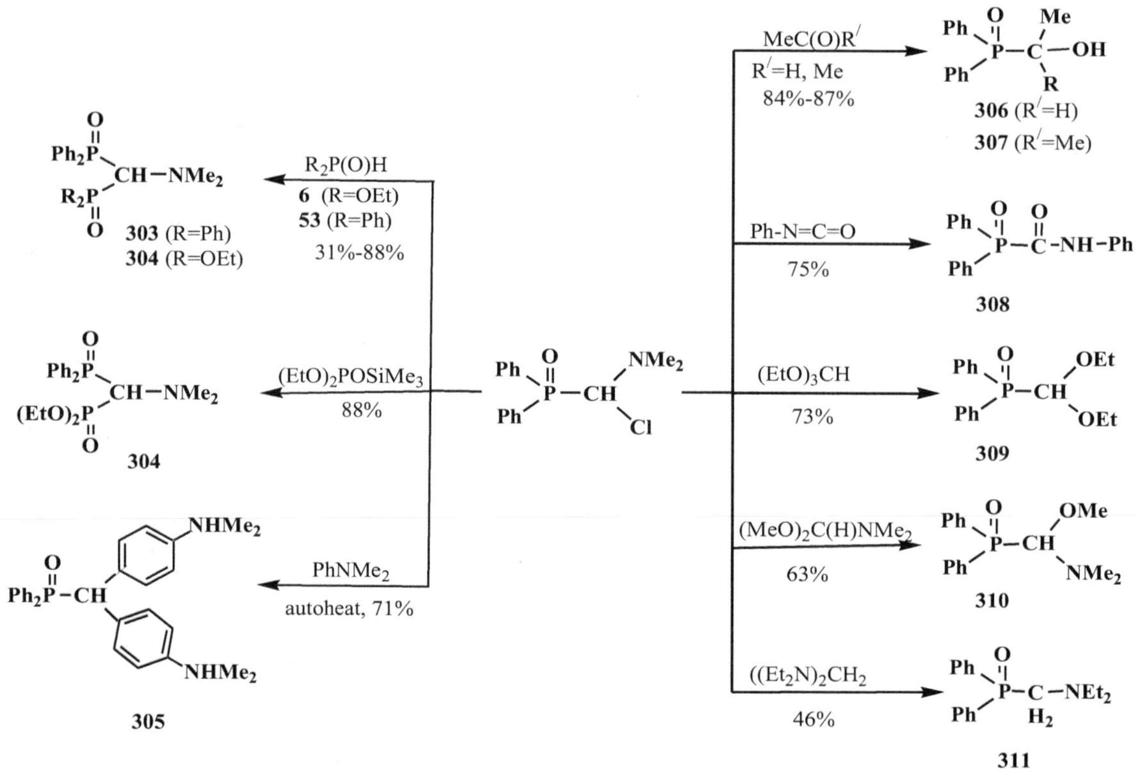

Scheme 104. Reactions of diphenyl[(*N,N*-dimethylamino)chloromethyl]phosphine oxide (**280**) with nucleophilic and electrophilic coreagents.

4.4. Synthesis and Chemical Properties of Diethyl [(N-Acylamino)bromomethyl]phosphonates **281**

Diethyl [(*N*-acylamino)bromomethyl]phosphonates **281** were prepared in high yields by the bromination of diethyl (*N*-acylamino)methylphosphonates **312** with N-bromosuccinimide (NBS) in carbon tetrachloride [164,171] (Scheme 105).

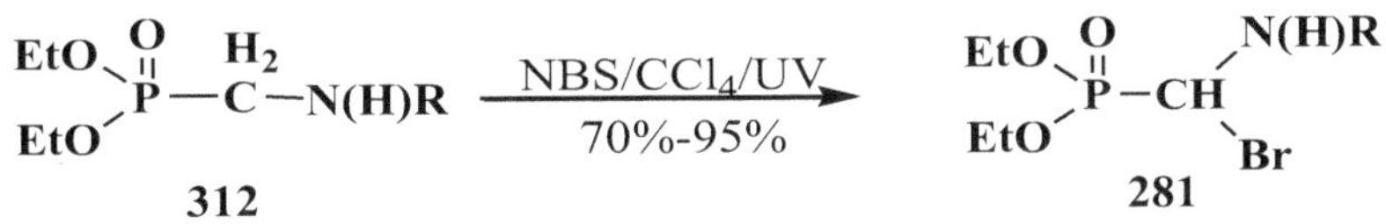

Scheme 105. Synthesis of diethyl [(*N*-acylamino)bromomethyl]phosphonates **281** where R = AlkC(O), ArC(O), AlkS$(O)_2$, ArS$(O)_2$.

To date diethyl [(*N*-acylamino)bromomethyl]phosphonates **281** remain poorly studied, and their chemical properties are insufficiently studied. Nonetheless, compounds **281**, structurally related to the phosphorylated Vilsmeier–Haak reagents, behave as electrophiles similarly to dialkyl [(*N,N*-dimethylamino)chloromethyl]phosphonates **171** [166]. They react readily with such nucleophiles as primary aliphatic and aromatic amines and secondary cyclic amines in the presence of ethyldiisopropylamine—Hünig's base as well as triethylamine (Scheme 106). Reaction products are diethyl [(*N*-acylamino)aminomethyl]phosphonates **313**, unsymmetrical amidoaminals of diethyl (formyl)phosphonate (**3**) [30] showing biological and pharmacological activity [172].

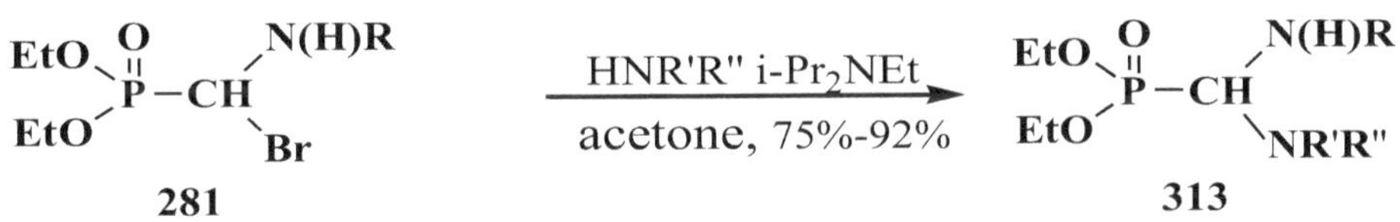

Scheme 106. Transformation of bromomethylphosphonates **281** into unsymmetrical amidoaminals **313**, where HNR′R″ are morpholine, piperidine, pyrrole, 2-aminomethyltetrahydrofuran, *N*-methylbenzylamine, and aniline.

The reaction of compounds **282** with sodium azide leads to diethyl [(*N*-acylamino)azidomethyl]-phosphonates **314**, valuable precursors in the synthesis of phosphinoyl-substituted 1,2,3-triazoles, the products of 1,3-dipolar addition to disubstituted alkynes [173,174]. A mixture of the resultant two regioisomers of diethyl [(*N*-acylamino)(1-(1,2,3-triazolyl))methyl]phosphonates (**315**) and diethyl [(*N*-acylamino)(2-(1,2,3-triazolyl))methyl]phosphonates (**316**) can be separated by chromatography on silica gel (Scheme 107).

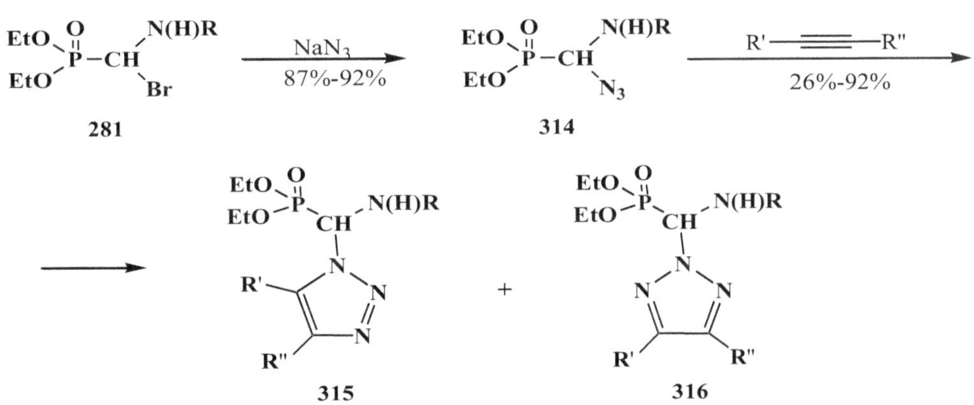

Scheme 107. Synthesis of (triazolylmethyl)phosphonates **315** and **316** from bromomethylphosphonates **281**, where R′, R″ = AlkC(O), ArC(O), AlkS$(O)_2$, ArS$(O)_2$.

Prolonged heating of acetonitrile solutions of **315** leads to migration of [(*N*-acylamino)(diethylphosphinoyl)]methyl group from the first to the second nitrogen atom (N1–N2 migration) of 1,2,3-triazole group to give **316**. In the presence of nucleophiles, azide ion or imidazole, the triazole fragment is displaced to yield diethyl [(*N*-acylamino)azidomethyl]phosphonates **314** or diethyl [(*N*-acylamino)(1-imidazolyl)methyl]phosphonates **317** (Scheme 108) [174].

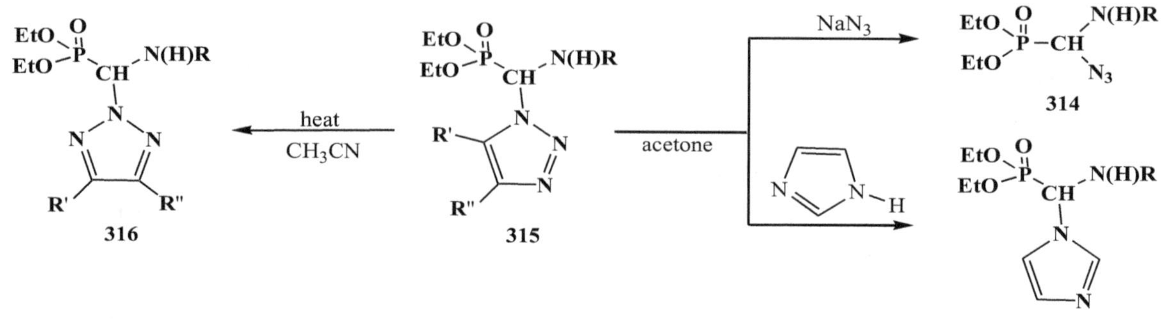

Scheme 108. Reaction replacement of 1,2,3-triazole group of compound **315** with nucleophiles to give compounds **314** and **317** and isomerisation of **315** into compound **316**.

The treatment of phosphonates **281** with Hünig's base i-Pr_2NEt [164] or triethylamine Et_3N [171] in tetrahydrofuran at −78 °C was shown to result in diethyl [(N-acyl)iminomethyl]phosphonates **318** in yields up to 98% (Scheme 109).

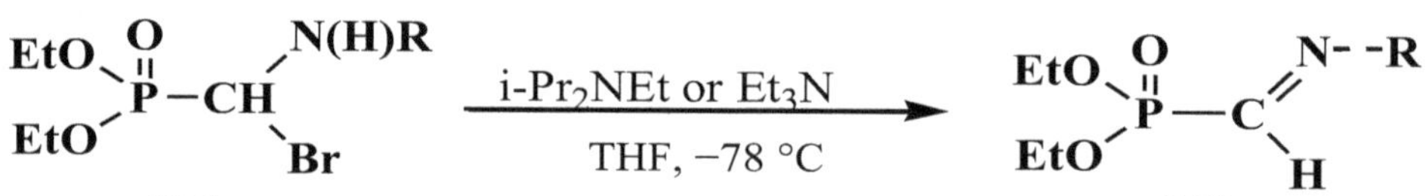

Scheme 109. Synthesis of diethyl [(*N*-acyl)iminomethyl]phosphonates (**318**) from phosphonates **281**.

5. Alkyl (dialkoxymethyl)phosphinates—H-Phosphinates 34. Syntheses and Chemical Properties

The synthesis of alkyl (dialkoxymethyl)phosphinates—H-phosphinates **34** in yields up to 96% by the reaction of hypophosphorous acid (**319**) with orthoformates **54** in the presence of catalytic amounts amounts of *p*-toluenesulfonic acid (p-$TolSO_3H$) [57,175] or trifluoroacetic acid (CF_3SO_3H) [58] was reported in 1977 (Scheme 110). The method is still used at present.

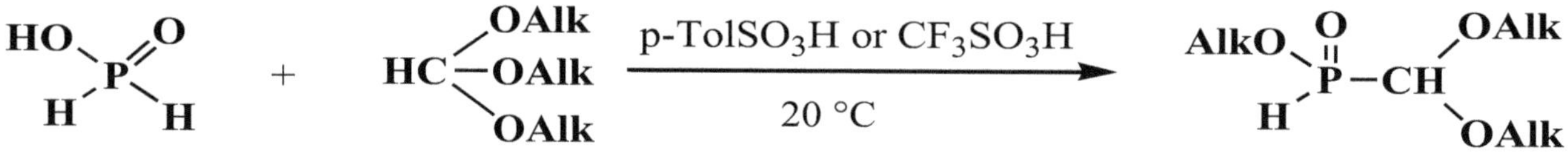

Scheme 110. Synthesis of alkyl (dialkoxymethyl)phosphinates **34**.

The chemical properties of phosphinates **34** were studied almost exclusively by the example of ethyl (diethoxymethyl)phosphinate (**86**). Ethyl (diethoxymethyl)phosphinate (**86**) retains properties typical for both phosphorus esters and hydrophosphoryl compounds and retains the general ability of phosphorylated formaldehyde acetals to undergo the cleavage of the phosphorus–carbon bond.

Phosphinate **86** is readily alkylated under the action of alkyl halides in the presence of bases: Na, NaH, BuLi (B^-) (via anion **320**) to give ethyl alkyl(diethoxymethyl)phosphinates **321** [58,176] (Scheme 111).

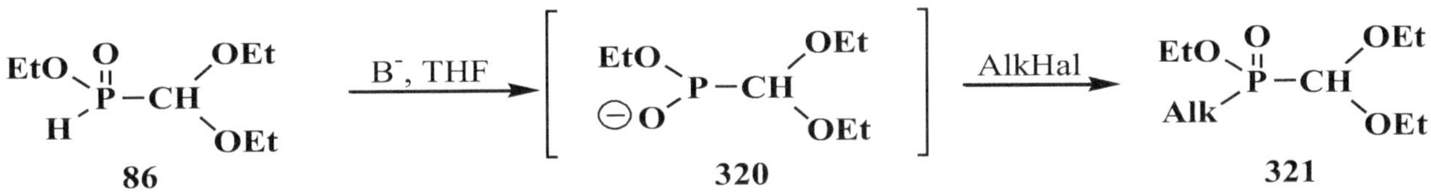

Scheme 111. Phosphinate **86** alkylation with alkylhalydes.

Compound **86** also readily undergoes addition to activated double bonds, or it can transform into three-coordinated phosphorus compounds [59,176,177], for example, in Scheme 112, compounds **322** and ethyl trimethylsilyl (diethoxymethyl)phosphinite (**323**), respectively.

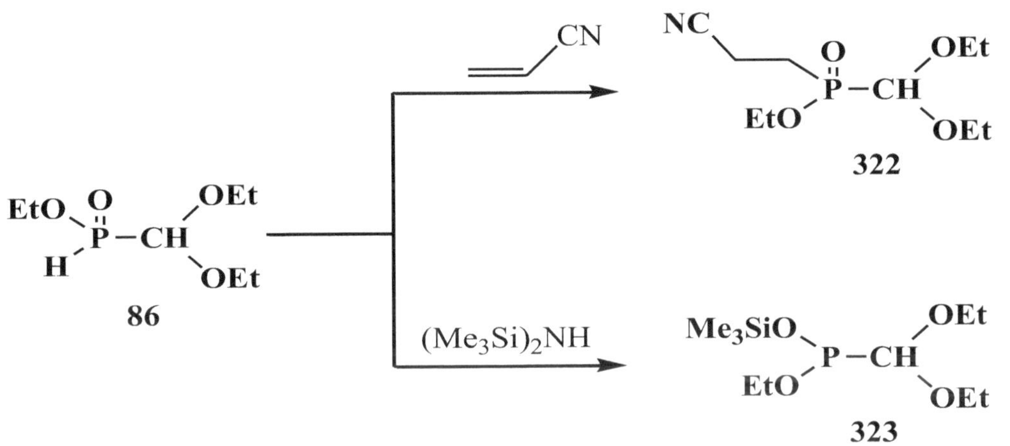

Scheme 112. Transformations of compound **86** into phosphinate **322** and a three-coordinated phosphorus compound (phosphinite **323**).

It also undergoes Todd–Atherton reaction with phenols and cross-coupling reactions with aryl bromides [106] in the presence of tetrakis(triphenylphosphine)palladium(0), $Pd(Ph_3P)_4$, see Scheme 17. In **86**, one of two P–H bonds of initial hypophosphorous acid **319** is protected by a diethoxymethyl group and can be restored in subsequent stages after deprotection [58,176]. When treated with two equivalents of organomagnesium or organolithium compounds, phosphinate **86** produces secondary phosphine oxides **324** [178], a hidden form of unstable primary phosphine oxides **325** apt to disproportionation [179], that can be easily obtained by subsequent acid hydrolysis. Further alkylation of secondary phosphine oxides **324** in the presence of a base leads to unsymmetrical tertiary phosphine oxides **326**, which in turn are the hidden form of unsymmetrical secondary phosphine oxides **328**. Similarly to primary phosphine oxides **325**, compounds **327** can be also obtained by subsequent acid hydrolysis of tertiary phosphine oxides **326**. Further oxidation leads to the corresponding phosphonic **328** and unsymmetrical phosphinic acids **329**. Unsymmetrical tertiary phosphine oxides **330** can be obtained by the following alkylation of secondary phosphine oxides **327** (Scheme 113).

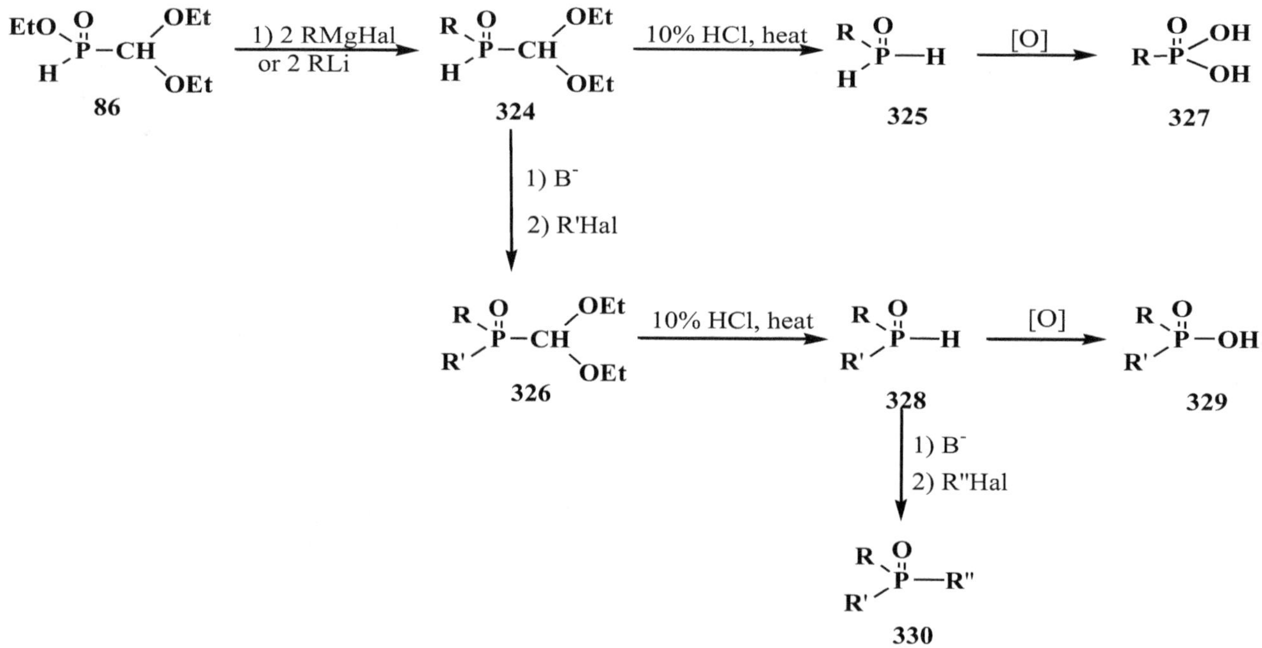

Scheme 113. Using phosphinate **86** for the synthesis of phosphine oxides **325**, **328**, **330**, phosphonic **328** and unsymmetrical phosphinic acids **329**.

Phosphonic and unsymmetrical phosphinic acids **328** and **329** can also be obtained by the scheme that begins with the alkylation reaction by introducing **86** in the reaction of one or sequentially both P–H bonds, thus solving the problem of selectivity [58,59,176,178], over phosphinates **331** and H-phosphinic acids **332**, respectively. The scheme allows one to avoid the use of highly toxic hypophosphorous acid in the syntheses [177] (Scheme 114).

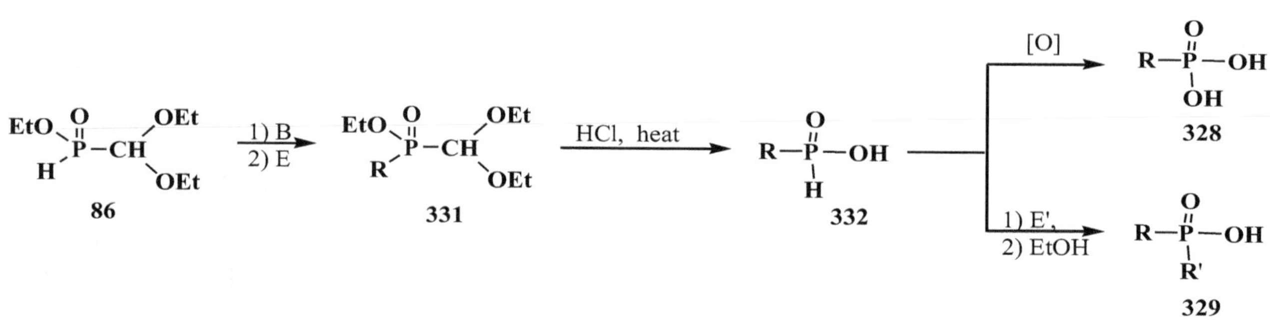

Scheme 114. Phosphonic **328**, H-phosphinic **332** and unsymmetrical phosphinic acids **329** syntheses starting with the reaction of **86** with electrophiles, where E = RHal or olefins with the activated double bond.

Due to the unique combination of chemical properties, phosphinate **86** is used in modern organic synthesis and synthesis of biologically active compounds. H-phosphinic analogs of natural (S-amino acids **333** were obtained with >95% enantiomeric excess by the reaction of **86** with chiral)-*N*-*tert*-butylsulfinylketimines (**334**) [180] (Scheme 115).

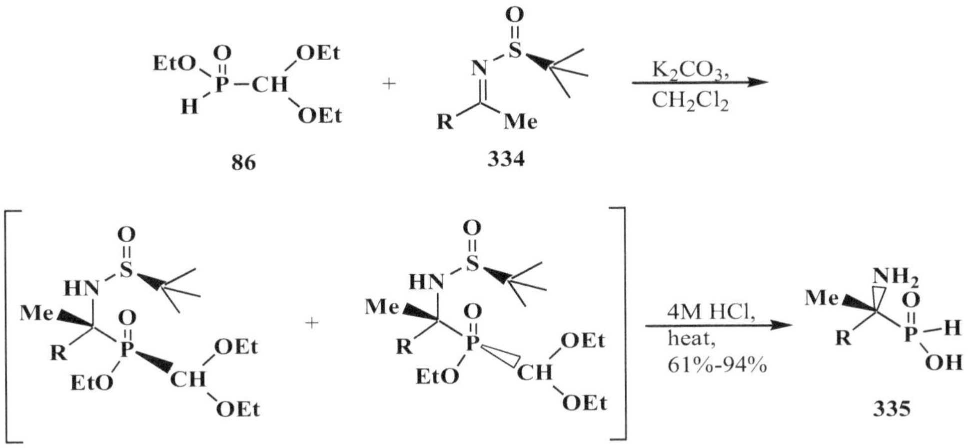

Scheme 115. Synthesis of H-phosphinic analogs of natural α-amino acids **333** by means of the reaction of phosphinate **86** with chiral (*S*)-*N*-*tert*-butylsulfinylketimines **334**.

3-Amino-2-fluoropropyl-*H*-phosphinic acid (**335**), a γ-aminobutyric acid (GABA) analog, a potential pharmaceutical for the treatment of central nervous system diseases, was prepared from compound **323** as the silylated form of ethyl (diethoxymethyl)phosphinate **86** [181] (Scheme 116).

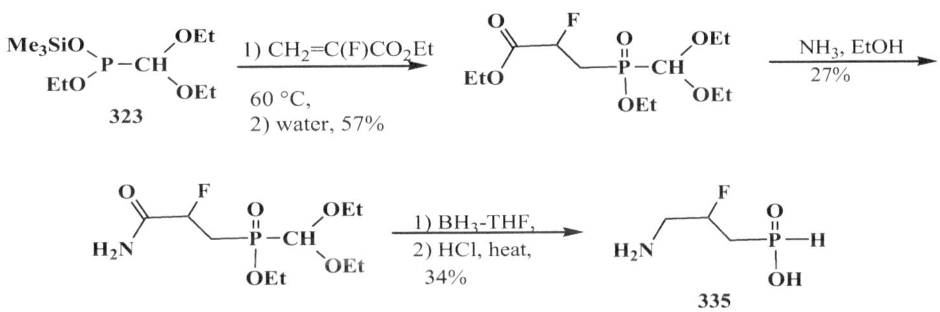

Scheme 116. Silylated compound **323** used for the synthesis of γ-aminobutyric acid analog **335**.

Disubstituted 3-aminopropyl-4-pyridylphosphinic acid (**336**), a potential neurotropic pharmaceutical and GABA antagonist, was obtained in a similar manner [182] (Scheme 117).

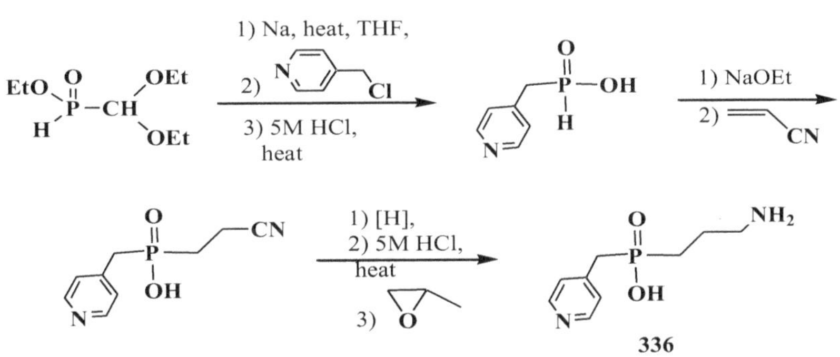

Scheme 117. Synthesis of potential neurotropic pharmaceutical, GABA antagonist **336**.

6. Phosphorylated Formaldehyde Hydrates—Geminal Diols 35. Syntheses and Chemical Properties

The chemistry of phosphorylated formaldehyde hydrates **35** began to develop since the mid 1990s when diethyl (dihydroxymethyl)phosphonate (**337**) was obtained in quantitative yield by the reaction of diethyl (diazomethyl)phosphonate (**338**) with 3,3-dimethyldioxirane—acetone peroxide at 20 °C [60,61] (Scheme 118).

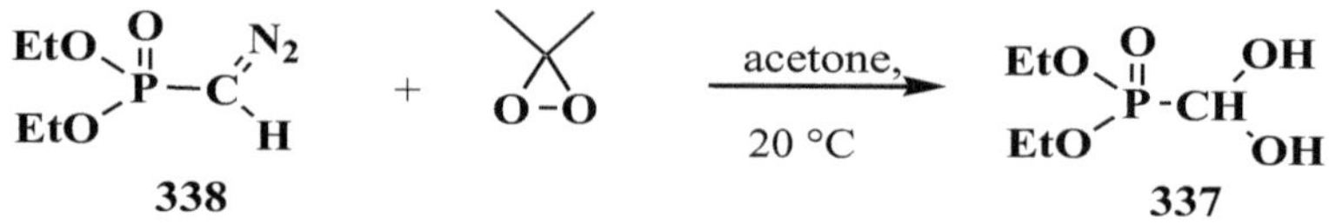

Scheme 118. Synthesis of diethyl (dihydroxymethyl)phosphonate (**337**) by the reaction of diazomethylphosphonate (**338**) with 3,3-dimethyldioxirane.

It was shown that **337** exists in an equilibrium with the hydrated form of diethyl formylphosphonate **3** [60,61] (Scheme 119).

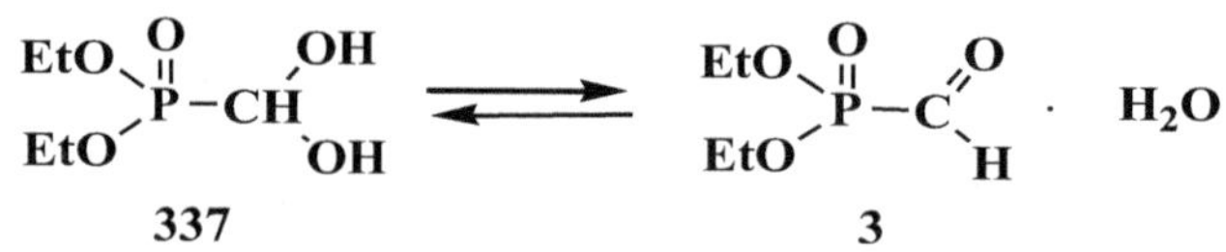

Scheme 119. Equilibrium between diethyl (dihydroxymethyl)phosphonate (**337**) and the hydrated form of diethyl formylphosphonate **3**.

The chemical properties of phosphonate **337** provide the possibility of considering it as a hidden form of phosphorylated formaldehyde. Compound **337** reacts with secondary amines R'_2NH with the cleavage of the phosphorus–carbon bond and formation of the corresponding *N,N*-disubstituted formamides and diethyl phosphite (**6**) [60,61] (it behaves as a source of formyl group $[C(O)H]^+$ cation, synthon type a^1 [183,184].

Compound **337** also shows properties of a typical aldehyde when reacted with primary amines RNH_2, trimethylsilyl cyanide Me_3SiCN, and hydroxylamine NH_2OH to give phosphorylated formaldimines **339**, formalcyanohydrin **340**, and formaldoxime **341**, respectively [60,61] (Scheme 120).

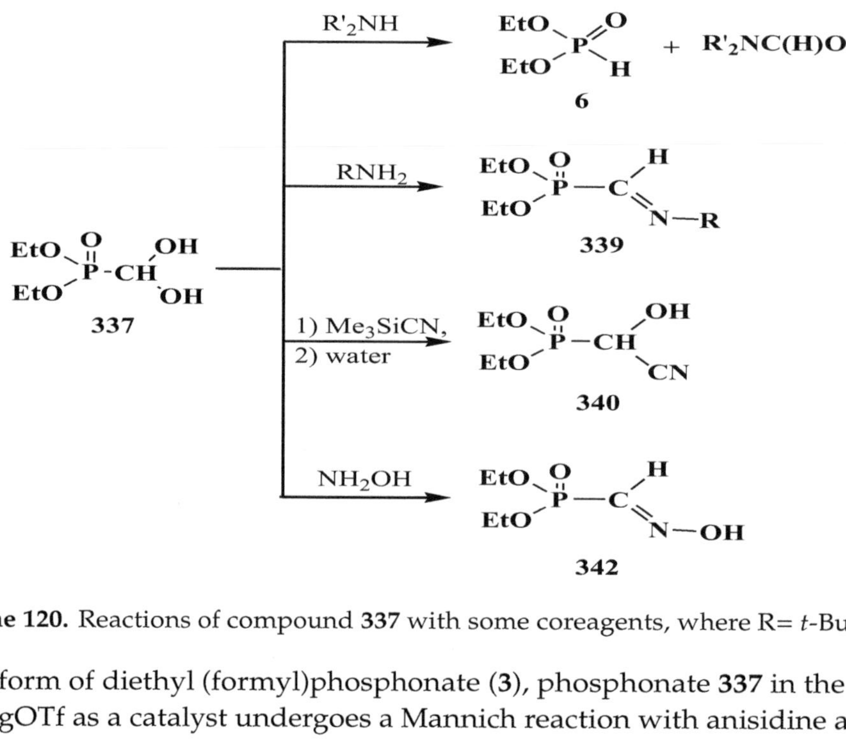

Scheme 120. Reactions of compound **337** with some coreagents, where R= *t*-Bu or Ph.

As a hidden form of diethyl (formyl)phosphonate (**3**), phosphonate **337** in the presence of silver trifluoroacetate AgOTf as a catalyst undergoes a Mannich reaction with anisidine and terminal alkyl-

or aryl-substituted alkynes. Reaction products, α-aminopropargylphosphonates **342**, are valuable precursors in the synthesis of α-aminophosphonic acids showing high biological activity as α-amino acid mimetics [63] (Scheme 121).

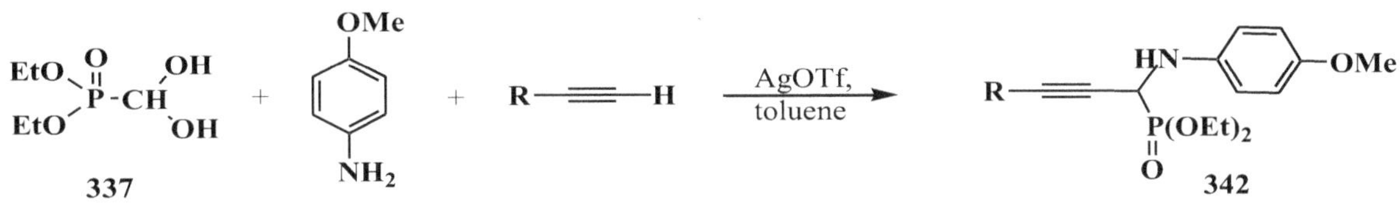

Scheme 121. Catalytical synthesis of α-aminopropargylphosphonates **342** by a Mannich reaction.

In the presence L-proline amide (**343**) as a catalyst, **337** undergoes an asymmetrical cross aldol condensation with aliphatic ketones. Diastereomerically pure α-hydroxyphosphonates **344**, precursors of α-hydroxyphosphonic acids, result from chirality induction with diastereomeric excesses greater that 99% [185] (Scheme 122).

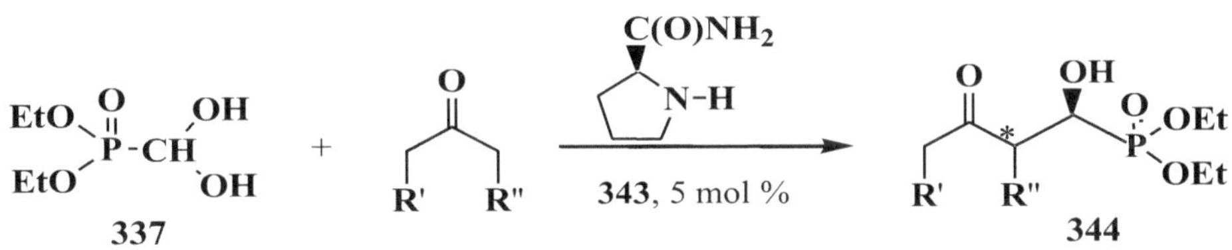

Scheme 122. Synthesis of diastereomeric α-hydroxyphosphonates **344** by asymmetrical cross-aldol condensation.

Like α-aminophosphonic acids, α-hydroxyphosphonic acids are also α-amino acid mimetics and exhibit high biological activity [185]. Similar results were obtained in the aldol condensation of ketones with racemic ethyl phenyl(dihydroxymethyl)phosphinate (**345**). In this case, the final products are α-hydroxyphosphinic acids **346** [62].

7. Conclusions

Phosphorylated formaldehyde derivatives, *i.e.*, acetals and related compounds, are a group of largely underinvestigated species. The experimental data accumulated since the early 1960s confirm that these compounds can be used in a wide variety of syntheses that have not been fully realized to date. For this reason, the growth of interest to these compounds will allow investigating their chemical properties in more detail and potentially enrich organic and organoelement chemistry with new synthetic methods.

References

1. Mitsuo, S.; Masaki, S.; Hikaru, Y.; Tsujiaki, H. Acylphosphonates: P-C Bond Cleavage of Dialkyl Acylphosphonates by Means of Amines. Substituent and Solvent Effects for Acylation of Amines. *J. Org. Chem.* **1980**, *45*, 4162–4167.
2. Miller, A.; Stewart, D. Reactions of Carbonyl Compounds with Tervalent Phosporus Reagents. Part 8.l, Acetyldiphenylphosphine Oxide. *J. Chem. Soc. Perkin Trans. I* **1977**, *17*, 1898–1901. [CrossRef]
3. Frey, G.; Lesiecki, H.; Lindner, E.; Vordermaier, G. Synthese und reaktives Verhalten von Acyldiorganylphosphanoxiden. *Chem. Ber.* **1979**, *112*, 763–772. [CrossRef]
4. Sekine, M.; Kume, A.; Nakajima, M.; Hata, T. A new method for acylation of enolates by means of dialkyl acylphosphonates as acylating agents. *Chem. Lett.* **1981**, *10*, 1087–1090. [CrossRef]

5. Shagidullin, R.R.; Plyamovatyi, A.K.; Mukhamadeeva, R.M.; Khairullin, V.K.; Pudovik, A.N. Fourier IR—Spectroscopic study of the mechanism of anomalous Shiff base reaction with phosphorus-containing CH—Acids. *Z. Obshch. Khim.* **1994**, *64*, 926–930.
6. Conti, P.; Pinto, A.; Tamborini, L.; Rizzo, V.; de Micheli, C. A regioselective route to 5-substituted pyrazole- and pyrazoline-3-phosphonic acids and esters. *Tetrahedron* **2007**, *63*, 5554–5560. [CrossRef]
7. Conti, P.; Pinto, A.; Tamborini, L.; Dunkel, P.; Gambaro, V.; Visconti, G.L.; de Micheli, C. A regioselective route to 5-substituted isoxazole- and isoxazoline-3-phosphonates. *Synthesis* **2009**, 591–596. [CrossRef]
8. Vasella, A.; Voeffray, R. Asymmetric Synthesis of α- Aminophosphonic Acids by Cycloaddition of *N*-Glycosyl-*C*- dialkoxyphosphonoylnitrones. *Helv. Chim. Acta* **1982**, *65*, 1953–1964. [CrossRef]
9. Moskva, V.V.; Mavrin, V.Y. *O,O*-Diethylformylphosphonate. *J. Gen. Chem. USSR* **1988**, *57*, 2492.
10. Iorga, B.; Eumery, F.; Mouriès, V.; Savignac, P. Phosphorylated aldehydes. *Tetrahedron* **1998**, *54*, 14637–14677. [CrossRef]
11. Groß, H. Zur Existenz von Estern der Formylphosphonsäure. *Z. Chem.* **1977**, *17*, 131–132.
12. Amsallem, D.; Gornitzka, H.; Baceiredo, A.; Bertrand, G. New types of stable aldehydes: Formylphosphane and formylphosphane oxide. *Angew. Chem. Int. Ed. Engl.* **1999**, *38*, 2201–2203. [CrossRef]
13. Razumov, A.I.; Liorber, B.G.; Moskva, V.V.; Sokolov, M.P. Phosphorylated aldehydes. *Russ. Chem. Rev.* **1973**, *42*, 538–550. [CrossRef]
14. Wagenknecht, J. An electrochemical method for the preparation of iminodimethylenediphosphonic Acid. *Synth. React. Inorg. Met. Org. Chem.* **1974**, *4*, 567–572. [CrossRef]
15. Firestone, R.A. Hydroxy and Imino Containing Phosphonic Acid Diesters. U.S. Patent 3784590, 8 January 1974.
16. Mührle, H.; Vetter, W. Reaktionsbeteiligung von Phosphoester-Nachbargruppen bei Aminodehydrierungen. *Z. Naturforsch. B* **1988**, *43*, 1663–1671.
17. Franczyk, T.S., II. Preparation of Formylphosphonic Acid from Tertiary Aminomethylphosphonic Acid N-oxides. U.S. Patent 6274760, 14 August 2001.
18. Costisella, B.; Gross, H. Synthese und NMR-Spektren offenkettiger und cyclischer Acetale von Formylphosphonsäureestern und Formylphosphinoxid. *J. Prakt. Chem.* **1977**, *319*, 8–16. [CrossRef]
19. Livantsov, M.V.; Proskurina, M.V.; Prishchenko, A.A.; Lutsenko, I.F. Synthesis and some properties of phosphorylated formals. *J. Gen. Chem. USSR* **1984**, *54*, 2504–2517.
20. Brunjes, M.; Kujat, C.; Monenschein, H.; Kirschning, A. Acylation of Alkyl Halides and Amino Aldehydes with Phosphane Oxide-Based d^1-Synthon. *Eur. J. Org. Chem.* **2004**, *5*, 1149–1160. [CrossRef]
21. Gross, H.; Keitel, I.; Costisella, B.; Mikołajczyk, M.; Midura, W. Zur Synthese von bis-alkylmercaptomethanphosphonsäuredialkylestern. *Phosphorus Sulfur Silicon Relat. Elem.* **1983**, *16*, 257–262. [CrossRef]
22. Młotkowska, B.; Gross, H.; Costisella, B.; Mikołajczyk, M.; Grejszczak, S.; Zatorski, A. Synthese offenkettiger und cyclischer *S,S*-Acetale von Formyl phosphonsäurestern. *J. Prakt. Chem.* **1977**, *319*, 17–22. [CrossRef]
23. Mikołajczyk, M.; Grejszczak, S.; Zatorski, A.; Młotkowska, B.; Gross, H.; Costisella, B. A new and general synthesis of ketene *S,S*- and *S,O*-Acetals based on the Horner-Wittig reaction. *Tetrahedron* **1978**, *34*, 3081–3088. [CrossRef]
24. Costisella, B.; Gross, H. 1-Dimethylamino-1-cyano methanphosphonsдurediethylester, ein neues Edukt zur Darstellung von Carbonsäuren, 1-Cyanoenaminen und Homoenolaten. *Tetrahedron* **1982**, *38*, 139–145. [CrossRef]
25. Costisella, B.; Gross, H. 1-Cyano-1-dimethylamino methanphosphonsäurediethylester-Lithium—Ein "Instant-Horner-Reagent". *Z. Chem.* **1987**, *27*, 143–144. [CrossRef]
26. Marrero, Y.; Harwood, L.M. Towards the total synthesis of colletofragaranes: Constructing the macrocyclic lactone by high pressure-mediated intramolecular Diels-Alder reaction. *Tetrahedron Lett.* **2009**, *50*, 3574–3576. [CrossRef]
27. Gross, H.; Costisella, B. Synthesis of carboxylic acids via PO-activated olefination of tetraethyl dimethilamino-methylene-diphosphonate. *Angew. Chem. Int. Ed. Engl.* **1968**, *7*, 391–392. [CrossRef]
28. Qian, D.Q.; Shi, X.D.; Cao, K.Z.; Liu, L.Z. The Synthesis and Reactivity of Alkylaminosubstitutedmethylenediphosphonates. *Heteroat. Chem.* **1999**, *10*, 271–276. [CrossRef]
29. McNulty, J.; Das, P. Development of one-pot method for the homologation of aldehides to carboxylic acids. *Tetrahedron* **2009**, *65*, 7794–7800. [CrossRef]

30. Gross, H.; Costisella, B. Über α-substituerte Phosphonate, XI. *Liebiegs Ann. Chem.* **1971**, *750*, 44–52. [CrossRef]
31. Gross, H.; Costisella, B. Derivate des Formylphosphonsäurediäthylestern. *J. Prakt. Chem.* **1969**, *311*, 925–929. [CrossRef]
32. Gross, H.; Costisella, B.; Haase, L. Zur Reaction von Dimethylformamidederivaten mit Phosphorigsäurediestern und Phosphinoxiden. *J. Prakt. Chem.* **1969**, *311*, 577–585. [CrossRef]
33. Risch, N.; Piper, S.; Winter, A.; Lefarth-Risse, A. An Efficient Synthesis of Novel α-Aminophosphonates Based on a Mannich-Type Reaction. *Eur. J. Org. Chem.* **2005**, *2*, 387–394. [CrossRef]
34. Schrader, T.; Steglich, W. Phosphorus analogs of amino acids. IV. Syntheses of unusual 1-aminophosphonic acids via Diels-Alder reactions of diethyl (*N*-acyliminomethyl)phosphonates. *Synthesis* **1990**, 1153–1156. [CrossRef]
35. Mikołajczyk, M.; Costisella, B.; Grejszczak, S.; Zatorski, A. Synthesis of *O,S*-thioacetals of formylphosphonates. *Tetrahedron Lett.* **1976**, *17*, 477–480. [CrossRef]
36. Kim, T.H.; Uh, D.Y. New synthetic method of *O,S*-thioacetals of formylphosphonates. *Tetrahedron Lett.* **1985**, *26*, 3479–3482.
37. Costisella, B.; Keitel, I. Synthese und Reaktionen von 1-Acetoxy-1-methylthio-methanphosphorylverbindungen. *Phosphorus Sulfur Silicon Relat. Elem.* **1988**, *40*, 161–165. [CrossRef]
38. Mavrin, V.Y.; Moskva, V.V. Reaction of trihetero-substituted carbonium salts with diethyl phosphite. *J. Gen. Chem. USSR* **1988**, *58*, 1930–1931.
39. Mavrin, V.Y.; Moskva, V.V. Condensation of diethyl phosphite with mixed derivatives of orthocarboxylic acid. *J. Gen. Chem. USSR* **1988**, *58*, 1490.
40. Gross, H.; Seibt, H. Synthese von α-chlor-α-methoxy-methanphosphonsäureestern und α-chlor-α-methylthio-methanphosphonsäureestern. *J. Prakt. Chem.* **1970**, *312*, 475–482. [CrossRef]
41. Petrov, K.A.; Chauzov, V.A.; Agafonov, S.V.; Pazhitkova, I.V. Methoxymethanediphosphonic acid esters. *J. Gen. Chem. USSR* **1980**, *50*, 1525–1526.
42. Galli, R.; Scaglioni, L.; Palla, O.; Gozzo, F. Synthesis of 2-alkylthyo(or trifluoromethylthio)-2-halogenothenyl deryvatives by of Wittig (under phase transfer conditions)or Wittig-Horner reactions. Applications in the field of pyretroids. *Tetrahedron* **1984**, *40*, 1523–1532. [CrossRef]
43. Kim, T.H.; Uh, D.Y. Synthesis of *S,S*-thioacetals of formylphosphonate from chloro(arylthio)methanephosphonate. *Synth. Commun.* **1988**, *18*, 1611–1614. [CrossRef]
44. Otten, P.A.; Davies, H.M.; van der Gen, A. A Horner-Wittig Synthesis of 1-Chlorovinyl Sulfoxides. *Tetrahedron Lett.* **1995**, *36*, 781–784. [CrossRef]
45. Otten, P.A.; Davies, H.M.; van der Gen, A. A Horner-Wittig Synthesis of 1-Chlorovinyl Sulfoxides. *Phosphorus Sulfur Silicon Relat. Elem.* **1996**, *109*, 449–452.
46. Dinizo, S.E.; Freerksen, R.W.; Rabst, W.E.; Watt, D.S. Synthesis of α-Alkoxyacrylonitriles Using Substituted Diethyl Cyanomethylphoaphonates. *J. Org. Chem.* **1976**, *41*, 2846–2849. [CrossRef]
47. Dinizo, S.E.; Freerksen, R.W.; Rabst, W.E.; Watt, D.S. A One-Carbon Homologation of Carbonyl Compounds to Carboxylyc Acids, Esters and Amides. *J. Am. Chem. Soc.* **1977**, *99*, 182–186. [CrossRef]
48. Diamond, P.M.; Dinizo, S.E.; Freerksen, R.W.; Curtis, R.; Haltiwanger, R.C.; Watt, D.S. Ferric Chloride-catalysed Conversion of α-t-butoxy- or α-acetoxy-acrylonitryles into Imides. *J. Chem. Soc. Chem. Commun.* **1977**, 298–299. [CrossRef]
49. Koidan, G.N.; Marchenko, A.P.; Oleinik, V.A.; Pinchuk, A.M. 2-Methyl-2-tribromophosphazopropane. *J. Gen. Chem. USSR* **1988**, *58*, 1304–1309.
50. Savignac, P.; Coutrot, P. Preparation of 1,1-dibromoalkanes by Halogen Exchange. *Synthesis* **1976**, 197–199. [CrossRef]
51. Majewski, P.; Koszuk, J.F. Synthesis of Dichloromethylphosphonates. *Phosphorus Sulfur Silicon Relat. Elem.* **2009**, *184*, 956–962. [CrossRef]
52. Bulpin, A.; Masson, S.; Sene, A. Reaction of Phosphonodithioformates with Nucleophilic Reagents; Potential synthetic Uses. *Phosphorus Sulfur Silicon Relat. Elem.* **1990**, *49–50*, 135–138. [CrossRef]
53. Costisella, B.; Oczegowski, S.; Gross, H. α-Substituierte phosphonate. 70 Untersuchungen über Methylthiomethanbisphosphorylderivate. *Phosphorus Sulfur Silicon Relat. Elem.* **1994**, *86*, 169–175. [CrossRef]
54. Binder, J.; Zbiral, E. A new procedure for homologation of carbonyl compounds to α-hydroxy-carboxylic esters by means of diethyl-[trimethylsilylethoxymethyl]phosphonate. *Tetrahedron Lett.* **1986**, *27*, 5829–5832. [CrossRef]

55. Dufrechon, S.; Combert, J.-C.; Malhiac, C.; Collignon, N. Efficient synthesis of α-enamino phosphonates in the series of piperidine and morpholine. *Phosphorus Sulfur Silicon Relat. Elem.* **1997**, *127*, 1–14. [CrossRef]
56. Mikołajczyk, M. α-Heterosubstituted phosphonates and phosphineoxides. *Pure Appl. Chem.* **1987**, *59*, 983–988. [CrossRef]
57. Gallagher, M.J.; Honegger, H. Dialkoxymethylation of phosphorus with trialkyl orthoformates: Reactions of phosphonic and phosphinic acids via their trivalent tautomers. *Tetrahedron Lett.* **1977**, *18*, 2987–2990. [CrossRef]
58. Coudray, L.; Montchamp, J.-L. Temporary Protection of H-Phosphinic Acids as a Synthetic Strategy. *Eur. J. Org. Chem.* **2009**, *27*, 4646–4654. [CrossRef]
59. Dingwall, Y.G.; Ehrenfreund, I.; Hall, R.G. Diethoxymethylphosphonites and phosphinates. Intermediates for the synthesis of α, β and γ-aminoalkylphosphinous acids. *Tetrahedron* **1989**, *45*, 3787–37808. [CrossRef]
60. Hamilton, R.; McKervey, M.A.; Rafferty, M.D.; Walker, B.J. The Reaction of Dimethyl Dioxirane with Diazomethylphosphonates; the First Synthesis of a Formylphosphonate Hydrate. *J. Chem. Soc. Chem. Commun.* **1994**, 37–38. [CrossRef]
61. Cairns, J.; Dunne, C.; Franczyk, T.S.; Hamilton, R.; Hardacre, C.; Stern, M.K.; Treacy, A.; Walker, B.J. The Synthesis and Chemistry of Formylphosphonate. *Phosphorus Sulfur Silicon Relat. Elem.* **1999**, *144*, 385–388. [CrossRef]
62. Samanta, S.; Perera, S.; Zhao, C.-G. Organocatalytic Enantioselective Synthesis of Both Diastereomers of α-Hydroxyphosphinates. *J. Org. Chem.* **2010**, *75*, 1101–1106. [CrossRef]
63. Dodda, R.; Zhao, C.-G. Silver(I) triflate-catalyzed direct synthesis of N-PMP protected α-aminopropargylphosphonates from terminal alkynes. *Org. Lett.* **2007**, *9*, 165–167. [CrossRef] [PubMed]
64. Hirai, T.; Han, L.-B. Palladium-Catalyzed Ibsertion of Isocyanides into P(O)-H Bonds: Selective Formation of phosphinoyl Imines and Bisphosphinoylaminomethanes. *J. Am. Chem. Soc.* **2006**, *128*, 4722–4723. [CrossRef] [PubMed]
65. Piotrowska, D.G. N-Substituted C-diethoxyphosphorylated nitrones as useful synthons for the synthesis of α-aminophosphonates. *Tetrahedron Lett.* **2006**, *47*, 5363–5366. [CrossRef]
66. Piotrowska, D.G.; Balzarini, J.; Glowacka, I.E. Design, synthesis antiviral and cytostatic evaluation of novel isoxalidine nucleotide analogues with a 1,2,3-triazole linker. *Eur. J. Med. Chem.* **2012**, *47*, 501–509. [CrossRef] [PubMed]
67. Seufert, D.; Marmor, R.S.; Hilbert, P. Some reactions of dimethilphosphono-substituted diazoalkanes. $(MeO)_2P(O)CR$ transfer to olefins and 1,3-dipolar additions of $(MeO)_2P(O)CR'$. *J. Org. Chem.* **1971**, *36*, 1379–1386.
68. Maehr, H.; Uskokovic, M.R.; Schaffer, C.P. Concise synthesis of dimethyl(2-Oxopropyl)phosphonate and homologation of aldehydes and to alkynes in a tandem process. *Synth. Commun.* **2009**, *39*, 299–310. [CrossRef]
69. Kosobokov, M.D.; Titnyuk, I.D.; Beletskaya, I.P. An expedient synthesis of diethyl diazomethylphosphonate. *Mendeleev Commun.* **2011**, *21*, 142–143. [CrossRef]
70. Neidlein, R.; Keller, H. Syntheses of 1-Benzyloxyaminoalkylphosphonates. *Heterocycles* **1993**, *36*, 1925–1932. [CrossRef]
71. Carter, W.A. Treatment of Human Viral Infection by Doublestranged RNA Combined with Viral Infections. Eur. Patent 286224, 12 October 1988.
72. Franczyk, T.S., II. Method for Preparing Formylphosphonic Acid. U.S. Patent 7294733, 13 November 2007.
73. Franczyk, T.S., II. Method of Making Phosphorus-containing Compounds and Products Thereof. U.S. Patent 6864218, 8 March 2005.
74. Oediger, H.; Lieb, F.; Disselnkotter, H. Phosphonoformaldehyde, a process for its preparation and its use as an intermediate product for the preparation of medicaments. U.S. Patent 4,348,332, 7 September 1982.
75. Livantsov, M.V.; Boiko, V.I.; Proskurina, M.V.; Lutsenko, I.F. Phosphorylation of orthoformates. *J. Gen. Chem. USSR* **1982**, *52*, 811.
76. Razumov, A.I.; Moskva, V.V. Reaction of dialkylphosphorous acids with orthoformic esters. *Z. Obshch. Khim.* **1964**, *34*, 3125–3126.
77. Razumov, A.I.; Moskva, V.V. Reaction of orthoformates with hydrogen phosphited and hydrogen phosphinites. *J. Gen. Chem. USSR* **1965**, *35*, 1599.

78. Beznosko, B.K.; Usanova, V.M.; Zhuravleva, L.V.; Kharitonov, A.V.; Bondarenko, N.A.; Yarkevich, A.N.; Antoshin, A.E.; Tsvetkov, E.N. Antiinflammatory and analgesic activity of quaternary phosphine oxides. *Pharm. Chem. J.* **1990**, *24*, 244–247. [CrossRef]
79. Gross, H.; Costisella, B. Reaktion von phosphoriger Säure bzw. deren Anhydrid mit Orthoameisensäureestern. *J. Prakt. Chem.* **1974**, *311*, 550–556. [CrossRef]
80. Gross, H.; Costisella, B. *O,O*-and S,S-Acetals of formylphosphoniumsalts. *J. Prakt. Chem.* **1978**, *320*, 128–132.
81. Gross, H.; Freiberg, J.; Costisella, B. Zur existanz von halogendialkoxyalkanen, eine einfache synthese von dialkoxymethanphosphonaten. *Chem. Ber.* **1968**, *101*, 1250–1256. [CrossRef]
82. Malenko, D.M.; Gololobov, Y.G. New method for synthesis of phosphorylated formals. *J. Gen. Chem. USSR* **1981**, *51*, 1214–1215.
83. Groß, H.; Costisella, B. Zur Umsetzung von Dimethylformamid-dimethyl-acetal mit Phosphortrichlorid. *Z. Chem.* **1970**, *10*, 404–405. [CrossRef]
84. Costisella, B.; Gross, H. Zur Reaktion von Orthoameisensäureestern mit Phosphor-III-Verbindungen. *J. Prakt. Chem.* **1969**, *311*, 571–576.
85. Krokhina, S.S.; Pyrkin, R.I.; Levin, Y.A.; Ivanov, B.E. Effect of triethyl othoformate on some trivalent phosphorus derivatives. *Russ. Chem. Bull.* **1968**, *17*, 1349. [CrossRef]
86. Prishchenko, A.A.; Livantsov, M.V.; Novikova, O.P.; Livantsova, L.I.; Petrosyan, V.S. Synthesis and reactivity of substituted α-carbonylphosphonites and their derivatives. *Heteroat. Chem.* **2012**, *23*, 352–372. [CrossRef]
87. Dietsche, W. Darstellung von C-phosphorylierten Formaldehydacetalen. *Liebigs Ann. Chem.* **1968**, *712*, 21–27. [CrossRef]
88. Moskva, V.V.; Maikova, A.I.; Razumov, A.I. Reaction of orthocarboxylates and acetals with trivalent phosphorus acid chlorides. *J. Gen. Chem. USSR* **1969**, *39*, 563–566.
89. Costisella, B.; Gross, H. Notiz zur Synthese unsymmetrisch substituirten derivate der diätoxymethanphosphonsäure. *J. Prakt. Chem.* **1977**, *319*, 343–346. [CrossRef]
90. Krutskaya, L.V.; Safiulina, O.Z.; Voronina, S.G.; Bryukhovetskaya, L.V.; Tsivunin, V.S. On the reaction of 2-chloro-1,3,2-dioxaphospholanes with 2-ethoxy-1,3-dioxolanes. *J. Gen. Chem. USSR* **1989**, *59*, 2408–2411.
91. Tsivunin, V.S.; Krutskii, L.N.; Ernazarov, M.; Kamai, G.K. Reaction of the diethylamido chloride of ethylphosphinous acid and ethyldichlorophosphine with orthoformic esters. *J. Gen. Chem. USSR* **1970**, *40*, 2551–2553.
92. Costisella, B.; Gross, H. Zur reaktion von P(III) amiden mit orthoameisensäureesternchloriden. *Phosphorus Sulfur Silicon Relat. Elem.* **1980**, *8*, 99–103. [CrossRef]
93. Młotkowska, B.; Costisella, B.; Gross, H. Eine einfache Methode zur Synthese cyclischer Acetale des Formylphosphonsäreesters. *J. Prakt. Chem.* **1974**, *316*, 913–916. [CrossRef]
94. Bennett, S.N.L.; Hall, R.G. New syntheses of arylphosphinic acids from the reaction of ethyl diethoxymethylphosphinate with aryl bromides and phenols. *J. Chem. Soc. Perkin Trans. I.* **1995**, 1145–1151. [CrossRef]
95. Costisella, B.; Gross, H. Zur reaktivität von Acetalen des Formylphosphonsäureesters. *J. Prakt. Chem.* **1971**, *313*, 265–276.
96. Niyazymbetov, M.S.; Costisella, B.; Kaitel, I.; Shvatrz, K.K. Cathode-catalyzed transesterification of alkylphosphonates. *Russ. Chem. Bull.* **1991**, *40*, 182–186. [CrossRef]
97. Mikołajczyk, M.; Graczyk, P.P.; Wiechorek, M.W.; Bujacz, G. A solution and solid state conformation of 2-diphenylphosphinoyl-1,3-dioxanes. The nature of O-C-P anomeric interactions. *Tetrahedron* **1992**, *48*, 4209–4230. [CrossRef]
98. Gross, H.; Böck, C.; Costisella, B.; Gloede, J. Entalkylierung von Phosphonoesthers mit labilen funktionellen Gruppen mittels Trimethylsilyl-Bromid. *J. Prakt. Chem.* **1978**, *320*, 344–350. [CrossRef]
99. Rol'nik, L.Z.; Pastushenko, E.V.; Livantsov, M.V.; Proskurnina, M.V.; Zlotskii, S.S.; Rakhmankulov, D.L. Free-radical reactions of dialkyl formylphosphonate phosphonacetals. *J. Gen. Chem. USSR* **1983**, *53*, 1104–1109.
100. Razumov, A.I.; Gurevich, P.A.; Liorber, B.G.; Borisova, T.B. A study in the series of phosphinic and phosphinous acid derivatives. *J. Gen. Chem. USSR* **1969**, *39*, 369–372.
101. Nguyen-Ba, P.; Lee, N.; Mitchell, H.; Chan, L.; Quimpere, M. Design and synthesis of a novel class nucleotide analogs with anti-HCMV activity. *Bioorg. Med. Chem. Lett.* **1998**, *8*, 3555–3560. [CrossRef]

102. Kruse, C.G.; Broekhof, N.L.Y.M.; Wijsman, A.; van der Gen, A. Synthetic application of 2-chloro-1,3-dithiane preparation of ketene dithioacetals. *Tetrahedron Lett.* **1977**, *18*, 885–888. [CrossRef]
103. Ishikawa, K.; Akiba, K.; Inamoto, N. Synthesis of 1, 4-benzodithiofulvenes via Wittig reaction. *Tetrahedron Lett.* **1976**, *17*, 3695–3698. [CrossRef]
104. Mikołajczyk, M.; Graczyk, P.P.; Wieczorek, M.W. Conformational preference in 1,3-dithianes containing 2-phosphoryl, -(thiophosphoryl), and (selenophosphoryl)groups. Chemical and crystallographic implications of the nature of the anomeric effect. *J. Org. Chem.* **1994**, *59*, 1672–1679. [CrossRef]
105. Mikołajczyk, M.; Łuczak, T.; Graczyk, P.P.; Wieczorek, M.W.; Błaszczyk, Y.; Bujacz, G.D.; Majzner, W.R. Solid state conformation of the anomeric effect in 2-phosphoryl, 2- thiophosphoryl and 2-selenophosphoryl-substituted 1,3-dithiolans. *J. Organomet. Chem.* **1997**, *536–537*, 355–360. [CrossRef]
106. Aggarwal, V.K.; Barnell, J.K.; Worrall, J.M.; Alexander, R. Highly diastereoselective epoxidation of ketene dithioacetal dioxides: A new approach to asymmetric synthesis of α-amino amides. *J. Org. Chem.* **1998**, *63*, 7128–7129. [CrossRef]
107. Mikołajczyk, M.; Grejszczak, S.; Chefchyńska, A.; Zatorski, A. Addition of elemental sulfur to phosphonate carbanions and its application for synthesis of α-phosphoryl organosulfur compounds, synthesis of oaromatic ketones. *J. Org. Chem.* **1979**, *44*, 2967–2972. [CrossRef]
108. Mikołajczyk, M.; Bałczewski, P.; Grejszczak, S. Sulphenylation of phosphonates. A facile synthesis of α-phosphoryl sulfides and *S,S*-acetals of oxomethanephosphonates. *Synthesis* **1980**, 127–129.
109. Grayson, J.I.; Warren, S. Acyl Anion Equivalents: Synthesis of Ketones and Enones from α-Phenylthioalkylphosphine Oxides. *J. Chem. Soc. Perkin Trans. I* **1977**, 2263–2272. [CrossRef]
110. Mikołajczyk, M.; Bałczewski, P. Synthesis and reactivity of diethyl(methylthio)(trimethylsilyl)methylphosphonate. *Synthesis* **1989**, 101–106. [CrossRef]
111. Makomo, H.; Masson, S.; Saquet, M. Reduction of phosphonodithioformates: Syntheses of α-phosphoryl thiols and hemidithioacetals. *Tetrahedron Lett.* **1994**, *50*, 10277–10288. [CrossRef]
112. Mikołajczyk, M.; Mikina, M.; Graczyk, P.P.; Bałczewski, P. Synthesis of dithio- and diselenoacetals of formylphosphonates. *Synthesis* **1996**, 1232–1238. [CrossRef]
113. Yuaristi, E.; Cordillo, B.; Valle, L. Relative reactivity of 2-diphenylphosphinoil- and 2-diphenylthiophosphinoil-2-[1,3]-dithianyllithium as reagents Wittig-Horner. *Tetrahedron* **1986**, *42*, 1963–1970. [CrossRef]
114. Yuaristi, E.; Valle, L.; Valenzuela, B.A.; Aguilar, M.A. S-C-P anomeric interactions. 4. Conformational analys of 2-(diphenylphosphinoil)-1,3-dithyane. *J. Am. Chem. Soc.* **1996**, *108*, 2000–2005. [CrossRef]
115. Iorga, B.; Mouriès, V.; Savigniac, P. Carbanionic displacement reactions of phosphorus. Synthesis and reactivity of 5,5-dimethyl-2-oxo-2-(1,1-dithian-2-yl)-1, 3, 2-dioxaphosphorinane. *Bull. Chem. Soc. Fr.* **1997**, *134*, 891–985.
116. El-Wareth, A.; Sarhan, A.O.; Murakami, M.; Izumi, T. Synthesis of functionalized compounds related to π-extended tetrathiafulvenes with quinonoidal bearing a ferrocene moiety. *Monatsh. Chem.* **2002**, *133*, 1055–1066.
117. Costisella, B.; Gross, H. 1-Trimethylammonium-1-diethylphosphono-1-cyanomethylid, ein stabiles N-Ylid. *J. Prakt. Chem.* **1987**, *324*, 545–549. [CrossRef]
118. Costisella, B. ^{1}H, ^{13}C, ^{31}P, ^{15}N, ^{17}O-NMR investigations of α-substituted aminophosphonates. *Phosphorus Sulfur Silicon Relat. Elem.* **1990**, *51/52*, 226. [CrossRef]
119. Prishchenko, A.A.; Livantsov, M.V.; Petrosyan, V.S. Reaction of some CH acids with dimethylformamide dimethylacetal. *J. Gen. Chem. USSR* **1993**, *63*, 1326–1327.
120. Gross, H.; Costisella, B.; Gnauk, T.; Brenneke, L. Derivate der Aminomethan-bis-phosphonsäure. *J. Prakt. Chem.* **1976**, *318*, 116–126. [CrossRef]
121. Nesterov, L.V.; Krepysheva, N.E.; Aleksandrova, N.A. Reaction of dimethylformamide dimethylacetal with silyl phosphites. *J. Gen. Chem. USSR* **1989**, *59*, 641.
122. Prishchenko, A.A.; Livantsov, M.V.; Boganova, N.V.; Zhutskii, P.V.; Lutsenko, I.F. Synthesis of tetraalkyl dimethylaminomethylenediphosphorus-containing acids. *J. Gen. Chem. USSR* **1989**, *59*, 2132–2133.
123. Prishchenko, A.A.; Livantsov, M.V.; Petrosyan, V.S. New types of aminomethyl organophosphorus compounds. *J. Gen. Chem. USSR* **1994**, *63*, 1181–1193.
124. Degenhardt, C.R. Use of tetraethyl dimethylaminomethylene diphosphonate in the synthesis of benzothiophene-2-acetic acid and other carboxylic acids. *Synth. Commun.* **1982**, *12*, 415–421. [CrossRef]

125. Gross, H.; Costisella, B. Bis(diethoxyphosphinyl)(trimethylammonio)-methylide, a Stable *N*-Ylide. *Angew. Chem. Int. Ed. Engl.* **1968**, *7*, 463. [CrossRef]
126. Gross, H.; Costisella, B.; Bürger, W. Synthese von Alkylphosphonbetainen und stabilen N-Ylyden aus α-Trimethylammoniummethylphosphonaten. *J. Prakt. Chem.* **1969**, *311*, 563–570. [CrossRef]
127. Monenschein, H.; Dräger, G. Alexander Jung, Kirschning, A. Asymmetric Nucleophilic Acylation of Aldehydes via 1, 1-Heterodisubstituted Alkenes. *Chem. A Eur. J.* **1999**, *5*, 2270–2280. [CrossRef]
128. Kirschning, A.; Kujat, C.; Luiken, S.; Schaumann, E. Small and Versatile—Formyl Anion and Dianion Equivalents. *Eur. J. Org. Chem.* **2007**, 2387–2400. [CrossRef]
129. Mikołajczyk, M.; Midura, W.; Grejszczak, S. Synthesis of mono- and 1, 4-dicarbonyl compounds based on the oxygenation of phosphonate carbanions. Synthesis of dihydrojasmone, allethone and methylenomycin B. *Tetrahedron Lett.* **1984**, *25*, 2489–2492. [CrossRef]
130. Morita, T.; Okamoto, Y.; Sackurai, H. Dealkylation reaction of acetals, phosphonate, and phosphate esters with chlorotrimethylsilane/metal halide reagent in acetonitrile, and its application to the synthesis of phosphonic acids and vinyl phosphonates. *Bull. Chem. Soc. Jpn.* **1981**, *54*, 267–273. [CrossRef]
131. Morita, T.; Okamoto, Y.; Sackurai, H. A convenient dealkylation of dialkyl phosphonates by chlorotrimethylsilane in the presence of sodium iodide. *Tetrahedron Lett.* **1978**, *28*, 2523–2526. [CrossRef]
132. Gross, H.; Keitel, I. Heterosubstituierte olefine durch horner-reaktion mit α-substituierten Phosphonaten. *Z. Chem.* **1982**, *22*, 117–126.
133. Van Schaik, T.A.M.; Henzen, A.V.; van der Gen, A. A Horner-Wittig solution to the synthesis of ketene, O, O-acetals. *Tetrahedron Lett.* **1983**, *24*, 1303–1306. [CrossRef]
134. Streitwieser, A., Jr.; Juaristi, E. Carbon acidity. Equilibrium ion pair acidies of some phosphorus-substituted carbon acids. *J. Org. Chem.* **1982**, *47*, 768–770. [CrossRef]
135. Afarinkia, K.; Faller, A.; Twist, A.J. A new synthesis of α-ketophosphonates. *Synthesis* **2003**, 357–360. [CrossRef]
136. Coutrot, P.; Grison, C.; Lecouvey, M. Preparation of the phosphonic acid analogue of 3-deoxy-D-manno-2-octulosonic acid (KDO). *Tetrahedron Lett.* **1996**, *37*, 1595–1598. [CrossRef]
137. Horner, L.; Hoffmann, H.; Wippel, H.S. Phosphinoxide als Olefinierungsreagenzien. *Chem. Ber.* **1958**, *91*, 61–63. [CrossRef]
138. Horner, L.; Hoffmann, H.; Wippel, H.S.; Klahre, G. Phosphinoxide als olefinierungsreagenzien. *Chem. Ber.* **1959**, *92*, 2499–2505. [CrossRef]
139. Wittig, G.; Schöllkopf, U. Über Triphenyl-phosphin-methylene als olefinbildende Reagenzien (I Mitteilung). *Chem. Ber.* **1954**, *87*, 1318–1330. [CrossRef]
140. Wardsworth, W.S.; Emmons, W.D. The Utility of Phosphonate Carbanions in Olefins Synthesis. *J. Am. Chem. Soc.* **1961**, *83*, 1733–1738. [CrossRef]
141. Martin, S.F. Synthesis of Aldehydes, Ketones, and Carboxylic Acids from Lower Carbonyl Compounds by C-C Coupling Reactions. *Synthesis* **1979**, 633–665. [CrossRef]
142. Korotchenko, V.N.; Nenaidenko, V.G.; Balenkova, E.S.; Shastin, A.V. Olefination of carbonyl compounds. The newest and classical methods. *Russ. Chem. Rev.* **2004**, *73*, 957–989. [CrossRef]
143. Costisella, B.; Keitel, I.; Gross, H. α-Phosphononoenamine und Acyl phosphonate durch Horner-Olefinierung. *Tetrahedron* **1981**, *37*, 1227–1232. [CrossRef]
144. Gross, H.; Costisella, B.; Schick, H. γ-Butyrolactone aus metallierten Enamine. *Tetrahedron* **1984**, *40*, 733–736.
145. O'Boyle, N.M.; Greene, L.M.; Bergin, O.; Fichet, J.-B.; McCabe, T.; Lloyd, D.G.; Zisterer, D.M.; Meegan, M.J. Synthesis, evaluation and structural studies of antiproliferative tubulin-targeting azetidin-2-ones. *Bioorg. Med. Chem.* **2011**, *19*, 2306–2325. [CrossRef]
146. Mikołajczyk, M.; Grzejszczak, S.; Zatorski, A.; Młotkowska, B. A new general synthesis of ketene thioacetals. *Tetrahedron Lett.* **1976**, *31*, 2731–2734. [CrossRef]
147. Mikołajczyk, M.; Bałczewski, P. Diverse reactivity of α-carbanions derived from α-phosphoryl dithioacetals and α-phosphoryl sulphides towards α, β-unsaturated carbonyl compounds. A general synthesis of conjugated ketene dithioacetals. *Tetrahedron* **1992**, *8*, 8697–8710. [CrossRef]
148. Ceruti, M.; Degani, I.; Fochi, R. A new synthetic application of 1, 2-benzodithiolium cations: Synthesis of aldehydes by 1-carbon homologation of carbonyl compounds. *Synthesis* **1987**, 79–83. [CrossRef]

149. Hanessian, S.; Maji, D.K.; Govindan, S.; Matera, R.; Tintelnot-Blomley, M. Substrate-Controlled and Organocatalytic Asymmetric Synthesis of Carbocyclic Amino Acid Dipeptide Mimetics. *J. Org. Chem.* **2010**, *75*, 2861–2876. [CrossRef] [PubMed]
150. Hanessian, S.; Desilets, D.; Rancourt, G.; Fortin, R. The total stereocontrolled synthesis of a chemical precursor to (+)-thienamycin. *A formal synthesis of the antibiotic. Can. J. Chem.* **1982**, *60*, 2292–2294.
151. Moore, A.J.; Bryce, M.R. Generation and trapping of phosphorus stabilized 4, 5-ethylenedithio-1,3-dithiol-2-ide carbanions: Synthesis of ethylenedithio-1,3-dithiafulvalenes. *Synthesis* **1991**, 26–28. [CrossRef]
152. Misaki, Y.; Nishikawa, H.; Kawakami, K.; Uehara, T.; Yamabe, T. Bis(2-methylidene-1,3-dithiolo[4,5-*d*])tetrathiafulvalene (BDT-TTF): A tetrathiafulvalene condensed with 1,3-dithiol-2-ylidene moieties. *Tetrahedron Lett.* **1992**, *33*, 4321–4324. [CrossRef]
153. Heynderick, A.; Kaou, A.M.; Moustrou, C.; Samat, A.; Guglielmetti, R. Synthesis and photochromic behaviour of new Dipyrrolylperfluorocyclopentenes. *New J. Chem.* **2003**, *27*, 1425–1432. [CrossRef]
154. Seebach, D.; Bürstinghaus, R. S-methyl thiocarboxylates from aldehydes and ketones through ketene thioacetals. Reductive nucleophile thiocarboxylation. *Synthesis* **1975**, 461–462. [CrossRef]
155. Theil, F.; Costisella, B.; Mahrwald, R.; Gross, H.; Schick, H.; Schwarz, S. Prostaglandins and Prostaglandin Intermediates. XVII. Synthesis of the Main Metabolite of the $PGF_{2\alpha}$ Analogue Cloprostenol. *J. Prakt. Chem.* **1986**, *325*, 435–440. [CrossRef]
156. Theil, F.; Costisella, B.; Groß, H.; Schick, H.; Schwarz, S. A Three-step Procedure for the Conversion of γ-Lactones into δ-Lactones. *J. Chem. Soc. Perkin Trans. I* **1987**, 2469–2472. [CrossRef]
157. Zimmerman, H.E.; Baker, M.R.; Bottner, R.C.; Morrissey, M.M.; Murphy, S. Photochemistry of some allenic counterparts of cyclohexenones and 2, 5-cyclohexadienones. *J. Am. Chem. Soc.* **1993**, *115*, 459–466. [CrossRef]
158. Monenschein, H.; Brünjes, M.; Kirschning, A. Lithiated dimethoxymethyl diphenyl phosphine oxide. A versatile formiate carbanion equivalent. *Synlett* **2002**, 525–527. [CrossRef]
159. Okada, H.; Mori, T.; Saikawa, Y.; Nakata, M. Formation of α-hydroxyketones via irregular Wittig reaction. *Tetrahedron Lett.* **2009**, *50*, 1276–1278. [CrossRef]
160. Kirschning, A.; Dräger, G.; Yung, A. A new asymmetric formylation of aldehydes. *Angew. Chem. Int. Ed. Engl.* **1997**, *36*, 253–255. [CrossRef]
161. Becker, H.; Sharpless, B. A new ligand class for the asymmetric dihydroxylation of olefins. *Angew. Chem. Int. Ed. Engl.* **1996**, *35*, 448–451. [CrossRef]
162. Wittenberg, R.; Beier, C.; Dräger, G.; Jas, G.; Jasper, C.; Monenschein, H.; Kirschning, A. Towards the total synthesis of tonantzitlolone—Preparation of key fragments and the complete carbon backbone. *Tetrahedron Lett.* **2004**, *45*, 4457–4460. [CrossRef]
163. Morgalyuk, V.P.; Strelkova, T.V.; Nifant'ev, E.E. Synthesis of polyfunctionalized methylphosphine oxides. *Russ. Chem. Bull.* **2012**, *61*, 380–385. [CrossRef]
164. Schrader, T.; Kober, R.; Steglich, W. Synthese von 1-aminophosphonsäure-derivaten über acyliminophosphonsäure-ester. *Synthesis* **1986**, 372–375. [CrossRef]
165. Costisella, B.; Keitel, I.; Ozegowski, S. α-Substituierte Phosphonate. 69. Diastereoselektivität bei der Knüpfung der Phosphor-kohlenstoffbindung. *Phosphorus Sulfur Silicon Relat. Elem.* **1993**, *84*, 115–120. [CrossRef]
166. Marson, C.M. Reactions of carbonyl compounds with (monohalo) methyleniminium salts (Vilsmeier reagents). *Tetrahedron* **1992**, *48*, 3659–3726. [CrossRef]
167. Winter, A.; Risch, N. The vinylogous mannich reaction: An efficient access to substituted nicotinonitriles. *Synlett* **2003**, *13*, 1959–1964. [CrossRef]
168. Morgalyuk, V.P.; Petrovskii, P.V.; Lysenko, K.A.; Nifant'ev, E.E. Synthesis of polyfunctionalyzed methylphosphine oxides. *Russ. Chem. Bull.* **2009**, *58*, 248–250. [CrossRef]
169. Morgalyuk, V.P.; Strelkova, T.V. Stepwise scheme for the reaction of chlorodiphenylphosphine with N,N-dialkylformamides in the presence of NaI. *Russ. J. Gen. Chem.* **2011**, *81*, 2096–2101. [CrossRef]
170. Morgalyuk, V.P.; Strelkova, T.V.; Nifant'ev, E.E. Reactivity of Chloro(dialkylamino)(diphenylphosphinoyl)methanes. *Bull. Chem. Soc. Jpn.* **2012**, *85*, 93–100. [CrossRef]

171. Schrader, T.; Steglich, W. Phosphoranaloge von Aminosäuren IV. Synthesen ungevönlicher 1-Aminophosphonsäuren über Diels-Alder-Reaktionen von (*N*-Acyliminomethyl)phosphonsäurendiethylestern. *Synthesis* **1990**, 1153–1156. [CrossRef]
172. Boukallaba, K.; Elachqar, A.; El Hallaoui, A.; Alami, A.; El Hajji, S.; Labriti, B.; Atmani, A. Synthesis of α-Heterocyclic α-Aminophosphonates, Part II: Morpholine, Piperidine, Pyrrolidine, Tetrahydrofurylmethylamine, *N*-Benzyl-*N*-Methylamine, and Aniline Derivatives. *Phosphorus Sulfur Silicon Relat. Elem.* **2007**, *182*, 1045–1052. [CrossRef]
173. Elachqar, A.; El Hallaouiq, A.; Roumestant, M.L.; Viallefont, P. Synthesis of Heterocyclic α-Aminophosphonic Acids. *Synth. Commun.* **1994**, *24*, 1279–1286. [CrossRef]
174. Achamlale, S.; Mabrouk, H.; Elachqar, A.; El Hallaoui, A.; El Hajji, S.; Alami, A.; Bellan, J.; Mazières, M.R.; Wolf, J.G.; Pierrot, M. Synthesis and Thermal Isomerization of Carboxylic and Phosphonic α-Aminoesters Substituted With a Triazole Ring. *Phosphorus Sulfur Silicon Relat. Elem.* **2007**, *182*, 357–367. [CrossRef]
175. Gallagher, M.J.; Honegger, H. Organophosphorus intermediates. VI. The acid-catalysed reaction of trialkyl orthoformates with phosphinic acid. *Aust. J. Chem.* **1980**, *33*, 287–294. [CrossRef]
176. Fougère, C.; Guénin, E.; Hardouin, J.; Lecouve, M. Rapid and Efficient Synthesis of Unsymmetrical Phosphinic Acids R′P(O)OHR″. *Eur. J. Org. Chem.* **2009**, 6048–6054. [CrossRef]
177. Baylis, E.K. 1,1-Diethoxyethylphosphinates and Phosphinites. Intermediates for the Synthesis Functional Phosphorus Acids. *Tetrahedron Lett.* **1995**, *36*, 9385–9388.
178. Hall, R.G.; Riebli, P. Preparation of New Phosphine Oxides: Synthesis of an Analogue of Muscarinic Antagonists. *Synlett* **1999**, 1633–1635. [CrossRef]
179. Buckler, S.A.; Epstein, M. The preparation and reactions of primary phosphine oxides. *Tetrahedron* **1962**, *3*, 1221–1230. [CrossRef]
180. Zhang, D.; Yuan, C. A concise and first synthesis of α-aminophosphinates with two stereogenic atoms leading to optically pure α-amino-H-phosphinic Acids. *Chem. A Eur. J.* **2008**, *14*, 6049–6052.
181. Alstermark, C.; Amin, K.; Dinn, S.R.; Elebring, T.; Fjellström, O.; Fitzpatrick, K.; Geiss, W.B.; Gottfries, J.; Guzzo, P.R.; Harding, J.P.; *et al.* Synthesis and pharmacological evaluation of novel γ-aminobutyric acid type B (GABAB) receptor agonists as gastroesophageal reflux inhibitors. *J. Med. Chem.* **2008**, *51*, 4315–4320. [CrossRef]
182. Froestl, W.; Mickel, S.J.; von Sprecher, G.; Diel, P.J.; Hall, R.G.; Maier, L.; Strub, D.; Melillo, V.; Baumann, P.A.; Bernasconi, R.; *et al.* Phosphinic acid analogues of GABA. 2. Selective, orally active GABAB antagonists. *J. Med. Chem.* **1995**, *38*, 3313–3331. [CrossRef]
183. Seebach, D. Methoden und möglichkeiten der nucleophilen acylierung. *Angew. Chem. Int. Ed. Engl.* **1969**, *81*, 690–700. [CrossRef]
184. Seebach, D. Methods of reactivity umpolung. *Angew. Chem. Int. Ed. Engl.* **1979**, *18*, 39–336. [CrossRef]
185. Dodda, R.; Zhao, C.-G. Organocatalytic highly enantioselective synthesis of secondary α-hydroxyphosphonates. *Org. Lett.* **2006**, *8*, 4911–4914. [CrossRef]

8

Effect of Mono- and Poly-CH/P Exchange(s) on the Aromaticity of the Tropylium Ion

Ankita Puri and Raakhi Gupta *

Department of Chemistry, The IIS University, Jaipur 302020, India; ankita.puri@iisuniv.ac.in
* Correspondence: raakhi.gupta@iisuniv.ac.in

Academic Editor: György Keglevich

Abstract: In view of the fact that the phosphorus atom in its low co-ordination state (coordination numbers 1 and 2) has been termed as the carbon copy, there have been attempts to investigate, theoretically as well as experimentally, the effect of the exchange(s) of CH- moiety with phosphorus atom(s) (CH/P) on the structural and other aspects of the classical carbocyclic and heterocyclic systems. Tropylium ion is a well-known non-benzenoid aromatic system and has been studied extensively for its aromatic character. We have now investigated the effect of mono- and poly-CH/P exchange(s) on the aromaticity of the tropylium ion. For this purpose, the parameters based on the geometry and magnetic properties, namely bond equalization, aromatic stabilization energies (ASE), Nucleus-Independent Chemical Shift (NICS) values, (NICS(0), NICS(1), $NICS(1)_{zz}$), proton nucleus magnetic resonance (^{1}H-NMR) chemical shifts, magnetic susceptibility exaltation and magnetic anisotropic values of mono-, di-, tri- and tetra-phosphatropylium ions have been determined at the Density Functional Theory (DFT) (B3LYP/6-31+G(d)) level. Geometry optimization reveals bond length equalization. ASEs range from −46.3 kcal/mol to −6.2 kcal/mol in mono- and diphospha-analogues which are planar. However, the ions having three and four phosphorus atoms lose planarity and their ASE values approach the values typical for non-aromatic structures. Of the three NICS values, the $NICS(1)_{zz}$ is consistently negative showing aromatic character of all the systems studied. It is also supported by the magnetic susceptibility exaltations and magnetic anisotropic values. Furthermore, ^{1}H-NMR chemical shifts also fall in the aromatic region. The conclusion that mono-, di-, tri- and tetra-phosphatropylium ions are aromatic in nature has been further corroborated by determining the energy gap between the Highest Occupied Molecular Orbital (HOMO) and Lowest Unoccupied Molecular Orbital (LUMO) (HOMO − LUMO gap), which falls in the range, ca. 3×10^{-19}–9×10^{-19} J. The systems having more than four phosphorus atoms are not able to sustain their monocyclic structure.

Keywords: phosphatropylium ions; aromaticity; NICS; magnetic susceptibility exaltation; HOMO − LUMO gap

1. Introduction

The concept of aromaticity is of fundamental importance for explaining the structure, stability and reactivity of many molecules, due to which it continues to attract attention of many eminent theoretical chemists and forms the basis of many interesting monographs [1–3], research papers [4–9] and thematic issues [10–12]. Hückel's predictions based on the molecular orbital theory that conjugated monocyclic planar systems containing (4n + 2) π electrons should be aromatic, i.e., conjugatively stabilized [13], was modified by Platt [14], who broadened the scope of the postulation by including neutral as well as charged polycyclic systems in this category. The term "antiaromatic" was used in substitution of "pseudoaromatic" after 1965 by Breslow for those compounds that exhibited significant destabilization characteristics and possessed (4n) π conjugated electrons [15]. The possibility of verifying theoretical predictions with experimental isolations is limited to aromatic compounds only. Post-Hückel, there have been many attempts to propose newer and more reliable indicators to determine aromaticity/antiaromaticity of a particular system or a class of compounds. Schleyer et al. [16] endeavoured to divide various theoretical parameters into four categories, each based on a particular property:

Structure—tendency towards bond length equalization and planarity (if applicable).
Energy—enhanced stability.
Reactivity—lowered reactivity, electrophilic aromatic substitution (if applicable).
Magnetic properties—proton nucleus magnetic resonance (^{1}H-NMR) chemical shift, magnetic susceptibility exaltation and anisotropies, nucleus-independent chemical shift (NICS), ring current plots.

In a recent review [17], geometry-based various aromaticity indices have been discussed elegantly. Bond length equalization has been perceived as one of the important manifestations of aromaticity, but it cannot be used as the sole criterion to characterize it unambiguously, as there are examples like borazine, which has equalized bond length but is not aromatic. The quantitative indices have been developed by Francois and Julg [18] and Bird [19], but extension of these indicators to the heteroaromatic compounds is not straightforward.

The aromatic stabilization energies (ASEs) and the enhanced resonance energies (REs) have been recognized to be the keystone of aromaticity since long ago. However, these energy estimates vary significantly, depending strongly on the equations used and on the choice of the reference molecules. As such, the other effects like hyperconjugative, etc. are also not taken into consideration or are neglected while calculating the ASEs and REs, and, therefore, in some specific cases, large discrepancy is noted. Schleyer et al. [20] have reported improved results by calculating isomerization stabilization energies (ISE) instead of ASEs.

The most reliable and commonly used indicators for validating aromaticity of different systems including the transition states are the magnetic criteria. The exalted diamagnetic susceptibility has been used to quantify aromaticity ever since it was proposed by Pascal in 1910 [21]. ^{1}H-NMR chemical shifts are also used frequently for characterizing aromatic compounds [22]. The effect of the diatropic ring current induced in an aromatic molecule by an external magnetic field is much stronger inside the ring in comparison to outside. Schleyer et al. in 1996 proposed another indicator of aromaticity, namely NICS based on the magnetic shielding at the center of the aromatic ring caused by the induced diatropic current [23]. Subsequently, it was refined and magnetic shielding 1Å above the center of the aromatic ring (NICS(1)), the zz component of NICS(1) ($NICS(1)_{zz}$), and dissected NICS were considered to be better indicators [16].

An impressive analogy between P=C and C=C double bonds in the molecules having several PC double bonds was emphasized for the first time by Appel [24] while describing Cope-rearrangement of several diphospha- and tetraphospa-hexadienes. Thereafter, similar conjugative ability of the P=C and C=C double bonds was reiterated by analyzing π-ionization energies and ring-fragmentation/isodesmic energies of a series of heterocyclic conjugated systems containing σ^2,λ^3-P atoms [25–28]. The analogy turned out to be so prolific that besides a few research papers [29–33] on this theme, a monograph with the title, "Phosphorus: The Carbon Copy" [34] has been published. The effect of the exchange of the CH moiety with phosphorus atom (CH/P exchange) on the aromaticity of the carbocyclic rings has been discussed in two recent reviews [35,36], which include detailed experimental and theoretical discussion of the three- to six-membered rings having one or more phosphorus atoms. It was concluded that due to the small angle at the P atom (ca. 95 degrees), aromaticity is sustained on CH/P exchange in the small rings, but in large rings, it results in reduced aromatic character and planarity is disturbed.

On the basis of the recent studies carried out by our research group [37,38], it was demonstrated that mono- and poly-phospha analogues of carbocyclic cations and anions, namely, cyclopropenium, cyclopropenide, cyclopentadienium and cyclopentadienide, display aromatic or antiaromatic character comparable with that of the respective carbocyclic ion. However, the results are not uniform in the case of antiaromatic systems.

The aromatic character of the tropylium cation and its monophospha- and other hetero-analogues has been probed earlier at the hybrid functional level using various structural and magnetic descriptors [39–41]. The tropylium cation was found to have a planar structure with completely delocalized 6π-electrons. The CH/P exchange in the tropylium cation was accompanied by lowering of the aromaticity only marginally.

As no attempt has been made so far to study the effect of more than one CH/P exchanges on the aromaticity of the tropylium cation, the present study concerns the investigation of the effect of mono- and poly-CH/P exchange(s) and to compare the results obtained on the basis of geometry, energetic and magnetic criteria. For this purpose, we computed the optimized geometries and calculated ASEs, ^{1}H-NMR chemical shifts, NICS(0), NICS(1), NICS(1)$_{zz}$, magnetic susceptibility exaltation and magnetic susceptibility anisotropy values and the energy difference of the highest occupied molecular orbital and lowest unoccupied molecular orbital (HOMO − LUMO energy gap) for the tropylium ion and its mono-, di-, tri- and tetra-phospha analogues at the Density Functional Theory (DFT) (B3LYP/6-31+G(d)) level. As discussed later, phosphatropylium ions having five or more phosphorus atoms do not sustain their monocyclic structures. These results are presented here. The corresponding anions will be discussed in a separate communication.

2. Results and Discussion

The following model cations (Scheme 1) were computed at the B3LYP/6-31+G(d) level to study the effect of the CH/P exchange on the aromaticity of the tropylium ion.

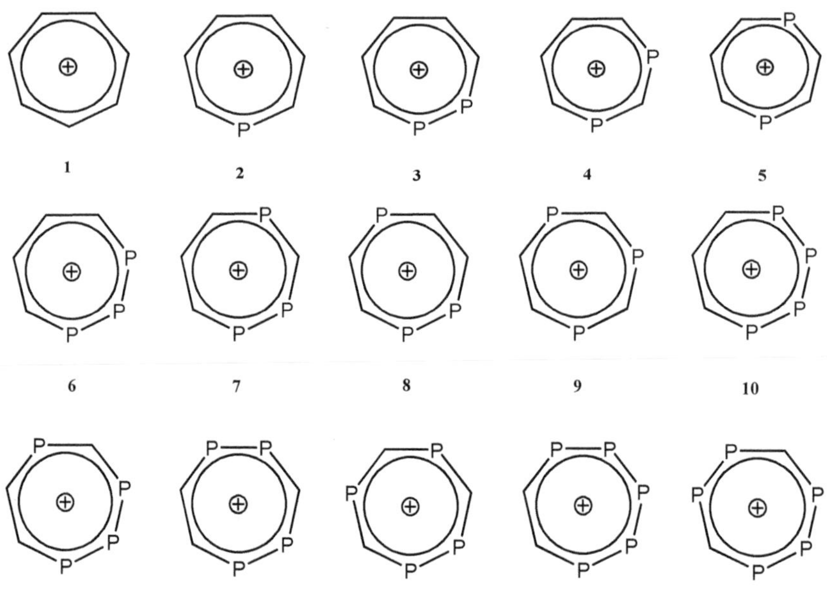

1 2 3 4 5 6 7 8 9 10 11 12 13 14 15

Scheme 1. Tropylium ion and its mono- and poly-phospha analogues.

2.1. Optimized Geometries

The Julg concept [18] is based on bond length equalization in the aromatic systems. However, it could not be extended to the heteroaromatic systems. To overcome this difficulty, the Bird index [19] was developed, but that too involved cumbersome calculations involving several aromaticity indices as constants. In the present study, our discussion is limited to the equalization or localization of the bonds without resorting to the calculation of the Julg or Bird indices.

The geometries of different systems optimized at the B3LYP/6-31+G(d) level along with bond lengths and Wiberg bond indices are reproduced in Figure 1.

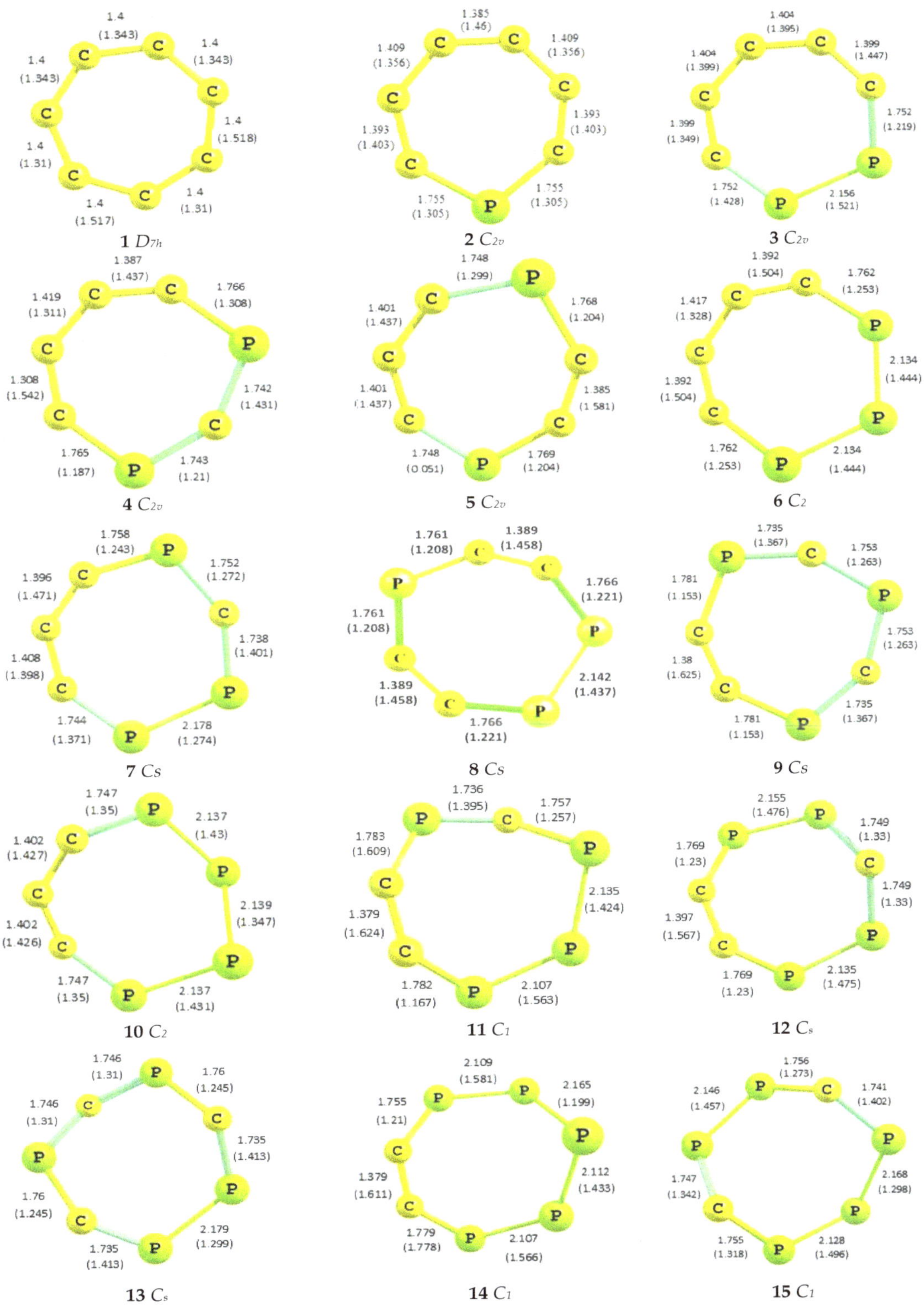

Figure 1. Optimized geometries of tropylium ion and its mono- and poly-phosha analogues at the B3LYP/6-31+G(d) level with bond distances (in Å) and Wiberg bond indices (in parenthesis). For better visualization, H atoms have been omitted.

The compounds **1–15** exhibit significant structural delocalization with CC, CP and PP bond lengths approaching the average of the respective double and single bond lengths of the prototype systems (Table 1).

Table 1. Average bond lengths (in Å) at the B3LYP/6-31+G(d) level.

Species	C-C (1.53 Å)	C=C (1.33 Å)	P-C (1.87 Å)	P=C (1.67 Å)	P-P (2.72 Å)	P=P (2.04 Å)
1	1.40	1.40	-	-	-	-
2	1.41	1.39	1.75	-	-	-
3	1.40	1.40	1.75	1.75	2.15	-
4	1.41	1.35	1.75	1.75	-	-
5	1.40	1.40	1.75	1.75	-	-
6	1.40	1.41	-	1.76	2.13	-
7	1.40	1.40	1.75	1.75	2.17	-
8	-	1.38	1.76	1.74	-	-
9	-	1.40	1.76	-	2.13	2.14
10	1.40	1.39	1.75	1.75	-	2.14
11	-	1.38	1.76	1.74	2.14	2.11
12	-	1.40	1.75	1.75	2.14	2.15
13	-	-	1.75	1.74	2.17	-
14	-	1.38	1.76	-	2.14	2.11
15	-	-	1.75	1.75	2.15	2.13

It is further noted that the planarity of the ring is not disturbed with one and two CH/P exchange(s) in the tropylium ion. Thus, systems **2–5** are planar with C_{2v} symmetry. It is in accordance with the earlier results, according to which incorporation of two σ^2, λ^3-P atoms in the five-membered ring reduces the ring strain and makes it planar even if a σ^3-P atom is also present in the α-position [42]. Likewise, pentaphosphole having a σ^3-P atom was found to be planar [43,44]. Experimentally also, the first aromatic 1*H*-1,2,4-triphosphole having a fully planar structure could be prepared [45]. Three or more CH/P exchanges in the tropylium cation, leading to **6**, **10**, **11**, **12**, **13**, **14** and **15** are accompanied by loss of planarity, with the exception of 1,2,4- (**7**) and 1,3,5-triphosphacycloheptatrienyl (**9**) cations, though this deviation is not much, as evident from the dihedral angles D_h C_a 30°.

As mentioned earlier, on making five or more CH/P exchanges in the tropylium ion, monocyclic structure is no more stable and it changes into bicyclic (**7P** atoms, **18**) or polycyclic (**5P** atoms, **16**; **6P** atoms, **17**) structures as shown in Figure 2.

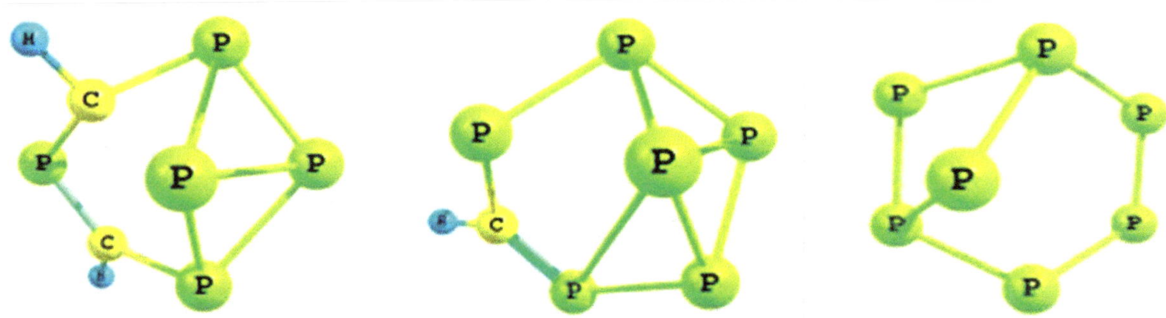

16 **17** **18**

Figure 2. 5P (**16**), 6P (**17**) and 7P (**18**) analogues of tropylium ion.

2.2. Frontier Molecular Orbitals

The tropylium cation (**1**) has doubly degenerate HOMOs, whereas, in the case of mono- and poly-phosphatropylium cations, the *p* orbital on the phosphorus atom(s) containing the lone pair constitutes the HOMO. As discussed later, the HOMO − LUMO gap in a molecule can be used as a measure of its aromatic character. In view of this, the frontier molecular orbitals (FMOs) of the tropylium cation (**1**) and two representative phosphatropylium cations (**2** and **3**) along with their energies are shown in Figure 3. It may be noted that, as compared to the tropylium cation, the HOMO − LUMO gap in the phosphatropylium cations is smaller.

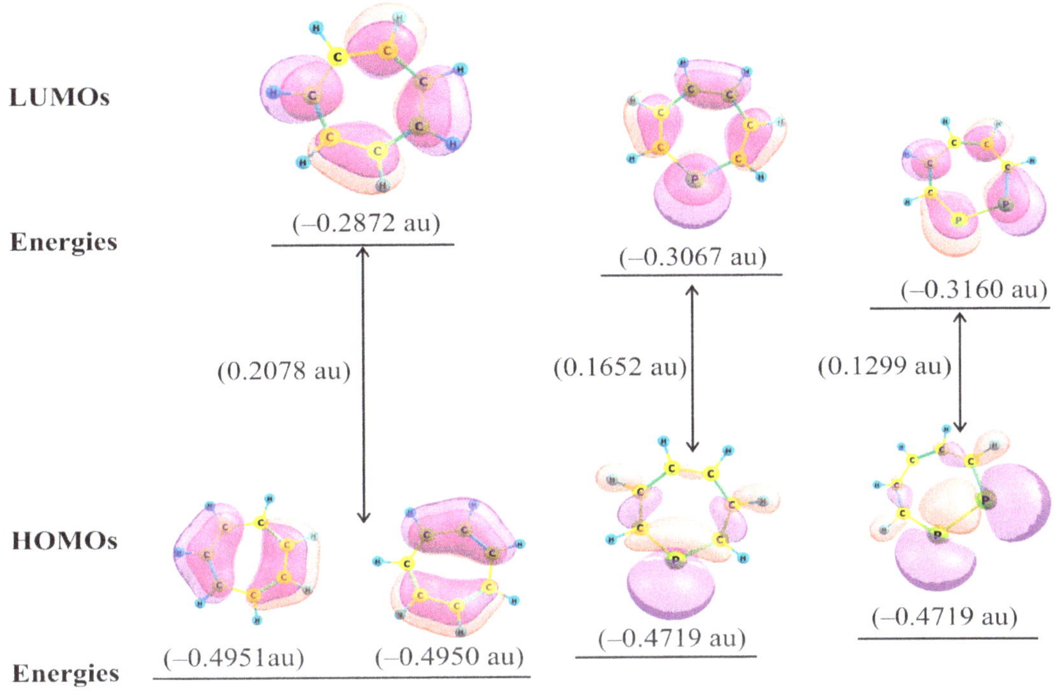

Figure 3. Frontier molecular orbitals (FMOs) of the tropylium cation (**1**); phosphatropylium cation (**2**) and 1,2-diphosphatropylium cation (**3**).

Further FMO analysis reveals that the HOMO-4 orbital of the species **1**, **2**, **3**, HOMO-5 orbital of **4**, **5**, **7**, **8** and HOMO-6 orbital of **9** are the delocalized orbitals of the π symmetry as in their non-phosha analogue i.e., compound **1** (Figure 4). This, in particular, indicates that the aromatic behaviour of these ions is comparable to that of the tropylium ion. However, in the case of the compounds **6**, **10**, **11**, **12**, **13**, **14** and **15**, loss of planarity of the ring reduces the interaction between some of the π orbitals, resulting in diminished delocalization of the π electrons.

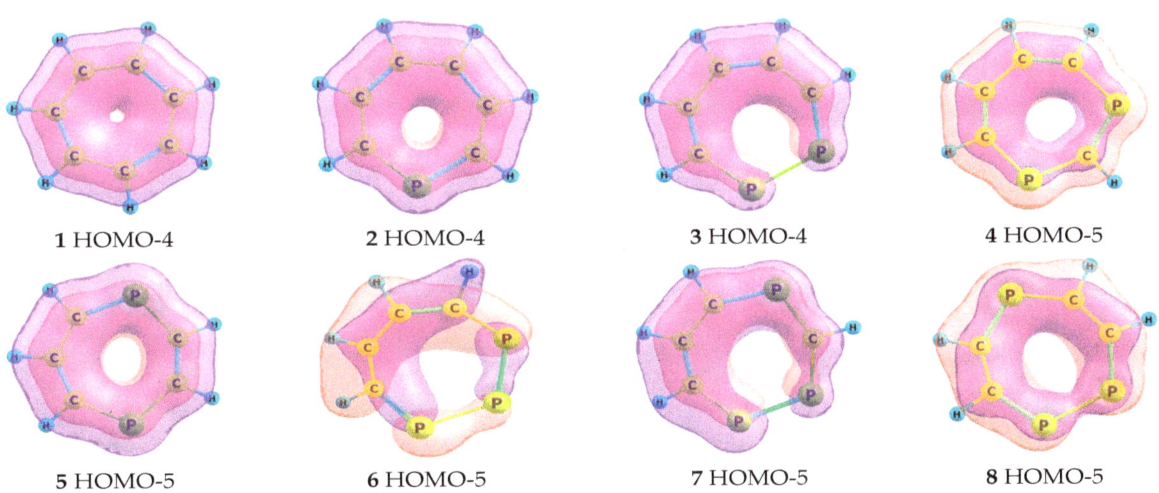

Figure 4. *Cont.*

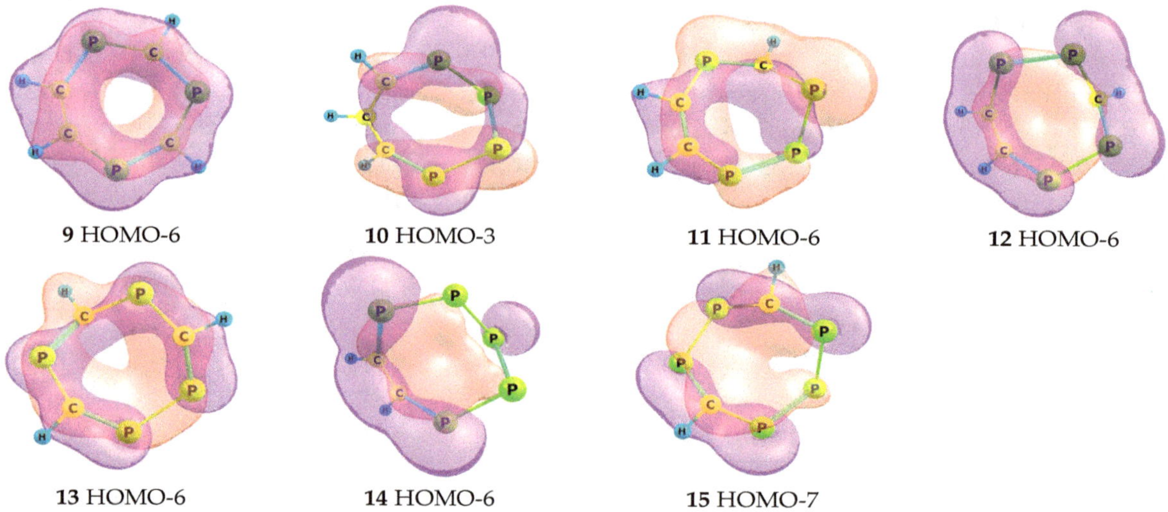

Figure 4. HOMOs of π-symmetry of the tropylium ion and its mono- and poly-phospha analogues at the density functional theory (DFT) (B3LYP/6-31+G(d)) level.

2.3. HOMO − LUMO Energy Gap

The correlation of the HOMO − LUMO energy gap with kinetic stability and chemical reactivity has been well established [46–54]. It can therefore be an important parameter to determine or explain aromaticity in molecules. A large HOMO − LUMO gap means high kinetic stability and low chemical reactivity associated with the aromatic behaviour. It can be well understood on the basis of the fact that it is energetically unfavourable to add electrons to a high-lying LUMO and to extract electrons from a low-lying HOMO to form the activated complex in a potential reaction [46].

This concept was for the first time used to rationalize aromatic character of the polycyclic aromatic hydrocarbons [55]. We extended application of this concept to the present study and have attempted to explain the aromatic behaviour of the tropylium cation and its phospha analogues on the basis of the HOMO − LUMO energy separation and chemical hardness. The chemical hardness has been calculated using Pearson formula (Equation (1)) [56,57]:

$$\eta = (\varepsilon\text{LUMO} - \varepsilon\text{HOMO}) / 2 \quad (1)$$

The values of the HOMO − LUMO energy gap ($\Delta\varepsilon$) and chemical hardness (η) are given in Table 2.

Table 2. Frontier molecular orbital energies and hardness of the tropylium ion and its mono- and poly-phospha analogues.

Species	HOMO (a.u.)	LUMO (a.u.)	LUMO − HOMO (a.u.)	$\Delta\varepsilon$ ($\times 10^{-19}$ J)	η ($\times 10^{-19}$ J)
1	−0.4951	−0.2872	0.2078	9.05	4.53
2	−0.4719	−0.3067	0.1652	7.19	3.60
3	−0.4459	−0.3160	0.1299	5.65	2.83
4	−0.4485	−0.3179	0.1306	5.65	2.84
5	−0.4650	−0.3176	0.1474	6.52	3.20
6	−0.4376	−0.3287	0.1089	4.74	2.37
7	−0.4372	−0.3259	0.1113	4.85	2.43
8	−0.4431	−0.3264	0.1167	5.09	2.55
9	−0.4384	−0.3258	0.1126	4.90	2.45
10	−0.4284	−0.3328	0.0956	4.16	2.08
11	−0.4332	−0.3341	0.0991	4.32	2.16
12	−0.4357	−0.3335	0.1022	4.45	2.27
13	−0.4331	−0.3261	0.1070	4.66	2.33
14	−0.4271	−0.3412	0.0859	3.75	1.87
15	−0.4272	−0.3394	0.0878	3.83	1.91

A low HOMO − LUMO energy gap (<2.08 × 10^{-19} J) was associated with the antiaromatic character of the dianions of the polycyclic hydrocarbons [58]. From these results, it was extrapolated that the aromatic systems would have a large HOMO - LUMO energy gap. It may be noted that the value of the HOMO − LUMO energy gap for **1** to **15** lies in the range of 3.75 × 10^{-19}–9.05 × 10^{-19} J, which indicates aromatic character of these systems. The decrease in HOMO − LUMO energy gap in the case of **6** and **10–15** can be attributed to the loss of planarity in these compounds.

It has been shown earlier that an excellent correlation exists between absolute hardness and aromaticity, both increasing almost parallel to each other [56]. In the present case, chemical hardness also varies parallel to the HOMO − LUMO energy gap vis-à-vis aromaticity as shown in Figure 5.

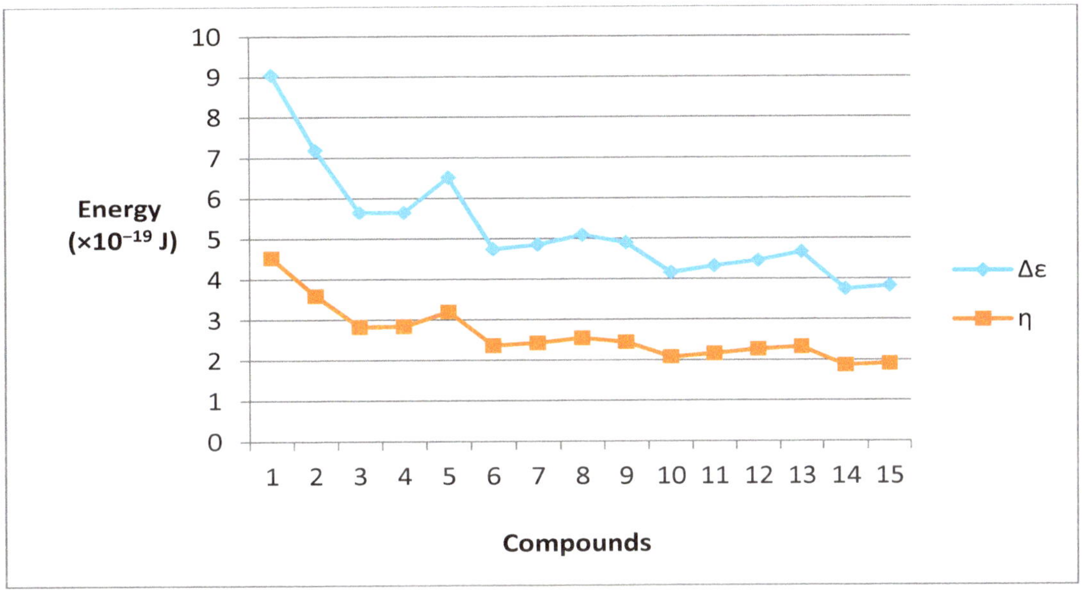

Figure 5. Variation of $\Delta_{HOMO-LUMO}$ and chemical hardness in tropylium cation and its mono- and poly-phospha analogues.

2.4. ASE

The ASEs of different systems were computed using the following homodesmotic reactions (Scheme 2).

The ASEs of the tropylium ion (**1**) and phosphatropylium ion (**2**) are in conformity with their aromatic character. However, on further CH/P exchange(s), the systems lose planarity as a result of which, ASEs do not change consistently. It is obvious that for the phosphatropylium ions having two or more phosphorus atoms, ASE is no more a reliable indicator of the aromaticity.

2.5. Magnetic Criteria

It is well known that a diatropic current is induced in an aromatic system when it is placed in an external magnetic field. This generates a secondary magnetic field causing a deshielding effect at the periphery and a shielding effect inside the ring. Based on this effect, three indicators, namely ^{1}H-NMR chemicals shifts [22], NICS [23] and magnetic susceptibility exaltation [59,60] have been used extensively to understand aromaticity. A closely related parameter, magnetic susceptibility anisotropy has also been used to validate aromatic character of different compounds. The aromatic character of the species **1–15** is discussed on the basis of all these criteria.

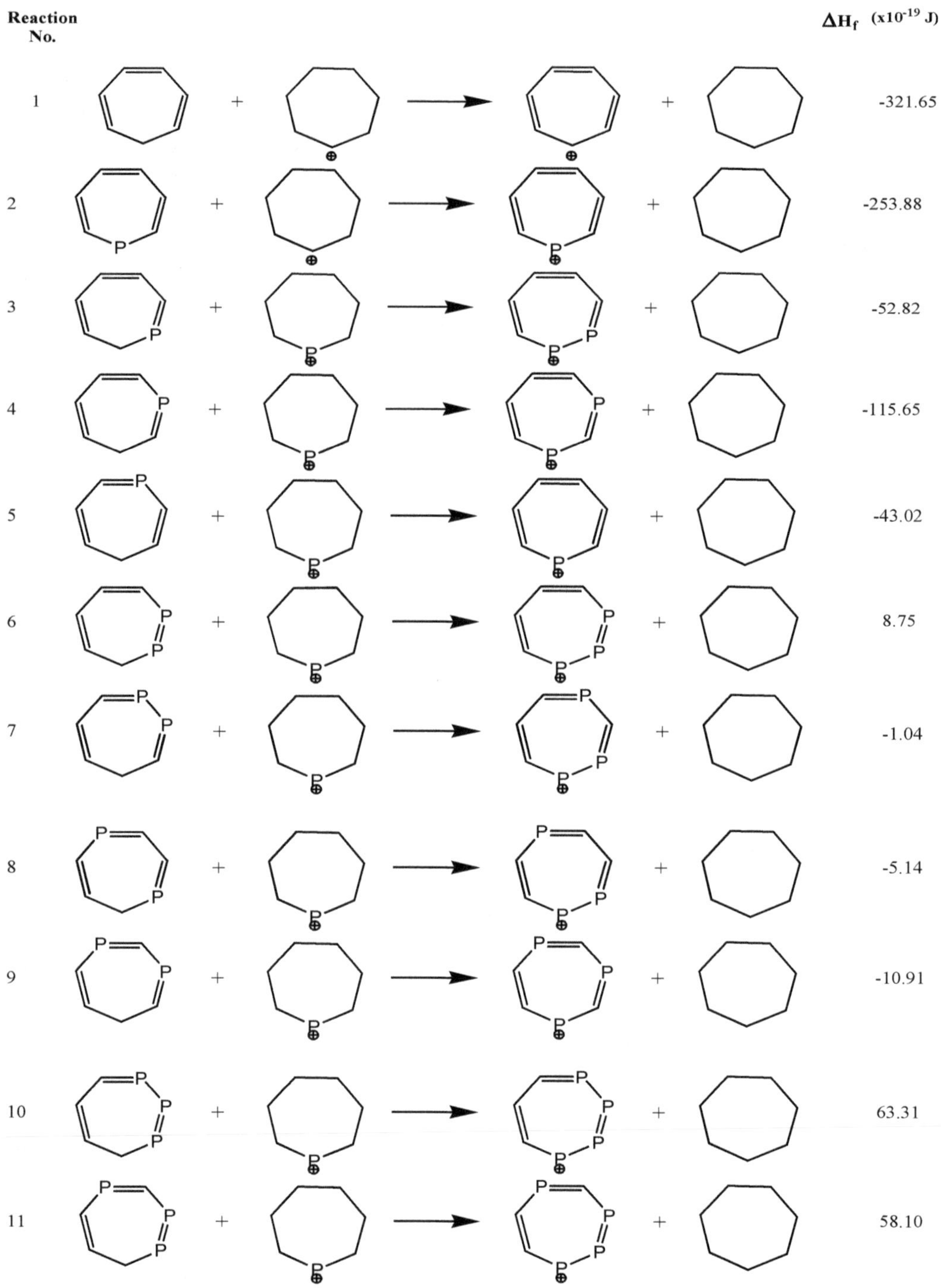

Scheme 2. Homodesmotic reactions for aromatic stabilization energy (ASE) determination.

2.5.1. ^{1}H-NMR Chemical Shifts

The most common measure of the ring current is the ^{1}H-NMR chemical shift, although Schleyer et al. reminded the chemical community of the limitations of proton shifts in the measurement of a ring current [16]. The chemical shift of a proton is the difference of the magnetic shielding of a reference (Tetramethylsilane in the present case) and the proton in question calculated at the same level of theory (Equation (2)):

$$\delta 1H = \sigma TMS - \sigma 1H \text{ ppm} \quad (2)$$

The ^{1}H-NMR chemical shift values of different species calculated are given in Table 3. It is noted that ^{1}H-NMR chemical shifts of protons in all of the species **1–15**, fall in the aromatic region, confirming their aromatic character. In tropylium ion (**1**), all protons are equivalent and the calculated chemical shift value is δ 9.3. It is noteworthy that on successive CH/P exchanges, the NMR shift values of the protons adjacent to the phosphorus atom(s) appear further downfield, which may be attributed to the stronger electron-acceptor character of the P=C- moiety.

Table 3. ^{1}H-NMR chemical shifts, nucleus-independent chemical shift (NICS), magnetic susceptibility exaltation, magnetic anisotropy and aromatic stabilization energy ASE values of tropylium ion and its phospha analogues.

Species	^{1}H-NMR Chem. Shift (δ)		NICS(0)	NICS(1)	NICS(1)$_{zz}$	Λ	χ_{anis}	ASE ($\times 10^{-19}$J)
1	H1	9.39	−6.30	−9.52	−26.42	−20.72	−79.89	−321.65
	H2	9.34						
	H3	9.38						
	H4	9.38						
	H5	9.35						
	H6	9.34						
	H7	9.36						
2	H2	11.14	−5.28	−8.72	−23.78	−34.05	−102.65	−253.88
	H3	9.39						
	H4	9.24						
	H5	9.24						
	H7	11.14						
	H6	9.39						
3	H3	11.53	−4.32	−7.78	−21.73	−32.72	−116.79	−52.82
	H4	9.31						
	H5	9.37						
	H6	9.31						
	H7	11.53						
4	H2	13.16	−4.19	−7.78	−20.70	−35.95	−105.56	−115.65
	H4	10.86						
	H5	9.18						
	H7	10.86						
	H6	9.15						
5	H2	11.11	−4.34	−7.92	−21.31	−36.66	−109.68	−43.02
	H3	11.11						
	H5	11.18						
	H6	9.24						
	H7	11.18						
6	H4	10.98	−4.69	−8.00	−21.26	−39.93	−121.50	8.75
	H5	9.36						
	H6	9.36						
	H7	10.98						
7	H3	13.59	−3.26	−6.97	−18.62	−33.49	−119.21	−1.04
	H5	11.19						
	H6	9.09						
	H7	11.56						
8	H3	11.27	−3.59	−7.31	−19.91	−35.64	−121.98	−5.14
	H4	11.18						
	H6	11.18						
	H7	11.27						
9	H2	13.02	−2.99	−6.75	−17.54	−37.45	−107.94	−10.91
	H4	13.02						
	H6	10.78						
	H7	10.78						
10	H5	10.84	−5.06	−8.06	−20.06	−44.93	−115.3	63.31
	H6	9.12						
	H7	10.84						
11	H4	12.73	−4.68	−8.13	−20.46	−45.94	−109.43	58.10
	H6	10.61						
	H7	10.45						
12	H3	12.31	−5.18	−8.96	-23.86	−47.65	−114.03	89.09
	H6	11.03						
	H7	11.03						
13	H3	13.06	−2.70	−6.99	−17.83	−38.43	−105.24	45.17
	H5	12.74						
	H7	13.06						
14	H7	11.06	−7.83	−9.79	−22.83	−54.31	−107.81	93.69
	H6	10.12						
15	H3	12.63	−3.25	−7.54	−19.96	−42.88	−105.24	173.95
		H7	11.54					

2.5.2. NICS Values

Among the widely used magnetic criteria for aromaticity and anti-aromaticity, NICS proposed by Schleyer and co-workers in 1996 is a simple efficient probe [23]. Subsequently, there have been many refinements and at present $NICS(1)_{zz}$ value is considered to be the most reliable indicator of aromaticity [61]. NICS(0) is defined as the negative value of the absolute shielding computed at the ring centre determined by the average of the heavy atoms coordinates in the ring. NICS(1) is the negative value of the absolute shielding measured 1 Å above the center of the ring, while $NICS(1)_{zz}$ is the out-of-plane component of the absolute shielding estimated in the same position as NICS(1). Rings with highly negative values of NICS are qualified as aromatic by definition, whereas those with positive values are anti-aromatic.

The calculated NICS(0), NICS(1) and $NICS(1)_{zz}$ tensor component values for all species are given in Table 3.

It is noteworthy that these values are negative ranging from −17.54 to −26.42, confirming the aromatic character of the species **1** to **15**.

2.5.3. Magnetic Susceptibility Exaltation

The magnetic susceptibility exaltation (Λ) evaluates the effect of a ring current by comparing the bulk magnetic susceptibility (χ) to the susceptibility of a localized ring system [59,60].

The magnetic susceptibility exaltation with CSGT (continuous set gauge transformations) method values is determined by subtracting the sum of the magnetic susceptibilities of the fragments of cyclic electron delocalization from the values of the respective species (Equation (3)):

$$\Lambda = \chi M - \chi M' \tag{3}$$

The Λ values calculated for different species are given in Table 3. As expected, the exaltation values (Λ) of all these compounds are negative ranging from −20.7 to −54.3 revealing their aromatic character.

2.5.4. Magnetic Anisotropy

Magnetic anisotropy is defined as the difference between out-of-plane and the average in-plane diamagnetic susceptibilities for a ring lying in the (xy) plane [62,63] (Equation (4)):

$$\Delta\chi = \chi zz - (1/2)\,[\chi xx + \chi yy] \tag{4}$$

The values of magnetic anisotropies so calculated are given in Table 3. It may be noted that these values are negative confirming aromatic character of the species.

3. Computational Method and Models

All calculations were carried out using a Gaussian 03 suite of programmes [64]. A diffuse function was added to the heavy atoms and geometry optimization of all the systems was done in the gas phase at the B3LYP/6-31+G(d) level of theory. Frequency calculations were done at the same level to determine zero-point energies and to characterize the energy minima.

^{1}H-NMR chemical shifts were calculated at the GIAO-B3LYP/6-311++G(d,p)//B3LYP/6-31+G(d) level. NICS values were calculated at the (3,+1) ring critical point of the electron density topology as defined by Bader [65] at the GIAO-B3LYP/6-311++G(d,p)//B3LYP/6-31+G(d) level. Magnetic susceptibilities were calculated at the csgt-B3LYP/6-31+G(d)//B3LYP/6-31+G(d) level and exaltation values were determined by subtracting the sum of the magnetic susceptibilities of the fragments (as explained in the supporting information) calculated at the same level from the value of the respective species.

4. Conclusions

On CH/P- exchange in the tropylium ion, aromaticity is sustained up to four carbon atoms. However, with further exchange(s), the system does not remain monocyclic, and it changes to bicyclic and polycyclic systems. Of the various indicators for the aromaticity, the ones based on the magnetic field effect, namely ^{1}H-NMR chemical shifts, NICS$(1)_{zz}$, magnetic susceptibility exaltation and magnetic anisotropies are more consistent and confirm the aromatic character of these species.

Acknowledgments: We are thankful to the authorities of The IIS University, Jaipur (India) for providing us with the computational facilities. Thanks are also due to Raj Kumar Bansal for helpful discussions. We would like to express our gratitude to the anonymous reviewer for making very constructive suggestions, thus contributing indirectly to the improvement of the paper.

Author Contributions: The research problem was conceptualized by the corresponding author (Raakhi Gupta), and the computational work was carried out by Ankita Puri.

References

1. Garratt, P.J. *Aromaticity*; Wiley-Inc.: Hoboken, NJ, USA, 1986.
2. Minkin, V.J.; Glukhovtsev, M.N.; Simkin, B.Y. *Aromaticity and Antiaromaticity; Electronic and Structural Aspects*; John Wiley & Sons, Inc.: New York, NY, USA, 1994.
3. Bergmann, E.D.; Pullman, B. Aromaticity, Pseudo-Aromaticity, Anti-Aromaticity, Proceedings of Jerusalem Symposium on Quantum Chemistry and Biochemistry, Jerusalem, Israel, 31 March–3 April 1970; Academic Press Inc.: Jerusalem, Israel, 1971.
4. Cyranski, M.K.; Krygowski, T.M.; Katritzky, A.R.; Schleyer, P.v.R. To what extent can aromaticity be defined uniquely. *J. Org. Chem.* **2002**, *67*, 1333–1338. [CrossRef] [PubMed]
5. Schleyer, P.v.R.; Jiao, H. What is aromaticity? *Pure Appl. Chem.* **1996**, *68*, 209–218. [CrossRef]
6. Schleyer, P.v.R.; Freeman, P.K.; Jiao, H.; Goldfuss, B. Aromaticity and Antiaromaticity in Five-Membered C_4H_4X Ring Systems: Classical and Magnetic Concepts May Not Be Orthogonal. *Angew. Chem. Int. Ed. Engl.* **1995**, *34*, 337–340. [CrossRef]
7. Mills, N.S.; Benish, M. The Aromaticity/Antiaromaticity Continuum. I. Comparison of the aromaticity of the dianion and the antiaromaticity of the dication of tetrabenzo[5.5]fulvalene via magnetic measures. *J. Org. Chem.* **2006**, *71*, 2207–2213. [CrossRef] [PubMed]
8. Katritzky, A.R.; Karelson, M.; Wells, A.P. Aromaticity as a Quantitative Concept. 6. Aromaticity Variation with Molecular Environment. *J. Org. Chem.* **1996**, *61*, 1619–1623. [CrossRef] [PubMed]
9. Katritzky, A.R.; Karelson, M.; Sild, S.; Krygowski, T.M.; Jug, K. Aromaticity as a Quantitative Concept. 7. Aromaticity Reaffirmed as a Multidimensional Characteristic. *J. Org. Chem.* **1998**, *63*, 5228–5231. [CrossRef]
10. Thematic issue on Aromaticity. *Chem. Rev.* **2001**, *101*, 1115–1566.
11. Thematic issue on Delocalization of pi- and sigma. *Chem. Rev.* **2005**, *105*, 3433–3947.
12. Thematic issue on Challenges in Aromaticity: 150 years after Kekule's benzene. *Chem. Soc. Rev.* **2015**, *44*, 6397–6643.
13. Hückel, E. Grundzüge der Theorie Ungesättiger und Aromatischer Verbindungen. *Z. Elektrochem. Angew. Phys. Chem.* **1937**, *43*, 752–788.
14. Platt, J.R. The box model and electron densities in conjugated systems. *J. Chem. Phys.* **1954**, *22*, 1448–1455.
15. Breslow, R. Aromatic character. *Chem. Eng. News* **1965**, *43*, 90–100. [CrossRef]
16. Schleyer, P.v.R.; Chen, Z.; Wannere, C.S.; Corminboeuf, C.; Puchta, R. Nucleus-independent chemical shifts (NICS) as an aromaticity criterion. *Chem. Rev.* **2005**, *105*, 3842–3888.
17. Krygowski, T.M.; Szatylowicz, H.; Stasyuk, O.A.; Dominikowska, J.; Palusiak, M. Aromaticity from the viewpoint of molecular geometry: Application to planar systems. *Chem. Rev.* **2014**, *114*, 6383–6422. [CrossRef] [PubMed]
18. Julg, A.; Francois, P. Researches sur la géométrie de quelques hydrocarbures non-alternants: Son influence sur les énergies de transition, une nouvelle définition de l'aromaticité. *Theor. Chim. Acta* **1967**, *8*, 249–259. [CrossRef]
19. Bird, C.W. A new aromaticity index and its application to five-membered ring heterocycles. *Tetrahedron* **1985**, *41*, 1409–1414. [CrossRef]

20. Schleyer, P.v.R.; Pülhofer, F. Recommendations for the evaluation of aromatic stabilization energies. *Org. Lett.* **2002**, *4*, 2873–2876. [CrossRef] [PubMed]
21. Pascal, P. Magnetochemical Researches. *Ann. Chim. Phys.* **1910**, *19*, 5–70.
22. Mitchell, R.H. Measuring aromaticity by NMR. *Chem. Rev.* **2001**, *101*, 1301–1315. [CrossRef] [PubMed]
23. Schleyer, P.v.R.; Maerker, C.; Dransfeld, A.; Jiao, H.; Hommes, N.J.R.v.E. Nucleus-independent chemical shifts: A simple and efficient aromaticity probe. *J. Am. Chem. Soc.* **1996**, *118*, 6317–6318. [CrossRef]
24. Appel, R. pπ-Double bonds between phosphorus and carbon—A challenge. *Pure Appl. Chem.* **1987**, *59*, 977–982. [CrossRef]
25. Nyulászi, L.; Veszprémi, T.; Réffy, J. A new look at the similarities of the conjugative ability and reactivity of P=C and C=C double bonding. *J. Phys. Chem.* **1993**, *97*, 4011–4015. [CrossRef]
26. Nyulászi, L.; Várnai, P.; Krill, S.; Regitz, M. Aromaticity of thia- and selena-phospholes: A photoelectron spectroscopic and quantum chemical study. *J. Chem. Soc. Perkin Trans.* **1995**, *2*, 315–318. [CrossRef]
27. Nyulászi, L.; Várnai, P.; Veszprémi, T. About the aromaticity of five-membered heterocycles. *J. Mol. Struct. THEOCHEM* **1995**, *358*, 55–61. [CrossRef]
28. Cloke, F.G.N.; Hitchcock, P.B.; Nixon, J.F.; Wilson, D.J.; Tabellion, F.; Fischbeck, U.; Preuss, F.; Regitz, M. Synthetic, structural and theoretical studies on new aromatic 1,2,4-diazaphosphole ring systems: Crystal and molecular structure of $P_2C_2Bu^t{}_2NPh$. *Chem. Commun.* **1999**. [CrossRef]
29. Mathey, F. Expanding the analogy between phosphorus-carbon and carbon-carbon double bonds. *Acc. Chem. Res.* **1992**, *25*, 90–96. [CrossRef]
30. Schmidpeter, A. Molecules that we made: An essay on phosphorus chemistry. *Heteroat. Chem.* **1999**, *10*, 529–537. [CrossRef]
31. Bansal, R.K.; Heinicke, J. Anellated heterophospholes and phospholides and analogies with related non-phosphorus systems. *Chem. Rev.* **2001**, *101*, 3549–3578. [CrossRef] [PubMed]
32. Simpson, M.C.; Protasiewicz, J.D. Phosphorus as a carbon copy and as a photocopy: New conjugated materials featuring multiply bonded phosphorus. *Pure Appl. Chem.* **2013**, *85*, 801–815. [CrossRef]
33. Bansal, R.K.; Gupta, R.; Maheshwari, P.; Kour, M. Analogy of phosphaalkenes and azaphospholes with their respective non-phosphorus analogues. *Curr. Org. Chem.* **2016**, *20*, 2099–2108. [CrossRef]
34. Dillon, K.B.; Mathey, F.; Nixon, J.F. *Phosphorus: The Carbon Copy*; Wiley: New York, NY, USA, 1988; pp. 203–226.
35. Nyulászi, L. Aromaticity of phosphorus heterocycles. *Chem. Rev.* **2001**, *101*, 1229–1246. [CrossRef] [PubMed]
36. Nyulászi, L.; Benkő, Z. Aromatic phosphorus heterocycles. In *Aromaticity in Heterocyclic Compounds*; Krygowski, T.M., Cyranski, M.K., Eds.; Springer-Verlag: Berlin/Heidelberg, Germany, 2009; Volume 19, pp. 27–81.
37. Gupta, R.; Bansal, R.K. Aromaticity/antiaromaticity of phospha-analogues of carbocyclic ions: A DFT investigation. *Comp. Theor. Chem.* **2016**, *1076*, 1–10. [CrossRef]
38. Gupta, R.; Maheshwari, P.; Kour, M. Reinvestigation of the aromaticity of mono- and polyphosphacyclopentadienide ions at the DFT level. *Comp. Theor. Chem.* **2015**, *1060*, 10–16. [CrossRef]
39. Ghiasi, R. Theoretical study of classical isomers tropylium, azatropylium, phosphatropylium, and arsatropylium cations: Structure, properties and aromaticity. *Main Group Chem.* **2008**, *7*, 147–154. [CrossRef]
40. Lin, L.; Lievens, P.; Nguyen, M.T. In search of aromatic seven-membered rings. *J. Mol. Struct.* **2010**, *943*, 23–31. [CrossRef]
41. Firouzi, R.; Ardani, S.S. Description of heteroaromaticity on the basis of π-electron density anisotropy. *Phys. Chem. Chem. Phys.* **2014**, *16*, 11538–11548. [CrossRef] [PubMed]
42. Nyulászi, L. Toward a planar σ^3–phosphorus. *J. Phys. Chem.* **1996**, *100*, 6194–6198. [CrossRef]
43. Nyulászi, L. Pentaphosphole: An aromatic ring with a planar σ^3–phosphorus. *Inorg. Chem.* **1996**, *35*, 4690–4693. [CrossRef]
44. Glukhovtsev, M.; Dransfeld, A.; Schleyer, P.v.R. Why pentaphosphole, P_5H, is planar in contrast to phosphole, $(CH)_4PH$. *J. Phys. Chem.* **1996**, *100*, 13447–13454. [CrossRef]
45. Cloke, F.G.N.; Hitchcock, P.B.; Hunnable, P.; Nixon, J.F.; Nyulászi, L.; Niecke, E.; Thelen, V. The First Delocalized Phosphole Containing a Planar Tricoordinate Phosphorus Atom: 1-[Bis (trimethylsilyl) methyl]-3, 5-bis (trimethylsilyl)-1,2,4-triphosphole. *Angew. Chem. Int. Ed.* **1998**, *37*, 1083–1086. [CrossRef]
46. Manolopoulos, D.E.; May, J.C.; Down, S.E. Theoretical studies of the fullerenes: C34 to C70. *Chem. Phys. Lett.* **1991**, *181*, 105–111. [CrossRef]

47. Haddon, R.C.; Fukunaga, T. Unified theory of the thermodynamic and kinetic criteria of aromatic character in the [4n + 2] annulenes. *Tetrahedron Lett.* **1980**, *21*, 1191–1192. [CrossRef]
48. Schmalz, T.G.; Seitz, W.A.; Klein, D.J.; Hite, G.E. Elemental carbon cages. *J. Am. Chem. Soc.* **1988**, *110*, 1113–1127. [CrossRef]
49. Zhou, Z.; Parr, R.G.; Garst, J.F. Absolute hardness as a measure of aromaticity. *Tetrahedron Lett.* **1988**, *29*, 4843–4846. [CrossRef]
50. Zhou, Z.; Parr, R.G. New measures of aromaticity: Absolute hardness and relative hardness. *J. Am. Chem. Soc.* **1989**, *111*, 7371–7379. [CrossRef]
51. Zhou, Z.; Parr, R.G. Activation hardness: New index for describing the orientation of electrophilic aromatic substitution. *J. Am. Chem. Soc.* **1990**, *112*, 5720–5724. [CrossRef]
52. Liu, X.; Schmalz, T.G.; Klein, D.J. Favorable structures for higher fullerenes. *Chem. Phys. Lett.* **1992**, *188*, 550–554. [CrossRef]
53. Parr, R.G.; Zhou, Z. Absolute hardness: Unifying concept for identifying shells and subshells in nuclei, atoms, molecules, and metallic clusters. *Acc. Chem. Res.* **1993**, *26*, 256–258. [CrossRef]
54. Aihara, J.; Oe, S.; Yoshida, M.; Osawa, E.J. Further test of the isolated pentagon rule: Thermodynamic and kinetic stabilities of C84 fullerene isomers. *Comput. Chem.* **1996**, *17*, 1387–1394. [CrossRef]
55. Aihara, J. Circuit resonance energy: A key quantity that links energetic and magnetic criteria of aromaticity. *J. Am. Chem. Soc.* **2006**, *128*, 2873–2879. [CrossRef] [PubMed]
56. Pearson, R.G. Absolute electronegativity and hardness correlated with molecular orbital theory. *Proc. Natl. Acad. Sci. USA* **1986**, *83*, 8440–8441. [CrossRef] [PubMed]
57. Pearson, R.G. Recent Advances in the Concept of Hard and Soft Acids and Bases. *J. Chem. Educ.* **1987**, *64*, 561–567. [CrossRef]
58. Minsky, A.; Meyer, A.Y.; Rabinovitz, M. Paratropicity and antiaromaticity: Role of the homo-lumo energy gap. *Tetrahedron* **1985**, *41*, 785–791. [CrossRef]
59. Dauben, H.J.; Wilson, J.D.; Laity, J.L. Diamagnetic susceptibility exaltation as a criterion of aromaticity. *J. Am. Chem. Soc.* **1968**, *90*, 811–813. [CrossRef]
60. Dauben, H.J.; Wilson, J.D.; Laity, J.L. Diamagnetic susceptibility in hydrocarbons. *J. Am. Chem. Soc.* **1969**, *91*, 1991–1998. [CrossRef]
61. Schleyer, P.v.R.; Jiao, H.J.; Hommes, N.J.R.v.E.; Malkin, V.G.; Malkina, O.L. An evaluation of the aromaticity of inorganic rings: Refined evidence from magnetic properties. *J. Am. Chem. Soc.* **1997**, *119*, 12669–12670. [CrossRef]
62. Benson, R.C.; Flygare, W.H. Molecular Zeeman effect of cyclopentadiene and isoprene and comparison of the magnetic susceptibility anisotropies. *J. Am. Chem. Soc.* **1970**, *92*, 7523–7529.
63. Schmalz, T.G.; Norris, C.L.; Flygare, W.H. Localized magnetic susceptibility anisotropies. *J. Am. Chem. Soc.* **1973**, *95*, 7961–7967. [CrossRef]
64. Frisch, M.J.; Trucks, G.W.; Schlegel, H.B.; Scuseria, G.E.; Robb, M.A.; Cheeseman, J.R.; Montgomery, J.A., Jr.; Vreven, T.; Kudin, K.N.; Burant, J.C.; et al. *Gaussian 03*, Revision B.05. Gaussian, Inc.: Wallingford, CT, USA, 2003.
65. Bader, R.F.W. *Atoms in Molecules: A Quantum Theory*; Clarendon Press: Oxford, UK, 1990.

9

Continuous Flow Alcoholysis of Dialkyl *H*-Phosphonates with Aliphatic Alcohols

Erika Bálint *, Ádám Tajti, Nóra Tóth and György Keglevich *

Department of Organic Chemistry and Technology, Budapest University of Technology and Economics, 1521 Budapest, Hungary; tajti.adam@mail.bme.hu (Á.T.); toth.nora@mail.bme.hu (N.T.)

* Correspondence: ebalint@mail.bme.hu (E.B.); gkeglevich@mail.bme.hu (G.K.)

Abstract: The continuous flow alcoholysis of dialkyl *H*-phosphonates by aliphatic alcohols in the absence of a catalyst was elaborated using a microwave (MW) reactor equipped with a flow cell. By the precise control of the reaction conditions, the synthesis could be fine-tuned towards dialkyl *H*-phosphonates with two different and with two identical alkyl groups. In contrast to the "traditional" batch alcoholysis, flow approaches required shorter reaction times, and the products became available at a larger scale.

Keywords: dialkyl *H*-phosphonates; alcoholysis; transesterification; microwave; continuous flow reactor

1. Introduction

Flow chemistry induced an inevitable revolution in the field of chemical transformations [1]. In modern chemistry, reactions have to be strictly controlled, which requires more and more precise synthetic techniques. In contrast to the "traditional" batch reactions, flow approaches create a significantly different processing environment, which enables a more efficient control of the reaction conditions (such as the reaction time or temperature). Due to this, flow protocols may be significantly better with respect to purity, selectivity and yields, as compared to batch approaches [2].

In microwave (MW) chemistry, the continuous flow technique represents a special importance. Although applications of the MW technique proved to be useful in many chemical transformations, the scale-up of MW-assisted reactions means a challenge due to the limited geometry of the MW devices [3–6]. One possibility to solve this problem is the use of continuous flow MW reactors, where the reaction mixture flows through an irradiated flow cell. The MW unit can be of a similar size used in batch mode. During the last decade, the advantages of the applications of continuous flow MW reactors were reported in certain transformations; however, in most cases, the usefulness of the occasional devices may be questionable, as non-professional MW reactors do not allow reproductions, and what is more important, the temperatures were not reported [7,8].

Esters represent a fundamental family among organic compounds. Esters of carboxylic acids may be important intermediates, solvents or products in organic chemistry [9].

The most common preparations of simple carboxylic esters comprise the acid-catalyzed reaction of a carboxylic acid with alcohol (Fischer esterification) and the reaction of an ester with alcohol (alcoholysis) [10]. Continuous flow Fischer esterifications may be carried out in systems containing a packed catalyst bed [11–20] or a heated coil [21,22], as well as in microreactors [23]. In a few cases, the esterifications were performed in continuous flow MW systems based on professional MW reactors [24–26] or in household MW ovens [27,28]. The continuous flow alcoholysis of carboxylic esters may also be performed in reactors equipped with a catalyst bed [29] or a heated coil [21,30]. The alcoholysis is also of great importance in biodiesel production, the area of which was recently summarized by Lee and co-workers [31].

Organophosphorus esters are also of great importance [32]. Dialkyl esters of phosphorous acid (dialkyl *H*-phosphonates) are widely applied building blocks in syntheses [33]. They are important starting materials of the Kabachnik–Fields condensations and the aza-Pudovik reactions [34] (towards α-aminophosphonates), the Pudovik reactions [35] (resulting in the formation of α-hydroxyphosphonates) and further organophosphorus transformations, such as the Hirao reaction [36] and the phospha-Michael addition [37]. Dialkyl *H*-phosphonates bearing different alkoxy groups on the phosphorus atom are valuable intermediates for P-chiral organophosphorus derivatives [38,39].

The industrial synthesis of dialkyl *H*-phosphonates is comprised of the reaction of phosphorus trichloride with alcohols (Scheme 1) [40–45]. Although this transformation is convenient, efficient and may be scaled up, it requires a solvent, and the liberating HCl decreases the atom efficiency. It is also a disadvantage that the synthesis of derivatives with different alkyl groups is not possible.

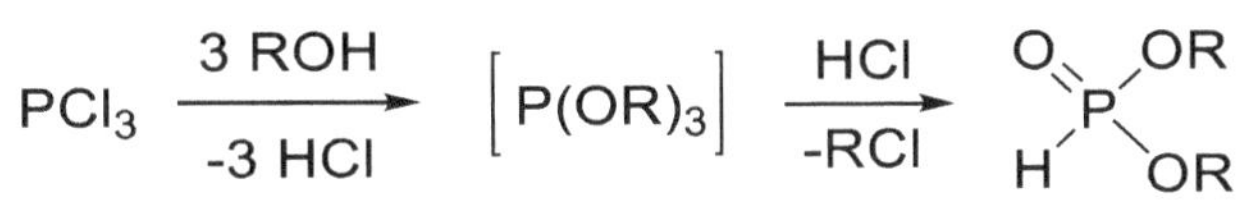

Scheme 1. Synthesis of dialkyl *H*-phosphonates from phosphorus trichloride.

Another possibility for the preparation of dialkyl *H*-phosphonates is alcoholysis (Scheme 2), which provides the products in good yields [46–54]. The only by-product is the leaving alcohol. The relatively high temperature and the need for the high excess of the alcohol mean disadvantages.

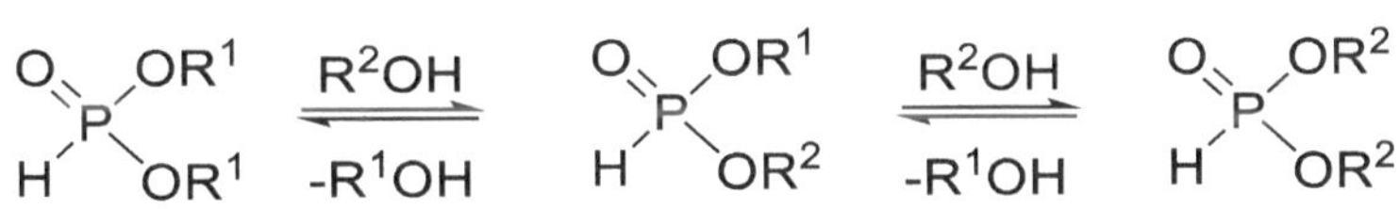

Scheme 2. Synthesis of dialkyl *H*-phosphonates by alcoholysis.

Besides the two main routes mentioned above, there are a few less important methods for the synthesis of dialkyl *H*-phosphonates. Such protocols are the oxidative reaction of elemental phosphorus with alcohols (Scheme 3a) [55–60] and the Ni-catalyzed oxidation of hypophosphorous esters (Scheme 3b) [61].

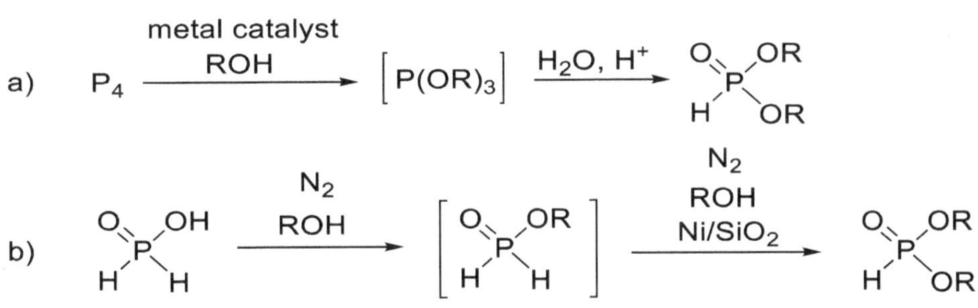

Scheme 3. Miscellaneous synthesis of dialkyl *H*-phosphonates.

The synthesis of dialkyl-*H*-phosphonates bearing different alkyl groups is not easy and requires special methods (Table 1).

Table 1. Synthetic methods for the preparation of dialkyl *H*-phosphonates bearing two different alkyl groups.

Entry	Reaction	Average Yield (%)	Flow Compatible	Ref.
1	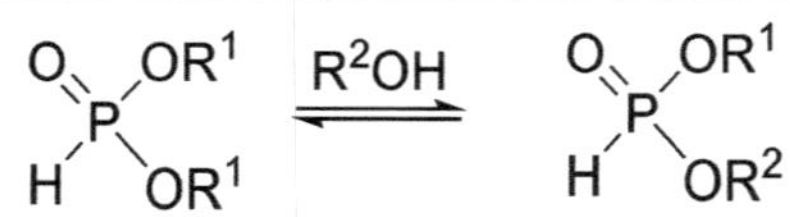	30–50	+	[46,50,52]
2	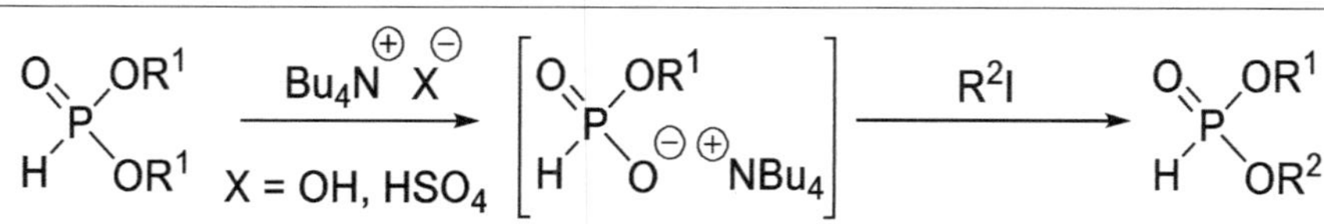	50–70	–	[62,63]
3	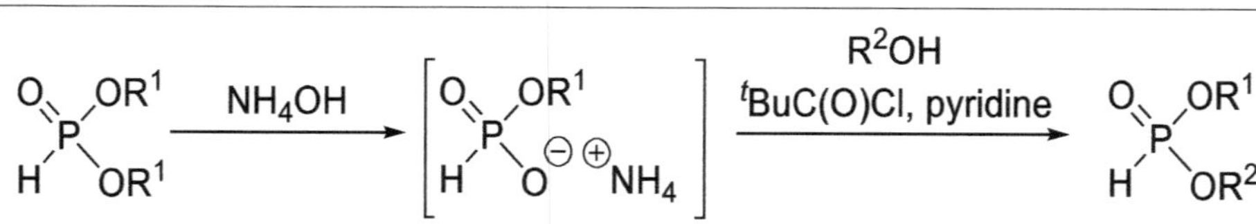	50–85	–	[64]
4	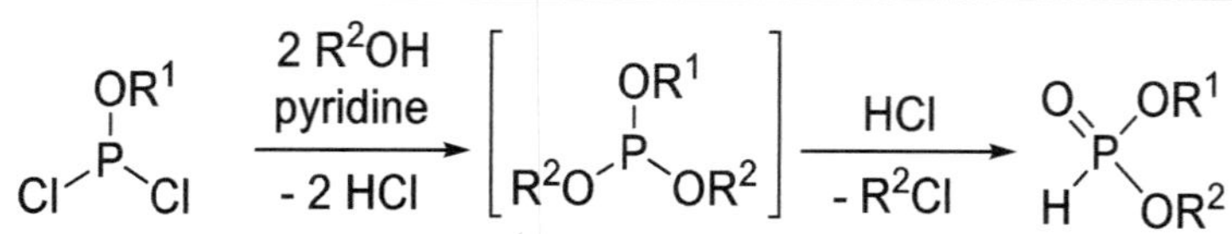	30–60	–	[65]

Phosphites with different alkyl groups may be synthesized by the partial alcoholysis of dialkyl *H*-phosphonates (Table 1, Entry 1). The reaction conditions must be controlled strictly to avoid complete transesterification.

Dialkyl *H*-phosphonates with two different alkyl groups may be obtained by the controlled O–C cleavage of dialkyl *H*-phosphonates (Table 2, Entries 2 and 3). In the first step of the first method, a phosphite-ammonium salt is prepared, which is then alkylated with alkyl iodides (Table 2, Entry 2), or with alcohols in the presence of pivaloyl chloride and pyridine (Table 2 entry 3) to furnish the target compounds selectively. A serious drawback of these transformations is the low atom efficiency.

Table 2. Alcoholysis of dialkyl *H*-phosphonates in a batch microwave (MW) reactor [33].

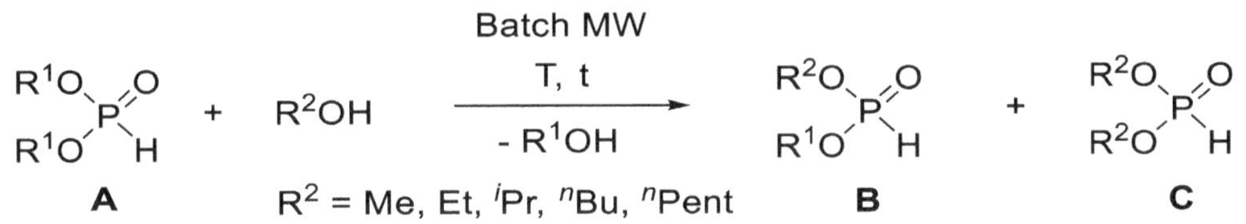

R^1 = Me, Et

Entry	R^1	R^2	R^2OH (equiv)	T (°C)	t (min)	Composition (%) [a]		
						A	B	C
1	Me	Et	25	100	120	22	**56**	22
2	Me	Et	50	175	40	0	4	**96**
3	Me	^{n}Bu	25	125	60	0	**40**	60
4	Me	^{n}Bu	50	150	60	0	0	**96**
5	Et	Me	25	125	120	49	**38**	13
6	Et	Me	50	175	40	0	21	**79**
7	Et	^{i}Pr	25	125	60	26	**57**	17
8	Et	^{i}Pr	50	175	40	0	6	**94**
9	Et	^{n}Bu	25	125	60	25	**54**	21
10	Et	^{n}Bu	50	175	40	0	2	**98**
11	Et	^{n}Pent	25	125	60	3	**52**	45
12	Et	^{n}Pent	50	175	40	0	8	**92**

[a] Analyzed by GC. Bold numbers indicate the target compounds in the reactions.

Dialkyl *H*-phosphonates bearing different alkoxy groups can also be obtained in the reaction of alkyl dichlorophosphites and alcohols. The primarily formed trialkyl phosphite is cleaved by a part of the hydrochloric acid formed (Table 2, Entry 4).

From these synthetic methods, the alcoholysis may be flow compatible. This solvolytic reaction may be carried out without any catalyst by fine-tuning the reaction conditions, to afford mixed or fully-transesterified dialkyl phosphites as the predominant component.

The alcoholysis of dialkyl *H*-phosphonates in a batch MW reactor was investigated by us previously (Table 2) [52]. The reaction of dialkyl *H*-phosphonates (**A**) with aliphatic alcohols under mild conditions (lower alcohol excess and lower temperature) afforded the derivatives with different alkyl groups (**B**) as the main products, while in the case of applying higher excess of alcohol and higher temperature, the fully-transesterified dialkyl *H*-phosphonates (**C**) predominated in the mixture.

In this paper, we aimed at elaborating an MW-assisted continuous flow method for the alcoholysis of dialkyl *H*-phosphonates. Our purpose was to find the optimum conditions for the formation of dialkyl *H*-phosphonates bearing different alkyl groups and for the fully-transesterified phosphites in a flow MW reactor. Furthermore, we wished to prepare new dialkyl *H*-phosphonates with two different alkyl groups, which may be valuable building blocks of chiral organophosphorus compounds.

2. Results and Discussion

In the first case, the alcoholysis of dimethyl *H*-phosphonate (DMP) (**1**) with *n*-butanol was investigated in the absence of any catalyst, in a CEM® (Matthews, NC, USA) MW reactor equipped with a commercially available CEM® continuous flow cell (Figures 1 and 2). The mixture of DMP (**1**) and a 25-fold excess of *n*-butanol was fed into the reactor by an HPLC pump at a flow rate of 0.1–1.4 mL/min (corresponding to residence times of 60–5 min, respectively). The temperature was monitored and controlled by an IR sensor. The mixture leaving the reactor was cooled down using a spiral-like cooler and was passed through a back pressure regulator operating at 250 psi. Consecutive fractions of the leaving mixture were analyzed by GC measurements in order to determine the composition and to identify the stationary operation.

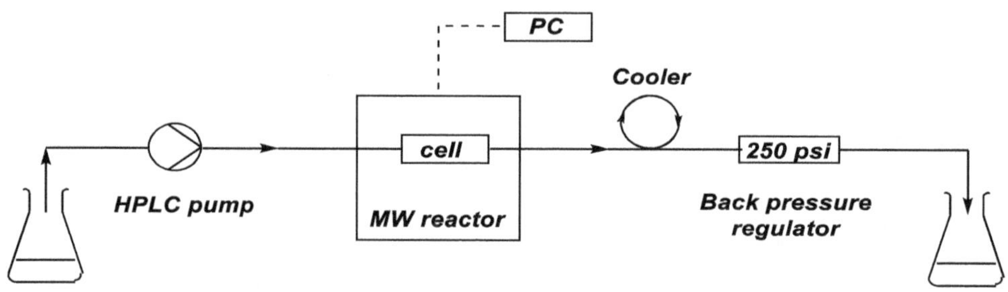

Figure 1. Schematic drawing of the continuous flow system developed.

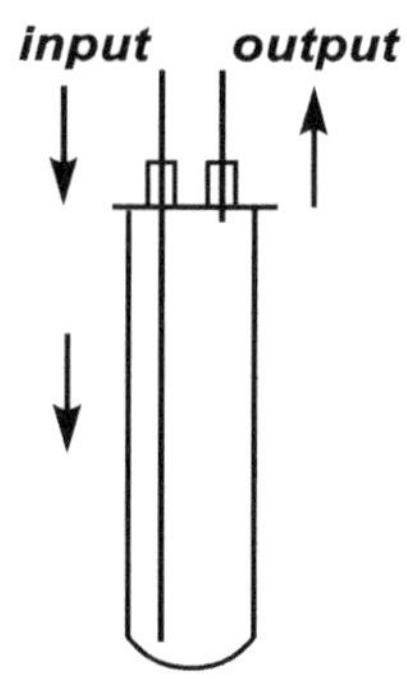

Figure 2. Sketch of the continuous flow cell.

The reaction conditions and the results are summarized in Table 3. To find the optimum conditions for the formation of *n*-butyl methyl *H*-phosphonate (nBMP) (**2a**) and *n*-dibutyl *H*-phosphonate (DnBP) (**3a**), the alcoholysis was carried out applying different temperatures and residence times. Performing the reaction at 100 °C, at a flow rate of 1.4 mL/min (with a residence time of 5 min), the conversion was only 26%, and the proportion of nBMP (**2a**) was 25%, along with 1% of DnBP (**3a**) (Table 3, Entry 1). Using a flow rate of 0.45 mL/min (with a residence time of 15 min), the starting DMP (**1**) was still the main component of the leaving mixture; however, the proportion of nBMP (**2a**) was 44% (Table 3, Entry 2). Increasing the residence time to 30 min (by decreasing the flow rate to 0.25 mL/min), the ratio of mixed *H*-phosphonate (**2a**) was 9% higher (53%) (Table 3, Entry 3). Applying a longer residence time of 45 or 60 min (flow rate of 0.15 or 0.10 mL/min), the proportion of phosphite **2a** was somewhat lower (50% or 47%), and the ratio of DnBP (**3a**) increased to 26% or 35%, respectively (Table 3, Entries 4 and 5). The maximum proportion of nBMP (**2a**) (53%) was obtained applying a reaction time of 30 min at 100 °C (Table 3, Entry 3). Increasing or decreasing the residence time at the same temperature resulted in lower ratio of product **2a** (Table 3, Entries 1, 2 and 4, 5, respectively). In a comparative thermal experiment at 100 °C at a residence time of 30 min, the conversion was only 15% (Table 3, Entry 6). Next, the effect of the reaction temperature was investigated at a residence time of 30 min (Table 3, Entries 6–8) (Figure 3). At 125 °C, the mixture was comprised of 7% of DMP (**1**), 43% of nBMP (**2a**) and

50% of DnBP (**3a**) (Table 3, Entry 7). Increasing the temperature to 150 °C, the conversion was complete, and the fully-transesterified product (**3a**) predominated in a proportion of 78% (Table 3, Entry 8). Performing the alcoholysis at 175 °C, the ratio of DnBP (**3a**) was 91% (Table 3, Entry 9). Carrying out the reaction at a longer residence time (45 min) and/or higher excess of the ^{n}BuOH (50 equiv), the composition did not change. Applying common heating at 175 °C, the reaction was not complete, and the ratio of product **3a** was 8% lower, than under MW conditions (Table 3, Entry 10). It can be observed that the difference between MW-assisted and conventionally-heated experiments was significant at 100 °C, while significantly smaller at 175 °C. After column chromatography, nBMP (**2a**) was obtained in a yield of 48%, while DnBP (**3a**) was isolated in a yield of 88% (Table 3, Entries 3 and 9).

Table 3. Continuous flow alcoholysis of DMP with ^{n}BuOH.

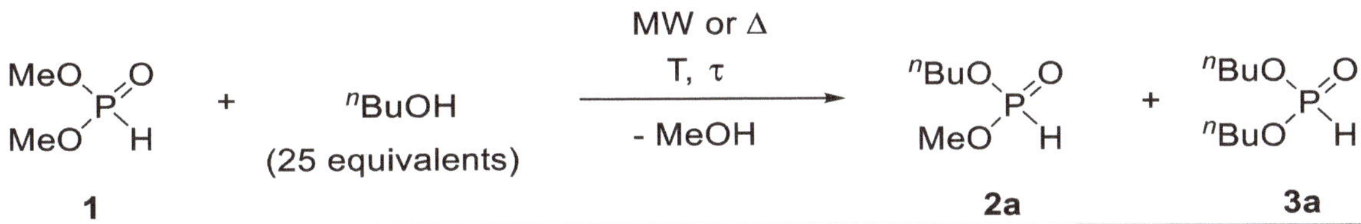

Entry	Mode of Heating	Power (W)	T (°C) [a]	Flow Rate (mL/min) [b]	τ (min)	Conversion (%)	Composition (%) [c]			Yield (%) [d]
							1	2a	3a	
1	MW	22	100	1.4	5	26	74	25	1	-
2	MW	14	100	0.45	15	52	48	44	8	-
3	MW	10	100	0.25	30	65	35	**53**	12	48 (**2a**)
4	MW	8	100	0.15	45	76	24	50	26	-
5	MW	5	100	0.10	60	83	18	47	35	-
6	Δ	-	100	0.25	30	15	85	15	0	-
7	MW	18	125	0.25	30	93	7	43	50	-
8	MW	38	150	0.25	30	100	0	22	78	-
9	MW	59	175	0.25	30	100	0	9	**91** [e,f]	88 (**3a**)
10	Δ	-	175	0.25	30	98	2	15	83	-

[a] The pressure was 17 bar; [b] Based on the HPLC pump; [c] the mixtures from the stationary operation were analyzed by GC; [d] after column chromatography; [e] No change on longer residence time (45 min); [f] No change on higher excess (50 equiv) of ^{n}BuOH. Bold numbers indicate the target compounds in the reactions.

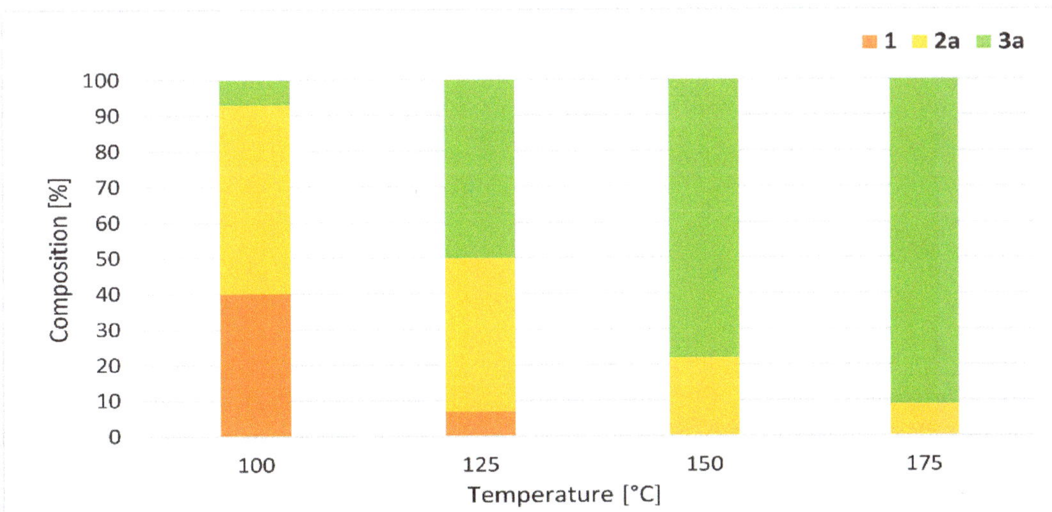

Figure 3. Composition of the reaction mixtures of the MW-assisted alcoholysis of DMP with ^{n}BuOH at different temperatures at a residence time of 30 min (Table 3 entries 3 and 7–9).

Comparative batch experiments were also carried out in the DMP (**1**)—^{n}BuOH model reaction at 100 °C. The alcohol excess (25 equiv) and the reaction time range studied (5–60 min) were the same as in the flow approaches. The batch results are listed in Table 4. Similarly to the flow experiments, the composition was highly reaction time-dependent.

Table 4. Comparative experiments for the alcoholysis of dimethyl *H*-phosphonate (DMP) with ^{n}BuOH in a batch MW reactor.

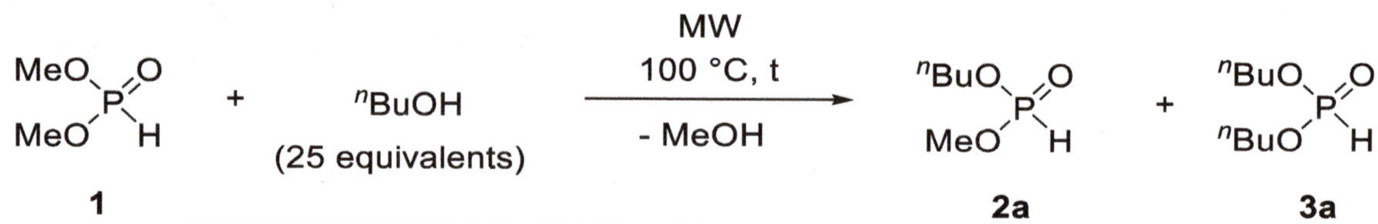

Entry	t (min)	Conversion (%)	Composition (%) [a]		
			1	2a	3a
1	5	24	76	20	4
2	15	46	54	37	9
3	30	59	41	47	12
4	45	67	33	49	18
5	60	73	26	45	29

[a] Based on GC.

Comparing the conversions of the flow and batch processes, the two plots show a similar shape; however, after a reaction time of 15 min, the batch conversions were 5–10% lower than the flow results (Figure 4).

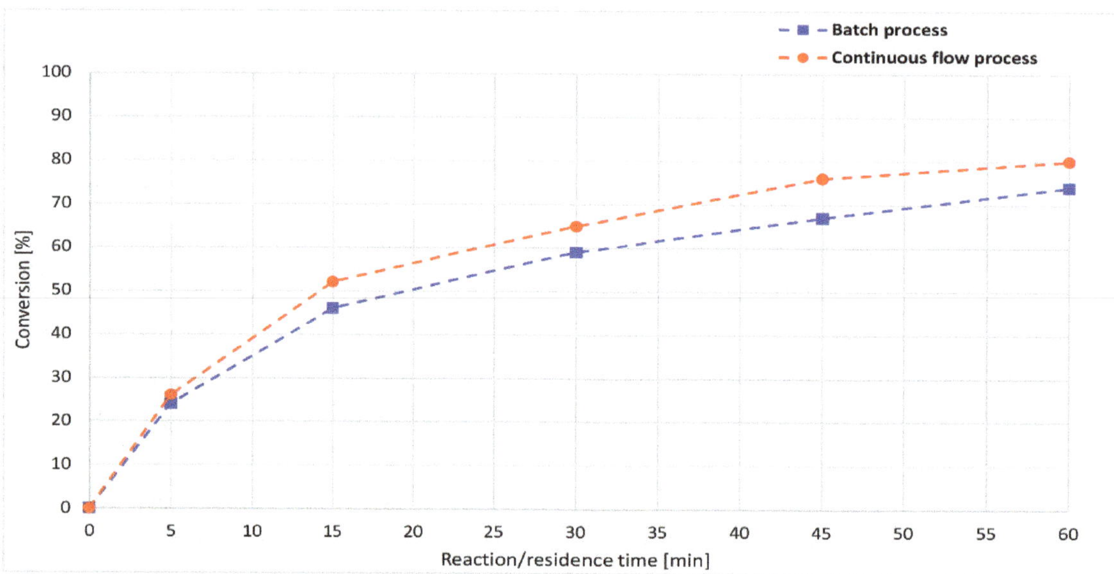

Figure 4. Reaction/residence time-dependence of the conversion of the alcoholysis of DMP with ^{n}BuOH at 100 °C.

A more significant difference may be observed if the ratio of product **2a** is examined separately (Figure 5). Maximum points are present on both curves: at 30 min and 53% in the flow experiments and at 45 min 49% in the case of the batch approaches. Based on the comparison, the flow reaction reaches the optimum point of phosphite **2a** under a shorter time, with a higher ratio.

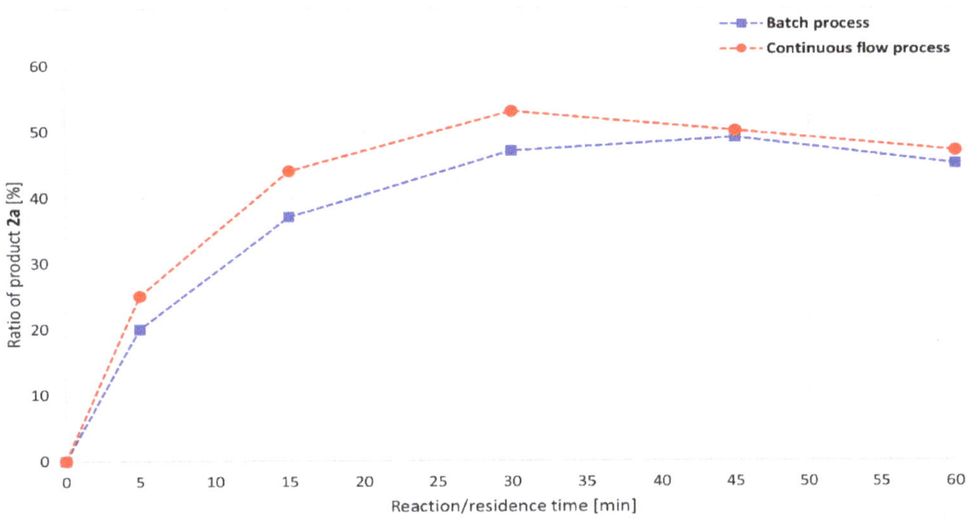

Figure 5. Reaction/residence time-dependence of the product **2a** ratio in the alcoholysis of DMP with ^{n}BuOH at 100 °C.

Based on the experiences above, the alcoholysis of DMP (**1**) was also carried out with *n*-propanol, *i*-butanol, as well as *n*-pentanol at 100 °C and 175 °C with a residence time of 30 min. In all cases, a mixture containing DMP (**1**) and 25 equivalents of the corresponding alcohol was passed through the continuous flow system. The data shown in Table 4 are concerned to the stationary operation. In the reaction of DMP (**1**) with ^{n}PrOH at 100 °C, the leaving mixture contained 45% of unreacted DMP (**1**), 46% of *n*-propyl methyl *H*-phosphonate (PrMP) (**2b**) and 9% of *n*-dipropyl *H*-phosphonate (DPrP) (**3b**) (Table 5, Entry 1). At a higher temperature of 175 °C, the fully-transesterified product (**3b**) was the major component (90%) (Table 5, Entry 2). The outcome of the alcoholysis of DMP (**1**) with ^{i}BuOH and ^{n}PentOH was similar (Table 5, Entries 3–6) However, in the case of the experiments carried out at 100 °C, the ratio of *i*-butyl methyl *H*-phosphonate (iBMP) (**2c**) and *n*-pentyl methyl *H*-phosphonate (PeMP) (**2d**) was above 50% (Table 5, Entries 3 and 5). After column chromatography, the corresponding mixed phosphonates (**2b**, **2c** and **2d**) were obtained in yields of 42–51% (Table 5, Entries 1, 3 and 5), and the fully-transesterified products (**3b**, **3c** and **3d**) were isolated in yields of 85–88% (Table 5, Entries 2, 4 and 6).

Table 5. Continuous flow alcoholysis of DMP with various alcohols.

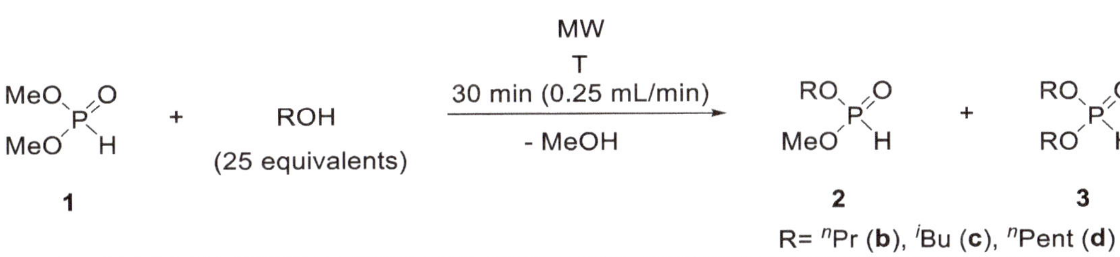

Entry	R	Power (W)	T (°C) [a]	Composition (%) [b]			Yield (%) [c]
				1	2	3	
1	^{n}Pr	10	100	45	**46**	9	42 (**2b**)
2		57	175	0	10	**90**	85 (**3b**)
3	^{i}Bu	10	100	38	**51**	11	46 (**2c**)
4		71	175	0	10	**90**	86 (**3c**)
5	^{n}Pent	10	100	32	**55**	13	51 (**2d**)
6		55	175	0	8	**92**	88 (**3d**)

[a] The pressure was 17 bar; [b] the mixtures from the stationary operation were analyzed by GC; [c] after column chromatography. Bold numbers indicate the target compounds in the reactions.

The next model was the continuous flow alcoholysis of diethyl *H*-phosphonate (DEP) (**4**) (Table 6). Based on the experiences from the transesterification of DMP (**1**), the reactions were carried out using 25 equivalents of the alcohols at a flow rate of 0.25 mL/min (or at a residence time of 30 min). First, the alcoholysis of DEP (**4**) by ^{n}BuOH was investigated. Performing the reaction at 100 °C, the unreacted DEP (**4**) was the main component along with 34% of *n*-butyl ethyl *H*-phosphonate (nBEP) (**5a**) and 8% of DBP (**3a**) (Table 6, Entry 1). At a higher temperature of 125 °C, the ratio of nBEP (**5a**) was 42%, and after column chromatography, compound **5a** could be isolated in a yield of 38% (Table 6, Entry 2). Increasing the temperature to 150 °C and 175 °C, the proportion of **5a** was decreased, while that of DBP (**3a**) increased (Table 6, Entries 3 and 4). In the latter case, the alcoholysis was complete, and the DBP (**3a**) was obtained in a yield of 85% (Table 6, Entry 4). The transesterification of DEP (**4**) by ^{n}PrOH at 125 °C took place similarly as the alcoholysis by ^{n}BuOH, and the *n*-propyl ethyl *H*-phosphonate (PrEP) (**5b**) was obtained in a yield of 36% (Table 6, Entries 2 and 5). At 175 °C, the conversion was not complete, and the resulting mixture contained only 74% of DPrP (**3b**) besides 4% of unreacted DEP (**4**) and 22% of PrEP (**5b**) (Table 6, Entry 6). To obtain a higher proportion of product **3b**, the temperature was increased to 200 °C. Hence, phosphite **3b** was formed in 84% (Table 6, Entry 7). In the case of ^{i}BuOH, the tendency was similar, and the highest amount of mixed phosphonate (**5c**) was formed at 125 °C, while at 175 and 200 °C, the fully-transesterified product (**3c**) predominated (Table 6, Entries 8–10). Using PentOH at 125 °C, the pentyl ethyl *H*-phosphonate (PeEP) (**5d**) was the main component in the departing mixture, from which it was isolated in a yield of 40% (Table 6, Entry 11). For the preparation of dipentyl *H*-phosphonate (DPeP) (**3d**), the application of 175 °C was enough, and compound **3d** was obtained in a yield of 80% (Table 6, Entry 12).

Table 6. Continuous flow alcoholysis of DEP with alcohols.

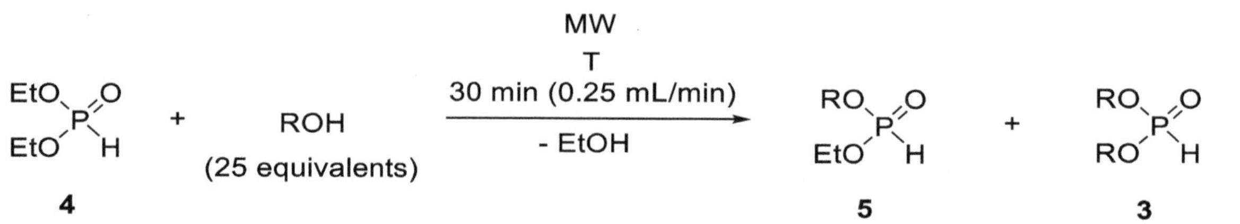

Entry	R	Power (W)	T (°C) [a]	Composition (%) [b]			Yield (%) [c]
				4	5	3	
1	^{n}Bu	10	100	58	34	8	-
2		17	125	44	**42**	14	38 (**5a**)
3		38	150	7	32	61	-
4		57	175	0	11	**89** [d,e]	85 (**3a**)
5	^{n}Pr	18	125	51	**41**	8	36 (**5b**)
6		54	175	4	22	74	–
7		83	200	2	14	**84** [d,e]	78 (**3b**)
8	^{i}Bu	21	125	21	**40**	39	36 (**5c**)
9		63	175	3	26	71	–
10		105	200	0	16	**84** [d,e]	77 (**3c**)
11	^{n}Pent	10	125	31	**44**	25	40 (**5d**)
12		39	175	2	13	**85** [d,e]	80 (**3d**)

[a] The pressure was 17 bar; [b] the mixtures from the stationary operation were analyzed by GC; [c] after column chromatography; [d] no change upon longer residence time (45 min); [e] no change upon higher excess (50 equiv) of ^{n}BuOH. Bold numbers indicate the target compounds in the reactions.

The catalyst-free alcoholysis of dialkyl *H*-phosphonates with various aliphatic alcohols was carried out efficiently in a continuous flow MW reactor. Applying a residence time of 30 min at different temperatures, the alcoholysis could be fine-tuned towards the dialkyl *H*-phosphonates with two different or with two identical alkyl groups. In the case of the reaction studied, the dimethyl phosphite proved to be the more reactive starting material. After almost complete conversions, the $(RO)_2P(O)H$ species formed by transesterification were obtained in yields above 85%. The ratio of (RO)(R'O)P(O)H derivatives with two different alkyl groups was around 50% in all mixtures, affording the target compounds in a yield range of ca. 40–50%. By the method developed, a series of mixed dialkyl *H*-phosphonates, which are valuable building blocks for the preparation of chiral organophosphorus derivatives, was synthesized.

3. Materials and Methods

3.1. General

GC measurements were performed on an HP5890 Series 2 GC-FID chromatograph, using a 15 m × 0.18 mm Restek, Rtx-5 column with a film layer of 0.20 μm. The temperature of the column was initially held at 40 °C for 1 min, followed by programming at 25 °C/min up to 300 °C and a final period at 300 °C (isothermal) for 10 min. The temperature of the injector was 290 °C and of the FID detector was 300 °C. The carrier gas was N_2.

GC-MS measurements were performed on an Agilent 6890 N-GC-5973 N-MSD chromatograph, using a 30 m × 0.25 mm Restek, Rtx-5SILMS column with a film layer of 0.25 μm. The initial temperature of the column was 45 °C for 1 min, followed by programming at 10 °C/min. up to 310 °C and a final period at 310 °C (isothermal) for 17 min. The temperature of the injector was 250 °C. The carrier gas was He, and the operation mode was splitless.

High resolution mass spectrometric measurements were performed using a TripleTOF 5600+ mass spectrometer in positive electrospray mode.

The ^{13}C- and ^{1}H-NMR spectra were obtained in $CDCl_3$ solution on a Bruker DRX-500 spectrometer operating at 125.7 and 500.1 MHz, respectively. The ^{13}C and ^{1}H chemical shifts are referred to TMS. ^{31}P-NMR spectra were obtained on a Bruker AV-300 spectrometer at 121.5 MHz. Chemical shifts are downfield relative to 85% H_3PO_4.

3.2. Equipment

The continuous flow reactions were performed in a self-developed continuous flow system comprising a 300-W CEM® Discover-focused microwave reactor equipped with a CEM® 10-mL Flow Cell Accessory continuous flow unit (irradiated volume 7 mL), a Gilson 305 HPLC pump, an HPLC backpressure regulator with a 250-psi (17.2 bar) cartridge and a cooler. Teflon® (Wilmington, DE, USA) PFA tubes with an outside diameter of 0.125" (3.175 mm) and an inside diameter of 0.064" (1.575 mm) were used. The exact lengths and volumes of each tube part are shown in Figure 6. All of the tubes, screws and ferrules applied were fully compatible with a regular HPLC system.

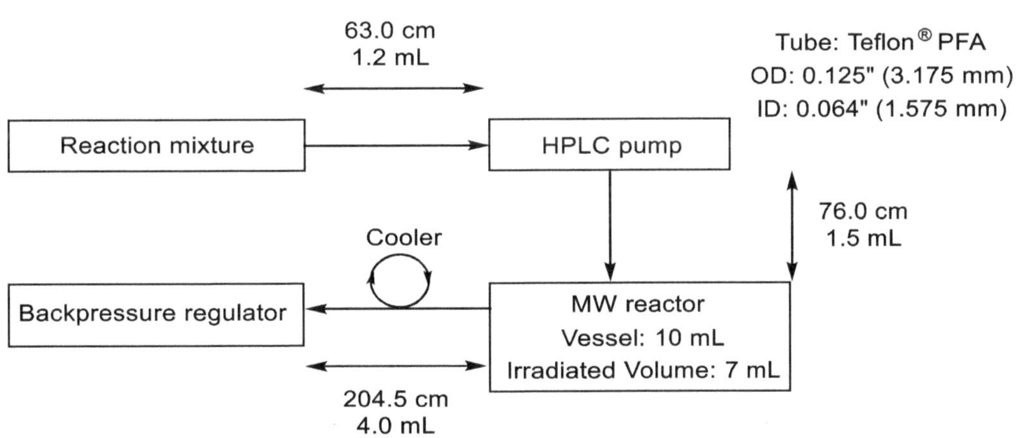

Figure 6. Design parameters of the continuous flow system.

3.3. General Procedure for the Continuous Flow Alcoholysis of Dialkyl H-Phosphonates

A mixture of 25.0 mmol of the dialkyl *H*-phosphonate (2.7 mL of dimethyl *H*-phosphonate, 3.9 mL of diethyl *H*-phosphonate) and 25 equivalents, 0.625 mol of alcohol (47 mL of *n*-propanol, 57 mL of *n*-butanol, 58 mL of *i*-butanol, 68 mL of *n*-pentanol) was homogenized by stirring for 5 min at 25 °C. The reactor was flushed with 20 mL of the mixture with a flow rate of 10 mL/min at 25 °C and 17 bar. Next, under the same pressure, the flow rate was set to the desired value, and the vessel was irradiated with a power of 45–300 W for 3–5 min, until the temperature reached the desired value, then the power was controlled automatically by the software of the MW reactor. The flow rate and the necessary power at stationary operation can be seen in Tables 3, 5 and 6. The operation was regarded as the steady state on the basis of the results of the GC measurements. The products were separated by column chromatography using silica gel as the absorbent and ethyl acetate as the eluent. Yields were calculated on the basis of the weights obtained after separation and evaporation taking into consideration the quantity of the dialkyl *H*-phosphonate fed in during a given time. The ^{31}P-NMR and mass data for the dialkyl *H*-phosphonates are listed in Table 7.

3.4. General Procedure for the Comparative Batch Experiments

A mixture of 0.5 mmol (0.05 mL) of dimethyl *H*-phosphonate and 25 equivalents, 12.5 mmol (1.1 mL) of *n*-butanol was heated at 100 °C in a vial in the MW reactor for 5–60 min as shown in Table 4. The volatile components were removed in vacuum, and the residual oil was analyzed by GC.

Methyl propyl H-phosphonate (**2b**): Yield: 42%; ^{31}P-NMR and HRMS (see Table 7); ^{13}C-NMR ($CDCl_3$) δ 10.0 (CH_3CH_2), 23.8 (J = 6.2, CH_2), 52.0 (J = 5.8, CH_3O), 67.1 (J = 6.0, CH_2O); ^{1}H-NMR ($CDCl_3$) δ 0.98 (t, J_{HH} = 7.4, 3H, CH_3CH_2), 1.65–1.82 (m, 2H, CH_2), 3.78, (d, J_{PH} = 11.9, 3H, CH_3O) 3.97–4.12 (m, 2H, CH_2O), 6.79 (d, J_{PH} = 695.4, 1H, PH).

i-Butyl methyl H-phosphonate (**2c**): Yield: 46%; ^{31}P-NMR and HRMS (see Table 7); ^{13}C-NMR ($CDCl_3$) δ 18.6 (J = 3.1, CH_3CH), 29.2 (J = 6.4, CH), 51.9 (J = 5.7, CH_3O), 71.8 (J = 6.3, CH_2O); ^{1}H-NMR ($CDCl_3$) δ 0.89 (d, J_{HH} = 6.8, 6H, CH_3CH), 1.84–1.96 (m, 1H, CH), 3.71 (d, J_{PH} = 11.9, 3H, CH_3O) 3.74–3.83 (m, 2H, CH_2O), 6.72 (d, J_{PH} = 695.8, 1H, PH).

Methyl pentyl H-phosphonate (**2d**): Yield: 50%; ^{31}P-NMR and HRMS (see Table 7); ^{13}C-NMR ($CDCl_3$) δ 13.9 (CH_3CH_2), 22.2 (CH_3CH_2), 27.6 ($CH_3CH_2CH_2$), 30.1 (J = 6.2, CH_2CH_2O), 51.9 (J = 5.7, CH_3O), 67.0 (J = 6.1, CH_2O); ^{1}H-NMR ($CDCl_3$) δ 0.91 (t, J = 6.9, 3H, CH_3CH_2), 1.29–1.44 (m, 4H, $CH_2CH_2CH_3$), 1.63–1.78 (m, 2H, CH_2CH_2O), 3.78, (d, J_{PH} = 11.9, 3H, CH_3O) 3.99–4.17 (m, 2H, CH_2O), 6.79 (d, J_{PH} = 695.2, 1H, PH).

Ethyl propyl H-phosphonate (**5b**): Yield: 30%; ^{31}P-NMR and HRMS (see Table 7); ^{13}C-NMR ($CDCl_3$) δ 10.0 ($CH_3CH_2CH_2$), 16.3 (J = 6.1, CH_3CH_2O), 23.8 (J = 6.4, $CH_3CH_2CH_2$), 61.8 (J = 5.8, CH_2CH_2O), 67.3 (J = 6.0, CH_3CH_2O); ^{1}H NMR ($CDCl_3$) δ 0.96 (t, J_{HH} = 7.4, 3H, $CH_3CH_2CH_2$), 1.35 (t, J_{HH} = 7.1, 3H, CH_3CH_2O), 1.63–1.78 (m, 2H, $CH_3CH_2CH_2$), 3.94–4.07, (m, 2H, CH_2CH_2O) 4.07–4.21 (m, 2H, CH_3CH_2O), 6.80 (d, J_{PH} = 692.4, 1H, PH).

i-Butyl ethyl H-phosphonate (**5c**): Yield: 36%; ^{31}P-NMR and HRMS (see Table 7); ^{13}C-NMR ($CDCl_3$) δ 16.3 (J = 6.2, CH_3CH_2), 18.7 (J = 1.7, CH_3CH), 29.1 (J = 6.5, CH), 61.8 (J = 5.7, CH_3CH_2O), 71.6 (J = 6.2, $CHCH_2O$); ^{1}H NMR ($CDCl_3$) δ 0.96 (d, J_{HH} = 6.7, 6H, CH_3CH), 1.37 (t, J_{HH} = 7.1, 3H, CH_3CH_2), 1.89–2.06 (m, 1H, CH), 3.77–3.91, (m, 2H, $CHCH_2O$) 4.09–4.22 (m, 2H, CH_3CH_2O), 6.82 (d, J_{PH} = 692.7, 1H, PH).

Table 7. ^{31}P-NMR and mass data for the dialkyl *H*-phosphonates prepared.

Compound	δ_P ($CDCl_3$)	δ_P [Ref]	$[M + H]^+_{found}$	$[M + H]^+_{requires}$
2a	9.3 (695.0)	9.3 [52]	153.0684	153.0681
2b	9.3 (695.4)	-	139.0522	139.0523
2c	6.7 (695.7)	-	153.0676	153.0681
2d	6.8 (695.2)	-	167.0833	167.0837
3a	7.8	7.9 [66]	195.1153	195.1150
3b	7.9	7.8 [67]	167.0833	167.0837
3c	8.1	8.0 [67]	195.1145	195.1150
3d	7.8	8.1 [68]	223.1462	223.1463
5a	7.6 (693.0)	7.6 [52]	167.0840	167.0837
5b	8.6 (692.4)	-	153.0673	153.0681
5c	7.8 (692.8)	-	167.0832	167.0837
5d	7.7 (692.1)	7.7 [52]	181.0992	181.0994

4. Conclusions

In summary, a MW-assisted continuous flow method was developed for the catalyst-free alcoholysis of dialkyl *H*-phosphonates. By the optimization of the reaction parameters, the alcoholysis was fine-tuned towards dialkyl *H*-phosphonates with two different or with two identical alkyl groups. The selectivity of the reaction was similar in the flow and batch approaches; however, the continuous protocol offered several advantages. The dialkyl *H*-phosphonates with two different alkyl groups can be obtained at a shorter reaction time and in a higher ratio. Furthermore, a somewhat higher productivity (0.8–1.0 g/h) can be attained by the continuous flow method. Five new dialkyl *H*-phosphonates with different alkyl groups were isolated and characterized, which may be valuable building blocks in the synthesis of chiral organophosphorus compounds.

Author Contributions: E.B., Á.T. and G.K. conceived of and designed the experiments. Á.T. and N.T. performed the experiments. E.B. and G.K. contributed reagents/materials/analysis tools. E.B., Á.T. and G.K. wrote the paper.

References

1. Plutschack, M.B.; Pieber, B.; Gilmore, K.; Seeberger, P.H. The Hitchhiker's guide to flow chemistry. *Chem. Rev.* **2017**, *117*, 11796–11893. [CrossRef] [PubMed]
2. Glasnov, T. *Continuous-Flow Chemistry in the Research Laboratory*; Springer International Publishing: Basel, Switzerland, 2016; ISBN 978-3-319-32194-3.
3. Bálint, E.; Keglevich, G. The Spread of the Application of the Microwave Technique in Organic Synthesis. In *Milestones in Microwave Chemistry*; Keglevich, G., Ed.; Springer: Basel, Switzerland, 2016; pp. 1–10. ISBN 978-3-319-30632-2.
4. de la Hoz, A.; Loupy, A. (Eds.) *Microwaves in Organic Synthesis*, 3rd ed.; Wiley: Weinheim, Germany, 2012; ISBN 978-3-527-65131-3.
5. Moseley, J.D. Microwave heating as a tool for process chemistry. In *Microwave Heating as a Tool for Sustainable Chemistry*; Leadbeater, N., Ed.; CRC Press: New York, NY, USA, 2010; pp. 105–147. ISBN 9781138111981.
6. Kappe, C.O.; Stadler, A.; Dallinger, D. *Microwaves in Organic and Medicinal Chemistry*, 2nd ed.; Wiley: Weinheim, Germany, 2012; Volume 52, ISBN 978-3-527-33185-7.
7. Estela, L.; Pouxb, M.; Benamaraa, N.; Polaerta, I. Continuous flow-microwave reactor: Where are we? *Chem. Eng. Process.* **2016**, *113*, 56–64. [CrossRef]
8. Baxendale, I.; Hayward, J.; Ley, S. Microwave reactions under continuous flow conditions. *Comb. Chem. High Throughput Screen.* **2007**, *10*, 802–836. [CrossRef] [PubMed]

9. Riemenschneider, W.; Bolt, H.M. Esters, Organic. In *Ullmann's Encyclopedia of Industrial Chemistry*, 7th ed.; Wiley: Weinheim, Germany, 2005; pp. 245–265. ISBN 978-3-527-30673-2.
10. Otera, J.; Nishikido, J. (Eds.) *Esterification: Methods, Reactions, and Applications*, 2nd ed.; Wiley: Weinheim, Germany, 2010; ISBN 978-3-527-32289-3.
11. Woodcock, L.L.; Wiles, C.; Greenway, G.M.; Watts, P.; Wells, A.; Eyley, S. Enzymatic synthesis of a series of alkyl esters using novozyme 435 in a packed-bed, miniaturized, continuous flow reactor. *Biocatal. Biotransform.* **2008**, *26*, 501–507. [CrossRef]
12. Junior, I.I.; Flores, M.C.; Sutili, F.K.; Leite, S.G.F.; Miranda, L.S.D.M.; Leal, I.C.R.; de Souza, R.O.M.A. Fatty acids residue from palm oil refining process as feedstock for lipase catalyzed monoacylglicerol production under batch and continuous flow conditions. *J. Mol. Catal. B Enzym.* **2012**, *77*, 53–58. [CrossRef]
13. Sutili, F.K.; Ruela, H.S.; Leite, S.G.F.; Miranda, L.S.D.M.; Leal, I.C.R.; de Souza, R.O.M.A. Lipase-catalyzed esterification of steric hindered fructose derivative by continuous flow and batch conditions. *J. Mol. Catal. B Enzym.* **2013**, *85–86*, 37–42. [CrossRef]
14. Patil, N.G.; Benaskar, F.; Rebrov, E.V.; Meuldijk, J.; Hulshof, L.A.; Hessel, V.; Schouten, J.C. Scale-up of Microwave Assisted Flow Synthesis by Transient Processing through Monomode Cavities in Series. *Org. Process Res. Dev.* **2014**, *18*, 1400–1407. [CrossRef]
15. Sutili, F.K.; Ruela, H.S.; Nogueira, D.D.O.; Leal, I.C.R.; Miranda, L.S.D.M.; de Souza, R.O.M.A. Enhanced production of fructose ester by biocatalyzed continuous flow process. *Sustain. Chem. Process.* **2015**, *3*, 6. [CrossRef]
16. Okuno, Y.; Isomura, S.; Sugamata, A.; Tamahori, K.; Fukuhara, A.; Kashiwagi, M.; Kitagawa, Y.; Kasai, E.; Takeda, K. Convenient and Simple Esterification in Continuous-Flow Systems using g-DMAP. *ChemSusChem* **2015**, *8*, 3587–3589. [CrossRef] [PubMed]
17. Koreniuk, A.; Maresz, K.; Odrozek, K.; Jarzebski, A.B.; Mrowiec-Bialon, J. Highly effective continuous-flow monolithic silica microreactors for acid catalyzed processes. *Appl. Catal. A* **2015**, *489*, 203–208. [CrossRef]
18. Baek, H.; Minakawa, M.; Yamada, Y.M.A.; Han, J.W.; Uozumi, Y. In-Water and Neat Batch and Continuous-Flow Direct Esterifcation and Transesterifcation by a Porous Polymeric Acid Catalyst. *Sci. Rep.* **2016**, *6*, 25925. [CrossRef] [PubMed]
19. Furuta, A.; Fukuyama, T.; Ryu, I. Efficient Flow Fischer Esterification of Carboxylic Acids with Alcohols Using Sulfonic Acid-Functionalized Silica as Supported Catalyst. *Bull. Chem. Soc. Jpn.* **2017**, *90*, 607–612. [CrossRef]
20. Iemhoff, A.; Sherwood, J.; McElroy, C.R.; Hunt, A.J. Towards sustainable kinetic resolution, a combination of bio-catalysis, flow chemistry and bio-based solvents. *Green Chem.* **2018**, *20*, 136–140. [CrossRef]
21. Razzaq, T.; Glasnov, T.N.; Kappe, C.O. Continuous-Flow Microreactor Chemistry under High-Temperature/Pressure Conditions. *Eur. J. Org. Chem.* **2009**, 1321–1325. [CrossRef]
22. Archambault, C.M.; Leadbeater, N.E. A benchtop NMR spectrometer as a tool for monitoring mesoscale continuous-flow organic synthesis: Equipment interface and assessment in four organic transformations. *RSC Adv.* **2016**, *6*, 101171–101177. [CrossRef]
23. Gumel, A.M.; Annuar, M.S.M. Thermomyces lanuginosus lipase-catalyzed synthesis of natural flavor esters in a continuous flow microreactor. *3 Biotech* **2016**, *6*, 24. [CrossRef] [PubMed]
24. Tajti, Á.; Tóth, N.; Bálint, E.; Keglevich, G. Esterification of benzoic acid in a continuous flow microwave reactor. *J. Flow Chem.* **2017**, *8*, 11–19. [CrossRef]
25. Krull, M.; Moschhaeuser, R. Continuous Method for Producing Esters of Aromatic Carboxylic Acids. U.S. Patent 0088918, 12 April 2012.
26. Cablewski, T.; Faux, A.F.; Strauss, C.R. Development and Application of a Continuous Microwave Reactor for Organic Synthesis. *J. Org. Chem.* **1994**, *59*, 3408–3412. [CrossRef]
27. Chen, S.-T.; Chiou, S.-H.; Wang, K.-T. Preparative scale organic synthesis using a kitchen microwave oven. *J. Chem. Soc. Chem. Commun.* **1990**, 807–809. [CrossRef]
28. Pipus, G.; Plazl, I.; Koloini, T. Esterification of benzoic acid in microwave tubular flow reactor. *Chem. Eng. J.* **2000**, *76*, 239–245. [CrossRef]
29. Asadi, M.; Hooper, J.F.; Lupton, D.W. Biodiesel synthesis using integrated acid and base catalysis in continuous flow. *Tetrahedron* **2016**, *72*, 3729–3733. [CrossRef]

30. Adeyemi, A.; Bergman, J.; Branalt, J.; Savmarker, J.; Larhed, M. Continuous Flow Synthesis under High-Temperature/High-Pressure Conditions Using a Resistively Heated Flow Reactor. *Org. Process Res. Dev.* **2017**, *21*, 947–955. [CrossRef]
31. Tran, D.-T.; Chang, J.-S.; Lee, D.-J. Recent insights into continuous-flow biodiesel production via catalytic and non-catalytic transesterification processes. *Appl. Energy* **2017**, *185*, 376–409. [CrossRef]
32. Corbridge, D.E.C. *Phosphorus: Chemistry, Biochemistry and Technology*, 6th ed.; CRC Press: New York, NY, USA, 2013; ISBN 978-1-439-84088-7.
33. Troev, K.D. *Chemistry and Application of H-Phosphonates*; Elsevier: Amsterdam, The Netherlands, 2006; ISBN 978-0-080-47649-0.
34. Bálint, E.; Tajti, Á.; Tripolszky, A. Synthesis of α-aminophosphonates by the Kabachnik–Fields reaction and by the Pudovik reaction. In *Organophosphorus Chemistry*; Keglevich, G., Ed.; Walter de Gruyter GmbH: Berlin, Germany, 2018; pp. 108–147. ISBN 978-3-11-053453-5.
35. Rádai, Z.; Kiss, N.Z.; Keglevich, G. Synthesis of α-hydroxyphosphonates, an important class of bioactive compounds. In *Organophosphorus Chemistry*; Keglevich, G., Ed.; Walter de Gruyter GmbH: Berlin, Germany, 2018; pp. 91–107. ISBN 978-3-11-053453-5.
36. Henyecz, R.; Keglevich, G. P-C couplings by the Hiaro reaction. In *Organophosphorus Chemistry*; Keglevich, G., Ed.; Walter de Gruyter GmbH: Berlin, Germany, 2018; pp. 158–178. ISBN 978-3-11-053453-5.
37. Enders, D.; Saint-Dizier, A.; Lannou, M.-I.; Lenzen, A. The phospha-Michael addition in organic synthesis. *Eur. J. Org. Chem.* **2006**, 29–49. [CrossRef]
38. Tajti, Á.; Bálint, E.; Keglevich, G. Synthesis of ethyl octyl α-aminophosphonate derivatives. *Curr. Org. Synth.* **2016**, *13*, 638–675. [CrossRef]
39. Bálint, E.; Tajti, Á.; Kalocsai, D.; Mátravölgyi, B.; Konstantin, K.; Czugler, M.; Keglevich, G. Synthesis and utilization of optically active α-aminophosphonate derivatives by Kabachnik-Fields reaction. *Tetrahedron* **2017**, *73*, 5659–5667. [CrossRef]
40. Gerrard, W. The interaction of n-butyl alcohol and the chlorides and oxychloride of phosphorus in the absence and in the presence of pyridine. *J. Chem. Soc.* **1940**, 1466–1469. [CrossRef]
41. Hardy, E.E.; Anniston, A.; Kosolapoff, G.M. Halogenated Compounds and Process for Making Same. U.S. Patent 2409039, 8 October 1946.
42. Foss, O. Di-O-alkylmonothiophosphates and the corresponding pseudohalogens. *Acta Chem. Scand.* **1947**, *1*, 8–31. [CrossRef] [PubMed]
43. Campbell, C.H.; Chadwick, D.H.; Kaufman, S. Continuous process for preparing dialkyl phosphites. *Ind. Eng. Chem. Res.* **1957**, *49*, 1871–1873. [CrossRef]
44. Mitschke, K.-H. Method for the Combined Production of Diethyl Phosphite and Ethyl Chloride. Application Number WO200424742 A1, 25 March 2004.
45. Kendall, A.J.; Salazar, C.A.; Martino, P.F.; Tyler, D.R. Direct conversion of phosphonates to phosphine oxides: An improved synthetic route to phosphines including the first synthesis of methyl JohnPhos. *Organometallics* **2014**, *33*, 6171–6178. [CrossRef]
46. Kosolapoff, G.M. Preparation of some mixed dialkyl phosphites. *J. Am. Chem. Soc.* **1951**, *73*, 4989. [CrossRef]
47. Fields, E. The synthesis of esters of substituted amino phosphonic acids. *J. Am. Chem. Soc.* **1952**, *74*, 1528–1531. [CrossRef]
48. Oswald, A.A. Synthesis of cyclic phosphorus acid esters by transesterification. *Can. J. Chem.* **1959**, *37*, 1499–1504. [CrossRef]
49. Kuskov, V.K.; Gradis, G.K. Reaction of diethyl phosphite with sodium alcoholates. *Dokl. Akad. Nauk SSSR* **1953**, *92*, 323–324.
50. Froneman, M.; Modro, T.A. The titanium-mediated transesterification of phosphorus esters. *Tetrahedron Lett.* **1988**, *27*, 3327–3330. [CrossRef]
51. Aitken, R.A.; Collett, C.J.; Mesher, S.T.E. Convenient preparation of long-chain dialkyl phosphates: Synthesis of dialkyl phosphates. *Synthesis* **2012**, *44*, 2515–2518. [CrossRef]
52. Bálint, E.; Tajti, Á.; Drahos, L.; Ilia, G.; Keglevich, G. Alcoholysis of dialkyl phosphites under microwave conditions. *Curr. Org. Chem.* **2013**, *17*, 555–562. [CrossRef]
53. Zarrougui, R.; Raouafi, N.; Lemordant, D. New series of green cyclic ammonium-based room temperature ionic liquids with alkylphosphite-containing anion: Synthesis and physicochemical characterization. *J. Chem. Eng. Data* **2014**, *59*, 1193–1201. [CrossRef]

54. Salin, A.V.; Il'in, A.V.; Shamsutdinova, F.G.; Fatkhutdinov, A.R.; Galkin, V.I.; Islamov, D.R.; Kataeva, O.N. Phosphine-catalyzed addition of P(O)-H compounds to ethyl phenylpropiolate. *Tetrahedron Lett.* **2015**, *56*, 6282–6286. [CrossRef]
55. Ernsberger, M.L.; Hill, J.W. Preparation of Organic Phosphorus Compounds, and in Particular, of Dialkyl Phosphites. U.S. Patent 2661364, 1 December 1953.
56. Budnikova, Y.G.; Kargin, Y.M. Electrosynthesis of aliphatic esters of phosphorus acids from white phosphorus in alcohol solutions, involving radical cations of phenothiazine and triarylamine. *Russ. J. Gen. Chem.* **1995**, *65*, 504–507.
57. Abdreimova, R.R.; Akbayeva, D.N.; Polimbetova, G.S.; Caminade, A.-M.; Majoral, J.-P. Chlorine free synthesis of organophosphorus compounds based on the functionalization of white phosphorus (P_4). *Phosphorus Sulfur Silicon Relat. Elem.* **2000**, *156*, 239–254. [CrossRef]
58. Trofimov, B.; Timokhin, B.; Gusarova, N.; Kazantseva, M. Golubin, Cu-catalyzed oxidative phosphorylation of alkanols with white phosphorus and H_2O_2. *Phosphorus Sulfur Silicon Relat. Elem.* **2002**, *177*, 2385–2390. [CrossRef]
59. Budnikova, Y.H.; Kafiyatullina, A.G.; Sinyashin, O.G.; Abdreimova, R.R. Electrochemical synthesis of phosphorus esters from white phosphorus in the presence of copper complexes and ethanol. *Russ. Chem. Bull.* **2003**, *52*, 929–938. [CrossRef]
60. Budnikova, Y.H.; Yakhvarov, D.G.; Sinyashin, O.G. Electrocatalytic eco-efficient functionalization of white phosphorus. *J. Organomet. Chem.* **2005**, *690*, 2416–2425. [CrossRef]
61. Fisher, H.C.; Prost, L.; Montchamp, J.L. Organophosphorus chemistry without PCl3: A bridge from hypophosphorous acid to *H*-phosphonate diesters. *Eur. J. Org. Chem.* **2013**, 7973–7978. [CrossRef]
62. Kluba, M.; Zwierzak, A. Alkylation of tetra-n-butylaminium alkyl hydrogen phosphites. A new route to mixed dialkyl phosphites. *Synthesis* **1978**, *2*, 134–137. [CrossRef]
63. Ilia, G.; Kurunczi, L. Synthesis of mixed alkylphosphites and alkylphosphates. *Phosphorus Sulfur Silicon Relat. Elem.* **2003**, *178*, 1513–1519. [CrossRef]
64. Berchel, M.; Haddad, J.; Le Corre, S.S.; Haelters, J.-P.; Jaffrès, P.-A. Synthesis of lipid-based unsymmetrical O,O-dialkylphosphites. *Tetrahedron Lett.* **2015**, *56*, 2345–2348. [CrossRef]
65. Guin, J.; Wang, Q.; van Gemmeren, M.; List, B. The catalytic asymmetric Abramov reaction. *Angew. Chem. Int. Ed.* **2014**, *53*, 1–5. [CrossRef] [PubMed]
66. Peng, W.; Schreeve, J.M. Rapid and high yield oxidation of phosphine, phosphite and phosphinite compounds to phosphine oxides, phosphates and phosphinates using hypofluorous acid–acetonitrile complex. *J. Fluorine Chem.* **2005**, *126*, 1054–1056. [CrossRef]
67. Santschi, N.; Togni, A. Electrophilic Trifluoromethylation of S-Hydrogen Phosphorothioates. *J. Org. Chem.* **2011**, *76*, 4189–4193. [CrossRef] [PubMed]
68. Kers, A.; Kers, I.; Stawinski, J.; Sobkowski, M.; Kraszewski, A. Studies on Aryl *H*-Phosphonates; Part 2: A General Method for the Preparation of Alkyl *H*-Phosphonate Monoesters. *Synthesis* **1995**, 427–430. [CrossRef]

10

Chiral Hypervalent, Pentacoordinated Phosphoranes

Dorota Krasowska [1,*,†], Jacek Chrzanowski [1,*,†], Piotr Kiełbasiński [1,†] and Józef Drabowicz [1,2,*,†]

1 Centre of Molecular and Macromolecular Studies, Polish Academy of Sciences, Sienkiewicza 112, 90-363 Lodz, Poland; piokiel@cbmm.lodz.pl
2 Department of Chemistry, Environment Protection and Biotechnology, Jan Dlugosz University in Czestochowa, Armii Krajowej Ave. 13/15, 42-200 Czestochowa, Poland
* Correspondence: dkrasowska@gmail.com (D.K.); jacekchrzanowski1@gmail.com (J.C.); drabow@gmail.com (J.D.)
† These authors contributed equally to this work.

Academic Editor: György Keglevich

Abstract: This review presents synthetic procedures applied to the preparation of chiral (mainly optically active) pentacoordinated, hypervalent mono and bicyclic phosphoranes. The mechanisms of their stereoisomerization and their selected interconversions are also presented.

Keywords: hypervalency; chirality; phosphoranes; Berry pseudorotation; turnstile rotation

1. Introduction

The phenomenon of chirality plays a very important role in organic chemistry and biological processes [1–5]. Therefore an easy access to chiral compounds, especially optically active species, constitutes a vital challenge in modern synthetic organic chemistry. At least two reasons are responsible for this fact. Firstly, single stereoisomers are used very often as active components in formulations which serve as drugs, food additives or substances having the requested flavor or fragrance. The second reason stems from their application as chiral catalysts in various asymmetric reactions [6,7]. Moreover, detailed mechanistic description of a number of reactions would not be possible without taking into account their stereochemical aspects. These can be studied only with the use of optically active species which allow simultaneous application of kinetic and polarimetric measurements carried out with the use of the suitably designated model substrates (enantiomerically or diastereomerically pure or enriched samples) containing an appropriate chirality element [8]. In the organophosphorus chemistry the phenomenon of chirality is very common among organic derivatives having different valency and/or coordination number [9,10]. Their optical activity can be related to the presence of the following elements of chirality:

(a) a stereogenic center (for trivalent, tricoordinated, tetravalent tetracoordinated, pentavalent tetracoordinated derivatives)
(b) "trigonal bipyramidal chirality"
(c) "tetragonal bipyramidal chirality"

Organophosphorus derivatives in which optical activity results from the presence of a phosphorus atom at a center of trigonal bipyramid, are commonly named phosphoranes [11,12]. They have the general formula $P(L_a)_2(L_e)_3$, where "a" means an axial position and "e" means an equatorial position (Figure 1).

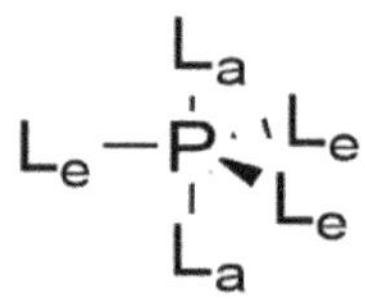

Figure 1. General formula of a phosphorus-containing trigonal bipyramid.

They constitute the family of hypervalent derivatives, showing "trigonal bipyramidal chirality" in which the central phosphorus atom has expanded its valence shell from 8 to 10 electrons (the concept was for the first time proposed by Musher in 1969) [13]. According to general systematic scheme proposed by Martin and coworkers [14,15] [N-P-L ($_{n}A_{m}B$) coding system, in which N stands for the number of valence electrons associated formally with a central phosphorus atom and L shows the number of ligands (A and B stand for the bonding element)]—this group should be considered as the family of 10-P-5 derivatives. The definition of hypervalency is also formally fulfilled by the corresponding phosphonium ylides (in which a phosphorus atom has also expanded its formal valence shell from 8 to 10 electrons). However, when one considers that the important resonance structures of these compounds are polarized from phosphorus to carbon it becomes evident that they represent 8-P-4 species (Figure 2). Therefore they do not meet the condition of hypervalency and they are excluded from this review.

Figure 2. Resonance structure of phosphonium ylides.

The nature of bonding of hypervalent derivatives is presented in details in a few chapters of *"The Chemistry of Hypervalent Compounds"* [16,17] and more recently has again been studied using the topological analysis of the electron localization function (ELF)" [18]. Their structure and reactivity are dominated by the weak three-center-four-electron (3c-4e) bond. Its weakness results from the fact that two of the four electrons that participate in the bonding are in a nonbonding molecular orbital.

Phosphoranes, like other compounds having a trigonal bipyramide (TBP) geometry can exist in an enantiomeric form when the number of different ligands is sufficiently large to induce chirality of such a structure. For 10-P-5 derivatives having the general structure with five substituents (Figure 1) bonded to the central phosphorus atom the numbers of achiral and chiral structures are given in Table 1. According to the Mutterties rule [19] in hypervalent structures the more electronegative ligands tend to occupy the apical positions whereas the lone electron pair "should be located at the equatorial position" [19]. Moreover, "five- and six-membered rings, which stabilize the hypervalent molecules, span both the axial and equatorial positions" [14].

Table 1. Number of chiral and achiral species of 10-P-5 (TBP geometry).

General Structure	Number of Different Structures and Their Chirality
PL_5	1 achiral
$PL^1{}_4L^2$	2 achiral
$PL^1{}_3L^2{}_2$	3 achiral
$PL^1{}_3L^2L^3$	2 achiral and 2 pairs of enantiomers
$PL^1{}_2L^2{}_2L^3$	2 achiral and 3 pairs of enantiomers
$PL^1{}_2L^2L^3L^4$	1 achiral and 6 pairs of enantiomers
$PL^1L^2L^3L^4L^5$	10 pairs of enantiomers

It is evident from this table that phosphoranes containing at least three different ligands can be chiral and for structures $PL^1{}_2L^2{}_2L^3$ chirality may appear in the symmetric bicyclic spiro derivatives

(Figure 3). Therefore, from the synthetic point of view, hypervalent structures exhibiting chirality due to the presence of one or two bidentate ligands should be more easily available than its acyclic analogues.

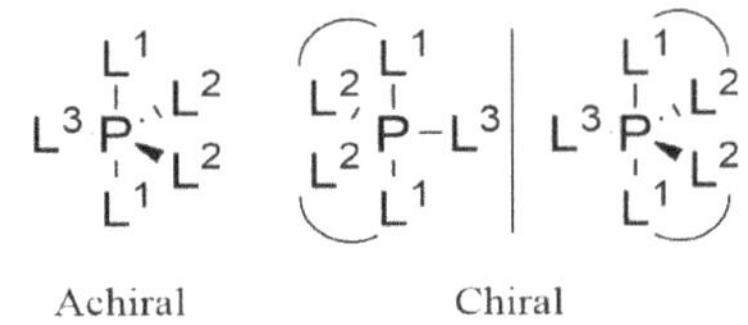

Figure 3. Chiral and achiral structure of hypervalent phosphoranes.

In hypervalent phosphoranes 10-P-5 the apical and equatorial ligands can be interchanged with each other according to the Berry pseudorotation mechanism (BPR) [20–22] (Scheme 1) or by the "turnstile" rotation mechanism (TR) proposed by Ugi and Ramirez [23–26] (Scheme 2). It should be noted here that the very recent DFT calculations of the fluxional behavior of experimentally known pentavalent molecules have suggested the equivalence of the turnstile rotation with the Berry pseudorotation. This suggestion is based on identification of three principal mechanisms by which the substituents interchange can be achieved (Berry pseudorotation, threefold cyclic permutation, and half-twist axial-equatorial interchange) [27].

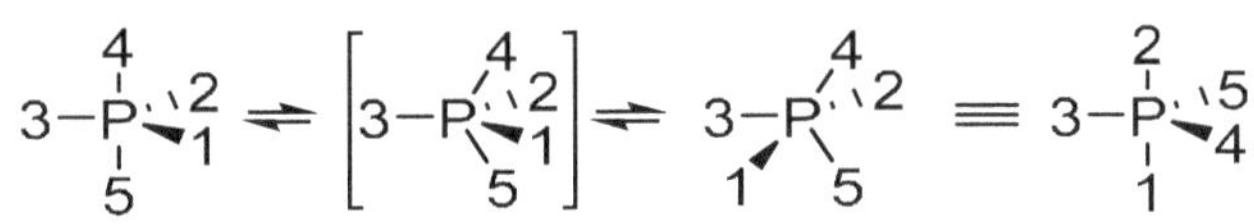

Scheme 1. Berry pseudorotation mechanism.

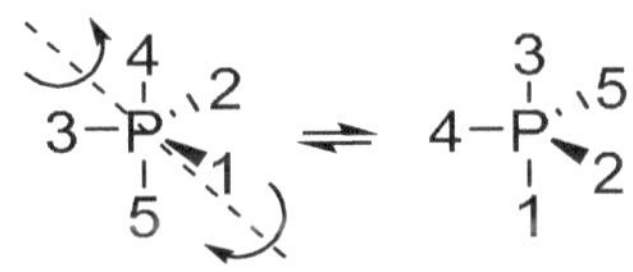

Scheme 2. "Turnstile" rotation mechanism (TR).

As early as 1949, Wittig and Rieber [28] reported isolation of pentaphenylphosphorane as the first example of a pentacoordinated organophosphorus derivative. However, a real opening in the hypervalent organophosphorus chemistry can be ascribed to a much later finding that the Ramirez reaction affords oxyphosphoranes by an oxidative addition of α-diketones to trivalent phosphines [29,30] and Westheimer's proposal that a pentacoordinate phosphorane (able to stereomutate by pseudorotation) is involved in the hydrolysis of phosphoric esters [31]. Since that time, studies devoted to the synthesis, structural determinations, stereochemistry and reactivity of this group of the organophosphorus derivatives, have been carried out in a number of academic and industrial laboratories. Their results are regularly published in renowned chemical journals and, fortunately for those who work on these topics, are regularly presented in the chapters devoted to penta- and hexacoordinated organophosphorus derivatives included in the subsequent volumes of RSC monographic series "*Organophosphorus Chemistry*" [32–43]. In this respect, a very recent review highlighting methods for the asymmetric synthesis of P-chiral pentacoordinated spirophosphoranes should also be mentioned [44].

2. Chiral Hypervalent Phosphoranes and Their Anions

2.1. Chiral Phosphoranes as Reactive Intermediates

In most cases the papers describing the involvement of unstable chiral phosphoranes as reactive intermediates tackle the problem of the mechanism of S_N-P reactions. This is connected with the fact that a nucleophilic substitution at phosphorus can occur either synchronously according to an S_N2-P mechanism or in a stepwise manner by an addition-elimination mechanism (A-E). The later involves formation of a phosphorane as an intermediate, by the addition of a nucleophile$_s$ (Nu) to a chiral substrate. It is now generally accepted that diaxial or diequatorial disposal of entering "Nu" and leaving "L" groups in a trigonal bipyramidal structure of a transient or intermediary phosphorane should lead to inversion of the configuration at phosphorus while the steric course of axial-equatorial substitution (which is possible if a phosphorane intermediate undergoes pseudorotation) is predicted to be a retention (Scheme 3).

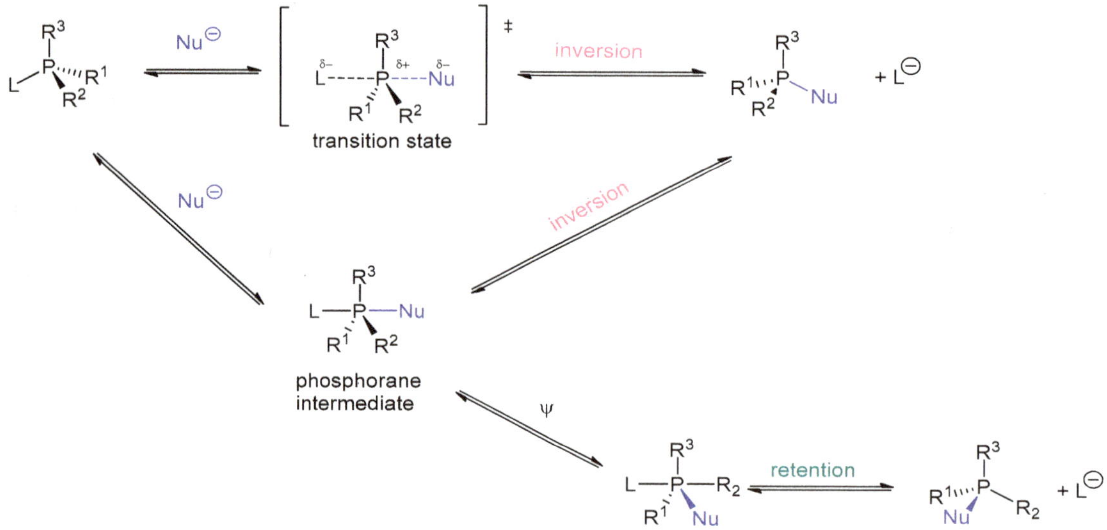

Scheme 3. Possible mechanism for nucleophilic substitution reaction at phosphorus.

The stereochemical consequences (retention or inversion) of such substitution reactions may be conveniently discussed with the aid of the Desargues–Levi graph [45], which originally was proposed by Mislow [46] for stereoisomerization of pentacoordinate phosphoranes.

The participation of such chiral phosphoranes **I–V** as intermediates and their pseudorotation was suggested by Juge and coworkers [47] to explain retention of configuration during the rearrangement of the lithium salts **2a–c** derived from the 2-bromophenylphosphinite boranes **1a–c** to the *o*-hydroxyphenylphosphine boranes **3a–c** induced by halogen–metal exchange (Scheme 4) [47].

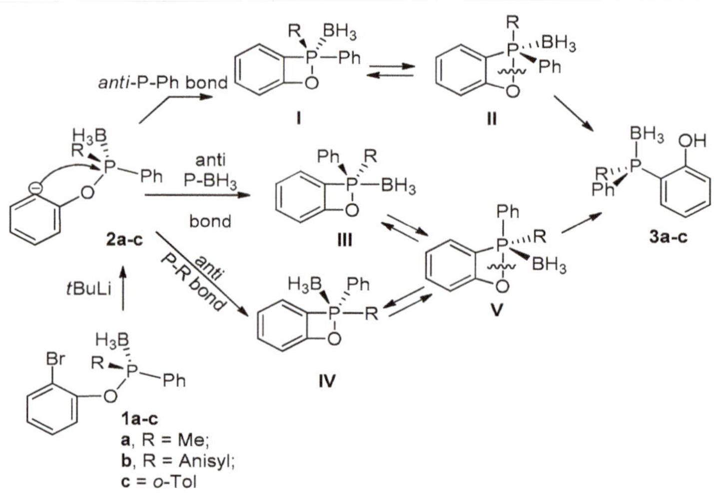

Scheme 4. The rearrangement of the lithium salts **2a–c** derived from the 2-bromophenylphosphinite boranes **1a–c** to the *o*-hydroxyphenylphosphine boranes **3a–c**.

Earlier the same group proposed that pseudorotation of the intermediate phosphoranes **VI, VII** was responsible for predominant retention of configuration during the regiospecific ring opening reaction of the diastereomerically pure dioxaphospholane borane **4** with organolithium reagents which afforded phosphinite boranes **7a,b** (Scheme 5) [48].

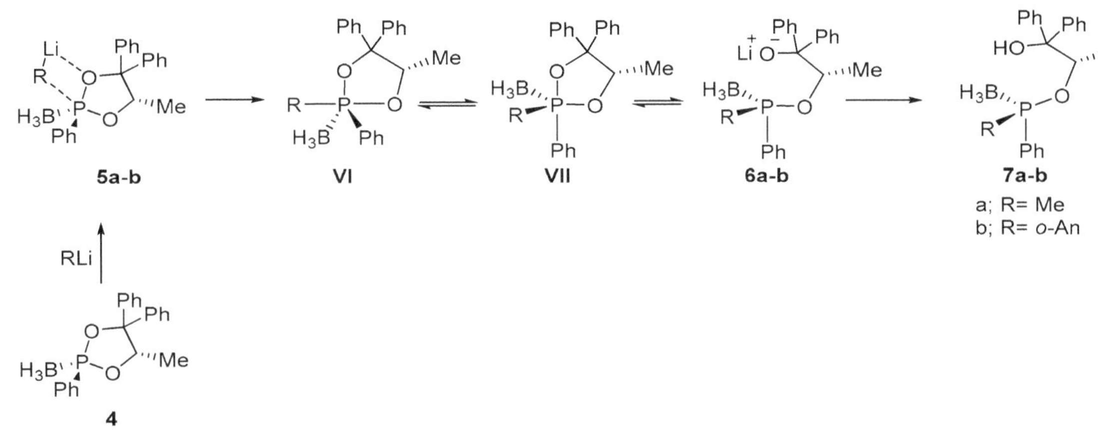

Scheme 5. Stereochemical outcome of the regiospecific ring opening reaction of the diastereomerically pure dioxaphospholane borane **4** with organolithium reagents.

As another example of such a transformation, hydrolysis of quasiphosphonium salts, *cis*- and *trans*-3-methoxy-2,2,6-trimethyl-3-phenyl-1,3-oxaphosphorinanium tetrafluoroborate salts **8a** and **8b** can be presented, by which the corresponding phosphine oxides **9a** and **9b** were formed with complete retention of configuration at phosphorus. The detailed NMR and stereochemical studies allow rationalization of the stereochemical outcome of this reaction in terms of the addition-elimination mechanism, A-E, involving two hydroxyphosphorane intermediates **10a** and **10b** (that are able to undergo pseudorotation) in which at least one oxygen is apical as shown in Scheme 6 [49].

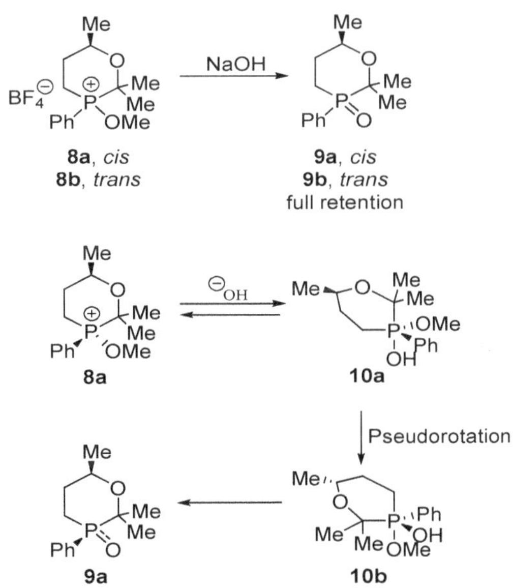

Scheme 6. The hydrolysis of quasiphosphonium salts **8a** and **8b** involving two hydroxyphosphorane intermediates **10a** and **10b**.

Similarly, the participation of the chiral hydridophosphorane **12** as an intermediate in the intramolecular transesterification of the 2-hydroxyphenyl phosphite **11** to the 3-hydroxypropyl phosphite **13** (Scheme 7) was supported by low temperature ^{31}P-NMR measurements [50].

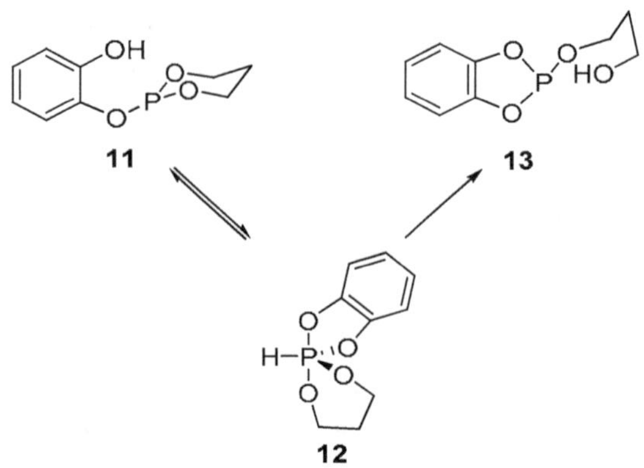

Scheme 7. The intramolecular transesterification of the 2-hydroxyphenyl phosphite **11**.

Earlier, the participation of chiral hydridophosphoranes was proposed in the intermolecular transesterification of chiral asymmetric phenyl phosphites [51] and in the addition of achiral alcohols to compounds having a stereogenic tricoordinated phosphorus atom [52] and very recently in the hydrolysis of a trinucleoside monophosphate by the intramolecular 2α-hydroxy group neighbouring the scissile phosphodiester linkage [53]. The regioselective formation of chiral 3-hydroxypropylphosphinates **17** from cyclic oxaphospholane **14** and Grignard reagents was explained by assuming that during the reactions a pentavalent TBP phosphorane intermediate **15** is formed, which could not undergo pseudorotation to generate another phosphorane **16** due to a high energetic barrier. Therefore, the endocyclic P–O bond in **15** is cleaved almost exclusively to form the phosphinates **17** in high yields (Scheme 8) [54].

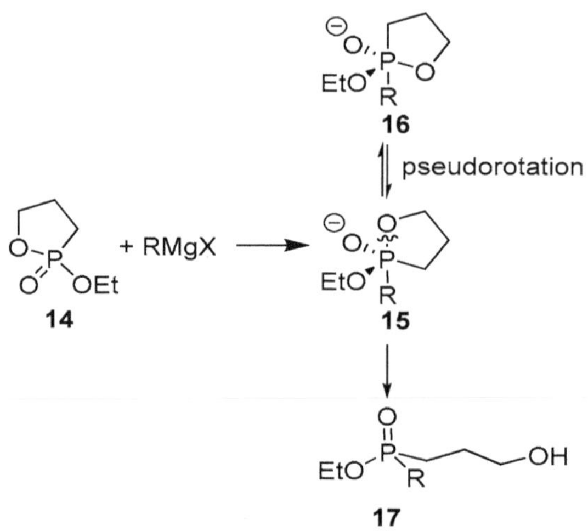

Scheme 8. Reaction of cyclic oxaphospholane **14** and Grignard reagents.

2.2. Chiral Phosphoranes and Their Anions as Isolable Species

As it was mentioned above, from the synthetic point of view, hypervalent structures exhibiting chirality due to the presence of one or two bidentate ligands should be more easily available than their acyclic analogues. Therefore, the majority of isolated hypervalent phosphoranes constitute mono- and especially bicylic derivatives in which a cyclic frame include pentacoordinated phosphorus atom bonded to a carbon and/or heteroatoms.

2.2.1. Chiral Monocyclic Phosphoranes and Their Anions as Isolable Species

The non-concerted [2 + 2] cycloaddition reaction of P-haloylides **18** with trifluorometyl ketones led to the stereoselective formation of monocyclic halophosphoranes **19** (Scheme 9) [55].

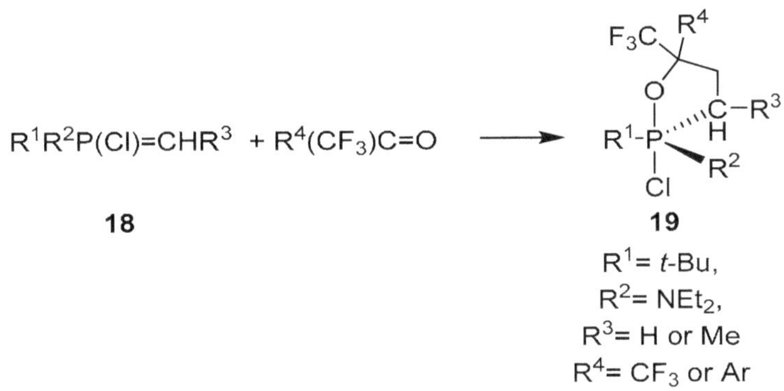

Scheme 9. Monocyclic halophosphoranes 19.

The monocyclic vinyl phosphoranes **21** were prepared by the reaction of cyclic phosphite **20** with dimethyl acetylenedicarboxylate in the presence of the appropriate phenol (Scheme 10) [56].

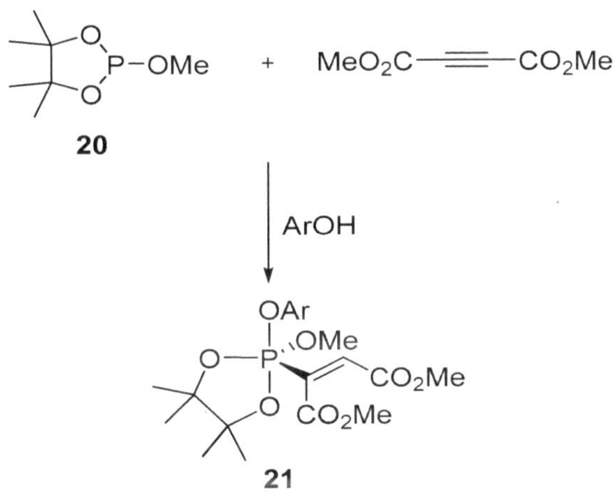

Scheme 10. Reaction of cyclic phosphite **20** with dimethyl acetylenedicarboxylate.

The formation of monocyclic phosphorane **23** was observed in the reaction of ylide **22** with hexafluoroacetone (Scheme 11) [57].

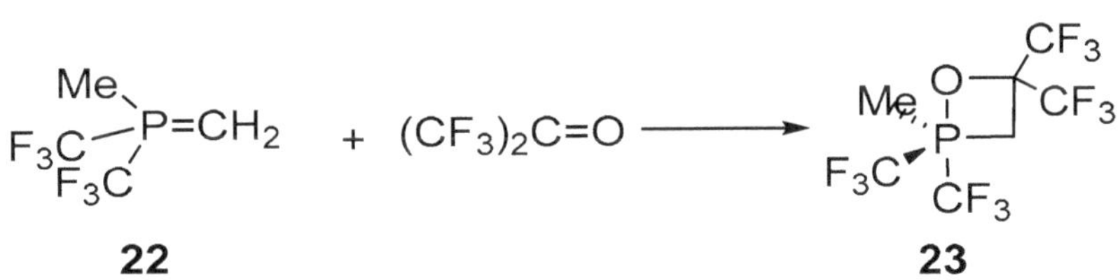

Scheme 11. Reaction of ylide **22** with hexafluoroacetone.

A similar reaction of ylide **24** with hexafluoroacetone gave monocyclic phosphorene **25** (Scheme 12) [57].

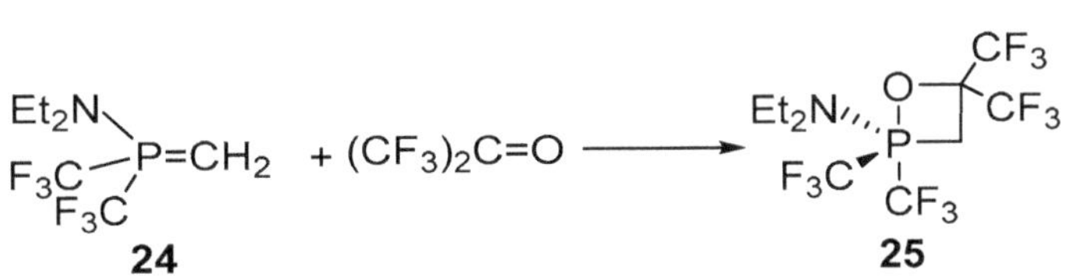

Scheme 12. Reaction of ylide **24** with hexafluoroacetone.

The tricoordinate organophosphorus derivatives **26a–c** were found to react with hexafluoroacetone to give chiral monocyclic phosphoranes **27a–c** (Scheme 13) in which no ligand exchange processes were observed at room temperature. However, the chiral phosphonic acid ester **29** was isolated upon hydrolysis of the chlorophosphorane **27a**, which proceeds most probably via hydroxyphosphorane **28** (Scheme 13) [58].

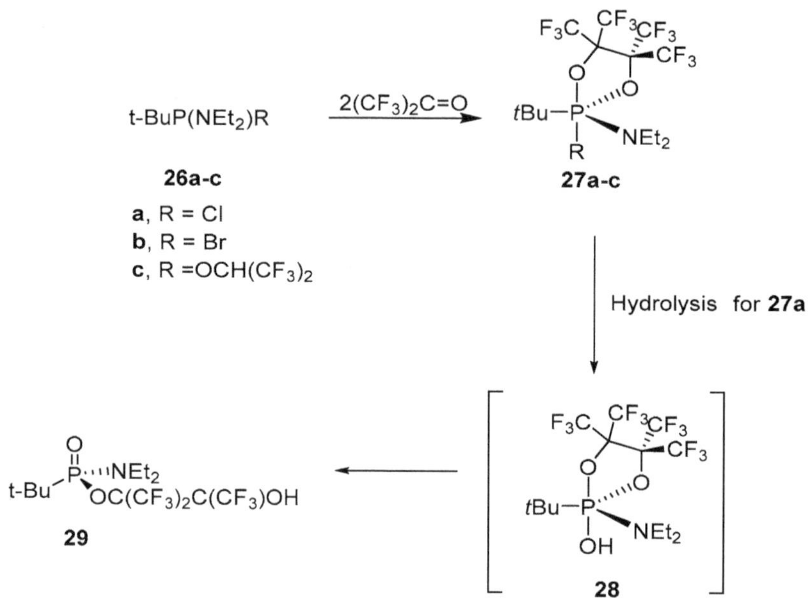

Scheme 13. Reaction of organophosphorus derivatives **26a–c** with hexafluoroacetone.

Preparation of monocyclic halophosphoranes **31a,b** and **31c** was based on the reaction of the tricoordinate phosphinites **30a,b** with benzyl chloride or benzyl bromide. The corresponding fluorophosphoranes **31d** or **31e** were isolated upon the chloride-fluoride or bromide-fluoride exchange reactions when the chloride **31a** or bromide **31c** were used as a substrate respectively (Scheme 14) [59].

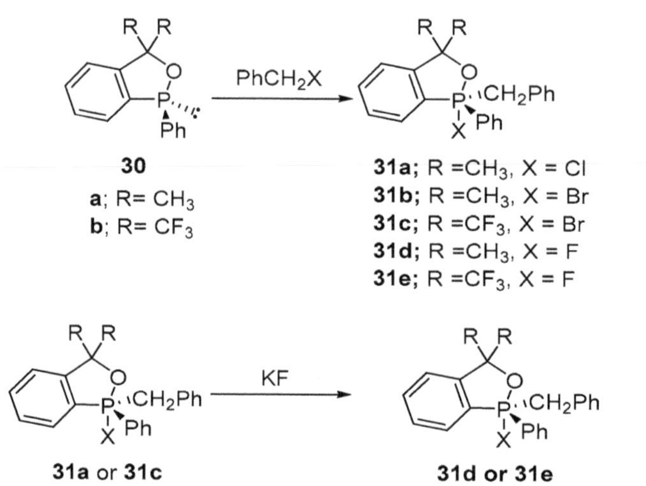

Scheme 14. Preparation of monocyclic halophosphoranes **31a–e**.

Monocyclic (ethylthio)fluorophosphorane **34** was isolated in quantitative yield upon fluoride anion abstraction from BF_4 counter anion of phosphonium salt **33** easily generated by the reaction of cyclic phosphinothionate **32** with triethyloxonium tetrafluoroborate. Defluorination of the fluorophosphorane **34** with trimethylsilyl trifluoromethanesulfonate gave phosphonium salt **35** [60] (Scheme 15).

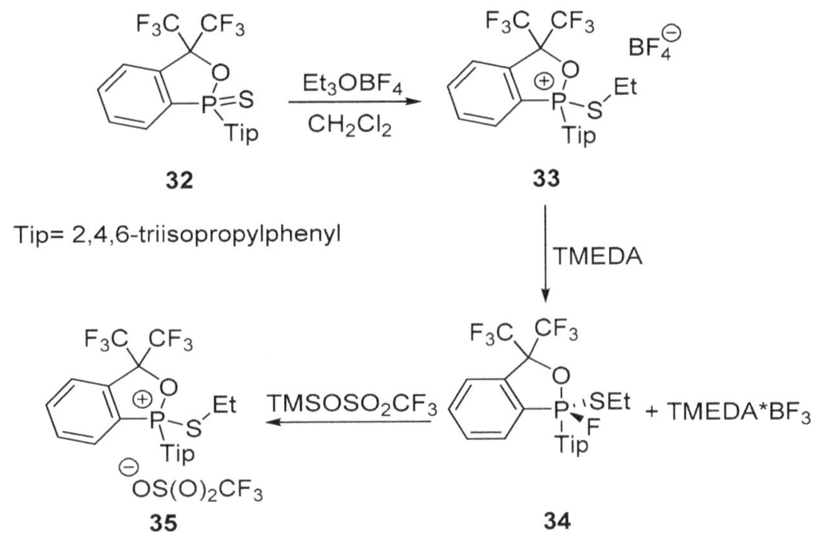

Scheme 15. Preparation of monocyclic (ethylthio)fluorophosphorane **34**.

2.2.2. Chiral Bicyclic Phosphoranes and Their Anions as Isolable Species

A relatively rich number of isolable bicyclic hypervalent phosphoranes have been well documented in the older and more recent chapters by Hellwinkel [61], Burger [62], Holmes [63], Husband and McNab [64], Burgada and Setton [65], Akiba [16], and Kawashima [66]. These chapters mention, more or less extensively, chiral bicyclic phosphoranes, but without deeper comments devoted to the phenomenon of chirality in such derivatives. Therefore, in this part of our compilation we are going to present, as comprehensively as possible, isolable hypervalent derivatives in which their chirality results from the presence of two heterocyclic units forming a P-spiro system. They are divided in accord with the commonly accepted Martin's N-P-L ($_nA_mB$) coding system, in which N stands for the number of valence electrons associated formally with a central phosphorus atom and L shows the number of ligands (A and B stand for the bonding element) [14].

10P-5C Phosphoranes

Hellwinkel as early as 1966 reported on the isolation of the first optically active pentaarylphosphorane. Initially, hexacoordinate phosphorane potassium salt **37** was synthesized in good yield in racemic form by the reaction of bis(biphenyl)phosphonium iodide **36** with 2,2′-dilithio-4-methyl-biphenyl. The optical resolution of the racemic salt **37** was based on its conversion into two diastereomeric ammonium salts by the treatment with *N*-methylammonium iodide derived from brucine. The isolation of the levorotatory diastereoisomer ($[\alpha]_{578}$ = −1200) was achieved in moderate yield by recrystallization of a crude crystalline fraction from acetone. The dextrorotatory diastereoisomer of the ammonium salt was isolated from a mother liquor in poor yield and with a lower optical purity ($[\alpha]_{578}$ = +986). The isolated diastereoisomers were reconverted into the enantiomeric potassium salts upon the treatment with potassium iodide in acetone followed by repeating crystallizations. Their maximum specific rotation reached values $[\alpha]_{578}$ = + and −1870. Acidification of optically active potassium salts of hexacoordinate phosphorus derivatives (+) and (−)-**37** with HCl in methanol/acetone solution provided the mixture of spirophosphoranes from which enantiomeric spirophosphoranes **38** having $[\alpha]_{578}$ = + and −94, respectively were isolated by multiple recrystallization (Scheme 16). Additionally, the structure of **38** as a racemate was confirmed by the alternative method for its preparation using biphenyl-biphenylyl-2-phosphine **39** as the starting material (Scheme 17) [67].

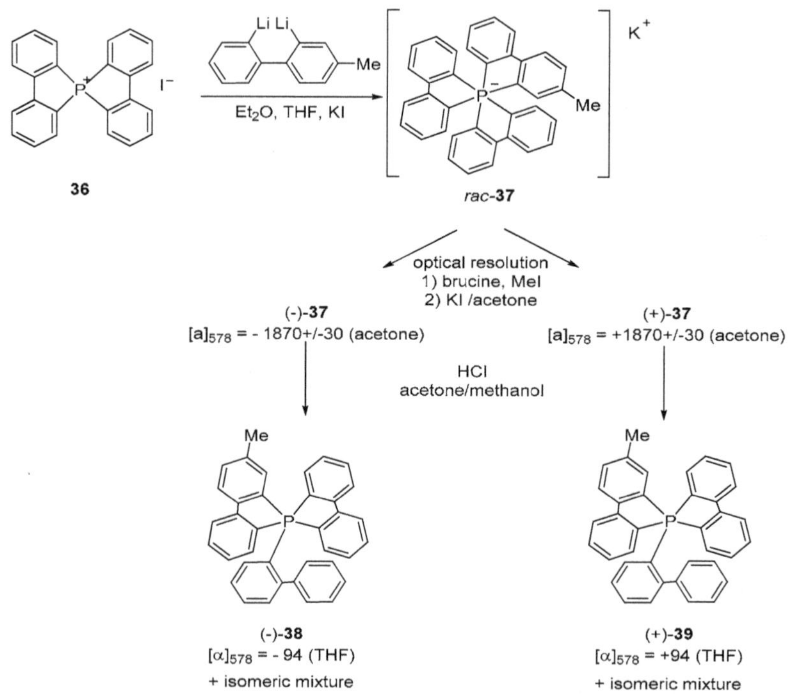

Scheme 16. Preparation of enantiomeric spirophosphoranes **38**.

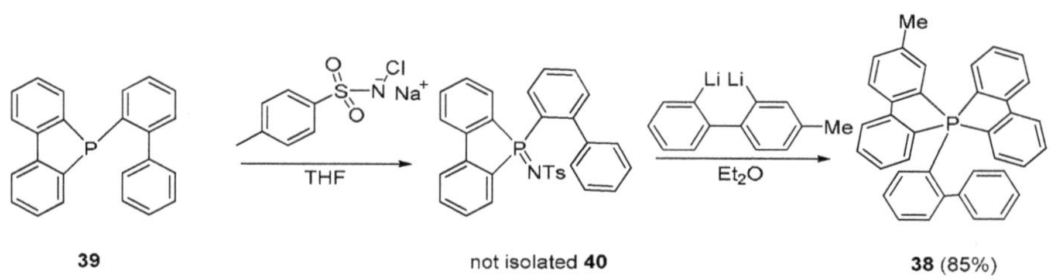

Scheme 17. Preparation of **38** using biphenyl-biphenylyl-2-phosphine **39**.

More recently the reaction of spirocyclotetraalkylphosphonium salts **41** with a series of organolithium reagents RLi, was found to give spirobicyclophosphoranes **42a–e** in good to low yields (Scheme 18). Single crystal X-ray analysis of **42a** showed a TBP geometry with the axial-equatorial rings and the methyl group in an equatorial position. Low temperature NMR measurements indicated that all the pentacoordinate structures **42** show fluxional behaviour in a solution with a very low pseudorotation barrier [68].

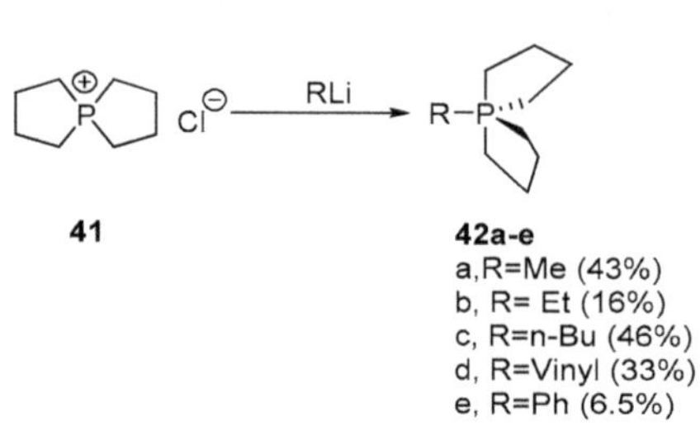

Scheme 18. Reaction of spirocyclotetraalkylphosphonium salts **41** with a series of organolithium reagents.

10P-4C-1O Phosphoranes

Oxaphosphetane (OPA) is a widely recognized intermediate formed via an irreversible [2 + 2] cycloaddition of an aldehyde or ketone with the phosphonium ylide in the Wittig reactions.

The stereoselectivity of the reaction is decided by the course of the first irreversible step. Then, the oxaphosphetane undergoes irreversible [2 + 2] *syn* cycloreversion to give an alkene and phosphine oxide. In most cases the diastereomeric ratio of the oxaphosphetane (OPA) intermediate exactly corresponds to the final alkene, except the cases when a stereochemical drift occurs in the formation of the alkene product. The process "stereochemical drift" refers to the nonstereospecific decomposition of the oxaphosphetane (OPA) intermediate in reactions of certain alkylides with certain aldehydes. The particular investigations on the stereochemical drift in the Wittig reactions (alkylidenephosphonium ylide with aldehydes) under Li-salt-free conditions were undertaken by D. G. Gilheany et al. [69], and based on Variable-Temperature NMR measurements. A particularly suitable candidate for the study of stereochemical drift was intermediate **44a** formed in the reaction of biphenylphenylethylidene-phosphonium ylide **43** with 2-bromobenzaldehyde. It is due to its stability at room temperature (Scheme 19). It undergoes significant stereochemical drift on decomposition in THF at reflux: A sample containing *cis*- and *trans*-**44a** in a ratio of 94:6 gave alkenes of a (*Z*)/(*E*) ratio 82:18 under reflux conditions. On the other hand, the decomposition conducted at 50 °C was not complete after 6d, but the maximum possible (*Z*)/(*E*) ratio that could have been achieved in this reaction was 88:12, revealing that a smaller amount of the stereochemical drift had occurred. A very similar experiment supported with VTNMR spectra analysis over the temperature range −20 to +40 °C on **44b** (produced in the reaction of biphenylphenylethylidenephosphonium ylide **43** and benzaldehyde) was performed. Oxaphosphetane intermediate **44b** was formed in a diastereomeric ratio 71:29 (*cis*/*trans*). Again, the ratio of OPA, ylide and phosphine oxide was invariant within the temperature range from −20 to 30 °C, which indicates that no decomposition to alkene and phosphine oxide occurred. On heating the OPA **44b** was decomposed to produce the alkene in 53:47 (*Z*)/(*E*) isomeric ratio. In the ^{31}P NMR spectra recorded for **44a** and **44b** in the presence of an excess aldehyde additional signals appeared in the pentavalent region (δ = −50 to −80 ppm). The observation was consistent with that reported earlier by Vedejs et al. [70]. Thus, the apparent interaction between OPA and aldehyde was expected to have an impact on the occurrence of the stereochemical drift.

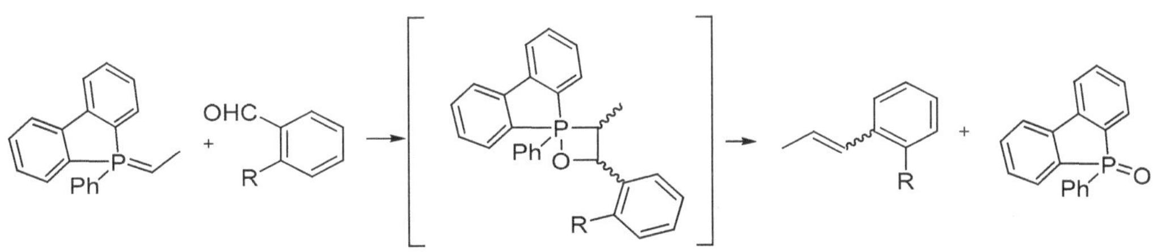

43

44a: R=Br
44b: R=H

Scheme 19. Reaction of biphenylphenylethylidene-phosphonium ylide **43** with benzaldehydes.

10P-3C-1N-1O Phosphoranes

In 1996 Kawashima and Okazaki succeeded in preparing pentacoordinate *N*-apical 1,2-azaphosphetidines, whose structures were fully characterized. They also investigated their thermolysis. The first observation on the occurrence of the *N*-equatorial pseudorotamers was also reported [71]. The synthetic pathway was as follows: the treatment of 2-(methylphenylphosphinyl)-α, α-bis(trifluoromethyl)phenylmethanol **45** with *n*-BuLi led to the generation of methylene carbanion species, which subsequently underwent nucleophilic addition to a Schiff base providing the β-aminoethyl derivative **46**. 1,2-Azaphosphetidine **47** was formed in 93% yield by intramolecular cyclization followed by dehydration under the Mitsunobu reaction condition (Scheme 20).

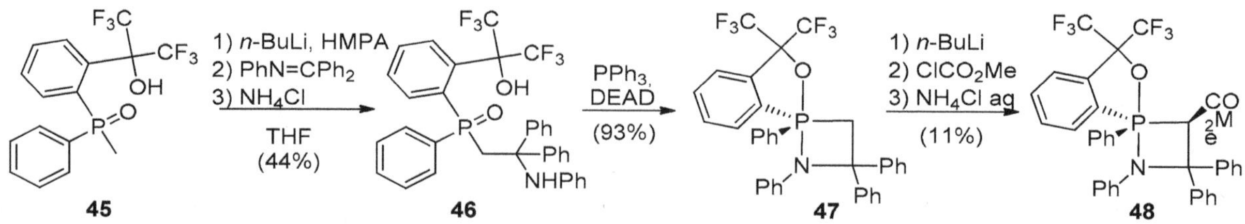

Scheme 20. Synthesis of 1,2-Azaphosphetidine **47** and **48**.

A typical chemical shift in the NMR spectra was observed for compounds with a TBP geometry in which an apical position was occupied with O or N atom. X-ray single crystal analysis revealed the distorted TBP geometry with phosphorus as a central core. Hydrolysis of **47** on silica gel resulted in its cycloreversion to the starting material **46**. Finally, 1,2-azaphosphetidine **47** was converted into the 3-methoxycarbonyl derivative **48**, likewise it had been reported, for the corresponding 1,2-λ^5-oxaphosphetanes [72,73]. In the ^{31}P-NMR spectrum recorded for the solution of **47** in deuterated toluene, apart from the appropriate resonance signal ($\delta_P = -29.9$), another upfielded peak ($\delta_P = -50.9$), was observed which was assigned to the *N*-equatorial pseudorotamer **49** (Scheme 21). A similar pseudorotamer ($\delta_P = -51$) was observed also in the case of **48**.

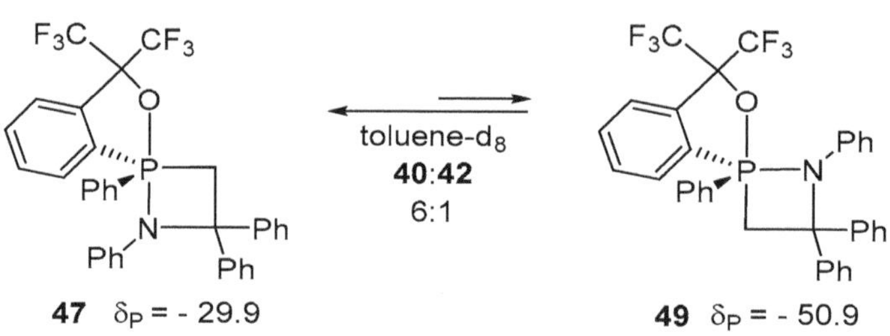

Scheme 21. The equilibrium between phosphoranes **47** and **49**.

Thermolysis of **47** or **48** proceeded with the quantitative formation of the corresponding alkenes and iminophosphorane (Scheme 22), which implies that azaphosphetidines should be considered as aza-Wittig intermediates.

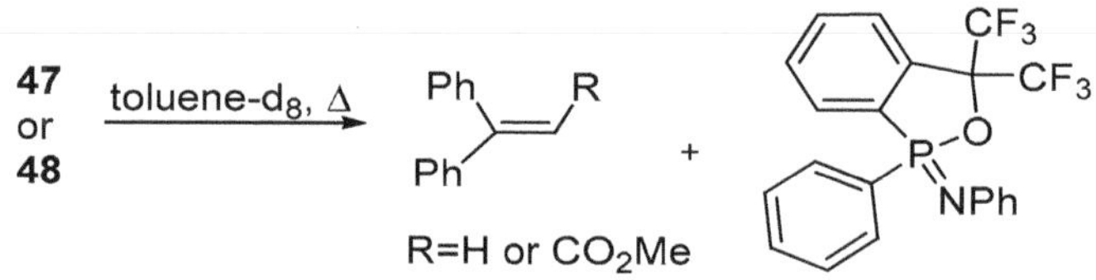

Scheme 22. Thermolysis of **47** or **48**.

The reaction of an iminophosphorane, the nitrogen analogue of a phosphonium ylide, with a reactive alkyne has been intensely studied because it constitutes a useful synthetic pathway for the corresponding α-iminoalkylidenephosphoranes via a [2 + 2]-cycloadduct. The intermediate had neither been observed nor isolated until the first stable pentacoordinate 1,2-λ^5-azaphosphetine **51** was synthesized by Kawashima et al. [74]. Its preparation was achieved by the cycloaddition reaction of the Martin ligand-based iminophosphorane **50** with an alkyne (Scheme 23). X-ray crystallography of **51** showed a distorted TBP with N and O atoms at the apical positions. The variable temperature ^{31}P NMR spectra of **51** in C_7D_8 or CD_3CN showed a shift to a lower field with decreasing temperature indicating that (**51**) was in an equilibrium with the corresponding ylide structure **52**.

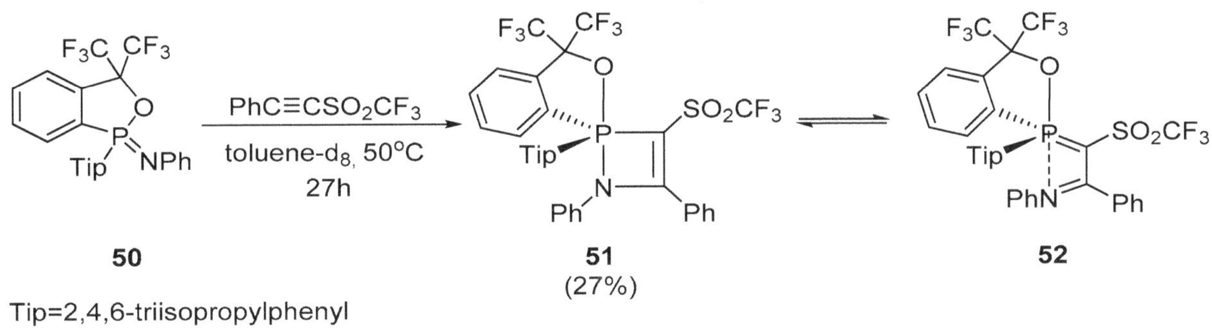

Scheme 23. Cycloaddition reaction of the iminophosphorane **50** with an alkyne.

The isomerization and thermal decomposition of isolable spiro-1,2-oxaphosphetanes has been studied in details with the aim to recognize a stereomutation mechanism of spiro-1,2-oxaphosphetanes, the Wittig intermediate products. Spiro-oxaphosphetanes **54–58**, stabilized by the presence of the *o*-benzamide moiety, were found to be good candidates for the studies [75]. The OPA's **54–56** were obtained according to the previously reported procedure [76] (Scheme 24). The reactions with the hindered ketone, L-(−)-camphor, gave products **57** and **58** (in a 45:55 ratio) arising from the endo attack at the CO group of L-camphor.

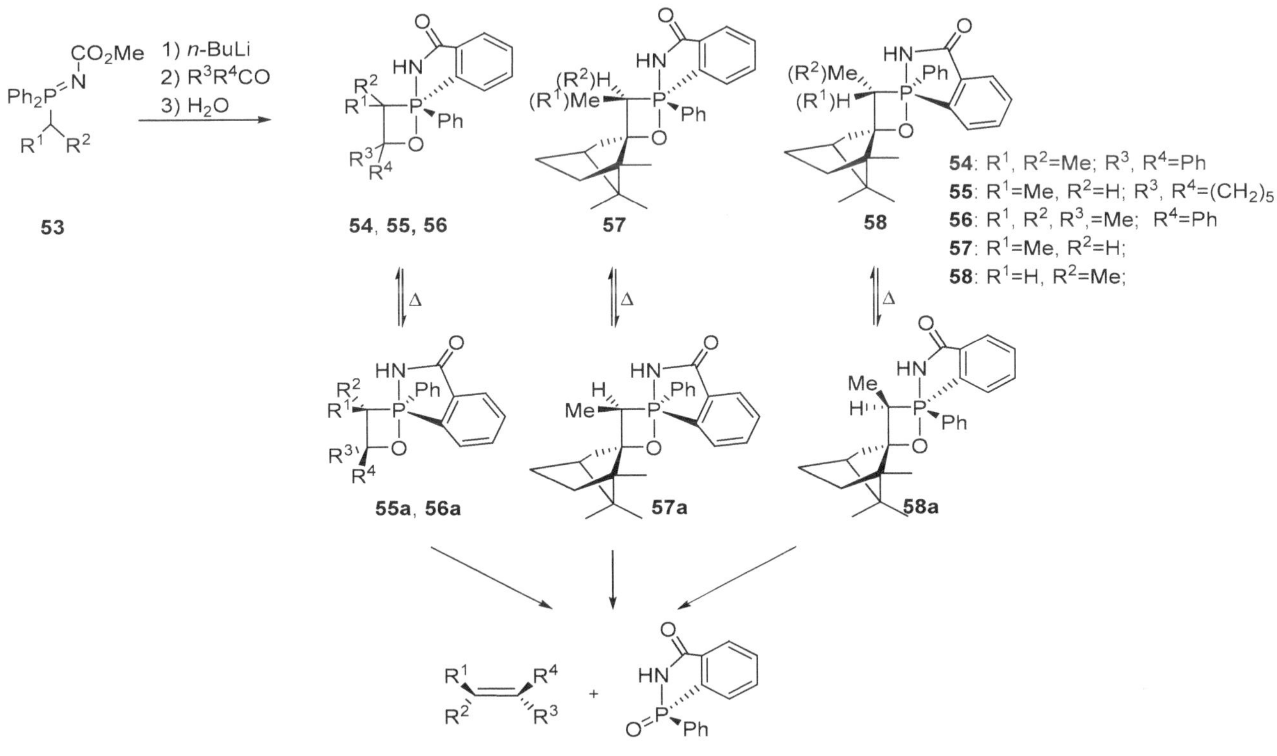

Scheme 24. Synthesis of spiro-oxaphosphetanes **54–58** and their thermal decompositions.

All thermal decompositions of OPAs **54–58** furnished the appropriate alkenes and a phosphinamide, quantitatively. On the thermolysis of **55–58**, partial isomerization to **55a–58a** was observed, which resulted in the inverted configuration at phosphorus atom. It has been found that oxaphosphetane decomposition took place in a single step via a polar transition state. For the first time the stereomutation through three possible mechanisms MB_2, MB_3, and MB_4 involving two, three, and four Berry pseudorotations (at phosphorus atom), respectively, supported by DFT calculations has been evidenced [75].

10P-3C-2O Phosphoranes

The first configurationally stable enantiomeric pair of 10-P-3C-2O phosphoranes with well-defined sole stereogenic centre at phosphorus atom was successfully obtained and characterized by Akiba et al. [77]. The synthesis of diastereomeric **60**-(R_P) and **60**-(S_P) was achieved via a facile alkylation of the in situ generated phosphoranide anion by (−)-menthyl chloroacetate in 87% yield (Scheme 25). The diastereomeric mixture could be resolved effectively by fractional crystallization to furnish **60**-(R_P) in 25% and **60**-(S_P) in 22% yield. Thus, the separation and the selection of suitable crystals allowed for the determination of absolute configurations for both diastereoisomers by X-ray diffraction. The alkylation of P–H phosphorane derivatives was recognized to proceed with complete retention of configuration at the phosphorus sterereogenic centre, most probably due to no detectable racemization at the step of phosphoranide intermediate formation.

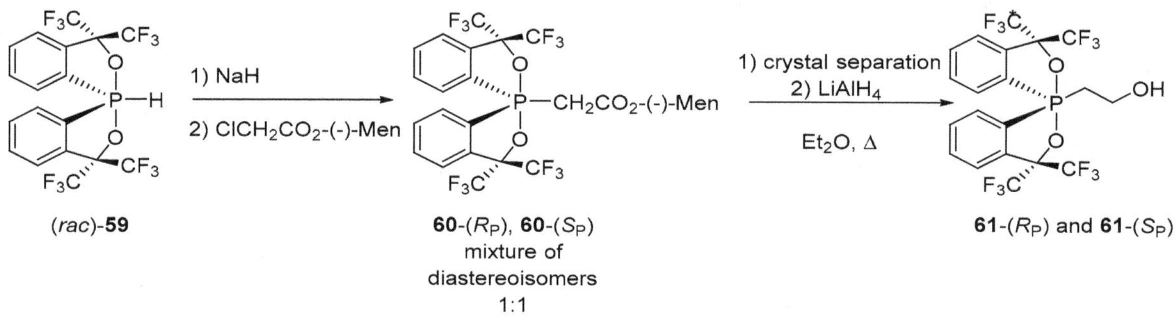

Scheme 25. Synthesis of diastereomeric phosphoranes **60** and enantiomeric phosphoranes **61**.

The removal of the menthyl unit by treating the diastereoisomers **60**-(R_P) and **60**-(S_P) with an excess of $LiAlH_4$ provided the enantiomerically pure phosphoranes **61**-(R_P) and **61**-(S_P), respectively, with a single chirality center at the pentacoordinated phosphorus atom. In order to verify enantiomeric purity of the obtained *P*-chiral 2-hydroxyethylphosphoranes **61**, they were converted into the (*R*)-(+) Mosher esters **62** using (*R*)-(+)-2-methoxy-2-(trifluoromethyl)phenylacetic acid chloride ((+)-MTPA-Cl)) (Scheme 26) [78].

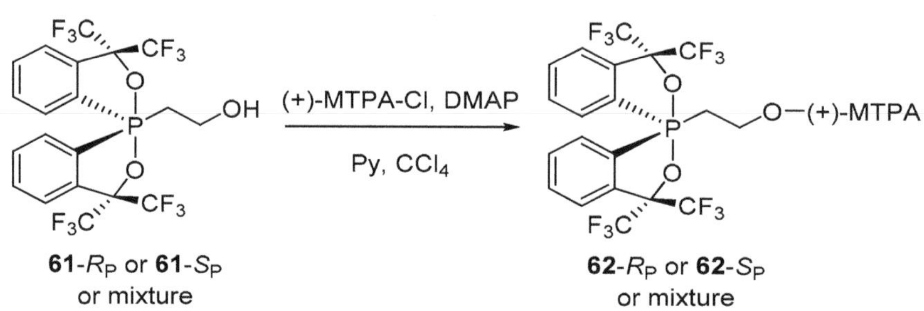

Scheme 26. Conversion of phosphoranes **61** to esters **62**.

Independently, the H-phosphorane **59** was converted (by the treatment with *n*-BuLi) into an anion, which was reacted with CH_2I_2 affording the alkylating product **63** in 82% yield. Subsequently α-iodomethyl phosphorane derivative was treated with α-methylbenzylamine to give a diastereomeric mixture of **64** in 56% yield (Scheme 27). The pure diastereoisomers were isolated by recrystallization and their absolute configuration was determined by X-ray analysis.

It was also found that deprotonation of the nitrogen-bonded proton in each single diastereomer **64** afforded enantiomeric phosphoroanidates **65**-(R_P) and **65**-(S_P) (via imine elimination). They, after acidification, gave the desired optically active P-H phosphoranes **59**-(R_P) and **59**-(S_P) (Scheme 28).

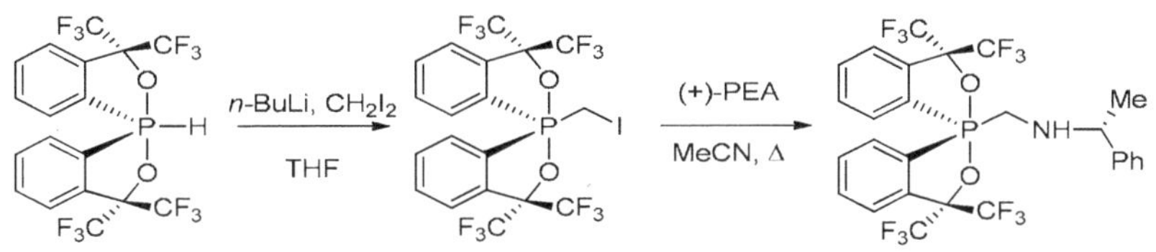

rac-**59** *rac*-**63** **64**-R_P and **64**-S_P mixture of diastereoisomers separated by crystallization

Scheme 27. Synthesis of diastereomeric phosphoranes **64**.

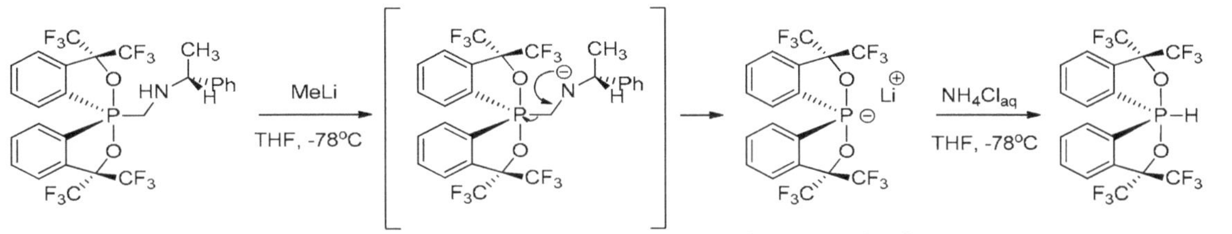

64-(R_P) or **64**-(S_P) **65**-(R_P) and **65**-(S_P) **59**-(R_P) (67%) **59**-(S_P) (87%)

Scheme 28. Conversion of diastereomeric phosphoranes **64** to enantiomeric phosphoranes **59**.

Later on, Akiba and co-workers reported the first example of an "anti-apicophilic" (*O-cis*) spirophosphoranes **67** [79–81], in which the oxygen atom occupies an equatorial position and the carbon atom is located in an apical position of a five membered ring, which is in contrast to the general concept of apicophilicity. According to the procedure shown in Scheme 29 (*O-cis*) phosphoranes **66** were obtained. The treatment of P–H phosphorane **59** with an excess of alkyl(aryl)lithium resulted in the formation of a dianionic intermediate which upon treatment with I_2, as an oxidizing agent, afforded P–I equatorial phosphorane. Its cyclization, initiated by the nucleophilic attack of the alkoxide anion at the phosphorous centre and the simultaneous extrusion of I^- produced *O-cis*-alkyl-or aryl spirophosphoranes **66**.

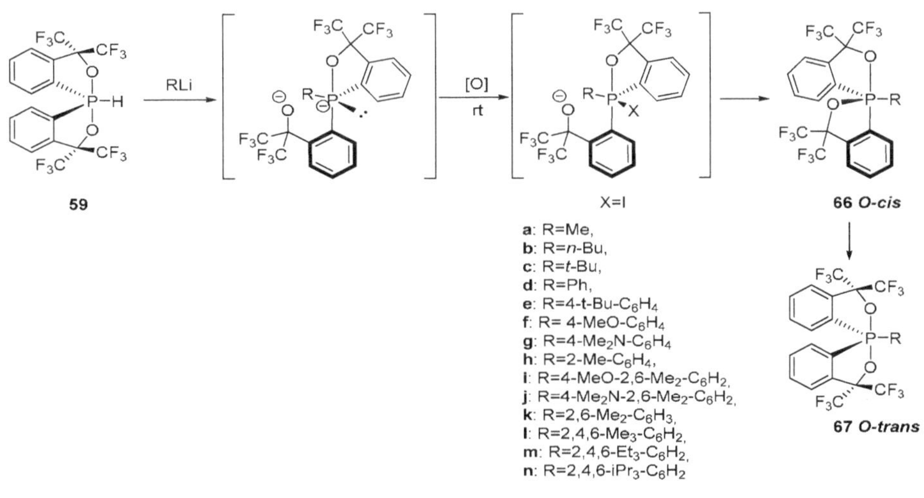

Scheme 29. Synthesis of (*O-cis*) spirophosphoranes **66** and their stereomutation to (*O-trans*) phosphoranes **67**.

The effects of substituents in the monodentate aryl ring of *O-cis* arylphosphoranes on the stereomutation process, along with a kinetic study on the stereomutation of *O-cis* **66k–n** to *O-trans* **67k–n** were studied in details. The exclusive generation of *O-cis* **66** could be observed for the alkyl derivatives **66a–c** right after oxidation, however, in the case of the phenyl derivative **66d** the

O-cis isomer was the minor product. The pseudorotation to *O-trans* **67d** isomer was complete in 30 min at room temperature, which could be explained by the higher apicophilicity of aryl groups in comparison to alkyl groups. For compounds **66e–g**, bearing electron donating groups (*t*Bu, OMe, NMe_2, respectively) in the 4-position of the monodentate aryl groups *O-cis* isomers were the major products. Finally, the use of bulky aryl organometallic reagents containing two substituents in *ortho* positions led to the exclusive formation of the *O-cis* isomers (**66i–n**), having the sufficient lifetimes for characterization. From the relative stability of **66i–n**, it is evident that steric effects contributed mainly to the stabilization against pseudorotation in these isomers that exhibited reversed apicophilicity. In the case of **66n**, pseudorotation was slow enough to allow for its isolation (purification on silica gel in 71% yield) and the crystal structure determination by X-ray diffraction. The differences in the reactivity of *O-cis*-**66b** and *O-trans*-**67b** toward nucleophiles were also indicated (Scheme 30). The reaction of *O-cis*-**66** with TBAF as a nucleophile afforded the hexacoordinated phosphate **68** with the newly formed P–F bond while the *trans*-isomer did not react. Rapid decomposition of **68** caused by the presence of traces of water was observed. *O-Cis* **66b** likewise was reacted with MeLi to give product **69**, while *O-trans*-**67b** treated with organolithium reagent remained unreacted under the same conditions; however by increasing MeLi molar equivalency and temperature the product **69** was successfully isolated.

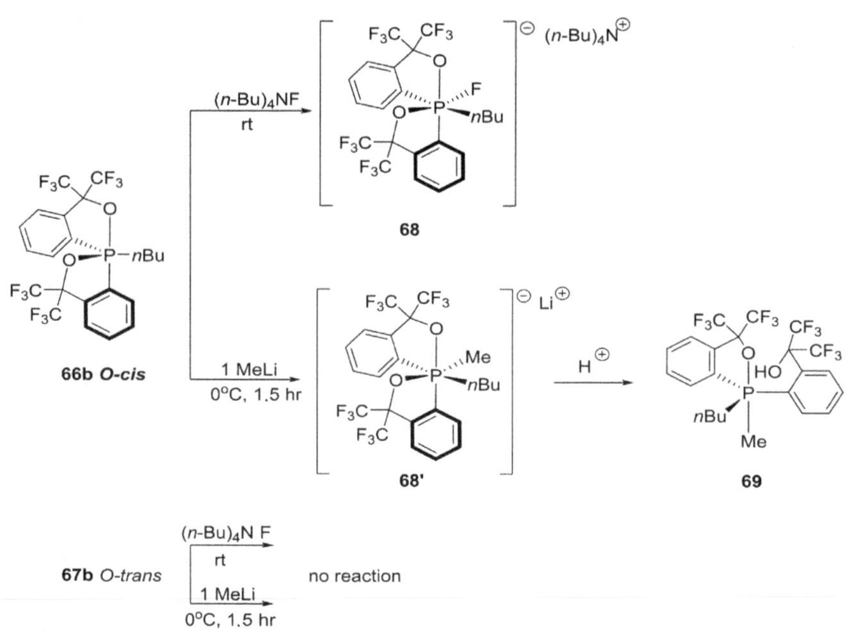

Scheme 30. Reactivity of *O-cis*-**66b** and *O-trans*-**67b** toward nucleophiles.

A three-component reaction of phosphoranide **70**, silane **71**, and THF was applied for the preparation of hypervalent compound **72** containing simultaneously pentacoordinated phosphorus and silicon atoms (Scheme 31) [82,83]. The ^{31}P-NMR spectrum indicated the presence of two diastereomeric products in a ratio of 1:1.

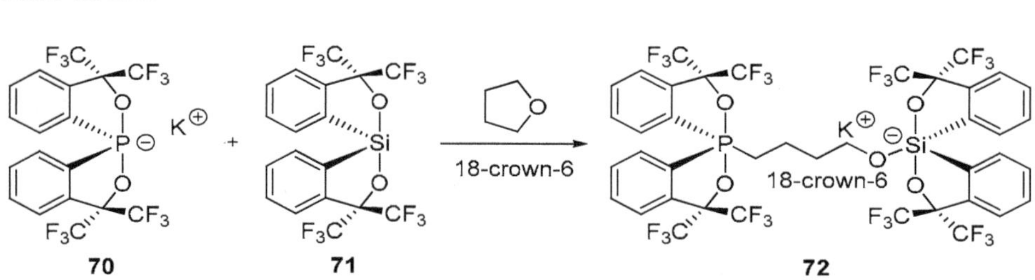

Scheme 31. Reaction of phosphoranide **70** with silane **71**.

The synthesis of other phosphoranylalkoxysilicates was achieved by the nucleophilic addition of phosphoranylalkoxides generated from hydroxy(alkyl)phosphoranes **61** and **73a–c** to silane **74** (Scheme 32). Hydroxy(alkyl)phosphoranes **61** and **73a–c** were prepared by deprotonation of hydrophosphorane **59** with DBU and subsequent treatment with an electrophile, such as hydrogen peroxide, formaldehyde, 2-bromoethanol, and 4-bromobutanol, respectively. Their treatment with KH in the presence of 18-crown-6 followed by the addition of silane **71** in THF afforded the appropriate phosphoranyloxysilicate **74a** (56%), and phosphoranylalkoxysilicates **74b,c** and **72** (53%–96%). Potassium silicates **74a–c** were obtained as a mixture of two diastereomers (in ratios of 66:34 for **74a**, 53:47 for **74b**, 50:50 for **74c**, respectively), as estimated by the integrals in the ^{31}P-NMR spectra [82,83].

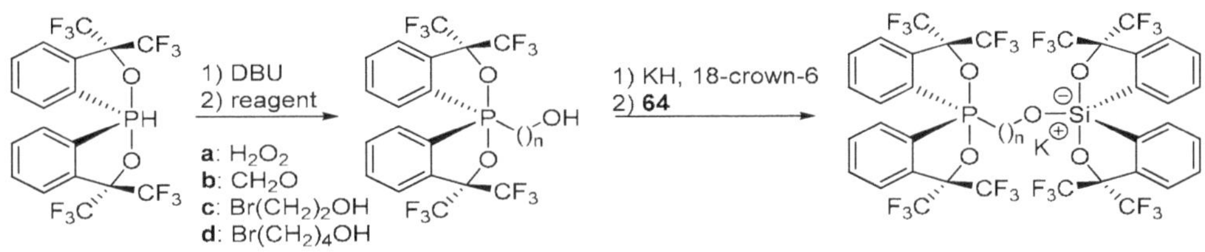

59

73a-d
a: n=0
b: n=1
c: n=4
61: n=2

74a-c,
a: n=0
b: n=1
c: n=2
65: n=4

Scheme 32. Reactivity of phosphoranes **59, 61** and **73**.

More recently, studies on spirophosphoranes, prepared from a new bulkier analogue of the Martin ligand containing two decafluoroetyl units, have been carried out by the Yamamoto group. This new ligand was expected to suppress BPR more efficiently than the Martin ligand [84]. The synthesis of P–H-spirophosphoranes **75a,b** was based on a two-step reaction involving dimetallation of perfluorocumyl alcohol derivatives leading to the corresponding dianion and its subsequent condensation with PCl_3, followed by acidification with 6 molar HCl. The *O*-equatorial spirophosphoranes **76a,b** were obtained by the reaction of **75a,b** with 3 molar equivalents of RLi followed by the oxidation with I_2. Their conversion to the corresponding *O*-apical stereoisomers **70a,b** proceeded in the solution at high temperatures. The stereomutation of **76a-1** to **77a-1** was found to proceed with first–order kinetics. The free energy of activation for stereomutation was higher by 3.6 kcal/mol in comparison with the CF_3 analogue indicating that steric hindrance of pentafluoroethyl group is the major factor for freezing pseudorotation. Recently, Jiang and coworkers [85] obtained a new stable anti-apicophilic phosphorene **76b** with four bulky *n*-C_3F_7 groups, which was converted to more stable *O*-apical phosphorane **77b** only by heating, (Scheme 33).

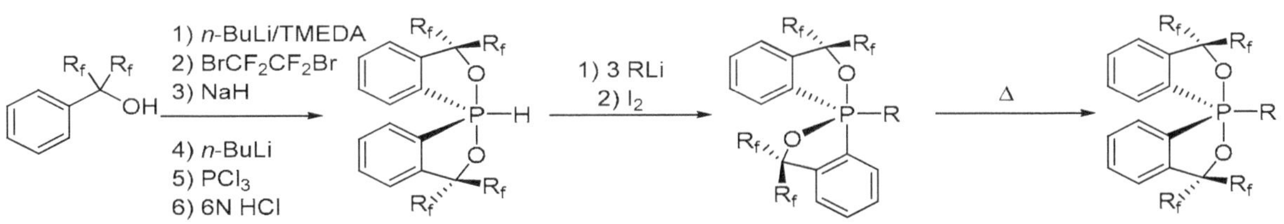

75a-b
a R_f= C_2F_5
b R_f= *n*-C_3F_7

76a-b *O-cis*
a R_f=C_2F_5
(76**a-1**: R=Me,
76**a-2**:R=n-Bu,
76**a-3**:R=t-Bu)
b R_f= *n*-C_3F_7, R=Me

77a-b *O-trans*
a R_f=C_2F_5
(77**a-1**: R=Me,
77**a-2**:R=n-Bu,
77**a-3**:R=t-Bu)
b R_f= *n*-C_3F_7, R=Me

Scheme 33. Synthesis of phosphoranes **76** and **77**.

The same research group elaborated the synthesis of a series of anti-apicophilic phosphoranes bearing a *para*-substituted aryl group ($-C_6H_4(p\text{-}X)$; X = H, CF_3, F, OMe) or a mesityl group as a more electronegative monodentate ligand than an alkyl group. The *O*-equatorial arylphosphoranes **69c–g** were successfully synthesized by the reaction of the P-H spirophosphorane **68a** with an excess of ArLi, followed by treatment with I_2 (Scheme 34) [86]. These phosphoranes were found to be stable at r.t. Their isolation indicates that the steric effect of the C_2F_5 groups for freezing the isomerization is remarkable. By heating in an organic solvent the *O*-equatorial arylphosphoranes **69c–g** were quantitatively converted into the more stable *O*-apical isomers **70c–g**. The effect of a solvent and a para-substituent on the rate of the isomerization were also investigated in order to provide an insight into the $\pi \rightarrow \sigma^*$ P–O interaction in the *O*-equatorial arylphosphoranes. The kinetic study showed a small para-substituent effect on the stereomutations. However, the multi-regression analysis for the para substituents revealed the 1.3 times greater contribution of the resonance effect on the on the isomerization rate in C_6D_6 than the inductive effect suggesting that the $\pi \rightarrow \sigma^*$ P–O stabilizing interaction in the *O*-equatorial isomer plays some role in the isomerization.

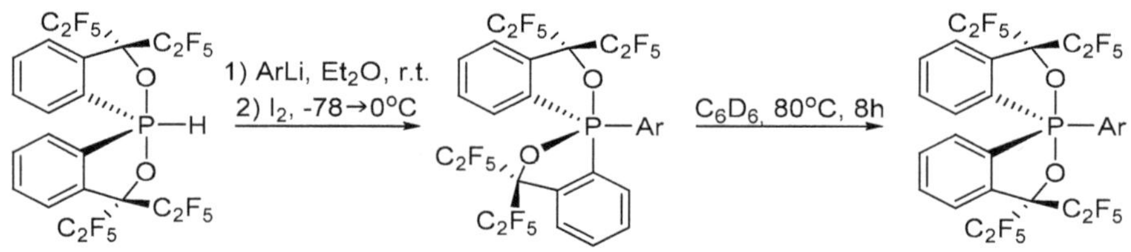

75a **76c-g** *O-cis* **77c-g** ***O-trans***

c: Ar=Ph
d: Ar=4-CF_3-C_6H_4
e: Ar=4-F-C_6H_4
f: Ar=4-MeO-C_6H_4
g: Ar=Mes

Scheme 34. Synthesis of phosphoranes **76** and their interconversion into **77**.

In this context it is interesting to note that the *O*-equatorial methylphosphorane **76a** treated with MeLi was converted to the more stable *O*-apical isomer **77a**, but on the contrary, the *O*-apical phosphorane did not react with MeLi at all (Scheme 35) [87].

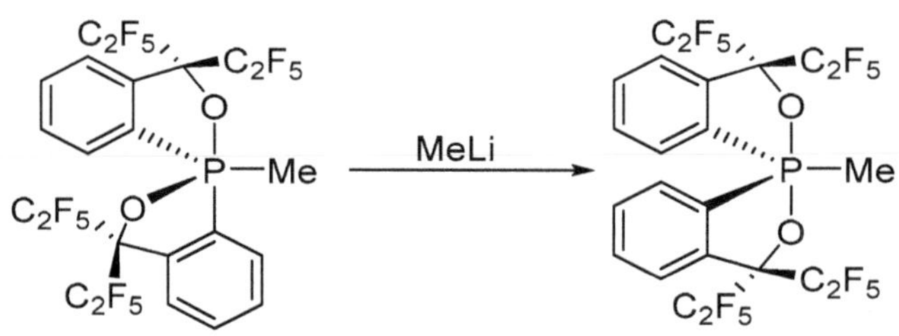

76a *O-cis* **77a** ***O-trans***

Scheme 35. Methyllithium induced epimerization of *O*-equatorial methylphosphorane **76a**.

Deprotonation of methylphosphorane **76a** at the methyl group using superbase (*t*-BuOK/*n*-BuLi) led to the corresponding α-anion, which upon treatment with electrophilic agents gave new phosphorane derivatives **78** and/or **79** (Scheme 36). The *O*-equatorial and *O*-apical phosphoranes having a β-hydroxyethyl group were synthesized according to this procedure using paraformaldehyde and applied as monodentate ligands [87].

Akiba et al. reported the preparation and characterization of an anti-apicophilic spirophosphorane bearing an oxaphosphetane ring and the Martin ligand [88]. Although the substitution pattern on the phosphorus atom slightly differs from that of oxaphosphetane in the typical Wittig reaction they can be considered as a model for the possible reactive intermediate for the Wittig reaction.

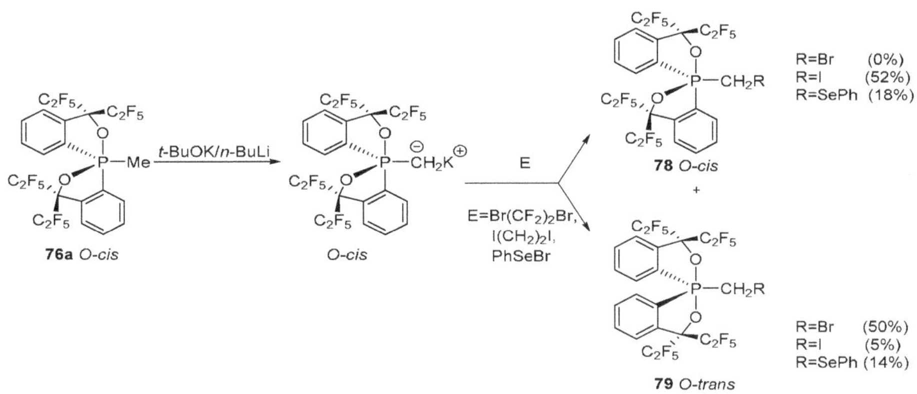

Scheme 36. Synthesis of phosphorane derivatives **78** and/or **79**.

The synthesis of isomeric mixture of oxaphosphetane **80a** and **80b** in 1:1 ratio was achieved by the oxidation procedure using *n*-BuLi followed by iodine in ether (Scheme 37). The sole *anti*-apicophilic **80a** isomer was isolated from the reaction mixture by crystallization upon addition of hexane. Under acidic conditions **80a** underwent complete stereomutation to **80b**. A similar conversion took place within minutes upon dissolution of **80a** in anhydrous $CDCl_3$ at room temperature. A ^{31}P-NMR measurement of the **80b** sample heated to 120 °C in *p*-xylene did not show the presence of **80a** in equilibrium, indicating that **80a** is thermodynamically much less favorable than its *O*-apical isomer. Moreover, heating a solid sample of **80a** at 120 °C for 5 min. gave only **80b** as a result of pseudorotation.

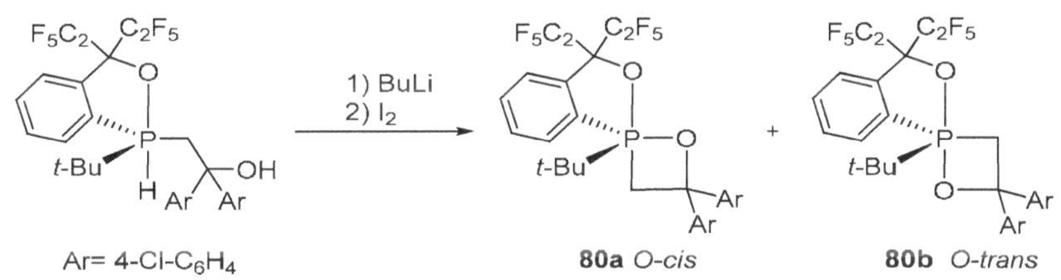

Scheme 37. Synthesis of isomeric mixture of oxaphosphetane **80a** and **80b**.

The synthesis of pentacoordinate phosphirenes **82a** and **82b** bearing a phenyl group on the phosphorus atom was achieved by the reaction of tricoordinate phosphirenes **81a** and **81b** with *o*-chloranil (Scheme 38) [89].

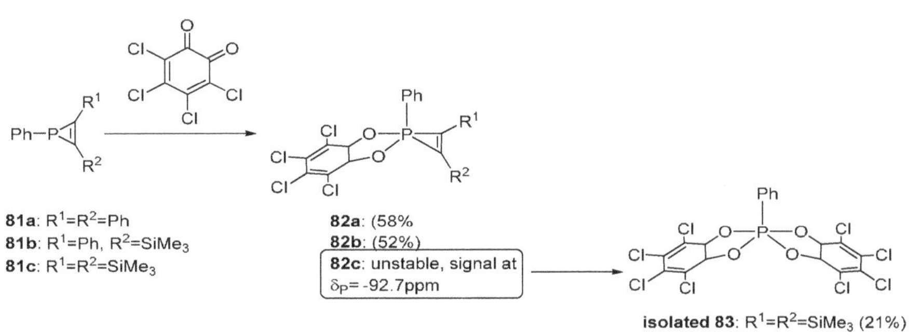

Scheme 38. Reaction of phosphirenes **81a** and **81b** with *o*-chloranil.

The formation of **82c** was evidenced by ^{31}P-NMR spectrum that showed a suitable signal at δ_P −92.7. It was converted rapidly to phosphorane **83** bearing two tetrachlorocatecholate ligands. For relatively more thermally stable phosphirene **82a** the crystal structure was determined by X-ray analysis. It showed that phosphirene **82a** has a highly distorted square pyramidal SP arrangement at the phosphorus atom as a core. The basal positions are occupied by two oxygen atoms from the tetrachlorocatecholate ligand and two carbon atoms belonging to the three-membered ring of phosphirene, when the phenyl group is located at the apical position. No decomposition was observed for **82a** after heating at 60 °C for 40 h. On the other hand, phosphirene **82b** was thermally less stable, and decomposed at room temperature to give the *o*-chloranil derived phenylphosphinite and phenyl(trimethylsilyl)acetylene. These results indicate that an attachment of a trimethylsilyl group to the endocyclic carbon lowers the thermal stability of the pentacoordinate phosphirenes.

10P-3C-1O-1Se; 10P-3C-1O-1S; 10P-3C-1O-1F; 10P-3C-1O-1H Phosphoranes

The synthesis of the first 1,2-σ^{5-}selenaphosphirane and 1,2-σ^5-thiaphosphirane involving a pentacoordinate phosphorus atom was based on the reaction of the Martin ligand based phosphorus ylide with elemental selenium [90], $CpTiSe_5$ [91] or elemental sulfur [92]. Treatment of ylide **84** with 1.6 equiv. of elemental selenium in THF at room temperature or with Cp_2TiSe_5 under the same conditions resulted in the formation of selenaphosphirane **85** as a solid in 96% and 85% yield, respectively (Scheme 37). Although the selenaphosphirane **85** is highly moisture-sensitive, its isolation was successful under an argon atmosphere and its structure was crystallographically characterized. Both, the solid-state $^{31}P\{^1H\}$ NMR spectrum and the $^{31}P\{^1H\}$ NMR spectrum of the C_6D_6 solution showed a single peak with chemical shifts at −26.1 and −26.6, respectively. This was considered as evidence for the pentacoordinate state of the phosphorus atom of the selenaphosphirane. It is interesting to note that in the $^{77}Se\{^1H\}$-NMR spectrum (the resonance signal at δ_{Se} 147.5 in C_6D_6), the coupling constant between the phosphorus and the selenium nuclei was not observed. This selenaphosphirane upon treatment with methyl triflate in $CDCl_3$ afforded the highly air-sensitive α-(methylseleno)-α-methylethyl phosphonium triflate **86** in 76% yield (Scheme 39). The formation of **86** was initiated by the electrophilic attack of methyl triflate on the negatively charged selenium atom in **85** and proceeded with the subsequent cleavage of the polarized P–Se bond. In a similar manner the thiaphosphirane **87** was prepared in 68% yield, by the reaction of the ylide **84** with 0.96 equivalent of elemental sulfur in THF at −30 °C for 7 h (Scheme 39). Thiaphosphirane **87** was found to be air-sensitive and its single crystal was obtained by recrystallization from hexane under an argon atmosphere. Its X-ray crystallographic analysis showed that the oxygen and sulfur atom are located at apical positions.

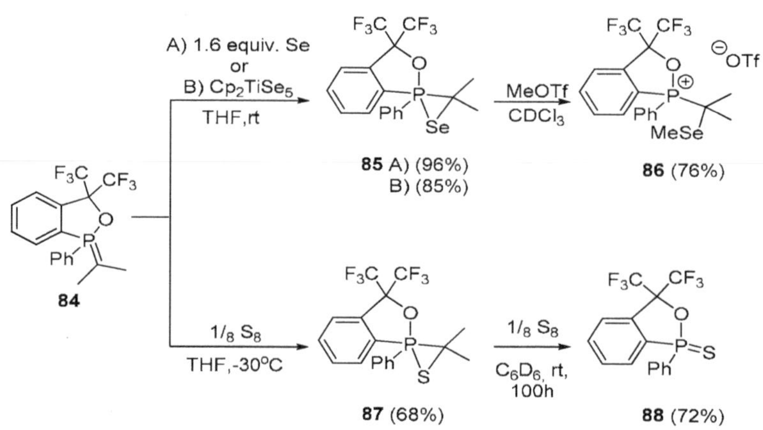

Scheme 39. Reactions of monocyclic phosphorus ylide **84**.

Thiaphosphirane **87** showed inertness toward triaryl- or trialkylphosphines. This is in a contrast to tri- and tetracoordinate thiaphosphiranes which underwent desulfurization to form phosphaalkene derivarives and phosphine sulfides [93]. When the thiaphosphirane **87** was treated with additional portion of elemental sulfur the corresponding phosphinothionate **88** was isolated in 72% yield. The independent experiment showed that the phosphorus ylide **84** treated with 2 equiv. of elemental sulfur provided phosphinothionate **88** in 31% yield.

10P-2C-2N-1O; 10P-2C-1N-2O Phosphoranes

The reactions of iminophosphorane bearing the Martin ligand and a bulky 2,4,6-triisopropyl-phenyl group with a ketone, an isothiocyanate and an alkyne has been found to proceed with the formation of the corresponding cycloadducts as novel heterocycles bearing pentacoordinated phosphorus atom [94,95]. Cyclic iminophosphinate **89** reacted with carbonyl compounds, phenyl isothiocyanate and phenylethynyl trifluoromethyl sulfone affording the appropriate cycloadducts: 1,3,2λ^5-oxazaphosphetidines **90a–c**, 1,3,2λ^5-diazaphosphetidine-4-thione **91**, 1,2λ^5-azaphosphetine **92**, respectively. The reaction with hexafluoroacetone proceeded at room temperature yielding cyclic phosphorane which could be even purified by silica gel chromatography. With dimethyl acetylenedicarboxylate (DMAD) and water the reaction provided 1,2λ^5-oxaphosphol-5(2*H*)-one **93** instead of the expected 1,2-azaphosphetine (Scheme 40).

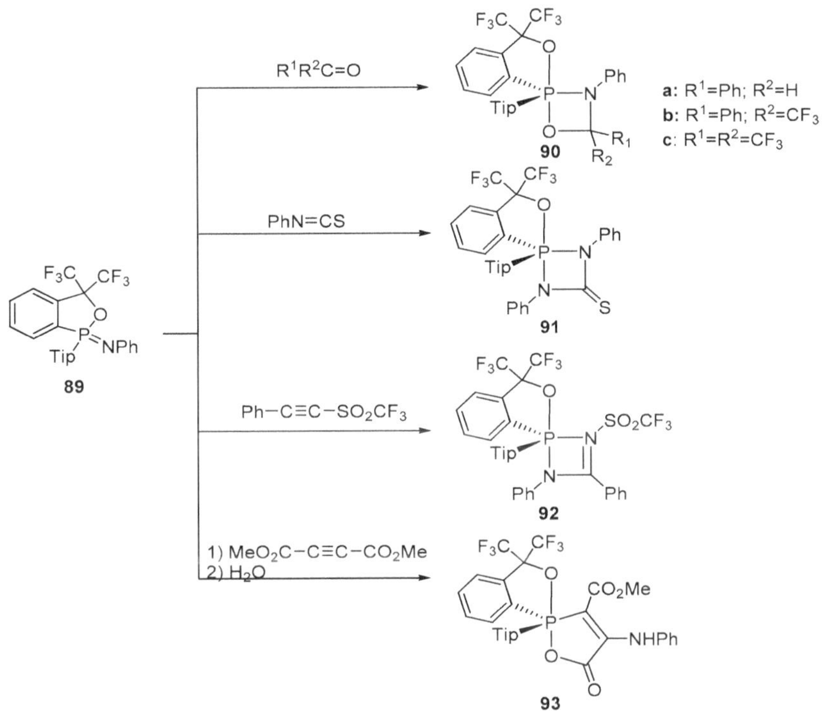

Scheme 40. Syntheses of bicyclic phosphoranes **90–93**.

Thermolysis of **90c** at 140 °C in a sealed tube gave the corresponding imine **94** and cyclic phosphinate **95** (Scheme 41), indicating that 1,3,2λ^5-oxazaphosphetidine **90c** is regarded as an intermediate of the aza-Wittig reaction [95].

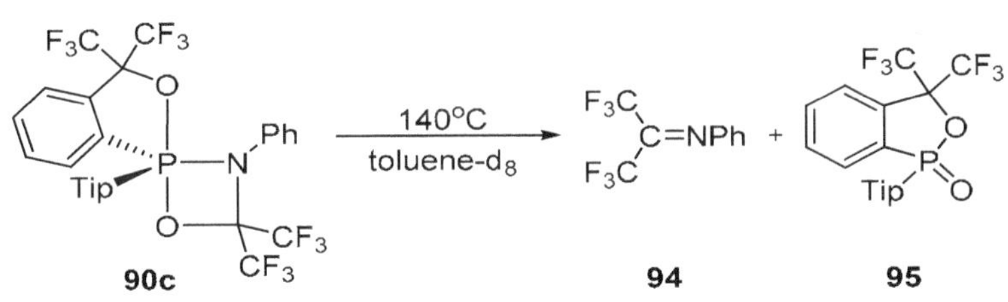

Scheme 41. Thermolysis of spirophosphorane **90c**.

The detailed studies on the formation of 1,2λ^5-oxaphosphol-5(2*H*)-one **93** were conducted on the basis of ^{31}P-NMR analysis [96]. In the course of the reaction a decrease of the resonance signal at +25.6 ppm along with an increase of the new one at +47.3 ppm corresponding to **96** was observed.

The latter was found to be unstable in the presence of moisture which resulted in the slow conversion to the final product. The proven synthetic pathway was concluded in Scheme 42.

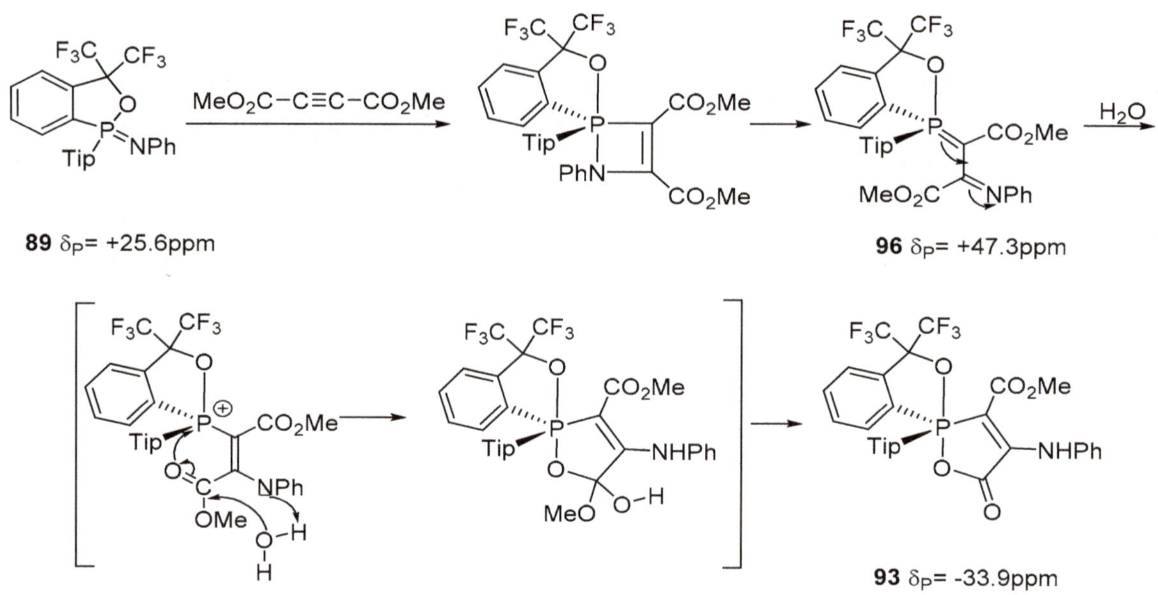

Scheme 42. The mechanism of formation of the bicyclic phosphorane **93** from cyclic iminophosphinate **89**.

10P-2C-2O-1Z (Z = N, O, etc.) Phosphoranes

It is already well known that the preparation of pentacoordinate phosphoranes can be accomplished by the reaction of phosphorus (III) compounds with ylidene derivatives of β-carbonyl compounds. This approach was exemplified by the reaction of 2-phenyl-1,3,2-benzodioxaphosphorin-4-one **97** with diethyl benzylidenemalonate **98** which occurred along two pathway, yielding pentacoordinate phosphorane **99** and phosphinate derivative **100** containing tetracoordinate phosphorus atom [97]. When the reaction was allowed to heat for a short time the formation of the seven-membered phosphonate derivative with relatively high stereoselectivity equal to 50% was observed (Scheme 43).

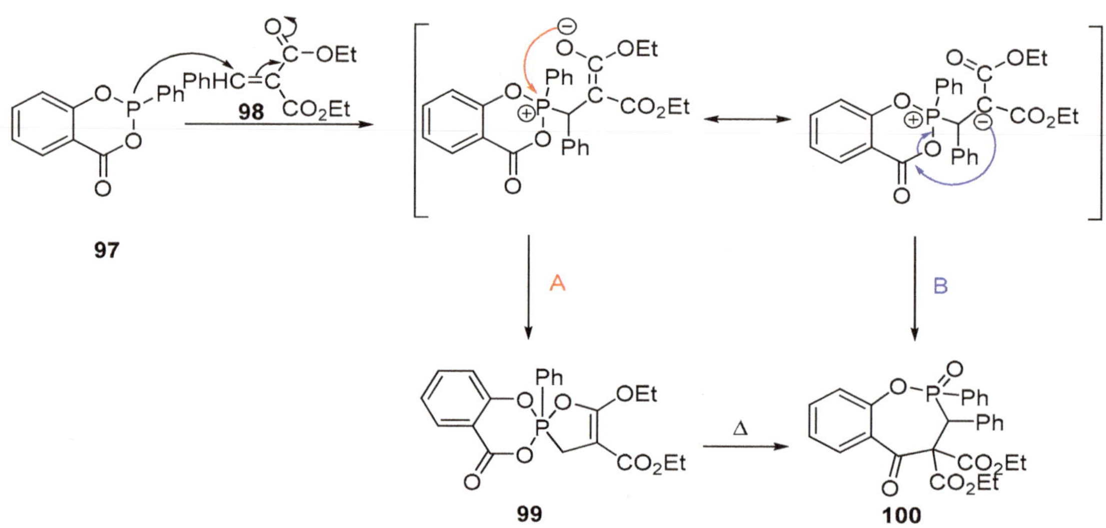

Scheme 43. Synthesis of pentacoordinate phosphorane **99** and phosphinate derivative **100**.

The plausible mechanism involves the nucleophilic attack of the phosphorus atom at the β-carbon atom in α,β-unsaturated carbonyl compounds to produce an intermediate with delocalized π-electrons along three atoms, as is presented by its two resonance structures. In case of synthetic pathway A the attack of negatively charged oxygen on phosphorus atom occurs with reversible formation of pentacoordinate phosphorane. The second synthetic pathway (B) assumes nucleophilic substitution at the carbonyl carbon atom to yield the seven-membered ring, 1,2-benzoxaphosphepine. In further investigations Akiba et al. [98] succeeded in preparing spirophosphoranes bearing Martin ligand units and a monodentate primary amino substituent. These *O*-equatorial isomers were found to be surprisingly stable, taking into account the fact that the equatorial substituents, which are more electronegative than carbon, are expected to facilitate pseudorotation. The *O*-equatorial aminophosphoranes **102** along with *O*-apical isomers **103** were obtained from the in-situ generated chlorophosphorane **101** which was allowed to react with the corresponding amines (Scheme 44). However, when methylamine or aniline were used, the corresponding *O*-equatorial isomer were not produced.

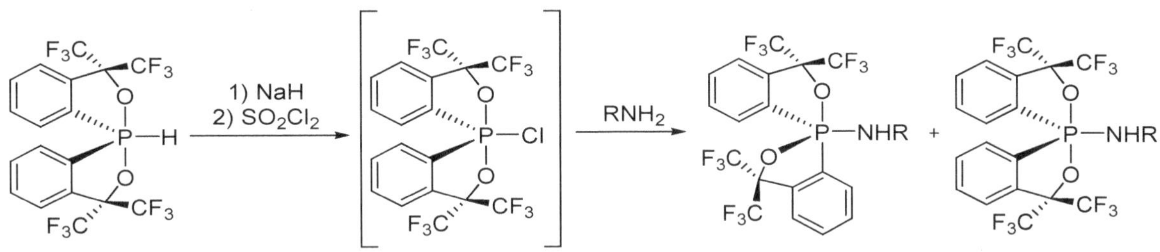

59 **101** **102** **103**

	102/103
a: R=Me,	2/98
b: R=*n*-Pr,	36/64
c: R=$PhCH_2$,	62/38
d: R=*i*-Pr,	80/20
e: R=*t*-Bu	98/2
f: R= Ph,	2/98

Scheme 44. Synthesis of aminophosphoranes **102** and **103**.

The *O*-equatorial phosphoranes **102** were stable at room temperature and could still be converted to their more stable *O*-apical pseudorotamers **103** when they were heated in a solution, indicating that the phosphoranes **102** are kinetic products, as in the case of *O*-equatorial alkylphosphoranes. The structures of **102c** and **102e** were determined by single-crystal X-ray analyses to be in the *O*-equatorial configuration. The unusual stability of the *O*-equatorial phosphorane **102** could be attributed to the stabilizing $n_N \rightarrow \sigma^*_{P\text{-}O}$ orbital interaction. More detailed studies on the experimental determination of the $n_N \rightarrow \sigma^*_{P\text{-}O}$ interaction energy of *O*-equatorial *C*-apical phosphorane supported with DFT calculations have been also reported providing an additional proof for this assumption.

Juge et al. described the formation of chiral bicyclic phosphoranes in the Arbuzov-type reaction of diastereomerically pure cyclic tricoordinated oxaphosphacycloalkanes and very reactive Koshland reagent I **104** [99]. They found that both enantiomers of 2-phenyl-1.3.2-oxazaphospholidine [(−)-**105** derived from (+)-ephedrine or (+)-**105** from (−)-ephedrine], treated with 2-hydroxy-5-nitrobenzyl bromide **104a** afforded the stable pentacoordinated phosphoranes [(−)-**106** from (−)-**105** or (+)-**106** from (+)-**105**] which were isolated by simple chromatography. The minor isomeric spirophosphoranes with the opposite configuration at the phosphorus atom were also formed (Scheme 45). 2-Hydroxy-5-nitrobenzyl halides **104a,b** reacted also with five- or six-membered cyclic tricoordinated phosphorus compounds (−)-**107** (derived from (+)-2-methyl-2,3-butanediol) or (−)-**110** (derived from (−)-chloramphenicol), yielding spirophosphoranes, **108** and **109** or **110**, respectively. However, these spirophosphoranes were too unstable to be isolated. Their formation was confirmed only by the ^{31}P-NMR spectra of a crude reaction product.

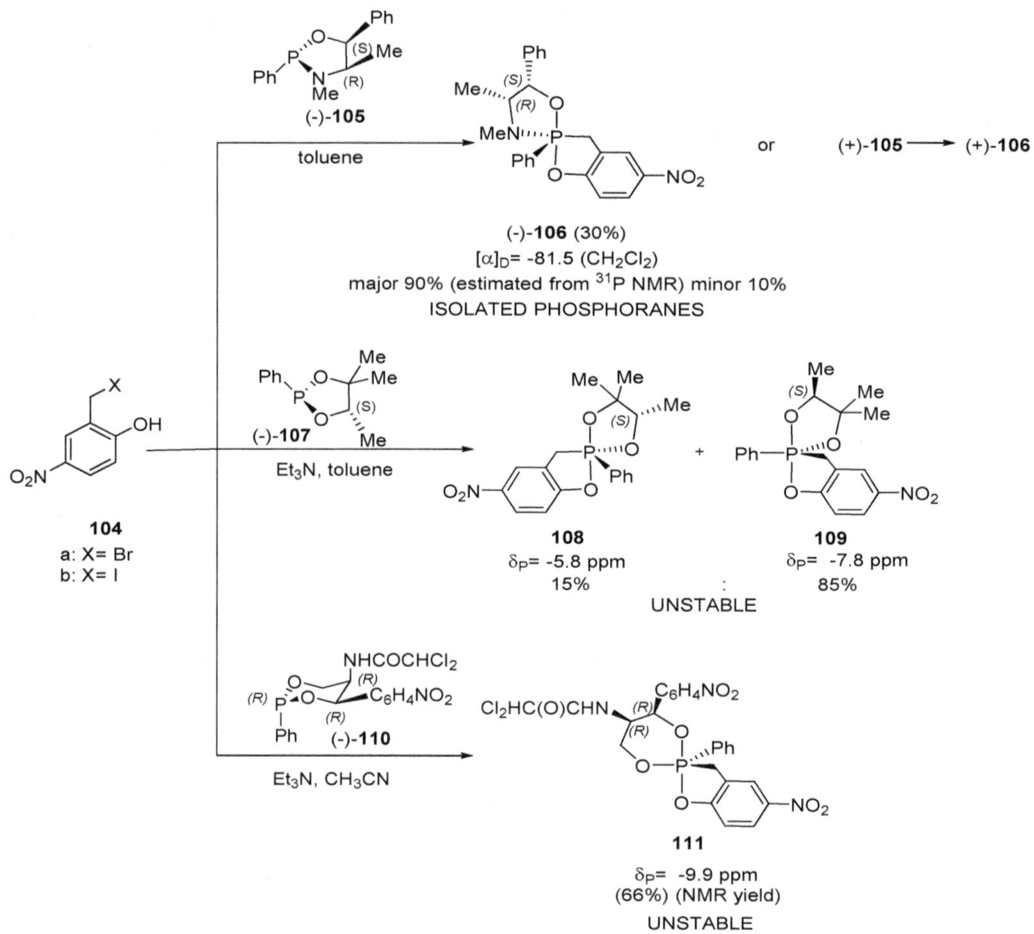

Scheme 45. Reactions of cyclic five or six-membered tricoordinated phosphorus compounds with 2-hydroxy-5-nitrobenzyl halides.

10P-1C-1N-3O Phosphoranes

The synthesis of a pentacoordinated phosphorane containing tricyclic system was reported by Sevenard et al. [100]. 2-Fluoroacetylcycloalkanones **111a,b** reacted with diethyl isocyanatophosphite diastereospecifically leading to phosphoranes **112a,b** via intermediate species formed by the addition of phosphorus to a trifluoromethyl substituted carbonyl atom which subsequently underwent two additional heterocyclizations (Scheme 46).

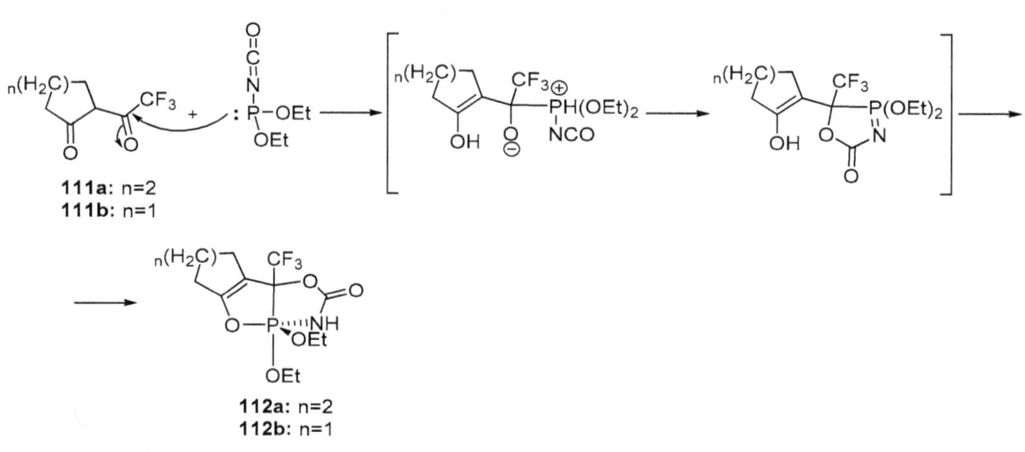

Scheme 46. Synthesis of bicyclic aminophosphoranes **112**.

The phosphorane derived from 2-trifluoroacetylcyclopentanone **112b** was moisture-sensitive, unlike the cyclohexane derivative **112a**, probably due to the greater ring strains. The proposed structure was strictly confirmed by means of NMR data on the basis of which it was found that carbon atom occupies an axial position and two annulated five membered rings are arranged in the equatorial-axial-equatorial order.

10P-1C-2N-2O Phosphoranes

Pudovik et al. reported the synthesis and crystal structure analysis of spirophosphoranes derived from benzoxazaphospholidines and/or 2-aminophenol [101]. It has been found that the condensation reaction of 2-aminophenol with chloromethylphosphonic dichloride proceeded with the formation of the expected benzoxazaphospholidine-2-oxide **99** along with the unexpected one: spirophosphorane **100** (Scheme 47).

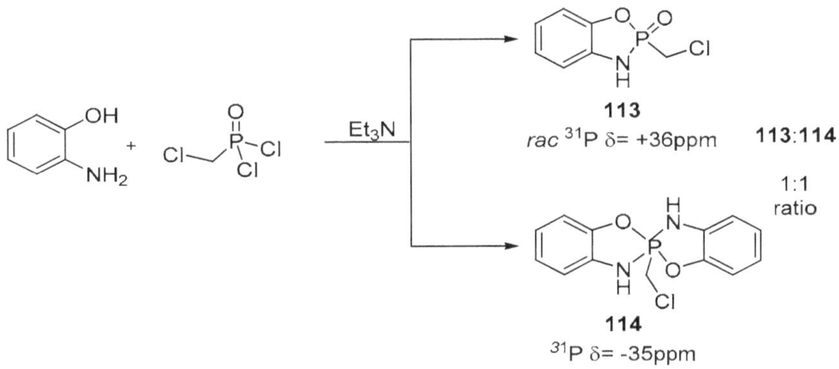

Scheme 47. Synthesis of bicyclic aminophosphorane **114**.

As indicated by the authors, the crucial step in a plausible pathway for their synthesis is an initial phosphorylation of a hydroxy group in 2-aminophenol with phosphonic dichloride leading to chloromethylphosphonochloridate intermediate (I). Subsequent nucleophilic attack of the amino group at the phosphorus atom leads to the cyclization product **99**. Simultaneously, the second pathway is realized when chloromethylphosphonochloridate intermediate reacts with the next molecule of the starting aminophenol to give the OH condensation product, which undergoes cyclization to provide unstable intermediate (II, III). It is then converted to the spirophosphorane as a result of intramolecular nucleophilic attack of the amino group at electrophilic phosphorus atom with the elimination of water (Scheme 48).

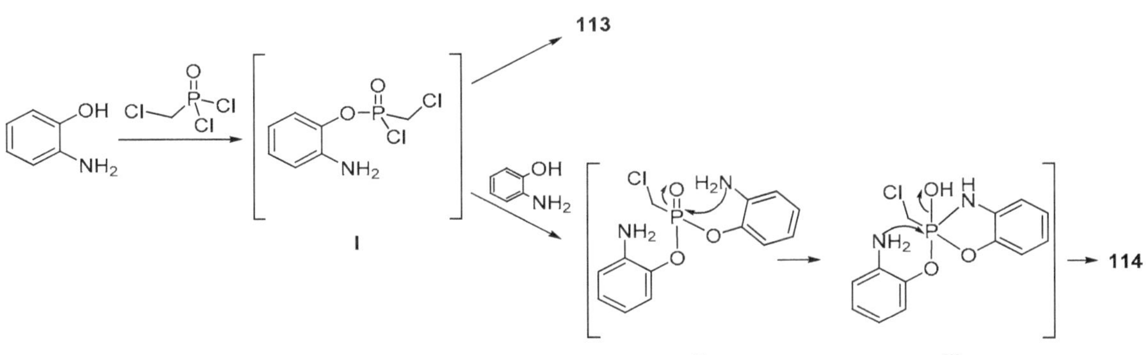

Scheme 48. The mechanism for the formation of spirophosphorane **114**.

The alkylation of 2-alkoxy-1,3,2-benzoxazaphospholidine with bis(dialkylamino)methanes afforded N-alkylated product **115** which underwent tautomerization to phosphonimidate **116** and further transformation to polycyclic phosphoranes **117** in 30 days [102]. It is worth to mention that both benzoxazaphospholidine **115** and polycyclic phosphorane **117** were used as phosphorylating agent for 2-aminophenol yielding spirophosphorane **118** (Scheme 49). The crystal and molecular structure of the synthesized spirophosphoranes **114** and **118** were determined by single-crystal X-ray diffraction [101,103,104].

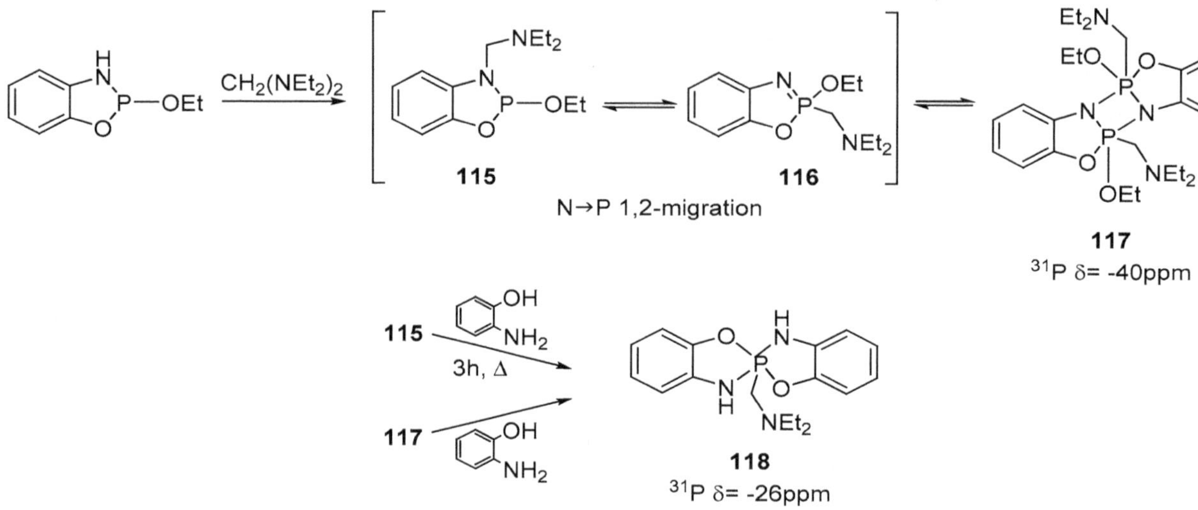

Scheme 49. Syntheses of bicyclic diaminophosphoranes **117** and **118**.

Amino alcohol based monocyclic hydrophosphorane **119** was prepared in the reaction of 2-aminophenol with tetraalkylphenylphosphonous diamide and used for the preparation of bicyclic spirophosphorane **120** [105] based on the oxidation procedure or on heating above its melting point (Scheme 50).

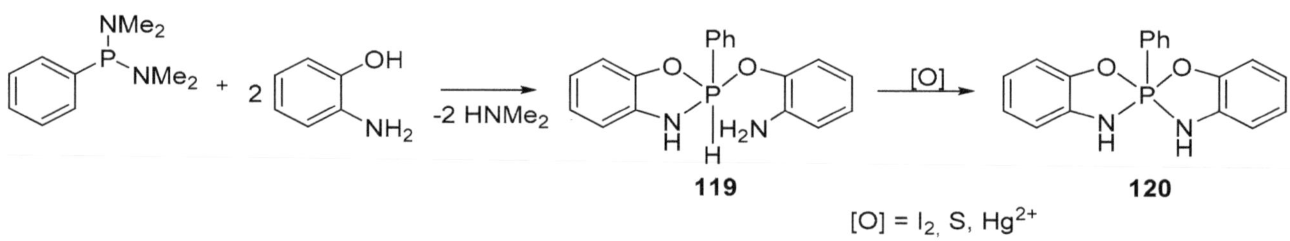

Scheme 50. Synthesis of bicyclic diaminophosphorane **120**.

The alternative approach to synthesize bicyclic pentacoordinate spirophosphoranes proposed by Gholivand et al. [106] is based on the condensation of simple hydrazides (i.e., benzhydrazide **123a** and 4-pyridinecarboxylic acid hydrazide **123b**) with phosphoryl reagents followed by a dehydration–cyclization rearrangement of the resulting phosphorylated hydrazides. Such spiro-bicyclophosphorane with a trigonal bipyramidal structure is formed if phosphoryl reagent contains at least two appropriate leaving groups such as chlorine atoms (in $PhPOCl_2$ or $POCl_3$). Benzhydrazide reacted with $POCl_3$ in refluxing acetonitrile to give the P–Cl phosphorane intermediate **121** which treated with the appropriate amine was converted to the P-N spirophosphorane **122a,b** (Scheme 51). The mechanism proposed for the formation of the intermediate involved β-amidic proton elimination in phosphorylated hydrazides **A** upon which the cyclization product with new C=N imine bond **B** was formed. The intermediate was finally converted into **121** by dehydration.

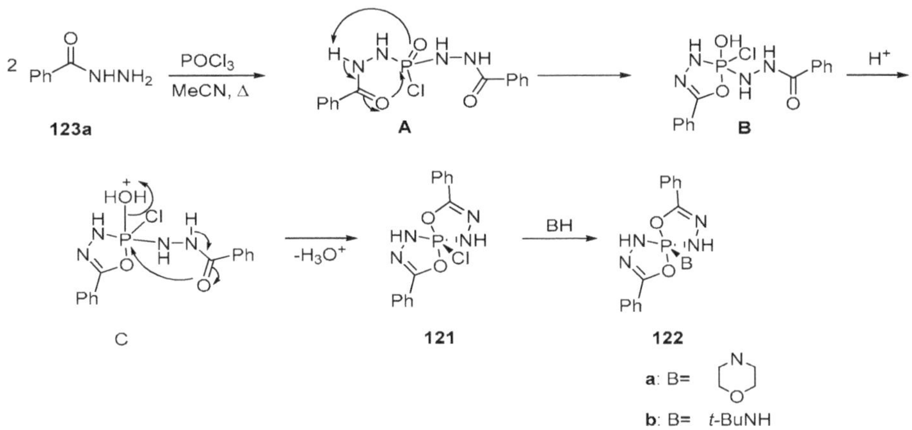

Scheme 51. Syntheses of bicyclic diaminophosphoranes **121** and **122**.

Similarly, several pentacoordinate phosphoranes **122c–g** were easily prepared in the reaction of benzhydrazide **123a** or 4-pyridinecarboxylic acid hydrazide **123b** with $PhPOCl_2$, $PhNHPOCl_2$ or $POCl_3$ in the presence of triethylamine [106] (Scheme 52).

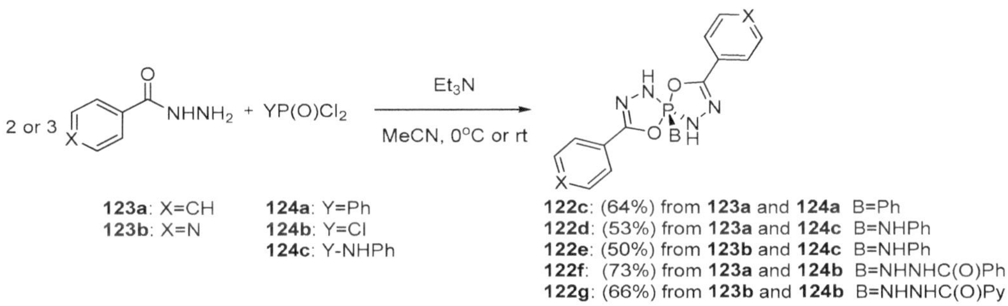

Scheme 52. Synthesis of bicyclic diaminophosphoranes **122**.

10P-1C-4O Phosphoranes

The reaction of (*E*)-bis-(2,4,6-tri-*t*-butylphenyl)diphosphane **125** with a fourfold excess of tetrachloro-*o*-benzoquinone **126** was found to give the chiral perchlorinated spirophosphorane **127**. It was suggested that the reaction, proceeding according to an electron transfer mechanism, occurred via **128** with a cleavage of the P-P bond and indeed, **127** was isolated in low yield even with two equivalents of **126** (Scheme 53) [107].

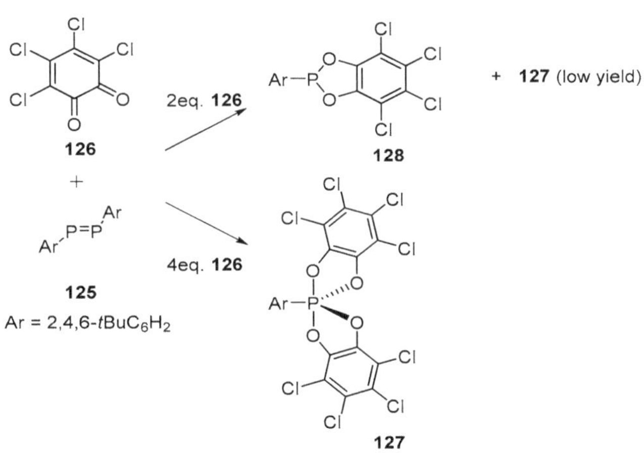

Scheme 53. Synthesis of bicyclic phosphoranes **127**.

Similarly, the reactions of tricoordinated organophosphorus precursors **129–131** with an appropriate equivalent of ethylene glycol was found to give the spirophosphorane **132**. All the three reactions occurred via the bis-ester **133** formed in a nucleophilic exchange reaction at the tricoordinated phosphorus center of precursors **129–131** [108] (Scheme 54).

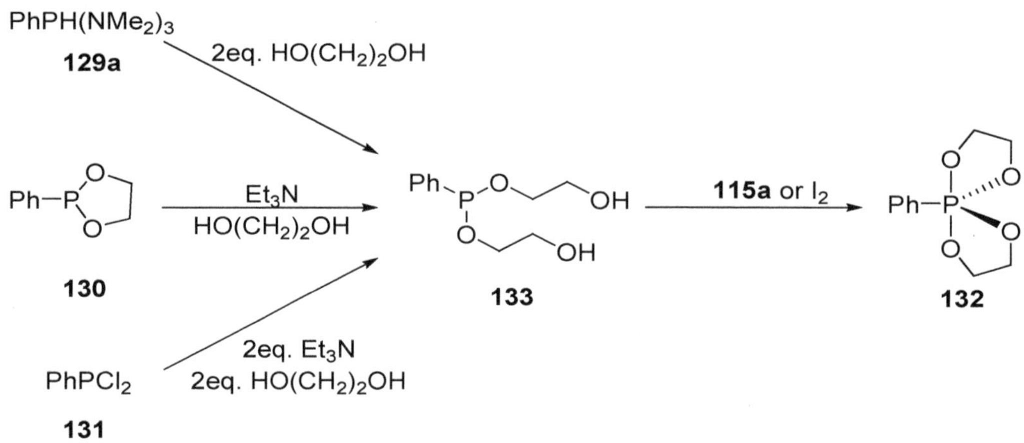

Scheme 54. Synthesis of bicyclic phosphorane **132**.

10P-1C-1N-2O-1H Phosphoranes

The reactions of tricoordinated phosphonous diamides **115a–d** with an appropriate equivalent of aminodiols **120** led to a series of chiral enantiomeric or diastereomeric bicyclic spirophosphoranes **121a–f** (Scheme 55) [109].

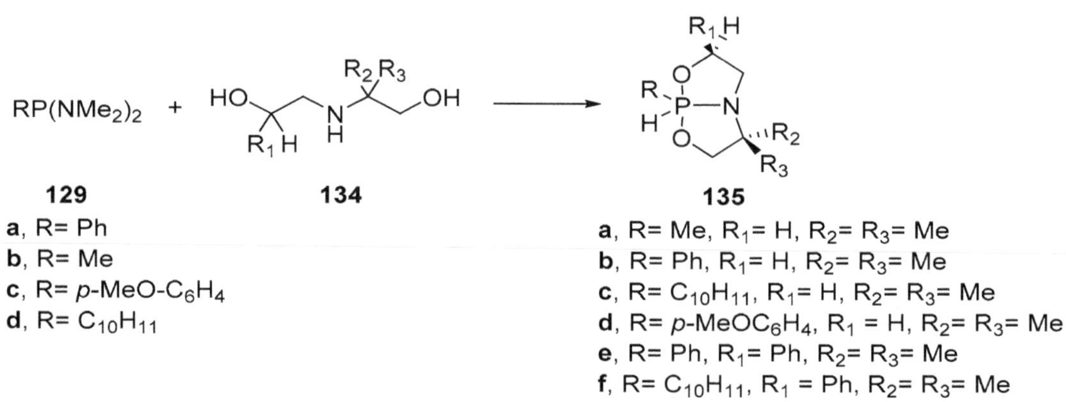

Scheme 55. Synthesis of bicyclic azaphosphoranes **135**.

10P-2N-2O-1H and 10P-2N-3O Phosphoranes

Chiral hydrospirophosphoranes derived from *L*-amino acids were for the first time obtained by Koenig et al. as early as 1979 [110]. More recently this approach based on the reaction of phosphorus trichloride with amino acids was used by Zhao and coworkers for the preparation of spirophosphoranes from enantiomerically pure valine **136a**, isoleucine **136b** and phenylalanine **136c**. [111–114]. The spirophosphoranes formed as mixtures of diastereoisomers were separated to the pure stereoisomers **137a–c** and **138a–c** by reverse-phase HPLC or recrystallization (Scheme 56).

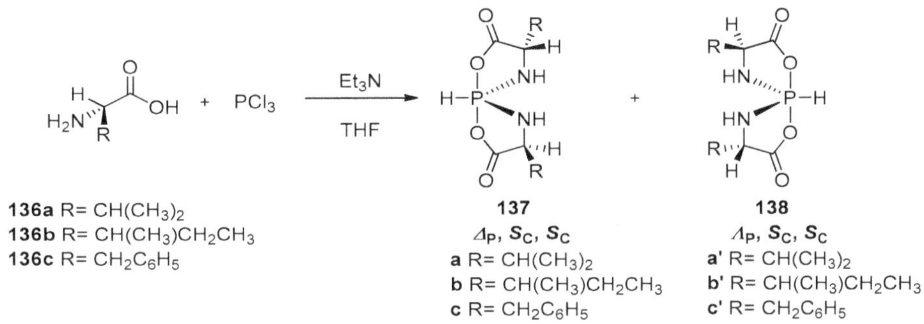

Scheme 56. Synthesis of hydrospirophosphoranes **137** and **138** derived from L-amino acids.

The pure stereoisomers **137** were used for the preparation of pentacoordinate spirophosphorane carbamates **139** [115] via the two-steps Atherton–Todd-type reaction (Scheme 57). Initially, the spirophosphoranes **137** reacted with CCl_4 yielding chlorospirophosphoranes **140** (with retention of configuration at the phosphorus atom). In a second step chlorophosphoranes, upon the reaction with carbamate anions, (formed in situ from CO_2 and secondary amines), afforded the carbamates **139** with inversion of configuration (Scheme 58).

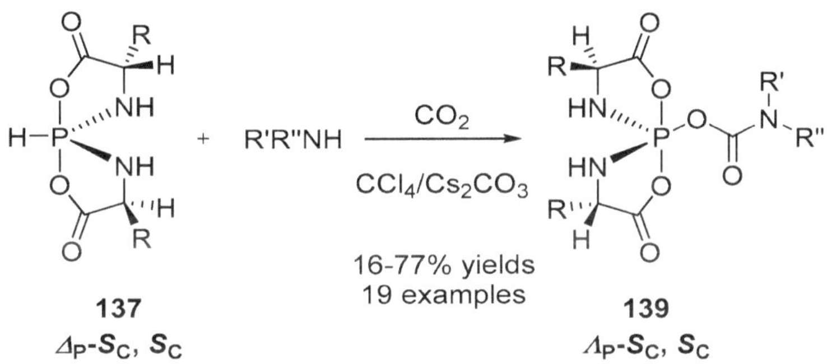

Scheme 57. Preparation of pentacoordinate spirophosphorane carbamates **139**.

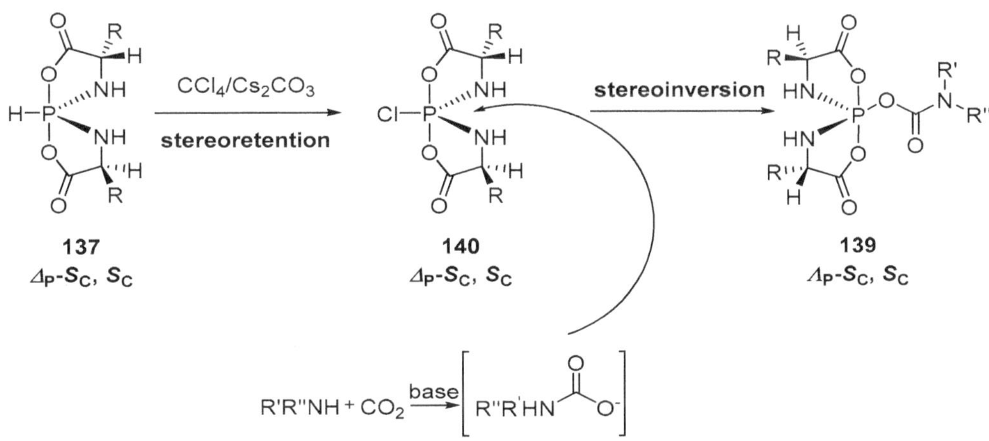

Scheme 58. The mechanism of formation spirophosphorane carbamates **139**.

Similarly, the spirophosphoranes **137a–e** and **138a–e** were used as substrates for the synthesis of pentacoordinate pyrospirophosphoranes containing a P–O–P bond **141–143** under modified Atherton–Todd conditions [116] (Scheme 59). It was found, upon optimization of the reaction conditions, that the spirophosphoranes **137a,b** gave the pure diasteroisomer of pyrospirophosphoranes **142a,b** whereas a mixture of diastereoisomers **141c–e** and **143c–e** were isolated if hydrospirophosphoranes **137c–e** were used as substrates.

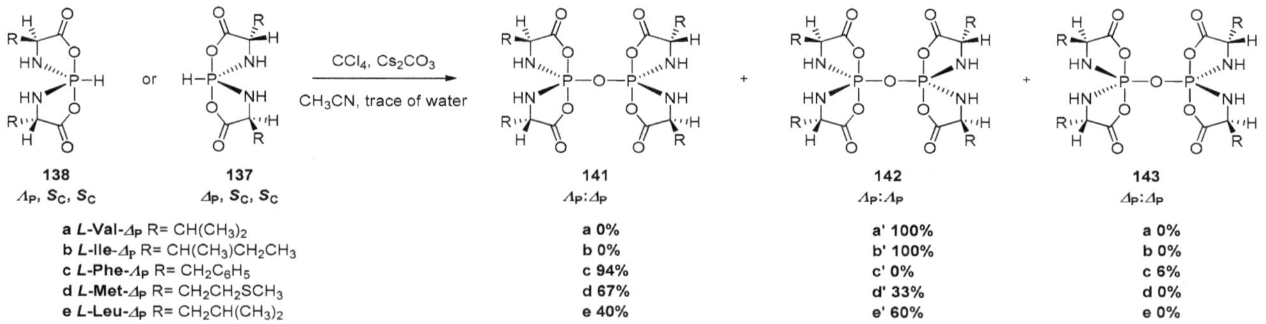

Scheme 59. Synthesis of pentacoordinate pyrospirophosphoranes **141–143**.

A series of the alkoxy spirophosphoranes **144,145** was also prepared by this approach (Scheme 60) but unfortunately as a mixture of diastereoisomers (in most cases in 1:1 ratio) [117].

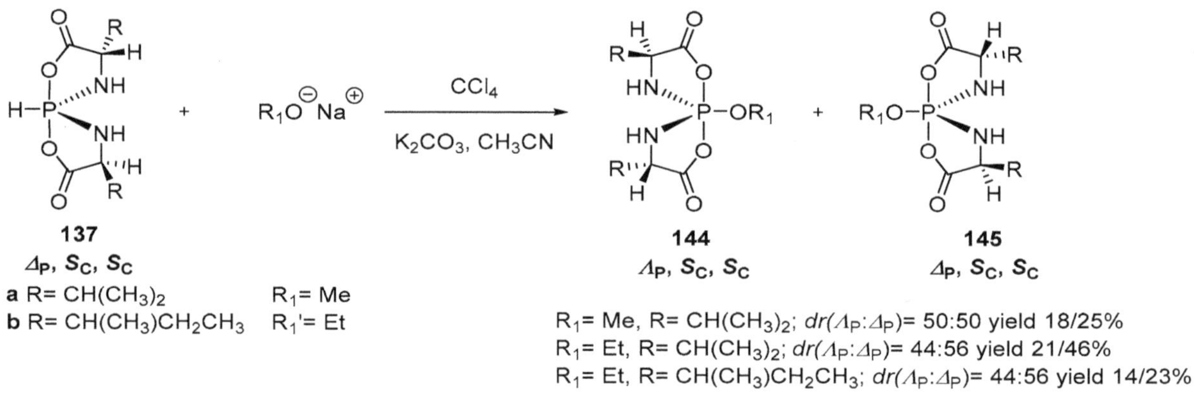

Scheme 60. Synthesis of alkoxy spirophosphoranes **144,145**.

Bicyclic hydrophosphoranes **147** and **148** were synthesized as a 3:2 mixture of two epimers by the reaction of tris-(*N,N*-diethyl)phosphorus amide with isoleucinol **146**. On the other hand, the tricyclic phosphorane **149** was isolated as a single stereoisomer in a similar reaction with the diaminodiole **151** (Scheme 61) [118]. The isolated spirophosphoranes **147–149** were used as chiral ligands in the Pd-catalyzed alkylation of 1,3-diphenyl allyl acetate with acceptable stereoselectivity (up to 74% *ee*) (Scheme 61) [119]. Moreover, these hydrophosphorane derivatives were used to obtain complexes with $[Pt(COD)Cl_2]$, $[Rh(CO)_2Cl]_2$ $[Rh(THF)_2(COD)]^+BF_4^-$.

Chiral triquinphosphoranes **152–154** were easily synthesized from chiral enantiopure diaminodiols with a C_2 symmetry axis. It was shown that their structure is best represented by two trigonal bipyramid structures (TBP) with opposite absolute configurations at the phosphorus atom, R_P and S_P, being in a fast equilibrium by a Berry pseudorotation process (Scheme 62) [120]. They reacted with various activated carbonyl compounds: trifluoroacetophenone, ketopantolactone and aromatic aldehydes to afford two diastereomeric hydroxyphosphoranes with a *de* up to 90% depending on the nature of the electrophile (Scheme 62). Diastereomeric mixture of hydroxyphosphoranes formed by the addition of ketopantolactone **155–156d–f** were quantitatively converted into alkoxyphosphoranes **157** with diastereomeric excesses decreasing from 86% to 8% for R = Me, from 90% to 50% for R = iPr and from 84% to 2% for R = Bn.

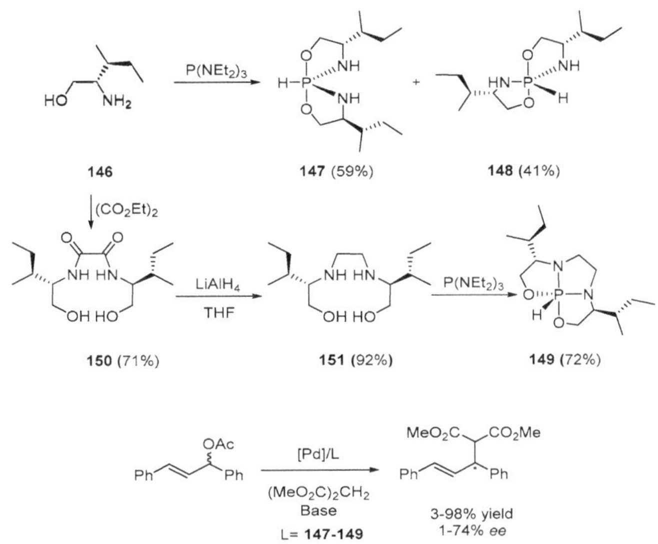

Scheme 61. Syntheses of bicyclic hydrophosphoranes **147–149** and their use as chiral ligands.

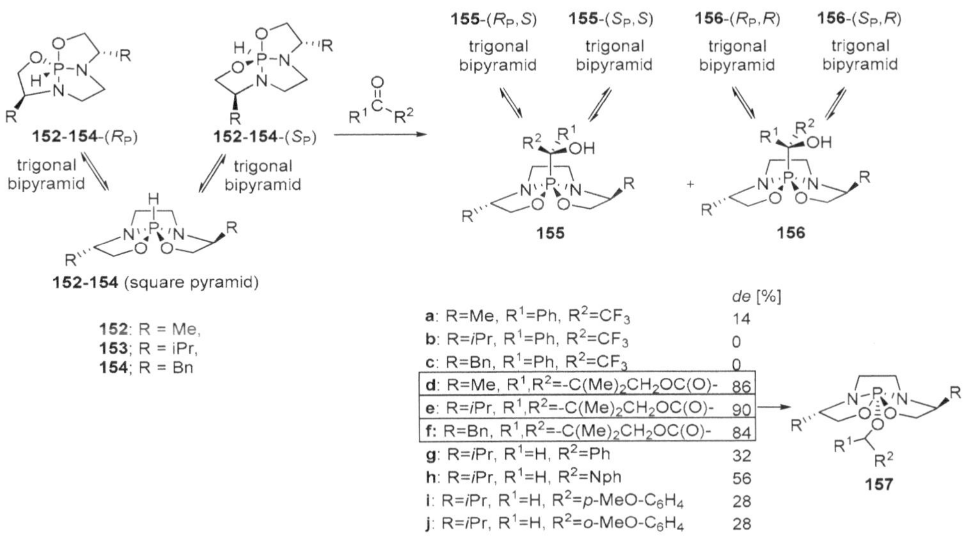

Scheme 62. Chiral triquinphosphoranes **152–154** and their reactions with various activated carbonyl compounds.

Similarly, the parent triquinphosphorane **158** was found to react with methyl, *n*-butyl, or *t*-butyl disulfide to produce the corresponding alkylthiophosphoranes **159a–c**, although *t*-butylthio-phosphorane **159c** was only a minor product and the major one was identified as the thiophosphoramide **160** (Scheme 63). Under irradiation at −50 °C *t*-butylthiophosphorane **159c** was produced in high yield, however, by increasing the reaction temperature its slow conversion to **160** was observed. The reactivity of the triquinphosphorane **158** with alkyl disulfides was compared both under irradiation and in the dark at room temperature. The reactivity increased significantly in the case of *n*-butyl disulfide and *t*-butyl disulfide when the reactions were conducted under irradiation. This observation led to the conclusion that this type of the reaction does involve free radical species [121].

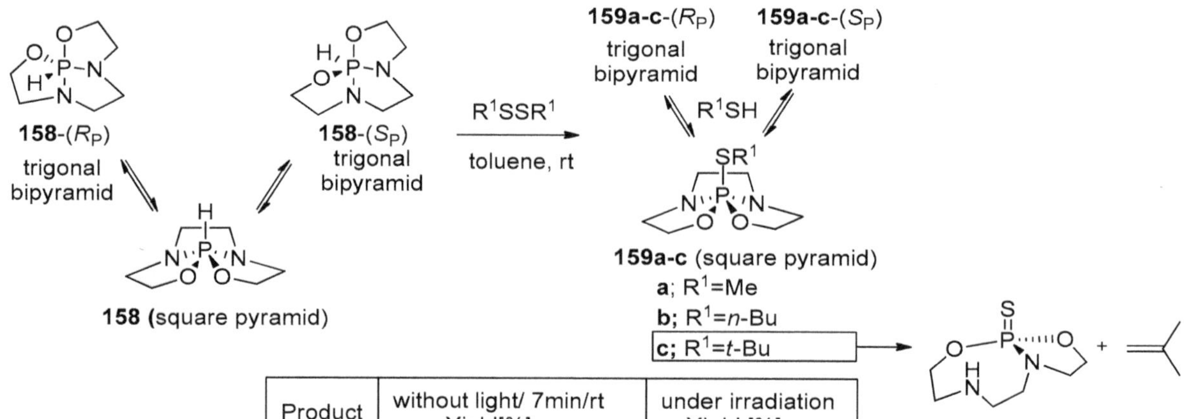

Product	without light/ 7min/rt Yield[%]	under irradiation Yield [%]
159a	100	100
159b	20	100
159c	0	80 (**159c+160**)

160

Scheme 63. Reaction of chiral hydrido-triquinphosphorane **158** with alkyl disulfides.

10P-1N-3O-1H and 10P-2N-2O-1H Phosphoranes

Three synthetic protocols were applied to the preparation of this type of hydrophosphoranes with the use of amidooxime derivatives as substrates [122]. The first procedure started with the reaction of amidooxime **161a–e** with PCl_3 leading to 5-chloro-1,2,4,5-oxadiazaphospholines **162a–e**. Their reactions with ethylene glycol or *ortho*-hydroxyphenol afforded the unsymmetrical spirophosphoranes **163** or **164** whereas the treatment with another dose of the amidooxime **161** led to symmetrically substituted spirophosphoranes **165a–e** (Scheme 64).

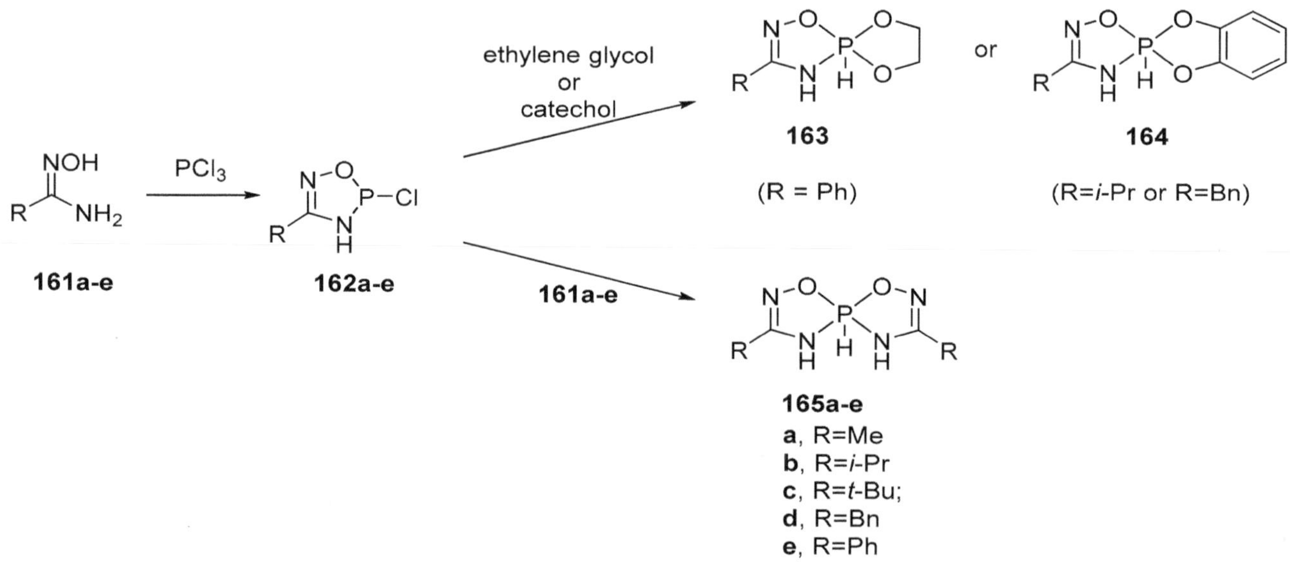

Scheme 64. Syntheses of bicyclic hydrophosphoranes **163–165**.

In the second approach, amidooxime **161a,c,d** upon reaction with 2-dimethylamino-1,3,2-dioxaphospholane **166** provided unsymmetrical spirophosphoranes **163a,c,d** and in the reaction with 2-dimethylamino-(4,5)benzo-1,3,2-dioxaphospholane **167** in acetonitrile yielded unexpected symmetrically substituted spirophosphoranes **165a–e** instead of unsymmetrically substituted phosphorane **164a–e** (Scheme 65).

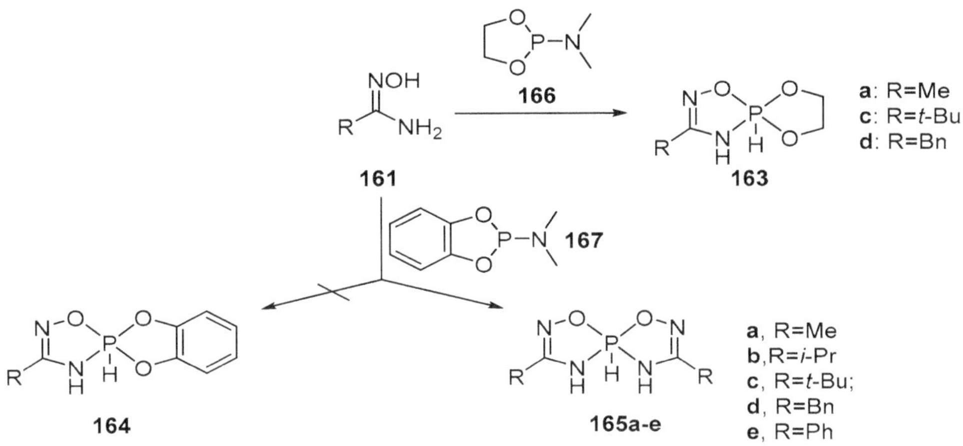

Scheme 65. Syntheses of bicyclic hydrophosphoranes **163** and **165**.

The third synthetic pathway involved the treatment of amidooximes **161b,d,e** with 2-chloro-(4,5)benzo-1,3,2-dioxaphospholane **168** leading, as it was expected, to the corresponding unsymmetrical spirophosphoranes **164**. The latter were converted into the corresponding unsymmetrical spirophosphorane **170** by transesterification with pinacol **169** taking place in the presence of triethylamine (Scheme 66).

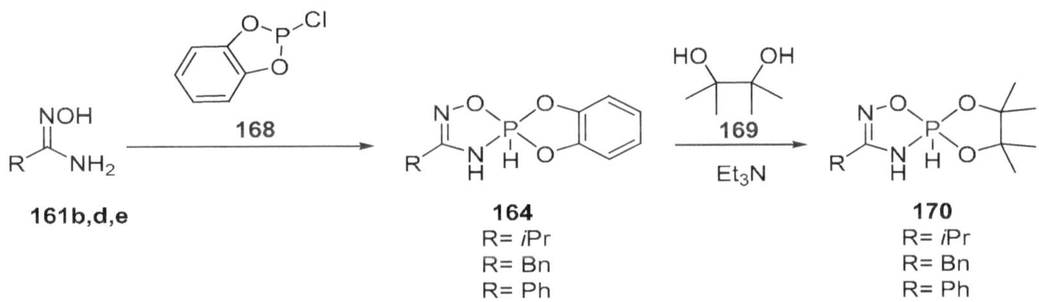

Scheme 66. Synthesis of bicyclic hydrophosphoranes **164** and **170**.

10P-2N-3O and 10P-1N-4O Phosphoranes

Amidooximes **161d,e** were used also as substrates for the preparation of the spirophosphoranes containing the P–OMe grouping **173, 175**. They were isolated upon application of the oxidation procedure carried out under basic conditions with iodine as an oxidant [122] (Scheme 67).

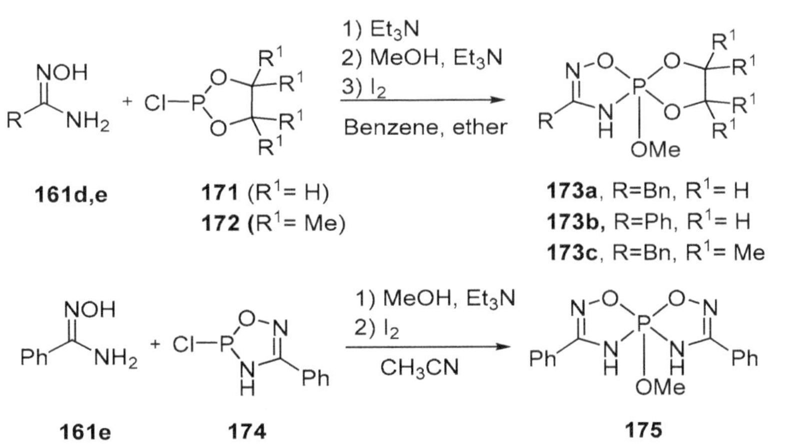

Scheme 67. Synthesis of bicyclic spirophosphoranes **173** and **175**.

O-Trimethylsilyl derivatives of spirophosphoranes **177** containing an amino acid residue were formed by the cyclisation of the corresponding cyclic amides **176** [123] (Scheme 68). The resulting spirophosphoranes **177** were used for the formation of peptides from amino acids such as histidine, serine, threonine and α-alanine but not β-alanine.

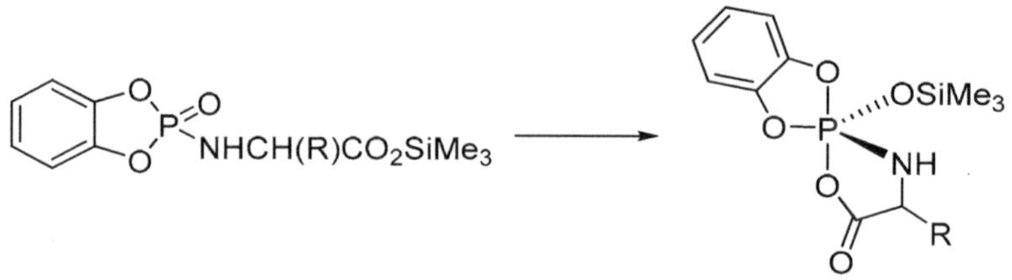

176 **177**

Scheme 68. Synthesis of bicyclic phosphoranes **177**.

Later on, spirophosphoranes **177** were also used as substrates in a new method for the solid phase synthesis of oligopeptides. This approach was based on their rapid reaction with hydroxymethyl polystyrene resin leading to unstable hydroxyphosphoranes **178** which after mild hydrolysis gave the phosphate **180** and a solid phase bound amino-acids **179**. Coupling of **179** with further phosphorane **177**, followed by hydrolysis afforded a solid phase bound dipeptide. The process was repeated to give oligopeptides. The final oligopeptide was liberated from the resin by treatment with TFA [124] (Scheme 69).

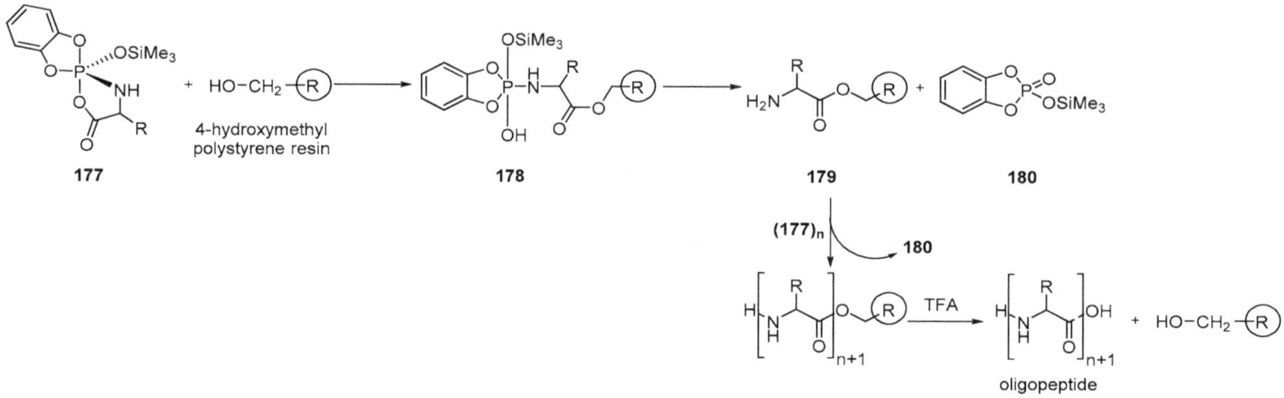

Scheme 69. Synthesis of oligopeptides with the use of spirophosphoranes **177** as substrates.

10P-4O-1H Phosphoranes

The synthesis and reactivity of the hydrophosphorane **181** derived from perfluoropinacol was described as early as 1983 [125]. It was found that it could be rapidly oxidized by dimethyl sulfoxide to form the hydroxyphosphorane **182**, and then silylated to the derivative **183**. Chlorine and bromine reacted with **181** to give the corresponding halospirophosphoranes **184a,b**. In this context it is interesting to note that the corresponding fluorospirophosphorane **184c** was obtained from the dichlorofluoro-perfluoropinacolophosphorane $FCl_2P(CH_2Ph)[OC(CF_3)C(CF_3)O]$ **185** and dilithium perfluoropinacolate $LiOC(CF_3)C(CF_3)OLi$ **186**. In the presence of triethylamine the hydrophosphorane **181** reacted with benzyl bromide or acetyl chloride to form the phosphoranes **187** or **188**. Moreover, the treatment of the hydrophosphorane **181** with trimethylphosphine gave a thermally unstable phosphonium salts **189**. Hydrophosphorane **181** undergoes fast hydrolysis affording phosphorous acid and perfluoropinacol [125] (Scheme 70).

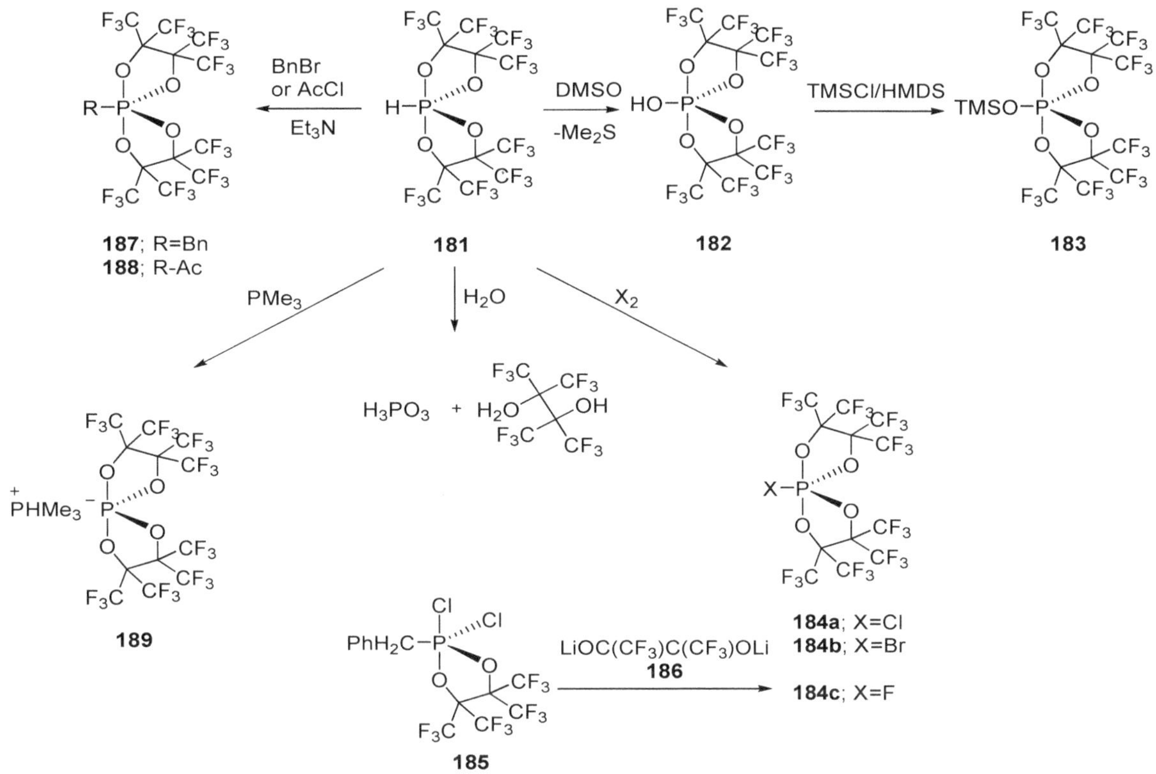

Scheme 70. Interconversions of spirophosphoranes **181** and **182**.

10P-4O-1Metal and 10P-3N-1O-1Metal Phosphoranes

Another group of chiral pentacoordinate organophosphorus compounds are metallophosphoranes **192a–e**. Nakazawa et al. [126] developed a synthetic method for the preparation of metallophosphoranes **192**, which was based on the nucleophilic attack of the, in situ generated, anion **191a–e** on the complexed trivalent phosphorus atom **190a,b** (Scheme 71). For example, ferrocenyl complex **190a** reacted with in situ generated anions **191a–e** to form chiral metallophosphoranes **192a–e**. The valency of the phosphorus atom increased from III to V without breaking the Fe-P bond and the nature of Fe-P bond changed from coordinative to covalent.

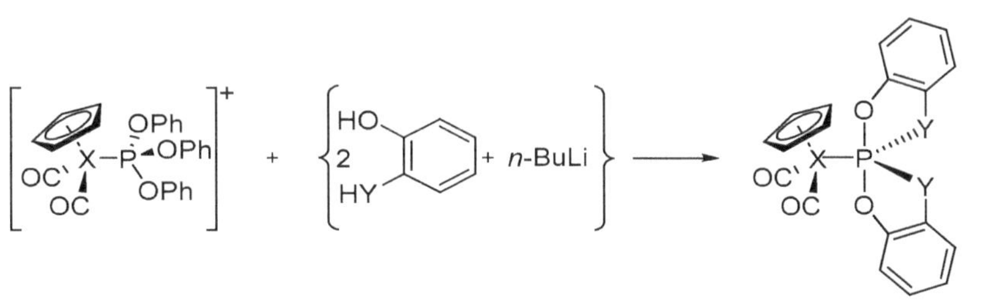

190
a, X= Fe
b, X= Ru

191
a, Y=O
b, Y=NH
c, Y=NMe

192
a, X= Fe, Y=O
b, X= Fe, Y= NH
c, X= Fe, Y= NMe
d, X= Ru, Y= O
e, X= Ru, Y= NH
f, X= Ru, Y= NMe

Scheme 71. Preparation of chiral metallophosphoranes **192**.

It is worthy to note that treatment of $HP(C_6H_4NH)_2$-**193** with *n*BuLi led to unexpected deprotonation on nitrogen atom (giving **194b**) instead on phosphorus (to afford **194a**). Moreover, an amide anion **194b** is in an equilibrium with phosphoranide **194a**. The reaction of the in situ generated anion **194a** with [Cp(CO)LFeCl] (**195a** or **195b**) gave a P-metallated phosphorane **196a** or **196b**. On the contrary, the reaction the equilibrated mixture of **194** with [Cp(CO)CoI$_2$]-**197** led to *N*-metallated compound **198**. Interestingly, in the reaction of the phosphoranide **194a** with MeI the P-methylated product **199** was formed. On the other hand, electrophiles such as Me_3SiCl, Me_3GeCl or Me_3SnCl, provided *N*-substituted products (**200a–c**) [127,128] (Scheme 72).

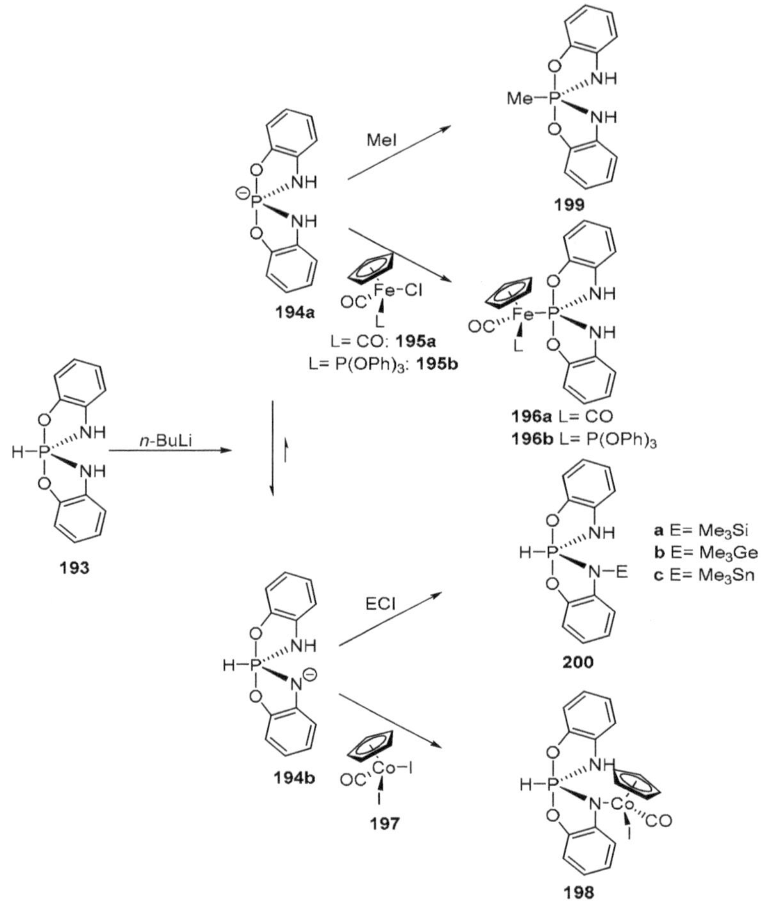

Scheme 72. Functionalization of hydrophosphorane **193** leading to phosphoranes **196** and **198–200**.

Moreover, it was found that in the presence of a base a migration of pentacoordinate phosphorane fragment occured from the transition metal in **196** to the carbon atom in the Cp ring in **201** [129] (Scheme 73).

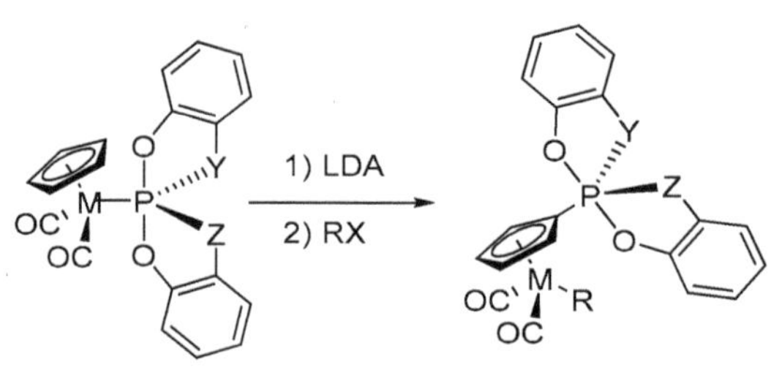

196
M= Fe or Ru

201
Y= NMe or O
Z= NMe or O
R= Me or CH_2Ph

Scheme 73. Alkylation of metallophosphoranes **196**.

10P-5O Phosphoranes

The hydroxyphosphorane **203** was prepared by N_2O_4 oxidation of the parent hydrophosphorane **202** (Scheme 74). Its single crystal analysis showed an almost perfect trigonal bipyramidal structure with the unit cell containing two molecules of the same helicity connected by H-bonds between the P–OH and carbonyl groups [130].

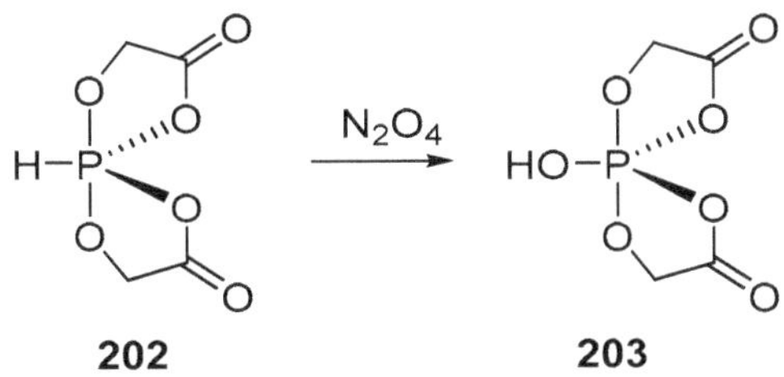

Scheme 74. Oxidation of hydrophosphorane **202**.

The triethylammonium salt of the hydroxyspirophosphorane **205** derived from *n*-butyl tartrate was prepared on treatment of the very acidic [pK_a = 4.4(DMSO)] monocyclic phosphorus ester **204** with triethylamine (Scheme 75) [131].

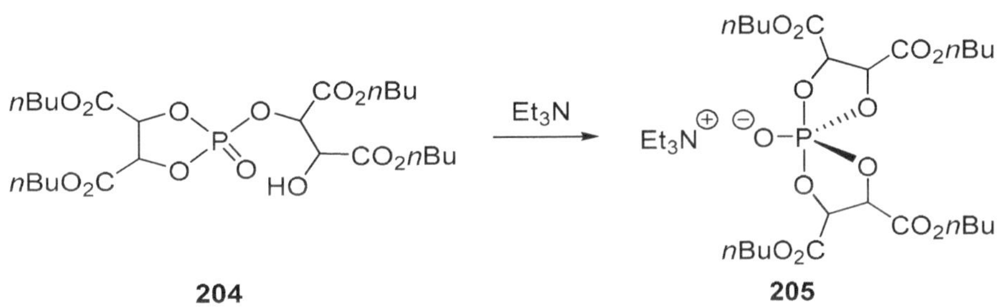

Scheme 75. Formation of salt **205**.

The radical reaction of the bicyclic hydrophosphorane **206** derived from glycol with ethyl vinyl ether provided the corresponding P–C spirophosphorane **207** along with the monocyclic phosphite **208**. The reaction of the hydrospirophosphorane **206** with ethylene glycol and enamine **209** afforded the pentaoxyspirophosphorane **210** with the simultaneous reduction of the enamine **209** to amine **211** [132] (Scheme 76).

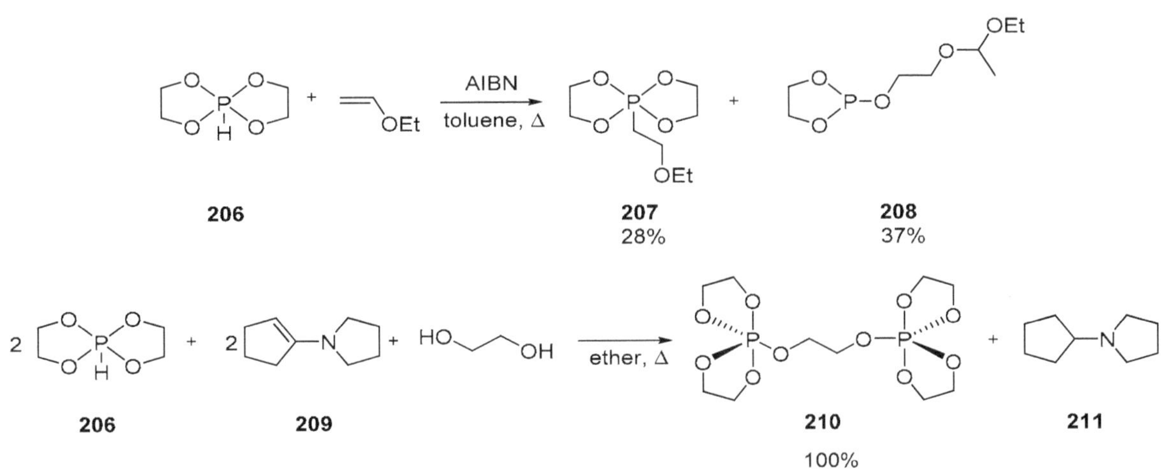

Scheme 76. Synthesis of spirophosphoranes **207** and **210**.

3. Conclusions

Enantiomerically enriched pentavalent phosphoranes constitute an interesting group of heterocyclic chiral auxiliaries. Their still limited applications in asymmetric synthesis are however hampered mainly by a limited access to diastereomerically pure and especially enantiomerically pure (or at least enantiomerically enriched) samples. It is our hope that this review, which describes briefly the basic procedures used for the preparation of these derivatives, their stereoisomerization mechanisms and their selected interconversions, will encourage wider interest in the synthesis, structural determinations and applications of these groups of chiral, hypervalent organophosphorus derivatives.

Acknowledgments: Preparation of this review was financially supported by the National Science Center, Poland fund awarded based on the decisions (grant Maestro UMO-2012/06/A/ST5/00227 to (Józef Drabowicz).

References

1. Eliel, E.L.; Wilen, S.H. *Stereochemistry of Organic Compounds*; John Wiley & Sons: New York, NY, USA, 1994.
2. Wolf, C. *Dynamic Stereochemistry of Chiral Compounds: Principles and Applications*; The Royal Society of Chemistry: Cambridge, UK, 2008.
3. Singh, J.; Hagen, T.J. Chirality and Biological Activity. In *Burger's Medicinal Chemistry and Drug Discovery*, 7th ed.; Abraham, D., Rotella, D., Eds.; John Wiley & Sons: New York, NY, USA, 2010; pp. 127–166.
4. Reddy, I.K.; Mehvar, R. *Chirality in Drug Design and Development*; CRC Press: New York, NY, USA; Basel, Switzerland, 2004.
5. Mori, K. Bioactive Natural Products and Chirality. *Chirality* **2011**, *23*, 449–462. [CrossRef] [PubMed]
6. Ojima, I. (Ed.) *Catalytic Asymmetric Synthesis*, 3rd ed.; Wiley-VCH: New York, NY, USA, 2010.
7. Christmann, M.; Brase, S. (Eds.) *Asymmetric Synthesis: More Methods and Applications*; Wiley-VCH: New York, NY, USA, 2012.
8. Bruckner, R. *Organic Mechanisms: Reactions, Stereochemistry and Synthesis*; Harmata, M., Ed.; Springer: Berlin/Heidelberg, Germany, 2010.
9. Montchamp, J.-L. Phosphorus Chemistry I: Asymmetric Synthesis and Bioactive Compounds. In *Topics in Current Chemistry*; Springer International Publishing: Cham, Switzerland, 2015; Volume 360.
10. Dutartre, M.; Bayardon, J.; Juge, S. Applications and stereoselective syntheses of P-chirogenic phosphorus compounds. *Chem. Soc. Rev.* **2016**, *45*, 5771–5794. [CrossRef] [PubMed]
11. International Union of Pure and Applied Chemistry (IUPAC). *Compendium of Chemical Terminology*, 2nd ed.; (the "Gold Book"); McNaught, A.D., Wilkinson, A., Eds.; Blackwell Scientific Publications: Oxford, UK, 1997.
12. Quin, L.D. *A Guide to Organophosphorus Chemistry*; John Wiley & Sons: New York, NY, USA, 2000.
13. Musher, J.I. The Chemistry of Hypervalent Molecules. *Angew. Chem. Int. Ed. Engl.* **1969**, *8*, 54–68. [CrossRef]
14. Perkins, C.W.; Martin, J.C.; Arduengo, A.J.; Lau, W.; Alergia, A.; Kochi, J.K. An Electrically Neutral sigma-Sulfuranyl Radical from the Homolysis of a Perester with Neighboring Sulfenyl Sulfur: 9-S-3 Species. *J. Am. Chem. Soc.* **1980**, *102*, 7753–7759. [CrossRef]
15. Hayes, R.A.; Martin, J.C. *Organic Sulfur Chemistry: Theoretical and Experimental Advances*; Bernardi, F., Csizmadia, J.G., Mangini, A., Eds.; Elsevier: Amsterdam, The Netherlands, 1985; Chapter 8; pp. 408–458.
16. Akiba, K. Hypervalent Compounds. In *Chemistry of Hypervalent Compounds*; Akiba, K., Ed.; Wiley-VCH: New York, NY, USA, 1999; Chapter 1; pp. 1–8.
17. Akiba, K. Structure and Reactivity of Hypervalent Organic Compounds: General Aspects. In *Chemistry of Hypervalent Compounds*; Akiba, K., Ed.; Wiley-VCH: New York, NY, USA, 1999; Chapter 2; pp. 9–47.
18. Noury, S.; Silyi, B.; Gillespie, J.R. Chemical Bonding in Hypervalent Molecules: Is the Octet Rule Relevant? *Inorg. Chem.* **2002**, *41*, 2164–2172. [CrossRef] [PubMed]
19. Mutterties, E.L.; Schunn, R.A. Pentacoordination. *Q. Rev.* **1966**, *20*, 245–299.
20. Berry, R.S. Correlation of Rates of Intramolecular Tunneling Processes, with Application to Some Group V Compounds. *J. Chem. Phys.* **1960**, *32*, 933–938. [CrossRef]
21. Berry, J.S. Time-Dependent Measurements and Molecular Structure: Ozone. *Rev. Mod. Phys.* **1960**, *32*, 447–454. [CrossRef]
22. Sheppard, W.A. Alkyl- and Arylsulfur Trifluorides. *J. Am. Chem. Soc.* **1962**, *84*, 3058–3063. [CrossRef]

23. Ugi, I.; Marquarding, D.; Klusacek, H.; Gillespie, P.; Ramirez, F. Berry Pseudorotation and Turnstile Rotation. *Acc. Chem. Res.* **1971**, *4*, 288–296. [CrossRef]
24. Gillespie, P.; Hoffmann, P.; Klusacek, H.; Marquarding, D.; Pfohl, S.; Ramirez, F.; Tsolis, E.A.; Ugi, I. Bewegliche Molekülgerüste—Pseudorotation und Turnstile-Rotation Pentakoordinierter Phosphorverbindungen und Verwandte Vorgänge. *Angew. Chem.* **1971**, *83*, 691–721. [CrossRef]
25. Ugi, I.; Marquading, D.; Klusacek, H.; Gokel, G.; Gillespie, P. Chemie und Logische Strukturen. *Angew. Chem.* **1970**, *82*, 741–771. [CrossRef]
26. Ramirez, F.; Ugi, I. Turnstile Rearrangement and Pseudorotation in the Permutational Isomerization of Pentavalent Phosphorus Compounds. *Adv. Phys. Org. Chem.* **1971**, *9*, 25–126.
27. Couzijn, E.P.A.; Slootweg, J.C.; Ehlers, A.W.; Lammertsma, K. Stereomutation of Pentavalent Compounds: Validating the Berry Pseudorotation, Redressing Ugi's Turnstile Rotation, and Revealing the Two- and Three-Arm Turnstiles. *J. Am. Chem. Soc.* **2010**, *132*, 18127–18140. [CrossRef] [PubMed]
28. Wittig, G.; Rieber, M. Darstellung und Eigenschaften des Pentaphenyl-phosphors. *Justus Liebigs Ann. Chem.* **1949**, *562*, 187–192. [CrossRef]
29. Ramirez, F. Condensations of carbonyl compounds with phosphite esters. *Pure Appl. Chem.* **1964**, *9*, 337–370. [CrossRef]
30. Ramirez, F. Recent developments in the chemistry of hydroxyphosphoranes. *Bull. Soc. Chem. Fr.* **1966**, *8*, 2443–2450.
31. Westheimer, F.H. Pseudo-rotation in the hydrolysis of phosphate esters. *Acc. Chem. Res.* **1968**, *1*, 70–78. [CrossRef]
32. Hall, C.D. Pentacoordinated and hexacoordinated compounds. In *Organophosphorus Chemistry*; Hutchinson, D.W., Walker, B.J., Eds.; The Royal Society of Chemistry: London, UK, 1986; Volume 16, pp. 51–78.
33. Hall, C.D. Pentacoordinated and hexacoordinated compounds. In *Organophosphorus Chemistry*; Walker, B.J., Hobbs, J.B., Eds.; The Royal Society of Chemistry: London, UK, 1988; Volume 19, pp. 47–69.
34. Hall, C.D. Pentacoordinated and hexacoordinated compounds. In *Organophosphorus Chemistry*; Walker, B.J., Hobbs, J.B., Eds.; The Royal Society of Chemistry: London, UK, 1991; Volume 21, pp. 51–72.
35. Hall, C.D. Pentacoordinated and hexacoordinated compounds. In *Organophosphorus Chemistry*; Allen, D.W., Tebby, J.C., Eds.; The Royal Society of Chemistry: London, UK, 2003; Volume 33, pp. 68–83.
36. Hall, C.D. Pentacoordinated and hexacoordinated compounds. In *Organophosphorus Chemistry*; Allen, D.W., Tebby, J.C., Eds.; The Royal Society of Chemistry: London, UK, 2006; Volume 35, pp. 265–303.
37. Tebby, J.C. Pentacoordinated and hexacoordinated compounds. In *Organophosphorus Chemistry*; Allen, D.W., Tebby, J.C., Eds.; The Royal Society of Chemistry: London, UK, 2007; Volume 36, pp. 184–196.
38. Roschenthaler, G.-V. Pentacoordinated and hexacoordinated compounds. In *Organophosphorus Chemistry*; Allen, D.W., Tebby, J.C., Eds.; The Royal Society of Chemistry: London, UK, 2009; Volume 38, pp. 318–331.
39. Pajkert, R.; Roschenthaler, G.-V. Pentacoordinated and hexacoordinated compounds. In *Organophosphorus Chemistry*; Allen, D.W., Tebby, J.C., Loakes, D., Eds.; The Royal Society of Chemistry: London, UK, 2010; Volume 39, pp. 290–307.
40. Pajkert, R.; Roschenthaler, G.-V. Pentacoordinated and hexacoordinated compounds. In *Organophosphorus Chemistry*; Allen, D.W., Tebby, J.C., Loakes, D., Eds.; The Royal Society of Chemistry: London, UK, 2011; Volume 40, pp. 297–315.
41. Pajkert, R.; Roschenthaler, G.-V. Pentacoordinated and hexacoordinated compounds. In *Organophosphorus Chemistry*; Allen, D.W., Tebby, J.C., Loakes, D., Eds.; The Royal Society of Chemistry: London, UK, 2012; Volume 41, pp. 322–348.
42. Pajkert, R.; Roschenthaler, G.-V. Pentacoordinated and hexacoordinated compounds. In *Organophosphorus Chemistry*; Allen, D.W., Tebby, J.C., Loakes, D., Eds.; The Royal Society of Chemistry: London, UK, 2013; Volume 42, pp. 197–215.
43. Pajkert, R.; Roschenthaler, G.-V. Pentacoordinated and hexacoordinated compounds. In *Organophosphorus Chemistry*; Allen, D.W., Tebby, J.C., Loakes, D., Eds.; The Royal Society of Chemistry: London, UK, 2014; Volume 43, pp. 348–365.
44. Kolodiazhnyi, O.I. Recent developments in the asymmetric synthesis of *P*-chiral phosphorus compounds. *Tetrahedron Asymmetry* **2012**, *23*, 1–46. [CrossRef]
45. Randić, M. Symmetry properties of graphs of interest in chemistry. II. Desargues Levi graph Quantum Chemistry. *Int. J. Quantum Chem.* **1979**, *15*, 663–682. [CrossRef]

46. DeBruin, K.E.; Naumann, K.; Zon, G.; Mislow, K. Topological representation of the stereochemistry of displacement reactions at phosphorus in phosphonium salts and cognate systems. *J. Am. Chem. Soc.* **1969**, *91*, 7031–7040. [CrossRef]
47. Moulin, D.; Bago, S.; Bauduin, C.; Darcel, C.; Juge, S. Asymmetric synthesis of P-stereogenic o-hydroxyarylphosphine (borane) and phosphine-phosphinite ligands. *Tetrahedron Asymmetry* **2000**, *11*, 3939–3956. [CrossRef]
48. Uziel, J.; Stéphan, M.; Kaloun, El B.; Genet, J.P.; Juge, S. Regio- and stereochemistry of nucleophilic attack at the P-chiral center of a dioxaphospholane-borane complex: A model of study for the P-O bond cleavage. *Bull. Soc.Chim. Fr.* **1997**, *134*, 379–389.
49. Lopez-Cortina, S.; Basiulis, D.I.; Marsi, K.; Munoz-Hernandez, M.A.; Fernandez-Zertuche, M. Synthesis of New 1,3-Oxaphosphorinanium Salts. Stereochemistry of Hydroxide-Induced Displacement of Methoxide Ion. *J. Org. Chem.* **2005**, *70*, 7473–7478. [CrossRef] [PubMed]
50. Van Lier, J.J.C.; Hermans, R.J.M.; Buck, H.M. Evidence in support of tetraoxaspirophosphorane intermediates with a six-membered ring and P[sbnd]H bond in intramolecular transesterification reactions. *Phosphorus Sulfur Silicon Relat. Elem.* **1984**, *19*, 173–188. [CrossRef]
51. Boisdon, M.T.; Malavaud, C.; Mathis, F.; Barrans, J. Investigations of phosphoranes with P–H bond during the reaction of aminophosphine with alcohols. *Tetrahedron Lett.* **1977**, *18*, 3501–3505. [CrossRef]
52. Chung, F.-L.; Earl, R.A.; Townsend, L.B. The novel synthesis of a [6:5:6] linear isoguanosine type tricyclic nucleoside using carbonyl sulfide. *Tetrahedron Lett.* **1980**, *21*, 1599–1602. [CrossRef]
53. Lönnberg, T.; Laine, M. Phosphorane intermediate vs. leaving group stabilization by intramolecular hydrogen bonding in the cleavage of trinucleoside monophosphates: Implications for understanding catalysis by the large ribozymes. *Org. Biomol. Chem.* **2010**, *8*, 349–356. [CrossRef] [PubMed]
54. Binyamin, I.; Meidan-Shani, S.; Ashkenaz, N. Synthesis of γ-hydroxypropyl P-chirogenic ($\pm$)-phosphorus oxide derivatives by regioselective ring-opening of oxaphospholane 2-oxide precursors. *Beilstein J. Org. Chem.* **2015**, *11*, 1332–1339. [CrossRef] [PubMed]
55. Kolodyazhnyi, O.I. 2-Halo-1,2λ^5-oxaphosphetanes. *Zh. Obsh Kh.* **1986**, *56*, 283–298.
56. Labaudiniere, L.; Burgada, R. Evolution thermique de vinylphosphoranes. *Tetrahedron* **1986**, *42*, 3521–3536. [CrossRef]
57. Dieckbreder, U.; Roschenthaler, G.-V.; Kolomeitsev, A.A. *P*-bis(trifluoromethyl) ylides: Synthesis and reactions. *Heteroat. Chem.* **2002**, *13*, 650–653. [CrossRef]
58. Francke, R.; Roschenthaler, G.-V.; Di Giacomo, R.; Dakternieks, D. Oxidative addition von hexafluoraceton an die phosphor(III)-verbindungen tBuP(X)NEt$_2$ (X = F, Cl, OCH(CF$_3$)$_2$. *Phosphorus Sulfur Silicon Relat. Elem.* **1984**, *20*, 107–115. [CrossRef]
59. Yamamoto, Y.; Nakao, K.; Hashimoto, T.; Matsukawa, S.; Suzukawa, N.; Kojima, S.; Akiba, K.-Y. Crystallographic and NMR studies on species intermediate between haloalkoxyphosphoranes and alkoxyphosphonium halides. *Heteroat. Chem.* **2011**, *22*, 523–530. [CrossRef]
60. Sase, S.; Kano, N.; Kawashima, T. Novel Synthetic Method of Fluorophosphoranes by Fluoride Ion Abstraction from Tetrafluoroborate. *Phosphorus Sulfur Silicon Relat. Elem.* **2002**, *177*, 2041. [CrossRef]
61. Hellwinkel, D. *Organic Phosphorus Chemistry*; Kosolapoff, G.M., Maier, L., Eds.; Wiley Interscience: New York, NY, USA, 1973; Volume 3, pp. 186–339.
62. Burger, K. *Organophosphorus Reagents in Organic Synthesis*; Cadogan, J.I.G., Ed.; Academic Press: London, UK, 1979.
63. Holmes, R.R. *Structure and Spectroscopy and Reaction Mechanisms in Pentacoordinated Phosphorus*; ACS Monograph Series 175 and 176; American Chemical Society: Washington, DC, USA, 1980; Volume 1–2.
64. Husband, J.B.; McNab, H. The thermolysis of pentacoordinate phosphorus heterocycles. *Phosphorus Sulfur Silicon Relat. Elem.* **1984**, *20*, 207–230. [CrossRef]
65. Burgada, R.R.; Setton, R. *The Chemistry of Organophosphorus Compounds*; Hartley, F.R., Ed.; Wiley-Interscience: Chichester, UK, 1994; Volume 3, pp. 185–277.
66. Kawashima, T. *Chemistry of Hypervalent Compounds*; Akiba, K.-Y., Ed.; Wiley: New York, NY, USA, 1999.
67. Hellwinkel, D. Die Stereochemie organischer Derivate des fünf-und sechsbindigen Phosphors, II. Über ein erstes optisch aktives Pentaarylphosphoran. *Chem. Ber.* **1966**, *99*, 3642–3659. [CrossRef]
68. Monkowius, U.N.; Mitzel, W.; Schier, A.; Schmidbaur, H. 5-Organyl-5-phosphaspiro[4.4]nonanes: A Contribution to the Structural Chemistry of Spirocyclic Tetraalkylphosphonium Salts and Pentaalkylphosphoranes. *J. Am. Chem. Soc.* **2002**, *124*, 6126–6132. [CrossRef] [PubMed]

69. Byrne, P.A.; Muldoon, J.; Ortin, Y.; Müller-Bunz, H.; Gilheany, D.G. Investigations on the Operation of Stereochemical Drift in the Wittig Reaction by NMR and Variable-Temperature NMR Spectroscopy of Oxaphosphetane Intermediates and Their Quench Products. *Eur. J. Org. Chem.* **2014**, 86–98. [CrossRef]
70. Vedejs, E.; Meier, G.P.; Snoble, K.A.J. Low-temperature characterization of the intermediates in the Wittig reaction. *J. Am. Chem. Soc.* **1981**, *103*, 2823–2831. [CrossRef]
71. Kawashima, T.; Soda, T.; Okazaki, R. Synthesis, Structure, and Thermolysis of N-Apical 1,2λ^5-Azaphosphetidines with a Pentacoordinate P Center and the First Observation of Their *N*-Equatorial Pseudorotamers. *Angew. Chem. Int. Ed. Engl.* **1996**, *35*, 1096–1098. [CrossRef]
72. Kawashima, T.; Kato, K.; Okazaki, R. Synthese, Struktur und Thermolyse eines 3-Methoxycarbonyl-1,2λ^5-oxaphosphetans. *Angew. Chem.* **1993**, *105*, 941–942. [CrossRef]
73. Kawashima, T.; Kato, K.; Okazaki, R. Synthesis, Structure, and Thermolysis of a 3-Methoxycarbonyl-1,2λ^5-oxaphosphetane. *Angew. Chem. Int. Ed. Eng.* **1993**, *32*, 869–870. [CrossRef]
74. Kano, N.; Kikuchi, A.; Kawashima, T. The first isolable pentacoordinate 1,2λ^5-azaphosphetine: Synthesis, X-ray crystallographic analysis, and dynamic behaviour. *Chem. Commun.* **2001**, 2096–2097. [CrossRef]
75. López, J.G.; Ramallal, A.M.; González, J.; Roces, L.; Garcia-Ganda, S.; Iglesias, M.J.; Oña-Burgos, P.; Ortis, F.L. Mechanisms of Stereomutation and Thermolysis of Spiro-1,2-oxaphosphetanes: New Insights into the Second Step of the Wittig Reaction. *J. Am. Chem. Soc.* **2012**, *134*, 19504–19507. [CrossRef] [PubMed]
76. García-López, J.; Peralta-Pérez, E.; Forcén-Acebal, A.; García-Granda, S.; López-Ortiz, F. Dilithiated phosphazenes: Scaffolds for the synthesis of olefins through a new class of bicyclic 1,2-oxaphosphetanes. *Chem. Commun.* **2003**, 856–857. [CrossRef]
77. Kojima, S.; Kajiyama, K.; Akiba, K.-Y. Characterization of Enantiomeric Pairs of Optically Active 10-P-5 Phosphoranes with Asymmetry Only at Phosphorus. *Bull. Chem. Soc. Jpn.* **1995**, *68*, 1785–1797. [CrossRef]
78. Kojima, S.; Kajiyama, K.; Akiba, K.-Y. Characterization of an optically active pentacoordinate phosphorane with asymmetry only at phosphorus. *Tetrahedron Lett.* **1994**, *35*, 7037–7040. [CrossRef]
79. Kajiyama, K.; Yoshimune, M.; Nakamoto, M.; Matsukawa, S.; Kojima, S.; Akiba, K.-Y. Highly Selective One-Pot Synthesis of Spirophosphoranes Exhibiting Reversed Apicophilicity by Oxidation of Dianions Generated from P–H Spirophosphorane. *Org. Lett.* **2001**, *3*, 1873–1875. [CrossRef] [PubMed]
80. Akiba, K.-Y.; Matsukawa, S.; Kajiyama, K.; Nakamoto, M.; Kojima, S.; Yamamoto, Y. Novel results obtained by freezing berry pseudorotation of phosphoranes (10-P-5). *Heteroat. Chem.* **2002**, *13*, 390–396. [CrossRef]
81. Kajiyama, K.; Yoshimune, M.; Kojima, S.; Akiba, K.-Y. A New Method for the Formation of Anti-apicophilic (*O-cis*) Spirophosphoranes—Kinetic Studies on the Stereomutation of *O-cis* Arylspirophosphoranes to Their *O-trans* Isomers. *Eur. J. Org. Chem.* **2006**, 2739–2746. [CrossRef]
82. Kano, N.; Miyaka, H.; Kawashima, T. Hypervalent Silicon and Phosphorus Atoms in Single Molecules: Synthesis and Properties of Phosphoranylalkoxysilicates and a Phosphoranyloxysilicate. *Chem. Lett.* **2007**, *36*, 1260–1261. [CrossRef]
83. Miyaka, H.; Kano, N.; Kawashima, T. Synthesis and Properties of Pentacoordinate Phosphorus Compounds Containing a Pentacoordinate Silicon Atom. *Phosphorus Sulfur Silicon Relat. Elem.* **2008**, *183*, 673–674. [CrossRef]
84. Jiang, X.-D.; Kakuda, K.; Matsukawa, S.; Yamamichi, H.; Kojima, S.; Yamamoto, Y. Synthesis and application of a bidentate ligand based on decafluoro-3-phenyl-3-pentanol: Steric effect of pentafluoroethyl groups on the stereomutation of *O*-equatorial C-apical spirophosphoranes. *Chem. Asian J.* **2007**, *2*, 314–323. [CrossRef] [PubMed]
85. Jiang, X.-D.; Matsukawa, S.; Kakuda, K.; Fukuzaki, Y.; Zhao, W-L.; Li, L.-S.; Shen, H.-B.; Kojima, S.; Yamamoto, Y. Efficient synthesis of tetradecafluoro-4-phenylheptan-4-ol by a Cannizzaro-type reaction and application of the alcohol as a bulky Martin ligand variant for a new anti-apicophilic phosphorane. *Dalton Trans.* **2010**, *39*, 9823–9829. [CrossRef] [PubMed]
86. Jiang, X.-D.; Matsukawa, S.; Yamamoto, Y. Synthesis, structure and isomerization of arylphosphoranes with anti-apicophilic bonding modes using a novel bidentate ligand with two C2F5 groups. *Dalton Trans.* **2008**, 3678–3687. [CrossRef] [PubMed]
87. Jiang, X.D.; Matsukawa, S.; Yamamichi, H.; Yamamoto, Y. Some Reactions of *O*-Equatorial Spirophosphoranes Bearing the Bidentate Ligand Based on Decafluoro-3-phenyl-3-pentanol. *Heterocycles* **2007**, *73*, 805–824.
88. Kojima, S.; Sugino, M.; Matsukawa, S.; Nakamoto, M.; Akiba, K.-Y. First Isolation and Characterization of an Anti-Apicophilic Spirophosphorane Bearing an Oxaphosphetane Ring: A Model for the Possible Reactive Intermediate in the Wittig Reaction. *J. Am. Chem. Soc.* **2002**, *124*, 7674–7675. [CrossRef] [PubMed]

89. Sase, S.; Kano, N.; Kawashima, T. Pentacoordinate 1H-Phosphirenes: Reactivity, Bonding Properties, and Substituent Effects on Their Structures and Thermal Stability. *J. Org. Chem.* **2006**, *71*, 5448–5456. [CrossRef] [PubMed]
90. Sase, S.; Kano, N.; Kawashima, T. Synthesis and Structure of the First 1,2σ^5-Selenaphosphirane. *J. Am. Chem. Soc.* **2002**, *124*, 9706–9707. [CrossRef] [PubMed]
91. Sase, S.; Kano, N.; Kawashima, T. Synthesis of the First Stable Pentacoordinate Selenaphosphirane. *Phosphorus Sulfur Silicon Relat. Elem.* **2002**, *177*, 2039–2040. [CrossRef]
92. Sase, S.; Kano, N.; Kawashima, T. Isolation of a Cyclic Intermediate in the Reaction of a Phosphorus Ylide with Elemental Sulfur: Synthesis, Structure, and Reactivity of a 1,2σ^5-Thiaphosphirane. *Chem. Lett.* **2004**, *33*, 1434–1435. [CrossRef]
93. Ken Hirotsu, K.; Akihiro Okamoto, A.; Kozo Toyota, K.; Masaaki Yoshifuji, M. X-ray structure of 3-diphenylmethylene-2-(2,4,6-tri-*t*-butylphenyl)thiaphosphirane 2-sulfide: The first thiaphosphirane with *exo*-methylene. *Heteroat. Chem.* **1990**, *1*, 251–254. [CrossRef]
94. Kano, N.; Xing, J.-H.; Kikuchi, A.; Kawa, S.; Kawashima, T. Synthesis of Four- and Five-Membered Heterocycles Derived from an Iminophosphorane. *Phosphorus Sulfur Silicon Relat. Elem.* **2002**, *177*, 1685–1687. [CrossRef]
95. Kano, N.; Xing, J.-H.; Kawa, S.; Kawashima, T. Synthesis, structure, and thermolysis of pentacoordinate 1,3,2λ^5-oxazaphosphetidines: The intermediates of aza-Wittig reactions. *Tetrahedron Lett.* **2000**, *41*, 5237–5241. [CrossRef]
96. Kano, N.; Xing, J.-H.; Kikuchi, A.; Kawashima, T. Formation and X-ray crystallographic analysis of a 1,2λ^5-oxaphosphol-5(2*H*)-one. *Heteroat. Chem.* **2001**, *12*, 282–286. [CrossRef]
97. Burnaeva, L.M.; Mironov, V.F.; Romanov, S.V.; Ivkova, G.A.; Shulaeva, I.L.; Konovalova, I.V. Reaction of 2-Phenyl-1,3,2-benzodioxaphosphinan-4-one with Diethyl Benzylidenemalonate. *Russ. J. Gen. Chem.* **2001**, *71*, 488–489. [CrossRef]
98. Adachi, T.; Matsukawa, S.; Nakamoto, M.; Kajiyama, K.; Kojima, S.; Yamamoto, Y.; Akiba, K.-Y.; Re, S.; Nagase, S. Experimental Determination of the $n_N \rightarrow \sigma^*_{P\text{-}O}$ Interaction Energy of *O*-Equatorial C-Apical Phosphoranes Bearing a Primary Amino Group. *Inorg. Chem.* **2006**, *45*, 7269–7277. [CrossRef] [PubMed]
99. Acher, F.; Juge, S.; Wakselman, M. Chiral bicyclic spirophosphoranes in an Arbuzov-type reaction. *Tetrahedron* **1987**, *43*, 3721–3728. [CrossRef]
100. Sevenard, D.V.; Kazimir, E.L.; Pashkevich, I.; Röschenthaler, G.-V. Polyfluoroacylcycloalkanones in reactions with selected phosphorus(III) compounds. *Heteroat. Chem.* **2002**, *13*, 97–107. [CrossRef]
101. Terent'eva, S.A.; Pudovik, M.A.; Gubaidullin, A.T.; Litvinov, I.A.; Pudovik, A.N. Synthesis and crystal and molecular structures of 2-diethylaminomethyl- and 2-chloromethyl-2,2′-spirobi[benzo-1,3,2-oxazaphospholines]. *Russ. J. Gen. Chem.* **2001**, *71*, 330–336. [CrossRef]
102. Terent'eva, S.A.; Pudovik, M.A.; Pudovik, A.N. Aminoalkylation of 2-substituted 4,5-benzo-1,3,2-oxazaphospholanes. *Zh. Obshch. Khim.* **1987**, *57*, 496–499.
103. Chekhlov, A.N.; Bovin, A.N.; Tsvetkov, E.N. Structure of 2-chloromethyl-2λ5–2,2′(3*H*,3′*H*)-spirobis[1,3,2-benzoxazaphosphole]. *Izv. Akad. Nauk. SSSR Ser. Khim.* **1991**, *7*, 1523–1526.
104. Terent'eva, S.A.; Nikolaeva, I.L.; Burilov, A.R.; Kharitonov, D.I.; Popova, E.V.; Pudovik, M.A.; Litvinov, I.A.; Gubaidullin, A.T.; Konovalov, A.I. Reaction of 2-Ethyl-1,3,2-bezoxaphospholine with Calix[4]resorcinarenes. *Russ. J. Gen. Chem.* **2001**, *71*, 389–395. [CrossRef]
105. Malavaud, C.; Barrans, J. Les phosphoranes monocycliques a ligand hydrogene intermediaires dans la synthese de spirophosphoranes. *Tetrahedron Lett.* **1975**, *16*, 3077–3080. [CrossRef]
106. Gholivand, K.; Mahzouni, H.R.; Molaei, F.; Kalateh, A.A. A phosphoryl to spiro-bicyclophosphorane transformation via β-amidic proton elimination in phosphorylated hydrazides. *Tetrahedron Lett.* **2012**, *53*, 5944–5947. [CrossRef]
107. Freytag, M.; Jones, P.G.; Schmutzler, R.; Yoshifuji, M. Reaction of (*E*)-bis-(2,4,6-tri-*t*-butylphenyl)diphosphane with tetrachloro-*o*-benzoquinone. *Heteroat. Chem.* **2001**, *12*, 300–308. [CrossRef]
108. Malavaud, C.; Charbonnel, Y.; Barrans, J. Synthese de P-methyl spirophosphoranes a partir de diols ou D-aminoalcohols. *Tetrahedron Lett.* **1975**, *16*, 497–498. [CrossRef]
109. Bonningue, C.; Brazier, J.F.; Houalla, D.; Osman, F.H. Phosphoranes par fermeture *trans* annulaire IV. Quelques Exemples de Phosphoranes Bicycliques Chiraux. *Phosphorus Sulfur Silicon Relat. Elem.* **1979**, *5*, 291–298. [CrossRef]

110. Koenig, M.; Munoz, A.; Garrigues, B.; Wolf, R. Spirophosphoranes a Liaison P–H D'α-Hydroxyacidies et de Tartrates D'Alcoyle: Preparation, Properties-Stereochimie. *Phosphorous Sulfur Relat. Elem.* **1979**, *6*, 435–451. [CrossRef]
111. Hou, J.-B.; Tang, G.; Guo, J.-N.; Liu, Y.; Zhang, H.; Zhao, Y.-F. Stereochemistry of chiral pentacoordinate spirophosphoranes correlated with solid-state circular dichroism and ^{1}H NMR spectroscopy. *Tetrahedron Asymmetry* **2009**, *20*, 1301–1307. [CrossRef]
112. Hou, J.-B.; Zhang, H.; Guo, J.-N.; Liu, Y.; Xu, P.-X.; Zhao, Y.-F.; Blackburn, G.M. Chirality at phosphorus in pentacoordinate spirophosphoranes: Stereochemistry by X-ray structure and spectroscopic analysis. *Org. Biomol. Chem.* **2009**, *7*, 3020–3023. [CrossRef]
113. Yang, G.; Xu, Y.; Hou, J.; Zhang, H.; Zhao, Y. Determination of the Absolute Configuration of Pentacoordinate Chiral Phosphorus Compounds in Solution by Using Vibrational Circular Dichroism Spectroscopy and Density Functional Theory. *Chem. Eur. J.* **2010**, *16*, 2518–2527. [CrossRef] [PubMed]
114. Yang, G.; Xu, Y.; Hou, J.; Zhang, H.; Zhao, Y. Diastereomers of the pentacoordinate chiral phosphorus compounds in solution: Absolute configurations and predominant conformations. *Dalton Trans.* **2010**, *39*, 6953–6959. [CrossRef] [PubMed]
115. Cao, S.; Gao, P.; Gua, Y.; Zhao, H.; Wang, J.; Liu, Y.; Zhao, Y. Unexpected Insertion of CO_2 into the Pentacoordinate P–N Bond: Atherton–Todd-Type Reaction of Hydrospirophosphorane with Amines. *J. Org. Chem.* **2013**, *78*, 11283–11293. [CrossRef] [PubMed]
116. You, X.H.; Qi, L.; Zheng, J.; Dai, W.; Guo, Y.C.; Zhao, Y.F.; Cao, S.X. Synthesis and Characterization of New Pyrospirophosphoranes Containing a P–O–P Bond by the Atherton–Todd Reaction. *Heteroat. Chem.* **2015**, *26*, 168–174. [CrossRef]
117. Dai, W.; Liu, Q.; You, X.; Zhou, Z.; Guo, Y.; Zhao, Y.; Cao, S. Synthesis and Characterization of Alkoxy Spirophosphoranes Prepared from Hydrospirophosphoranes and Sodium Alcoholates. *Heteroat. Chem.* **2016**, *27*, 63–71. [CrossRef]
118. Bondarev, O.G.; Mikhel, I.S.; Tsarev, P.V.; Petrovskii, P.V.; Davankov, V.A.; Gavrilov, K.N. Platinum and rhodium complexes with isoleucinol-based bi- and tricyclic hydrophosphoranes. *Russ. Chem. Bull. Int. Ed.* **2003**, *52*, 116–121. [CrossRef]
119. Gavrilov, K.N.; Polosukhin, A.I.; Bondarev, O.G.; Lyubimov, S.E.; Lyssenko, K.A.; Petrovski, P.; Davankov, V.A. Synthesis and properties of pentacoordinated phospha derivatives of *iso*-leucinol. A rare example of using of hydrophosphoranes as ligands in asymmetric catalysis. *J. Mol. Catal. A Chem.* **2003**, *196*, 39–53. [CrossRef]
120. Marchi, C.; Buono, G. Asymmetric Addition of Chiral Triquinphosphoranes on Activated Carbonyl Compounds. *Tetrahedron Lett.* **1999**, *40*, 9251–9254. [CrossRef]
121. Marchi, C.; Buono, G. Alkylthiylation of triquinphosphoranes by disulfides: An entry to chiral thiatriquinphosphoranes. *Tetrahedron Lett.* **2000**, *41*, 3073–3076. [CrossRef]
122. Lopez, L.; Fabas, C.; Barrans, J. Heterocycles phosphores derivant des amidoximes. II. Composes du phosphore penta et hexacoordine. *Phosphorus Sulfur Silicon Relat. Elem.* **1979**, *7*, 81–87. [CrossRef]
123. Zhao, Y.-F.; Han, B.; Chen, J.; Jiang, Y. Penta-Coordinate Phosphorus Compounds and Biochemistry. *Phosphorus Sulfur Silicon Relat. Elem.* **2002**, *177*, 1391–1396. [CrossRef]
124. Li, Z.; Fu, H.; Gong, H.; Zhao, Y. Convenient Solid Phase Synthesis of Oligopeptides Using Pentacoordinated Phosphoranes with Amino Acid Residue as Building Blocks. *Bioorg. Chem.* **2004**, *32*, 170–177. [CrossRef] [PubMed]
125. Schenthaler, G.-V.R.; Bohlen, R.; Storzer, W.; Sopchik, A.E.; Bentrude, W.G. Reaktionen eines Tetraalkoxyhydro-spirophosphorans. *Z. Anorg. Allg. Chem.* **1983**, *507*, 93–99. [CrossRef]
126. Nakazawa, H.; Kubo, K.; Miyoshi, K. Metallaphosphorane Chemistry: Preparations, Structures, and Reactivities. *Bull. Chem. Soc. Jpn.* **2001**, *74*, 2255–2267. [CrossRef]
127. Nakazawa, H.; Ogawa, T.; Kawamura, K.; Miyoshi, K. Syntheses of Metallaphosphorane Complexes and Berry Pseudorotation. *Phosphorus Sulfur Silicon Relat. Elem.* **2002**, *177*, 2163–2164. [CrossRef]
128. Nakazawa, H.; Kawamura, K.; Ogawa, T.; Myoshi, K. Syntheses, structure, and Berry pseudorotation of ruthenium-, iron-, and cobalt-phosphorane complexes. *J. Organomet. Chem.* **2002**, *646*, 204–211. [CrossRef]
129. Kubo, K.; Nakazawa, H.; Kawamura, K.; Mizuta, T.; Miyoshi, K. Migration Reaction of a Hypervalent Fragment: Base-Induced Migration of a Phosphorane Fragment from Iron to the Cyclopentadienyl Ring in $Cp(CO)_2Fe\{P(OC_6H_4Y)(OC_6H_4Z)\}$ (Y, Z = NMe, NH, O). *J. Am. Chem. Soc.* **1998**, *120*, 6715–6721. [CrossRef]

130. Munoz, A.; Gornitzka, H. Phosphorus, Structure Cristalline D'un Hydroxyspirophosphorane Modele D'intermediaire de Reactions Impliquant Des Esters Cycliques de L'acide Phosphorique. *Phosphorus Sulfur Silicon Relat. Elem.* **2003**, *178*, 55–60. [CrossRef]
131. Boyer, D.G.; Boisdon, M.-T.; Rochal, A.; Munoz, A. Equilibre Ester Phosphorique, Hydroxyphosphorane Role des Liaisons Hydrogene, Acidite de Bronsted. *Phosphorus Sulfur Silicon Relat. Elem.* **2003**, *178*, 2117–2125. [CrossRef]
132. Laurenco, C.; Burgada, R. Mecanisme de formation et de transformation des spirophosphoranes-IX: Reactions des spirophosphoranes a liaison P–H avec les ethers vinyliques et les enamines. *Tetrahedron* **1976**, *22*, 2253–2255. [CrossRef]

Permissions

All chapters in this book were first published in MDPI; hereby published with permission under the Creative Commons Attribution License or equivalent. Every chapter published in this book has been scrutinized by our experts. Their significance has been extensively debated. The topics covered herein carry significant findings which will fuel the growth of the discipline. They may even be implemented as practical applications or may be referred to as a beginning point for another development.

The contributors of this book come from diverse backgrounds, making this book a truly international effort. This book will bring forth new frontiers with its revolutionizing research information and detailed analysis of the nascent developments around the world.

We would like to thank all the contributing authors for lending their expertise to make the book truly unique. They have played a crucial role in the development of this book. Without their invaluable contributions this book wouldn't have been possible. They have made vital efforts to compile up to date information on the varied aspects of this subject to make this book a valuable addition to the collection of many professionals and students.

This book was conceptualized with the vision of imparting up-to-date information and advanced data in this field. To ensure the same, a matchless editorial board was set up. Every individual on the board went through rigorous rounds of assessment to prove their worth. After which they invested a large part of their time researching and compiling the most relevant data for our readers.

The editorial board has been involved in producing this book since its inception. They have spent rigorous hours researching and exploring the diverse topics which have resulted in the successful publishing of this book. They have passed on their knowledge of decades through this book. To expedite this challenging task, the publisher supported the team at every step. A small team of assistant editors was also appointed to further simplify the editing procedure and attain best results for the readers.

Apart from the editorial board, the designing team has also invested a significant amount of their time in understanding the subject and creating the most relevant covers. They scrutinized every image to scout for the most suitable representation of the subject and create an appropriate cover for the book.

The publishing team has been an ardent support to the editorial, designing and production team. Their endless efforts to recruit the best for this project, has resulted in the accomplishment of this book. They are a veteran in the field of academics and their pool of knowledge is as vast as their experience in printing. Their expertise and guidance has proved useful at every step. Their uncompromising quality standards have made this book an exceptional effort. Their encouragement from time to time has been an inspiration for everyone.

The publisher and the editorial board hope that this book will prove to be a valuable piece of knowledge for researchers, students, practitioners and scholars across the globe.

List of Contributors

Sophie Wendels, Thiebault Chavez, Martin Bonnet, Khalifah A. Salmeia and Sabyasachi Gaan
Additives and Chemistry Group, Advanced Fibers, Empa, Swiss Federal Laboratories for Materials Science and Technology, Lerchenfeldstrasse 5, 9014 St. Gallen, Switzerland

Dambarudhar Parida and Sandro Lehner
Additives and Chemistry, Advanced Fibers, Empa Swiss Federal Laboratories for Materials Science and Technology, Lerchenfeldstrasse 5, 9014 St. Gallen, Switzerland

Antonia Neels
Center for X-ray Analytics, Swiss Federal Laboratories for Materials Science and Technology, Empa, Überlandstrasse 129, 8600 Dübendorf, Switzerland

Daniel Rentsch
Laboratory for Functional Polymers, Empa Swiss Federal Laboratories for Materials Science and Technology, Überlandstrasse 129, 8600 Dübendorf, Switzerland

Jarosław Lewkowski, Maria Rodriguez Moya and Marta Chmielak
Department of Organic Chemistry, Faculty of Chemistry, University of Łódź, Tamka 12, 91-403 Łódź, Poland

Diana Rogacz, Kamila Lewicka and Piotr Rychter
Institute of Chemistry, Environmental Protection and Biotechnology, Jan Długosz University in Częstochowa, 13/15 Armii Krajowej Av., 42-200 Częstochowa, Poland

Anastasy O. Kolodiazhna and Oleg I. Kolodiazhnyi
Institute of Bioorganic Chemistry and Petrochemistry, National Academy of Sciences of Ukraine, 1, Murmanska Street, Kiev 02094, Ukraine

Raj K. Bansal, Raakhi Gupta and Manjinder Kour
Department of Chemistry, The IIS University, Jaipur 302020, India

Gerhard Hägele
Institute of Inorganic Chemistry and Structural Chemistry, Heinrich-Heine-University Düsseldorf, Universitätsstraße 1, D-40225 Düsseldorf, Germany

Vasily P. Morgalyuk
Laboratory of Organophosphorus Compounds, Nesmeyanov Institute of Organoelement Compounds, Russian Academy of Sciences, Vavilova str., 28, Moscow 119991, Russia

Ankita Puri
Department of Chemistry, The IIS University, Jaipur 302020, India

Erika Bálint, Ádám Tajti, Nóra Tóth and György Keglevich
Department of Organic Chemistry and Technology, Budapest University of Technology and Economics, 1521 Budapest, Hungary

Dorota Krasowska, Jacek Chrzanowski and Piotr Kiełbasiński
Centre of Molecular and Macromolecular Studies, Polish Academy of Sciences, Sienkiewicza 112, 90-363 Lodz, Poland

Józef Drabowicz
Centre of Molecular and Macromolecular Studies, Polish Academy of Sciences, Sienkiewicza 112, 90-363 Lodz, Poland
Department of Chemistry, Environment Protection and Biotechnology, Jan Dlugosz University in Czestochowa, Armii Krajowej Ave. 13/15, 42-200 Czestochowa, Poland

Index

R

S

T

U

V

X

www.ingramcontent.com/pod-product-compliance
Lightning Source LLC
LaVergne TN
LVHW080929210926
848485LV00004B/758